AF685451

David Hilbert (1862–1943)

Eberhard Zeidler

Nonlinear Functional Analysis and its Applications

II/A: Linear Monotone Operators

Translated by the Author and by Leo F. Boron†

With 45 Illustrations

Springer-Verlag
New York Berlin Heidelberg
London Paris Tokyo

Eberhard Zeidler
Sektion Mathematik
Karl-Marx-Platz
7010 Leipzig
German Democratic Republic

Leo F. Boron†
Department of Mathematics
University of Idaho
Moscow, ID 83843
U.S.A.

Mathematics Subject Classification (1980): 46xx

Library of Congress Cataloging in Publication Data
(Revised for vol. 2 pts. A-B)
Zeidler, Eberhard.
Vol. 2, pts. A-B: Translated by the author and Leo L. Boron.
Vol. 3: Translated by Leo L. Boron.
Vol. 4: Translated by Juergen Quandt.
Includes bibliographies and indexes.
Contents: 1. Fixed point theorems — 2. pt A. Linear monotone operators. Pt. B Nonlinear operators — [etc.] — 4. Applications to mathematical physics.
1. Nonlinear functional analysis. I. Title.
QA321.5.Z4513 1985 515.7 83-20455

Printed on acid-free paper
Previous edition, *Vorlesungen über nichtlineare Funktionalanalysis*, Vols. I–III, published by BSB B. G. Teubner Verlagsgesellschaft, 7010 Leipzig, Sternwartenstrasse 8, Deutsche Demokratische Republik.

© 1990 by Springer-Verlag New York Inc.
All rights reserved. This work may not be translated or copied in whole or in part without the written permission of the publisher (Springer-Verlag, 175 Fifth Avenue, New York, New York 10010, U.S.A.), except for brief excerpts in connection with reviews or scholarly analysis. Use in connection with any form of information storage and retrieval, electronic adaptation, computer software, or by similar or dissimilar methodology now known or hereafter developed is forbidden.

Typeset by Asco Trade Typesetting Ltd., Hong Kong.
Printed and bound by Edwards Brothers, Inc., Ann Arbor, Michigan
Printed in the United States of America.

9 8 7 6 5 4 3 2 1

ISBN 0-387-96802-4 Springer-Verlag New York Berlin Heidelberg
ISBN 3-540-96802-4 Springer-Verlag Berlin Heidelberg New York

To the memory of my parents

Preface to Part II/A

A theory is the more impressive,
the simpler are its premises,
the more distinct are the things it connects,
and the broader is its range of applicability.
Albert Einstein

This is the second of a five-volume exposition of the main principles of nonlinear functional analysis and its numerous applications to the natural sciences and mathematical economics. The presentation is self-contained and accessible to a broader audience of mathematicians, natural scientists, and engineers. The basic content can be understood even by those readers who have little or no knowledge of linear functional analysis. The material of the five volumes is organized as follows:

Part I: Fixed-point theorems.
Part II: Monotone operators.
Part III: Variational methods and optimization.
Parts IV/V: Applications to mathematical physics.

The main goals of the work are discussed in detail in the Preface of Part I. A Table of Contents of Parts I through V can be found on page 871 of Part I. The emphasis of the treatment is based on the following considerations:

(a) Which are the basic, guiding concepts, and what relationship exists between them?
(b) What is the relationship between these ideas and the known results of classical analysis and of linear functional analysis?
(c) What are some typical applications?

The present Part II is divided into two subvolumes:

Part II/A: Linear monotone operators.
Part II/B: Nonlinear monotone operators.

These two subvolumes form a *unit*. They consist of the following sections:

introduction to the subject;
linear monotone problems;
generalization to nonlinear stationary problems;
generalization to nonlinear nonstationary problems;
general theory of discretization methods.

The numerous applications concern differential equations and integral equations, as well as numerical methods to their solution.

The Appendix, the Bibliography, and the Index material to Parts II/A and II/B can be found at the end of Part II/B.

The modern theory of linear partial differential equations of elliptic, parabolic, or hyperbolic type is based on the so-called *Hilbert space methods*. In this connection, boundary value problems and initial value problems are transformed into operator equations in Hilbert space. The solutions of these operator equations correspond to *generalized solutions* of the original classical problems. Here, the generalized solutions live in so-called *Sobolev spaces*. Roughly speaking, Sobolev spaces consist of functions which have sufficiently reasonable generalized derivatives.

The theory of *monotone operators* generalizes the Hilbert space methods to *nonlinear* problems. We want to emphasize that the Hilbert space methods and the theory of monotone operators are connected with the *main streams* of mathematics. They are closely related to Hilbert's rigorous justification of the *Dirichlet principle*, and to the 19th and 20th problems of Hilbert which he formulated in his famous Paris lecture in 1900. The relevant historical background will be discussed in Chapter 18. From the physical point of view, the Hilbert space methods and the more general theory of monotone operators are based on the fundamental concept of *energy*. Roughly speaking, Sobolev spaces can be regarded as spaces of functions which correspond to physical states of finite energy. We will show that, in our century, the notion of monotone operators played, both implicitly and explicitly, a fundamental role in the development of the calculus of variations, in the theory of linear and nonlinear partial differential equations, and in numerical analysis.

In order to help the reader understand the basic ideas, the first chapters of this volume serve as an elementary introduction to the modern functional analytic theory of *linear* partial differential equations. In particular, Chapter 18 contains an elegant functional analytic justification of the Dirichlet principle, based on a generalization of the classical Pythagorean theorem to Hilbert spaces. An introduction to the theory of *Sobolev spaces* can be found in Chapter 21. Experience shows that students frequently have trouble with the technicalities of Sobolev spaces. In Chapter 21, for the benefit of the reader, we choose an approach to Sobolev spaces which is as elementary as possible. To this end, we first prove all the embedding theorems in an extremely simple manner in $\mathbb{R}^1$, before passing to $\mathbb{R}^n$. For the convenience of the reader, the basic properties of the *Lebesgue integral* are summarized in the Appendix to Part II/B. In this connection, we choose the simplest definition of the Lebesgue integral. In contrast to other definitions of the Lebesgue integral, our definition also applies immediately to functions with values in B-spaces. Such functions are needed in connection with evolution equations. Moreover, in Chapter 18 we discuss a number of important principles which are frequently used in modern analysis, for example, the smoothing principle via mollifiers, the localization principle via partition of unity, the extension principle, and the completion principle.

The *basic ideas* and basic principles of the theory of nonlinear monotone operators are discussed in detail in a special section at the beginning of Part II/B. Any reader who wishes to learn about *nonlinear* monotone operators as quickly as possible may immediately begin reading Part II/B.

A reference of the form $A_1(20)$ and $A_2(20)$ is to formula (20) in the Appendix of Part I and II/B, respectively; while (18.20) refers to formula (20) in Chapter 18. Omission of a chapter number means that the formula is in the current chapter. The References to the Literature at the end of each chapter are of the following form: Krasnoselskii (1956, M, B, H), etc. The name and the year relate to the Bibliography at the end of Part II/B. The letters stand for the following:

M: monograph;
L: lecture notes;
S: survey article;
P: proceedings;
B: extensive bibliography in the work cited;
H: comments on the historical development of the subject contained in the work cited.

A List of Symbols may be found at the end of Part II/B. We have tried to use generally accepted symbols. A few peculiarities, introduced to avoid confusion, are described in the remarks introducing the List of Symbols.

Basic material on linear functional analysis may be found in the Appendix to Part I.

The theory of monotone operators is related to the simple fact that the *derivative* f' of a *convex* real function f is a *monotone* function. However, it is quite remarkable that the idea of the monotone operator allows many diversified applications. For example, there are *applications* to the following topics:

(i) variational problems and variational inequalities;
(ii) nonlinear elliptic, parabolic, and hyperbolic partial differential equations;
(iii) nonlinear integral equations;

(iv) nonlinear semigroups;
(v) nonlinear eigenvalue problems;
(vi) nonlinear Fredholm alternatives;
(vii) mapping degree for noncompact operators;
(viii) numerical methods such as the Ritz method (e.g., the method of finite elements), the Galerkin method, the projection–iteration method, the difference method, and the Kačanov method for conservation laws and variational inequalities.

Concerning time-dependent problems we emphasize both the *Galerkin method* and the method of *semigroups*. We also discuss in detail the fact that the theory of monotone operators generalizes both the theory of *bounded* and *unbounded* linear operators. To this end we develop, in Chapters 18 and 22 through 24, the theory of linear partial differential equations based on bounded linear operators, and in Chapter 19, we study in detail the elegant method of the Friedrichs extension for unbounded linear operators and its applications to variational problems and to linear and semilinear elliptic, parabolic, and hyperbolic equations, as well as applications to the semilinear Schrödinger equation. As we shall show in Parts IV and V, unbounded linear operators play a decisive role in quantum mechanics and quantum field theory (elementary particle physics). In contrast to this, for example, bounded linear operators are related to elasticity and hydrodynamics.

At the center of the theory of monotone operators there stands the notion of the *maximal monotone* operator, which generalizes both the theory of bounded and unbounded linear monotone operators. The theory of maximal monotone operators will be studied in detail in Chapter 32.

A number of diagrams contained in the text should help the reader to discover interrelationships between different topics. In particular, we recommend Figure 27.1 in Section 27.5 of Part II/B where the reader may find interrelationships between many important operator properties in nonlinear functional analysis. A list of all these schematic overviews can be found at the end of Part II/B. A list of the basic theorems and of the basic definitions can also be found there.

In Part I we studied equations involving compact operators. The decisive advantage of the theory of monotone operators is that it is also applicable to *noncompact* operators. Along with abstract existence theorems we also stress the methods of *numerical* functional analysis. Chapters 20 through 22 (resp. Chapters 33 through 35) may serve as an introduction to linear (resp. nonlinear) numerical functional analysis. For example, in terms of numerical functional analysis, monotone operators allow us to justify the following fundamental principle: Consistency and stability imply convergence. In this connection, the general notion of A-proper maps is crucial.

The connection between the theory of monotone operators and general variational methods will be studied in detail in Part III. In Parts IV and V we will consider applications of the theory of monotone operators to interesting problems in *mathematical physics*. For example, the theory of monotone operators plays an important role in elasticity, hydrodynamics (the Navier–Stokes equations), gas dynamics (subsonic flow), and semiconductor physics.

I hope that the reader will enjoy discovering a number of interesting interrelationships in mathematics.

Leipzig
Spring 1989

Eberhard Zeidler

Contents (Part II/A)

Contents (Part II/B)

INTRODUCTION TO THE SUBJECT

Each progress in mathematics is based on the discovery of stronger tools and easier methods, which at the time makes it easier to understand earlier methods. By making these stronger tools and easier methods his own, it is possible for the individual researcher to orientate himself in the different branches of mathematics.

The organic unity of mathematics is inherent in the nature of this science, for mathematics is the foundation of all exact knowledge of natural phenomena.

David Hilbert (1900)
(Paris lecture)

For me, as a young man, Hilbert became the kind of mathematician which I admired, a man with an enormous power of abstract thought, combined with a fully developed sense for the physical reality.

Norbert Wiener (1894–1964)

Hilbert always emphasized that mathematics is a unity, that its different parts are in permanent interaction with each other and with the natural sciences, and that this interaction not only provides the key to an understanding of the nature of mathematics, but also the best cure against a splitting into different and unrelated parts—a danger which, in our time of huge qualitative growth and alarming specialization of mathematical research, must always be kept in mind.

Pavel Sergeevič Aleksandrov (1971)

In the modern theory of partial differential equations, *generalized solutions* play a fundamental role. We want to explain briefly the mathematical and physical background of the notion of a generalized solution. To this end we consider the first boundary value problem for the Poisson equation

$$-\Delta u = f \quad \text{on } G, \tag{1a}$$

$$u = g \quad \text{on } \partial G. \tag{1b}$$

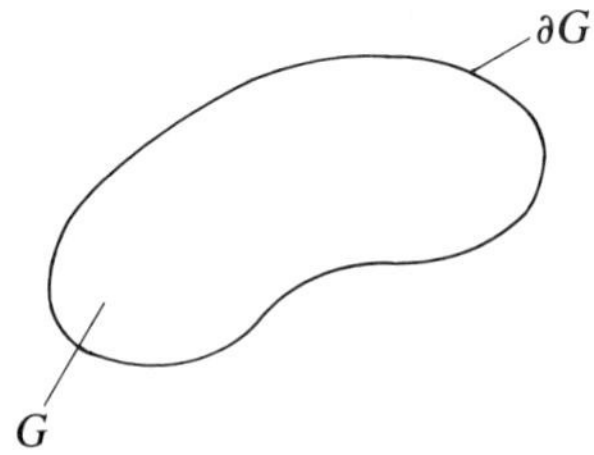

Figure 18.1

Here, G is a bounded region in $\mathbb{R}^3$, and ∂G denotes the boundary of G (Fig. 18.1).

Let $x = (\xi, \eta, \zeta)$ and

$$\Delta u = u_{\xi\xi} + u_{\eta\eta} + u_{\zeta\zeta}.$$

The functions f and g are given. We seek the function u. Problem (1) is closely related to the variational problem

$$\begin{gathered}\int_G (u_\xi^2 + u_\eta^2 + u_\zeta^2 - 2fu)\,dx = \min!,\\ u = g \quad \text{on } \partial G.\end{gathered} \tag{2}$$

In addition, let us consider the potential

$$u(x) = \frac{1}{4\pi}\int_G \frac{f(y)}{|x-y|}\,dy. \tag{3}$$

Up to a multiplicative constant, u represents the potential of a mass distribution on $\bar{G}$ which has density f.

The following observations are important.

(i) *Classical solutions.* For sufficiently *smooth* functions f and g and sufficiently smooth boundary ∂G, the original problem (1) has a classical solution. More precisely, if

$$f \in C^\alpha(\bar{G}), \qquad g \in C^{2,\alpha}(\partial G), \qquad \partial G \in C^{2,\alpha}, \qquad 0 < \alpha < 1,$$

then (1) has a unique classical solution u, where $u \in C^{2,\alpha}(\bar{G})$.

(ii) *Generalized solutions.* Counterexamples show that equation (1a) need *not* have a classical solution u in case $f: \bar{G} \to \mathbb{R}$ is *continuous.* This surprising fact depends essentially on the properties of the potential u in (3). If the density $f: \bar{G} \to \mathbb{R}$ is sufficiently smooth, say f is continuously differentiable, then the potential u in (3) is a classical solution of (1a). However, if $f: \bar{G} \to \mathbb{R}$ is only continuous, then u in (3) need not have second derivatives. Hence, u is *not* a classical solution of (1a).

Consequently, if $f: \bar{G} \to \mathbb{R}$ is continuous, then we have the following situation. The function u in (3) has a well-defined physical meaning, namely, u represents the potential of a mass distribution with continuous density f. But

u is *not* necessarily a solution of the differential equation (1a). Therefore it is reasonable to introduce the notion of a generalized solution of (1a). To this end we multiply (1a) by $\varphi \in C_0^\infty(G)$ and, using *integration by parts*, we obtain

$$-\int_G u\Delta\varphi\, dx = \int_G f\varphi\, dx \qquad \text{for all} \quad \varphi \in C_0^\infty(G) \tag{4}$$

(cf. Section 18.2). By definition, a function u is called a *generalized* solution of equation (1a) iff relation (4) holds. It can be shown that, in this sense, the potential u in (3) is a generalized solution of (1a) if f is continuous on $\bar{G}$.

Obviously, classical solutions of (1a) are also generalized solutions of (1a).

(iii) *Variational problems, generalized solutions, and the idea of completion.* Let us now consider the variational problem (2), which corresponds to the famous Dirichlet principle. In Chapter 18 we will show that each sufficiently smooth solution u of (2) is also a classical solution of the original problem (1). But the point is that the minimum problem (2) need *not* have a solution u which has classical first derivatives. This fact caused enormous trouble in the mathematics of the nineteenth century, related to the justification of the Dirichlet principle. This will be discussed in Section 18.10.

In order to understand the typical difficulties, let us first consider the following simple minimum problem

$$F(x) = \min!, \qquad x \in [a, b], \tag{5}$$

where $F\colon [a, b] \to \mathbb{R}$ is a continuous function on the compact interval $[a, b]$. The famous Weierstrass theorem tells us that problem (5) has always a solution (Fig. 18.2). Suppose now that the unique solution $x = \alpha$ of (5) is an *irrational* number, and suppose that we consider the modified problem

$$F(x) = \min!, \qquad x \in \mathbb{Q} \cap [a, b], \tag{5*}$$

where $\mathbb{Q}$ denotes the set of rational numbers. Then problem (5*) has *no* solution. Consequently, mathematicians, who do not know irrational numbers, cannot prove the Weierstrass existence theorem for (5). We now want to show that the variational problem (2) corresponds to a similar situation. For simplicity, let $g = 0$. We set

$$F(u) = \int_G (u_\xi^2 + u_\eta^2 + u_\zeta^2 - 2fu)\, dx$$

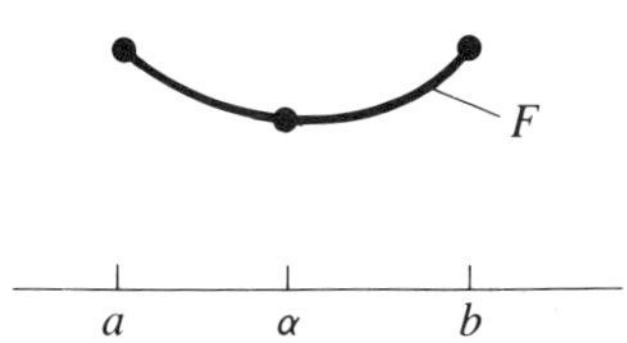

Figure 18.2

and

$$M = \{u \in C^1(\bar{G}): u = 0 \text{ on } \partial G\}.$$

Then, problem (2) reads as follows:

(D*) $$F(u) = \min!, \qquad u \in M.$$

This classical variational problem need not have a solution. Roughly speaking, problem (D*) corresponds to (5*). However, as we shall show in Section 18.10, if we introduce the *Sobolev space* $\mathring{W}_2^1(G)$, then we obtain the modified problem

(D) $$F(u) = \min!, \qquad u \in \mathring{W}_2^1(G),$$

which has a unique solution u if $\int_G f^2\,dx < \infty$.

By definition, this solution u is called a *generalized* solution of the original boundary value problem (1) with $g = 0$. In addition, u is also a generalized solution of the Poisson equation (1a) in the sense of (4). Roughly speaking, we obtain the following:

The introduction of Sobolev spaces corresponds to the introduction of real numbers via irrational numbers in classical mathematics.

To explain this, we start with the abstract problem

(P*) $$F(u) = \min!, \qquad u \in M.$$

In order to solve (P*) we consider the modified problem

(P) $$F(u) = \min!, \qquad u \in N,$$

where N is a completion of M, i.e., the set N is obtained from the set M by adding appropriate "ideal" elements. For example, in the case of problem (5*), we complete the set of rational numbers to the set of real numbers by adding irrational numbers. In the case of problem (D*), we complete the set

$$M = \{u \in C^1(\bar{G}): u = 0 \text{ on } \partial G\}$$

to the Sobolev space $N = \mathring{W}_2^1(G)$ by adding appropriate functions which have generalized first-order derivatives. The Sobolev space $\mathring{W}_2^1(G)$ is a Hilbert space equipped with the scalar product

$$(u|v) = \int_G (u_\xi v_\xi + u_\eta v_\eta + u_\zeta v_\zeta)\,dx.$$

The precise definition of $\mathring{W}_2^1(G)$ will be given in Section 18.10. Roughly speaking, the Sobolev space $\mathring{W}_2^1(G)$ is the smallest Hilbert space with the scalar product $(u|v)$ which contains the set of functions C_0^∞. Note that $C_0^\infty(G) \subseteq M$.

(iv) *Approximation methods.* In order to obtain approximate solutions for the original boundary value problem (1), one frequently uses the so-called *Ritz method.* The basic idea of this method is the following. Instead of solving the variational problem

$$\text{(D)} \qquad \int_G (u_\xi^2 + u_\eta^2 + u_\zeta^2 - 2fu)\,dx = \min!, \qquad u \in \mathring{W}_2^1(G),$$

we consider the approximate problem

$$\text{(D}_n\text{)} \qquad \int_G (u_\xi^2 + u_\eta^2 + u_\zeta^2 - 2fu)\,dx = \min!, \qquad u \in X_n,$$

where X_n is a finite-dimensional subspace of $\mathring{W}_2^1(G)$, i.e., X_n consists of all functions u of the form

$$u = c_1 w_1 + c_2 w_2 + \cdots + c_n w_n$$

with $w_j = 0$ on ∂G for all j, where the so-called basic functions $w_1, \ldots, w_n$ are fixed. This way the approximate problem (D_n) is reduced to the determination of the unknown real coefficients $c_1, \ldots, c_n$. Let u and u_n denote the solutions of (D) and (D_n), respectively. Generally, it is *not* possible to prove the convergence of this Ritz method in the sense of pointwise convergence:

$$\lim_{n\to\infty} u_n(x) = u(x) \qquad \text{for all} \quad x \in G.$$

However, in Chapter 22 we will prove that the sequence (u_n) converges to u in the Sobolev space $\mathring{W}_2^1(G)$, i.e.,

$$\lim_{n\to\infty} (u_n - u | u_n - u) = 0.$$

Explicitly, this means that

$$\lim_{n\to\infty} \int_G \left(\frac{\partial u_n}{\partial \xi} - \frac{\partial u}{\partial \xi}\right)^2 + \left(\frac{\partial u_n}{\partial \eta} - \frac{\partial u}{\partial \eta}\right)^2 + \left(\frac{\partial u_n}{\partial \zeta} - \frac{\partial u}{\partial \zeta}\right)^2 dx = 0.$$

Consequently, Sobolev spaces also play an important role in modern numerical analysis. For example, it is not necessary to use smooth basis functions $w_1, \ldots, w_n$ in (D_n) above; it is sufficient to use *piecewise-smooth* functions. This is the basic idea of the important method of *finite elements*, where $w_1, \ldots, w_n$ are piecewise polynomial functions.

Basic Strategy

The basic strategy of the modern theory of partial differential equations is the following:

(S1) *We prove the existence of generalized solutions.*

(S2) *We show that the generalized solutions are even classical solutions if the data of the problem are sufficiently smooth* (*method of regularization*).

In (S1) we use general results from functional analysis. The analytic substance of the existence proofs in (S1) is concentrated in the so-called *Sobolev embedding theorems* which correspond to inequalities for integrals. The proofs of the Sobolev embedding theorems are based on the Hölder inequality. The Sobolev embedding theorems generalize the famous classical inequalities of Poincaré and Friedrichs. In (S2) we need more sophisticated analytical methods which are based on the specific properties of the problems under consideration.

Along with the Ritz method and the Galerkin method, difference methods represent a universally applicable method for the numerical solution of partial differential equations. This will be considered in Chapter 20. In Chapters 34 and 35, we shall construct a general theory of discretization methods for nonlinear problems.

Figure 18.3 shows important interrelationships which will be studied in Chapters 18 and 19. There are two options for giving an *introductory lecture on Hilbert space methods* and their applications to integral equations and

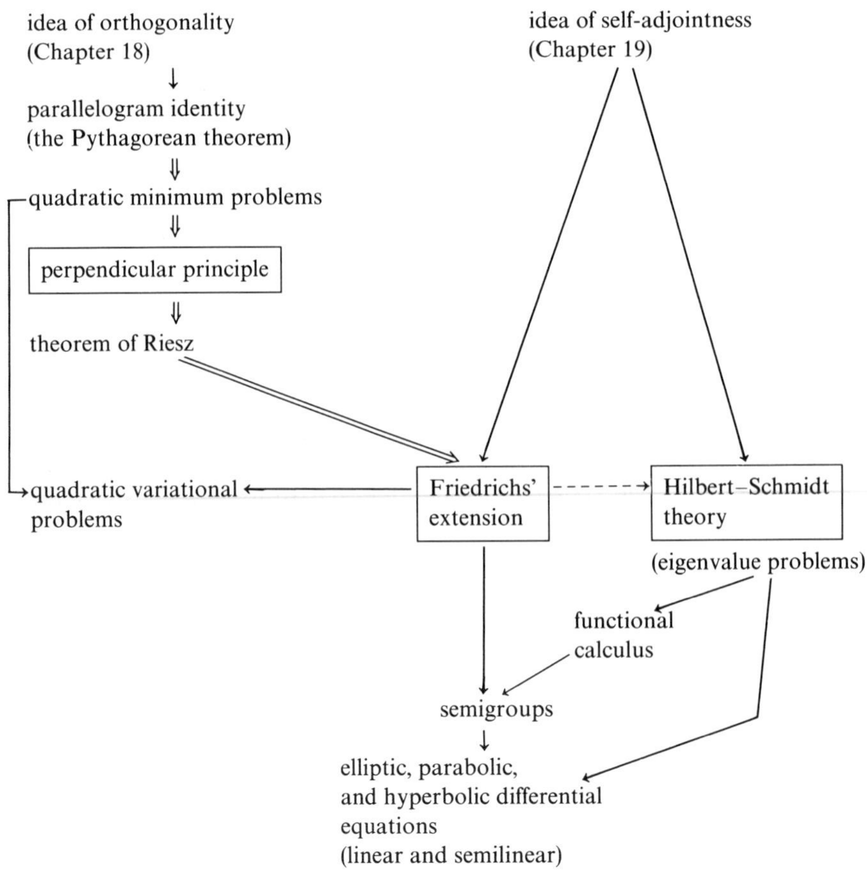

Figure 18.3

partial differential equations:

(i) Chapters 18 and 19;
(ii) Chapters 18, 21, and 22.

Approach (ii) is simpler than (i), since we only work with *bounded* operators. However, approach (i), based on *unbounded* operators, gives more insight. Note that unbounded operators are indispensable in quantum theory. If there is enough time in an introductory lecture, then one can add Chapters 23 and 24 to (ii) concerning existence theorems for parabolic and hyperbolic equations via the Galerkin method.

In Chapters 25 through 36, we generalize the basic principles for linear operators considered in Chapters 18 through 24 to nonlinear operators. An *introductory lecture on nonlinear monotone operators* can be based on Chapters 25 through 27.

Hilbert Spaces

For the convenience of the reader, we recall some basic facts on Hilbert spaces.

Scalar product. Let X be a linear space over $\mathbb{K} = \mathbb{R}, \mathbb{C}$ (cf. $A_1(22)$ in the Appendix of Part I). By definition, a scalar product on X is a function $(u, v) \mapsto (u|v)$ from $X \times X$ into $\mathbb{K}$ which has the following three properties for all $u, v, w \in X$ and all $\lambda, \mu \in \mathbb{K}$:

(i) $(u|\lambda v + \mu w) = \lambda(u|v) + \mu(u|w)$.
(ii) $\overline{(u|v)} = (v|u)$.
(iii) $(u|u) > 0$ iff $u \neq 0$.

Here, the bar denotes the conjugate complex number. From (i) and (ii) it follows that

$$(\lambda v + \mu w|u) = \bar{\lambda}(v|u) + \bar{\mu}(w|u).$$

Let $u, v \in X$. Then u is called *orthogonal* to v iff

$$(u|v) = 0.$$

Pre-Hilbert spaces. By definition, a pre-Hilbert space X is a linear space together with a scalar product. We set

$$\|u\| = (u|u)^{1/2}.$$

The convergence

$$u_n \to u \qquad \text{as} \quad n \to \infty$$

is defined by

$$\|u_n - u\| \to 0 \qquad \text{as} \quad n \to \infty.$$

A sequence (u_n) is called a *Cauchy sequence* iff, for each $\varepsilon > 0$, there exists an

$n_0(\varepsilon)$ with

$$\|u_n - u_m\| < \varepsilon \qquad \text{for all} \quad n, m \geq n_0(\varepsilon).$$

Hilbert spaces. By definition, a Hilbert space is a pre-Hilbert space with the additional property that each Cauchy sequence is convergent.

Hilbert spaces are briefly called H-spaces.

The Schwarz inequality and its consequences. The most important inequality in an H-space X is the so-called *Schwarz inequality*

$$|(u|v)| \leq \|u\| \, \|v\| \qquad \text{for all} \quad u, v \in X.$$

This follows from

$$0 \leq (u - \lambda v | u - \lambda v) = \|u\|^2 - \lambda(u|v) - \bar{\lambda}(v|u) + |\lambda|^2 \|v\|^2$$

with $\lambda = \|u\|^2/(u|v)$ in case $(u|v) \neq 0$.

It follows from the Schwarz inequality that

$$\begin{aligned}\|u + v\|^2 = (u + v | u + v) &= \|u\|^2 + (u|v) + (v|u) + \|v\|^2 \\ &\leq (\|u\| + \|v\|)^2.\end{aligned}$$

This yields the *triangle inequality*

$$\|u + v\| \leq \|u\| + \|v\| \qquad \text{for all} \quad u, v \in X.$$

Moreover, the *generalized triangle inequality*

$$| \, \|u\| - \|v\| \, | \leq \|u \pm v\| \leq \|u\| + \|v\| \qquad \text{for all} \quad u, v \in X$$

follows from

$$\|u\| - \|v\| = \|(u - v) + v\| - \|v\| \leq \|u - v\|.$$

If $u_n \to u$ as $n \to \infty$, then

$$\|u_n\| \to \|u\| \qquad \text{as} \quad n \to \infty,$$

since $| \, \|u_n\| - \|u\| \, | \leq \|u_n - u\| \to 0$. Finally,

$$u_n \to u \qquad \text{and} \qquad v_n \to v \qquad \text{as} \quad n \to \infty$$

implies

$$(u_n | v_n) \to (u|v) \qquad \text{as} \quad n \to \infty.$$

This follows from

$$\begin{aligned}|(u_n|v_n) - (u|v)| &= |(u_n|v_n - v) + (u_n - u|v)| \\ &\leq |(u_n|v_n - v)| + |(u_n - u|v)| \leq \|u_n\| \, \|v_n - v\| + \|u_n - u\| \, \|v\|.\end{aligned}$$

Equivalent scalar products. Let X be an H-space over $\mathbb{K}$ with the scalar product $(\cdot|\cdot)$, and let $(\cdot|\cdot)_*$ be another scalar product on X. By definition, $(\cdot|\cdot)$ and $(\cdot|\cdot)_*$ are called *equivalent* iff the corresponding norms

$$\|u\| = (u|u)^{1/2} \qquad \text{and} \qquad \|u\|_* = (u|u)_*^{1/2}$$

are equivalent, i.e., there exist positive constants c and d such that

$$c\|u\| \leq \|u\|_* \leq d\|u\| \qquad \text{for all} \quad u \in X.$$

In this case, X is also an H-space with respect to $(\cdot|\cdot)_*$.

In the following let X and Y be H-spaces over $\mathbb{K} = \mathbb{R}, \mathbb{C}$, and let M be a subset of X.

Bounded sets. The set M is called bounded iff there is a real number r such that

$$\|u\| \leq r \qquad \text{for all} \quad u \in M.$$

Open and closed sets. The set M is called open iff for each $u \in M$ there is a number $r > 0$ such that the ball $\{v \in X: \|v - u\| < r\}$ belongs to M.

The set M is called closed iff for each sequence (u_n) in M,

$$u_n \to u \qquad \text{as} \quad n \to \infty \qquad \text{implies} \qquad u \in M.$$

The closure $\overline{M}$ of M contains exactly all the elements u of X with the property that there is a sequence (u_n) in M such that $u_n \to u$ as $n \to \infty$.

Dense sets. The set M is called dense iff $\overline{M} = X$. Obviously, M is dense iff, for each $u \in X$ and each $\varepsilon > 0$, there is an element v in M with $\|v - u\| < \varepsilon$.

An H-space X is called *separable* iff X contains an at most countable dense set.

Compact sets. The set M is called relatively compact iff each sequence in M has a convergent subsequence.

The set M is called compact iff it is relatively compact and closed, i.e., each sequence (u_n) in M has a convergent subsequence $u_{n'} \to u$ as $n \to \infty$ where $u \in M$.

Linearity. The set M is called a linear subspace of X iff

$$u, v \in M \qquad \text{and} \qquad \alpha, \beta \in \mathbb{K} \qquad \text{imply} \qquad \alpha u + \beta v \in M.$$

The operator $A: D(A) \subseteq X \to Y$ is called linear iff $D(A)$ is a linear subspace of X and for all $u, v \in D(A)$ and $\alpha, \beta \in \mathbb{K}$,

$$A(\alpha u + \beta v) = \alpha A u + \beta A v.$$

The linear operator $A: D(A) \subseteq X \to Y$ is called bounded iff there is a real number c such that

$$\|Au\| \leq c\|u\| \qquad \text{for all} \quad u \in D(A).$$

If the linear operator $A: X \to Y$ is bounded, then we define

$$\|A\| = \sup_{\|u\| \leq 1} \|Au\|.$$

This implies

$$\|Au\| \leq \|A\| \|u\| \qquad \text{for all} \quad u \in X.$$

In this connection note the linearity of A.

Antilinearity. The operator $A: X \to Y$ is called antilinear iff, for all $u, v \in X$ and $\alpha, \beta \in \mathbb{K}$,

$$A(\alpha u + \beta v) = \bar{\alpha} Au + \bar{\beta} Au.$$

Here, the bar denotes the conjugate complex number. If X and Y are real H-spaces, then antilinear and linear operators coincide.

Continuity. The operator $A: X \to Y$ is called continuous iff, as $n \to \infty$,

$$u_n \to u \qquad \text{implies} \qquad Au_n \to Au.$$

We want to prove the following:

A linear operator $A: X \to Y$ is bounded iff it is continuous.

First suppose that A is bounded. Then it follows from $u_n \to u$ as $n \to \infty$ that

$$\|Au_n - Au\| = \|A(u_n - u)\| \le \|A\| \, \|u_n - u\|$$

and hence $Au_n \to Au$ as $n \to \infty$, i.e., A is continuous.

Conversely, suppose that A is continuous. If A is not bounded, then there is a sequence (u_n) such that

$$\|u_n\| = 1 \qquad \text{and} \qquad \|Au_n\| \ge n \qquad \text{for all } n.$$

Set $v_n = u_n/\|Au_n\|^{1/2}$. Then, as $n \to \infty$,

$$\|v_n\| = 1/\|Au_n\|^{1/2} \to 0$$

and $\|Av_n\| = \|Au_n\|^{1/2} \to \infty$. This contradicts the continuity of A.

Compactness. The operator $A: X \to Y$ is called compact iff it is continuous and it maps bounded sets into relatively compact sets. Thus, if A is compact, then each bounded sequence (u_n) contains a subsequence $(u_{n'})$ such that $(Au_{n'})$ is convergent.

Linear functionals. A linear continuous map $f: X \to \mathbb{K}$ is called a linear continuous functional on X. The set of all the linear continuous functionals on X is denoted by X^*. Consequently, we have $f \in X^*$ iff $f: X \to \mathbb{K}$ is linear and there is a real number c such that

$$|f(u)| \le c\|u\| \qquad \text{for all} \quad u \in X.$$

We set

$$\|f\| = \sup_{\|u\| \le 1} |f(u)|.$$

Instead of $f(u)$ we also write $\langle f, u \rangle$.

In the Appendix of Part I, the interested reader will find the relationship between the notions introduced above and general notions in topology.

The notion of an abstract Hilbert space was first introduced by von Neumann (1929). In his basic papers around 1906, Hilbert used the space l^2 consisting of all sequences $x = (\xi_i)$ with $\xi_i \in \mathbb{K}$ for all i and

$$\sum_{i=1}^{\infty} |\xi_i|^2 < \infty.$$

Here the scalar product is given by

$$(x|y) = \sum_{i=1}^{\infty} \bar{\xi}_i \eta_i.$$

The space l^2 generalizes the classical Euclidean space $\mathbb{R}^n$ to infinite dimensions.

A Look at the History of Hilbert Spaces

The following quotations should help the reader to understand the birth of the notion of Hilbert space.

> Further considerations of the subject led me to the conclusion that the systematic construction of a general theory of integral equations is of the utmost importance for the complete field of analysis and, in particular, for the expansion of functions in infinite series, linear differential equations, analytic functions, potential theory, and the calculus of variations.
>
> David Hilbert (1912)
> (From the Preface to his monograph on linear integral equations.)

In order to understand the great achievement of Hilbert (1862–1943) in the field of analysis, it is necessary to first comment on the state of analysis at the end of the nineteenth century. After Weierstrass (1815–1897) had made sure of the foundations of the complex function theory, and it had reached an impressive level, research switched to *boundary value problems*, which first arose in physics. The work of Riemann (1826–1866) on complex function theory, however, had shown that boundary value problems have great importance for pure mathematics as well. Two problems had to be solved:

(i) the problem of the existence of a potential function for given boundary values; and
(ii) the problem of eigenoscillations of elastic bodies, for example, string and membrane.

The state of the theory was bad at the end of the nineteenth century. Riemann had believed that by using the *Dirichlet principle*, one could deal with these problems in a simple and uniform way. After Weierstrass' substantial *criticism* of the Dirichlet principle in 1870, special methods had to be developed for these problems. These methods, by C. Neumann, Schwarz, and Poincaré, were very elaborate and still have great aesthetic appeal today; but because of their variety they were confusing, although at the end of the nineteenth century, Poincaré (1854–1912), in particular, endeavoured with great astuteness to standardize the theory. There was, however, a lack of "simple basic facts" from which one could easily get complete results without sophisticated investigations of limiting processes.

Hilbert first looked for these "simple basic facts" in the calculus of variations. He considered so-called regular variational problems which satisfy the Legendre condition. In 1900 he had an immediate and great success: he succeeded in *justifying* the Dirichlet principle.

While Hilbert used variational methods, the Swedish mathematician Fredholm

(1866–1927) approached the same goal by developing Poincare's work by using linear integral equations. In the winter semester 1900/1901 Holmgren, who had come from Upsala (Sweden) to study under Hilbert in Göttingen, held a lecture in Hilbert's seminar on Fredholm's work on linear integral equations which had been published the previous year. This was a decisive day in Hilbert's life. He took up Fredholm's new discovery with great zeal, and combined it with his variational methods. In this way he succeeded in creating a uniform theory which solved problems (i) and (ii) above.

In 1904 Hilbert's first note on the "Foundations of a General Theory of Linear Integral Equations" was published in the *Göttinger Nachrichten.* These results were based on lectures which Hilbert had held from the summer of 1901 onwards. Fredholm had proved the existence of solutions for linear integral equations of the second kind. His result was sufficient to solve the boundary value problems of potential theory. But Fredholm's theory did not include the eigenscillations and the expansion of arbitrary functions with respect to eigenfunctions. Only Hilbert solved this problem by using finite-dimensional approximations and a passage to the limit. In this way he obtained a generalization of the classical principal-axis transformation for symmetric matrices to infinite-dimensional matrices. The symmetry of the matrices corresponds to the symmetry of the kernels of integral equations, and it shows that the kernels appearing in oscillation problems are indeed symmetrical.

From our point of view today, Hilbert's paper of 1904 appears clumsy, compared to the elegance of Erhard Schmidt's method (1907a, b) which he developed in his dissertation written while a student of Hilbert in Göttingen. But the first step had been made. In the same year, 1904, Hilbert, in his second note, was able to apply his theory to general Sturm–Liouville problems. His third note in 1905 contained a very important result. Of the great problems which Riemann had posed with the complex function theory, there was still one left open: the proof of the existence of differential equations with a prescribed monodromy group. Hilbert solved this problem by reducing it to the determination of two functions which are holomorphic in both the interior and exterior of a closed curve, and whose real and imaginary parts satisfy appropriate linear relations on the curve (the Hilbert–Riemann problem[1]). The solution to this problem is a classic example for the axiomatics of limiting processes demanded by Hilbert. No concrete limiting processes are used, but everything results from the existence of the Green function for the interior and the exterior of the closed curve, and from the Fredholm alternative which says that either the homogeneous integral equation has a nontrivial solution or the inhomogeneous integral equation has a solution.

Hilbert soon noticed that limits are set to the method of integral equations. In order to overcome these limits he created, in his fourth and fifth notes in 1906, the general theory of quadratic forms of an infinite number of variables. Hilbert believed that with this theory he had provided analysis with a general basis which corresponded to an *axiomatics of limiting processes.* The further development of mathematics has proved him to be right.

Otto Blumenthal (1932)

The above quotation constitutes an extract from Hilbert's biography, which can be found in Volume 3 of his *Collected Works.* We recommend that the

[1] Today, Riemann–Hilbert problems play a fundamental role in the theory of solitons. See Faddeev and Takhtadjan (1987).

reader has a look at these *Collected Works*. Every single one of Hilbert's mathematical works and essays is a masterpiece.

> In the fall of 1926, the young John von Neumann (1903–1957) arrived at Göttingen to take up his duties as Hilbert's assistant. These were the hectic years during which quantum mechanics was developing at breakneck speed, with a new idea popping up every few weeks from all over the horizon. The theoretical physicists who were developing the new theory were groping for adequate mathematical tools, trying in succession infinite matrices without any considerations of convergence, differential operators, "continuous matrices" (whatever that might mean), etc. As late as 1924, most physicists did not even know that a finite matrix was! It finally dawned upon them that their "observables" had properties which made them look like Hermitian operators in Hilbert space, and that by an extraordinary coincidence, the "spectrum" of Hilbert (a name which he had apparently chosen from a superficial analogy) was to be the central conception in the explanation of the "spectra" of atoms. It was therefore natural that they should enlist Hilbert's help in trying to put some mathematical sense in their computations. With the assistance of Nordheim and von Neumann, Hilbert first tried integral operators in L_2, but that needed the use of the Dirac "δ-function," a concept which was for the mathematicians of that time self-contradictory. Von Neumann therefore resolved to try another approach.
>
> Jean Dieudonné (1981)

> Stimulated by an interest in quantum mechanics, John von Neumann began the work in operator theory which he was to continue as long as he lived. Most of the ideas essential for an abstract theory had already been developed by the Hungarian mathematician Frigyes Riesz (1880–1956), who had established the spectral theory for bounded symmetric operators in a form very much like that now regarded as standard. Von Neumann saw the need to extend Riesz's treatment to *unbounded* operators and found a clue to doing this in Carleman's highly original book on integral operators with singular kernels. The result was a paper von Neumann submitted for publication to the *Mathematische Zeitschrift* but later withdrew. The reason of this withdrawal was that in 1928 Erhard Schmidt (1876–1959) and I, independently, saw the role which could played in the theory by the concept of the adjoint operator, and the importance which should be attached to self-adjoint operators. When von Neumann learned from Professor Schmidt of this observation, he was at once able to rewrite his paper in a much more satisfactory and complete form, giving a full spectral theory for all closed symmetric operators as well as for the self-adjoint operators. This he did by abandoning Carleman's method, which he had been able to apply only by use of a transfinite induction, and introducing the Cayley transform, which served to reduce the theory of unbounded symmetric operators to that of bounded isometric operators. Incidentally, for permission to withdraw the paper without penalty, when it was already in page proof, the publisher exacted from Professor von Neumann a promise to write a book on quantum mechanics. The book soon appeared and has become one of the classics of modern physics, particularly valued for its analysis of quantum statistics (*Mathematische Grundlagen der Quantenmechanik*, Springer-Verlag, 1932).
>
> Marshall Harvey Stone (1970)

A great master of mathematics passed away when David Hilbert died in Göttingen on February 14, 1943, at the age of eighty-one. In retrospect, it seems to us that the era of mathematics upon which he impressed the seal of his spirit, and which is now sinking below the horizon, achieved a more perfect balance than has prevailed before or since, between the mastering of single concrete problems and the formation of general abstract concepts. Hilbert's own work contributed not a little to bringing about this happy equilibrium, and the direction in which we have since proceeded can in many instances be traced back to this impulse. No mathematician of equal stature has risen from our generation ....

Hilbert was singularly free from national and racial prejudices; in all public questions, be they political, social, or spiritual, he stood forever on the side of freedom, frequently in isolated opposition against the compact majority of his environment. He kept his head clear and was not afraid to swim against the current, even amidst the violent passions aroused by the First World War (1914–1918) that swept so many other scientists off their feet. It was not mere chance, when the Nazis "purged" the German universities in 1933 and their hand fell most heavily on the Hilbert school, that Hilbert's most intimate collaborators left Germany either voluntarily or under the pressure of Nazi persecution. He himself was too old, and stayed behind; but the years after 1933 became for him years of ever-deepening tragic loneliness.

Hermann Weyl (1944)

CHAPTER 18

Variational Problems, the Ritz Method, and the Idea of Orthogonality

> When we do scientific work, we must often step down from our high horse of grand principles, and dig in the dirt with our noses. When we achieve our purpose, we cover the tracks of our efforts in order to appear as gods of clear thought.
>
> Albert Einstein

In Sections 18.1 through 18.5 of this chapter we consider a number of important concrete examples in order to explain the following:

(i) The connection between variational problems and boundary value problems for elliptic partial differential equations.
(ii) The equivalence between classical and generalized solutions in case the solutions are sufficiently smooth.
(iii) The Ritz method for the construction of approximate solutions.

This should help the reader to recognize the simple basic ideas behind the general theory. In this connection, a fundamental role is played by:

(a) the formula of integration by parts; and
(b) the variational lemma.

A general convergence proof for the Ritz method will be given in Section 22.1.

In the middle of this chapter there stands an elegant functional analytic justification of the famous *Dirichlet principle.* Figure 18.4 shows the logical structure of our existence proof.

(i) In Section 18.8 we use the parallelogram identity in order to prove very simply a general existence theorem for quadratic minimum problems. The parallelogram identity is a generalization of the classical Pythagorean theorem.

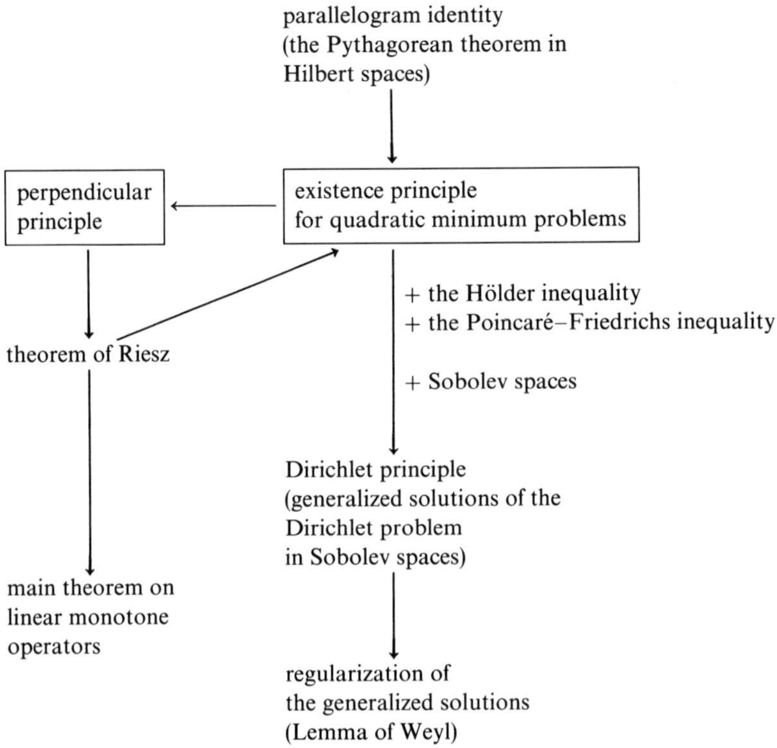

Figure 18.4

(ii) The functional analytic justification of the Dirichlet principle is a special case of the general existence theorem in (i). To this end, we need the following three analytical tools: the Hölder inequality, the Poincaré–Friedrichs inequality, and Sobolev spaces (Sections 18.9 and 18.10).

(iii) In Section 18.11 we show that the following three principles are mutually equivalent: the existence principle for quadratic minimum problems, the perpendicular principle, and the Riesz theorem.

The perpendicular principle says that there exists a perpendicular from each point of an H-space to a closed linear subspace. Consequently, we obtain the following:

The functional analytic justification of the Dirichlet principle is based on the idea of orthogonality.

Therefore, the Dirichlet principle is closely related to the classical Pythagorean theorem.

There are ideas in mathematics which remain eternally young and which lose nothing of their intellectual freshness, even after thousands of years.

Mathematicians of the Pythagorean school in ancient Greece attributed the Pythagorean theorem to the master of their school, Pythagoras of Samos (circa 560 B.C.–480 B.C.). It is said that Pythagoras sacrified one hundred oxen to the gods in gratitude. In fact, this theorem was already known in Babylon at the time of King Hammurabi (circa 1728 B.C.–1686 B.C.). Presumably, however, it was a mathematician of the Pythagorean school who first proved the Pythagorean theorem. This theorem appears as Proposition 47 in Book I of Euclid's *Elements* (300 B.C.).

The theory of Hilbert spaces is the abstract and very efficient formulation of the idea of orthogonality. It seems that this idea has deep roots in our real world. In Part V we shall show that Hilbert spaces represent the right mathematical tool in order to describe the strange nature of quantum phenomena. In nature we observe a duality between waves and particles, which is typical for quantum processes. For example, light possesses such a dual structure. In order to formulate this duality in terms of mathematics, one uses the theory of Hilbert spaces. The main idea is that the quantization of particle theories and wave theories lead to equivalent theories in H-spaces.

In Sections 18.7 and 18.19 we discuss briefly the historical development of the Dirichlet principle, and of the 19th and 20th problems of Hilbert and their relationship to the modern theory of monotone operators.

> One cannot comprehend what it is one possesses if one has not understood what one's predecessors possessed.
>
> Johann Wolfgang von Goethe (1749–1832)

> Mathematics has the advantage over many other fields of knowledge, such as history, for example, that one can sometimes distinguish between "true" and "false." Because of this mathematics pays dearly for this advantage in that it is the most remote of all human things. One best notices the mathematician's longing for humanity and history in the introductions to mathematical writings.
>
> Wilhelm Blaschke (1942)

18.1. The Space $C_0^\infty(G)$ and the Variational Lemma

We first list a number of function spaces.

Definition 18.1. Let G be a nonempty open set in $\mathbb{R}^N$, $N \geq 1$.

(a) $C^k(G)$ is the set of all real functions $u: G \to \mathbb{R}$ that have continuous partial derivatives of orders $m = 0, 1, \ldots, k$ on G. We understand the derivative of order $m = 0$ to be the function itself.
(b) $C^k(\bar{G})$ is the set of all $u \in C^k(G)$ for which all partial derivatives of orders $m = 0, \ldots, k$ can be extended continuously to $\bar{G}$.
(c) We denote briefly by $C(G)$ (resp. $C(\bar{G})$) the continuous real functions on G (resp. $\bar{G}$) instead of $C^0(G)$ (resp. $C^0(\bar{G})$).
(d) If $u \in C^k(G)$ (resp. $u \in C^k(\bar{G})$) for all $k = 0, 1, \ldots$, then we write $u \in C^\infty(G)$ (resp. $u \in C^\infty(\bar{G})$).

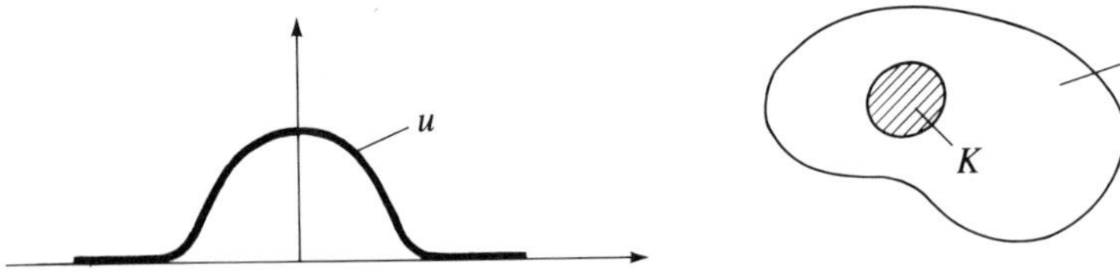

(a) $u \in C_0^\infty(\mathbb{R})$ (b)

Figure 18.5

(e) $C_0^\infty(G)$ is the set of all functions $u \in C^\infty(G)$ which vanish identically outside a compact subset K of G that depends on u (Fig. 18.5).

The following important result will be used extraordinarily often.

Proposition 18.2 (Variational Lemma). *Let G be a nonempty open set in $\mathbb{R}^N$, $N \geq 1$. Let $u \in L_2(G)$ and suppose that*

$$\int_G uv\,dx = 0 \qquad \text{for all} \quad v \in C_0^\infty(G). \tag{6}$$

Then we obtain

$$u(x) = 0 \qquad \text{almost everywhere on } G.$$

The notion "almost everywhere" will be used very frequently in this volume. We recall this in the Appendix, $A_2(2)$.

PROOF. We set $X = L_2(G)$ and use the well-known fact that $C_0^\infty(G)$ is dense in X. Hence there exists a sequence (u_n) in $C_0^\infty(G)$ with

$$u_n \to u \quad \text{in } X \qquad \text{as} \quad n \to \infty.$$

From (6) we obtain $(u|u_n) = 0$ for all n. Letting $n \to \infty$ this implies $(u|u) = 0$, and hence $u = 0$.

The density of $C_0^\infty(G)$ in X will be proved in Section 18.14 as a special application of smoothing operators. □

If $u \in C(G)$ in Proposition 18.2, then obviously we obtain $u(x) = 0$ for all $x \in G$.

In Sections 18.2 through 18.5 we shall make the following assumption:

G is a bounded region in $\mathbb{R}^N$ with $N \geq 1$. The boundary is piecewise smooth, i.e., to be precise, $\partial G \in C^{0,1}$ holds.[1] (7)

We gave the exact definition of "$\partial G \in C^{0,1}$" in Section 6.2. Intuitively, we think of this to mean all reasonable regions where the boundary can also have

[1] Throughout this volume, condition (7) means that G is a bounded open interval if $N = 1$.

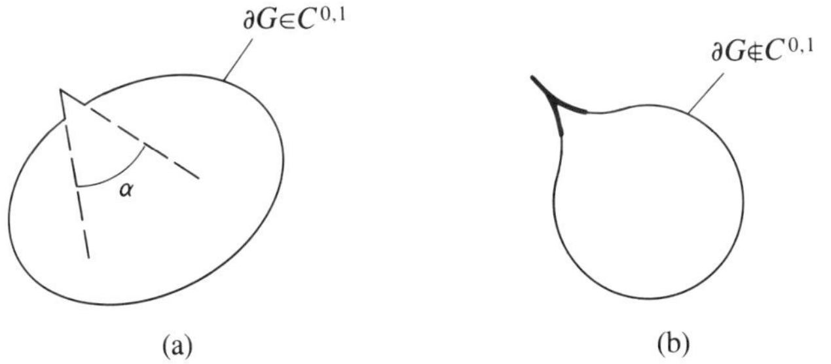

Figure 18.6

a finite number of vertices and edges with nonzero angles (Fig. 18.6(a)). Zero angles, as in Figure 18.6(b), are not allowed. Recall from Part I that a *region* is a nonempty open connected set.

Let $x = (\xi_1, \ldots, \xi_N) \in \mathbb{R}^N$ and

$$D_i = \partial/\partial\xi_i.$$

We denote the surface integral by

$$S = \int_{\partial G} g\, dO.$$

Here, for $N = 2$, S is equal to the contour integral over the boundary curve ∂G, where this curve is oriented in the mathematically positive sense. For $N = 1$ and $G =]a, b[$ we have $S = g|_a^b$, i.e., $S = g(b) - g(a)$. The *Laplacian* Δ is defined by

$$\Delta u = \sum_{i=1}^{N} D_i^2 u.$$

18.2. Integration by Parts

The integration by parts formula is the key to the modern theory of partial differential equations.

The classical integration by parts formula

$$\int_a^b u'v\, dx = uv|_a^b - \int_a^b uv'\, dx \tag{8a}$$

is valid in $\mathbb{R}^1$, where $-\infty < a < b < \infty$. The natural generalization to functions $u: G \subset \mathbb{R}^N \to \mathbb{R}$ reads as follows for $i = 1, \ldots, N$:

$$\int_G (D_i u)v\, dx = \int_{\partial G} uvn_i\, dO - \int_G uD_i v\, dx. \tag{8b}$$

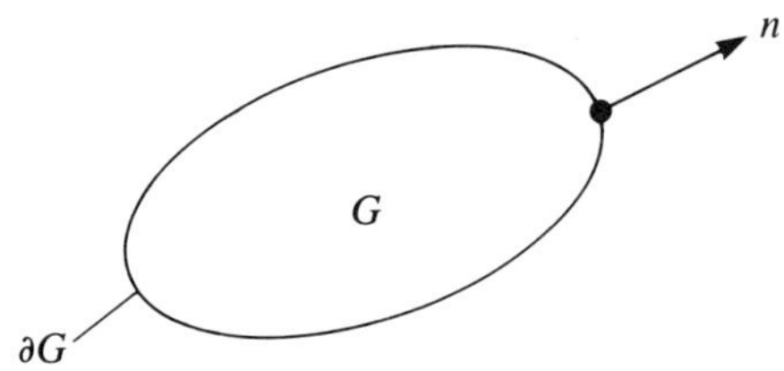

Figure 18.7

Here $n = (n_1, \ldots, n_N)$ denotes the outer unit normal to the boundary ∂G (Fig. 18.7). At the boundary points, in which no outer normal is defined, i.e., in vertices and edges, let $n_i = 0$, $i = 1, \ldots, N$. We denote the surface differential by dO. For $N = 1$, (8a) and (8b) coincide.

Proposition 18.3 (Integration by Parts). *Equation* (8b) *is valid for all* $u, v \in C^1(\bar{G})$ *in case* G *satisfies assumption* (7).

If u or v belongs to $C_0^\infty(G)$, then these functions vanish on ∂G; therefore, the boundary integral drops out in (8b).

Corollary 18.4. *For* (8b) *to hold, it suffices that* u, v *are elements of Sobolev spaces, to be introduced later. To be precise,*

$$u \in W_p^1(G), \qquad v \in W_q^1(G), \quad 1 < p < \infty, \qquad p^{-1} + q^{-1} = 1$$

must hold.

A proof of the well-known Proposition 18.3 can be found, for example, in Nečas (1967, M), p. 121. Corollary 18.4 follows by means of a passage to the limit, since $C^1(\bar{G})$ is dense in $W_p^1(G)$ and $W_q^1(G)$.

Definition 18.5. We understand the (outer) normal derivative of u at $x \in \partial G$ to be

$$\frac{\partial u(x)}{\partial n} = \sum_{i=1}^{N} n_i D_i u(x). \tag{9}$$

For $N = 1$ and $G =]a, b[$ we have

$$\frac{\partial u(a)}{\partial n} = -u'(a), \qquad \frac{\partial u(b)}{\partial n} = u'(b).$$

If we write the one-dimensional integration by parts formula (8a) in the form

$$\int_a^b (uv)'\, dx = uv|_a^b,$$

then this is precisely the fundamental theorem of calculus, which expresses the relationship between tangent and area discovered by Newton and Leibniz. The famous Gauss theorem

$$\int_G D_i(uv)\,dx = \int_{\partial G} uvn_i\,dO$$

generalizes this relationship to higher dimensions. But this is precisely the integration by parts formula (8b). We shall see later on that the integration by parts formula is the key to generalized derivatives, Sobolev spaces, distributions, and generalized solutions of partial differential equations. Thus, this formula is a cornerstone of modern analysis.

18.3. The First Boundary Value Problem and the Ritz Method

18.3a. Equivalent Problems

We want to investigate the connection between the following problems for the unknown function u. Let $M = \{v \in C^1(\bar{G})\colon v = 0 \text{ on } \partial G\}$.

(A) *Variational problem*

$$\begin{gathered}\int_G \left(\frac{1}{2}\sum_{i=1}^N (D_iu)^2 - uf\right)dx = \min!, \qquad u \in C^1(\bar{G}),\\ u = g \quad \text{on } \partial G.\end{gathered} \tag{10}$$

(B) *Generalized boundary value problem*

$$\begin{gathered}\int_G \left(\sum_{i=1}^N D_iuD_iv - fv\right)dx = 0 \qquad \text{for all} \quad v \in M,\\ u = g \quad \text{on } \partial G.\end{gathered} \tag{11}$$

(C) *Boundary value problem*

$$\begin{aligned}-\Delta u &= f \quad \text{on } G,\\ u &= g \quad \text{on } \partial G.\end{aligned} \tag{12}$$

Equation (12) is called the *Euler equation* to the original variational problem (10).

Proposition 18.6. *Let $f \in C(\bar{G})$ and $g \in C(\partial G)$. Assume that the region G satisfies* (7). *Then the following is valid*:

(a) *For $u \in C^2(\bar{G})$, the variational problem* (10), *the generalized boundary value problem* (11), *and the boundary value problem* (12) *are mutually equivalent.*

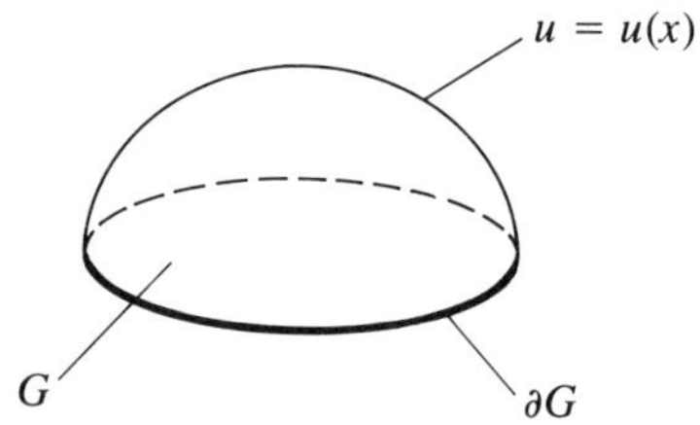

Figure 18.8

(b) *For $u \in C^1(\bar{G})$, the variational problem* (10) *and the generalized boundary value problem* (11) *are equivalent.*

In order to associate a physical picture with this problem, let G be a region in $\mathbb{R}^2$. Then we can interpret $u(x)$ as the vertical displacement of a membrane at the point x under the influence of an external force f (e.g., the force of gravity). The boundary condition "$u = g$ on ∂G" means that u is fixed at the boundary. Figure 18.8 shows the case $g \equiv 0$. Problem (10) corresponds to the principle of minimal potential energy, i.e., the membrane realizes the state of minimal potential energy for fixed boundary values.

The following argument (I) is *typical* for the calculus of variations. In Section 18.18 we shall show that the following implications (10) ⇒ (11) ⇒ (12) can be generalized in a straightforward manner to general variational problems.

PROOF. We set

$$F(u) = \int_G \left(\frac{1}{2} \sum_{i=1}^N (D_i u)^2 - fu \right) dx.$$

The decisive *trick* of the calculus of variations consists of reducing variational problems to extremal problems for real functions. To this end, we define the real function

$$\varphi(t) = F(u + tv) \qquad \text{for all} \quad t \in \mathbb{R}$$

and fixed $v \in M$, i.e.,

$$\varphi(t) = \int_G \left(\frac{1}{2} \sum_{i=1}^N (D_i(u + tv))^2 - f(u + tv) \right) dx.$$

Ad(a). Let $u \in C^2(\bar{G})$.

(I) (10) ⇒ (11). Suppose that u is a solution of the original variational problem (10), i.e.,

$$F(u) = \min!, \qquad u = g \quad \text{on } \partial G, \qquad u \in C^1(\bar{G}). \tag{10*}$$

Recall that $M = \{v \in C^1(\bar{G}): v = 0 \text{ on } \partial G\}$. Then for all $v \in M$, $t \in \mathbb{R}$, we have

$$u + tv = g \quad \text{on } \partial G, \qquad u + tv \in C^1(\bar{G}),$$

i.e., the functions $u + tv$ are *allowed* in the competition in (10*). Hence the real function φ has a minimum at $t = 0$; thus, we obtain the *key condition*

$$\varphi'(0) = 0.$$

This implies

$$\int_G \left(\sum_{i=1}^N D_i u D_i v - fv \right) dx = 0 \qquad \text{for all} \quad v \in M, \tag{11*}$$

which is precisely the generalized boundary value problem (11).

(II) (11) ⇒ (12). By integration by parts, it follows from (11*) that

$$\int_G (-\Delta u - f) v \, dx = 0 \qquad \text{for all} \quad v \in M,$$

in particular for all $v \in C_0^\infty(G)$. Application of the variational lemma (Proposition 18.2) yields

$$-\Delta u - f = 0 \quad \text{on } G.$$

Since $u = g$ on ∂G, we obtain the boundary value problem (12).

(III) (12) ⇒ (11). We multiply (12) by $v \in M$ and integrate over G. Then integration by parts yields (11).

(IV) (11) ⇔ (10). Since φ is a quadratic function, the following is valid:

$$\varphi \text{ has a minimum at } t = 0 \qquad \text{iff} \qquad \varphi'(0) = 0.$$

Recall that φ depends on v. Hence the original variational problem (10) is equivalent to

$$\varphi'(0) = 0 \qquad \text{for all} \quad v \in M.$$

But this is identical to (11).

Ad(b). In the proof of (10) ⇔ (11) we have only used $u \in C^1(\bar{G})$. □

In addition, we define

$$\delta^k F(u; v) = \varphi^{(k)}(0)$$

and call $\delta^k F(u; v)$ the kth *variation* of the variational integral F at the point u in the direction of v. In particular, we obtain

$$\delta F(u; v) = \int_G \left(\sum_{i=1}^N D_i u D_i v - fv \right) dx,$$

$$\delta^2 F(u; v) = \int_G \sum_{i=1}^N (D_i v)^2 \, dx.$$

Thus the generalized boundary value problem (11) is equivalent to the vanishing of the first variation, i.e.,

$$\delta F(u; v) = 0 \qquad \text{for all} \quad v \in M.$$

18.3b. The Ritz Method

We now explain, with the Ritz method for the approximate solution of (10) and hence of (12), a basic general approximation method for the solution of variational problems. The basic idea can be formulated briefly as follows:

(R) *For a given variational problem one varies only over finite-dimensional subsets that satisfy the side conditions.*

The advantage of this method consists in that one can *reduce* variational problems in function spaces to variational problems for real functions with finitely many variables. As a simple example, we first consider the minimum problem

$$f(\xi_1, \dots, \xi_N) = \min! \tag{13}$$

for the real function $f: \mathbb{R}^N \to \mathbb{R}$. An approximate solution can be obtained by varying over only special coordinates, for instance, one can choose $\xi_1, \dots, \xi_k$ to be free and set $\xi_{k+1} = \cdots = \xi_N = 0$. The variational problem (10) transpires, in contrast to (13), in an infinite-dimensional function space. The Ritz method for (10) is based on the following two conditions, in accordance with (R):

(i) One varies only over all real linear combinations of finitely many fixed functions.
(ii) All these linear combinations fulfill the boundary condition $u = g$ on ∂G, which corresponds to the side condition in (R).

As a formal simplification, we first assume that $g \equiv 0$ and discuss at the end of this section a general method for reducing problems with inhomogeneous side conditions to problems with homogeneous side conditions, by means of a simple subtraction trick. In order to formulate (i), we choose fixed functions $w_1, \dots, w_n$ and make the trial

$$u_n = \sum_{k=1}^{n} c_{kn} w_k$$

for the approximate solution u_n, with the unknown *real coefficients* $c_{1n}, \dots, c_{nn}$. In order to satisfy condition (ii) with $g \equiv 0$, we require

$$w_k = 0 \quad \text{on } \partial G \qquad \text{for all } k.$$

Then the boundary condition

$$u_n = 0 \quad \text{on } \partial G$$

is automatically fullfilled for all u_n. According to (i), we now replace the original variational problem (10),

$$\int_G \left(\frac{1}{2} \sum_{i=1}^{N} (D_i u)^2 - uf \right) dx = \min!,$$

$$u = 0 \quad \text{on } \partial G,$$

by the following *approximate problem*:

$$\int_G \left(\frac{1}{2} \sum_{i=1}^{N} (D_i u_n)^2 - uf \right) dx = \min!, \qquad c_n \in \mathbb{R}^n, \tag{14}$$

where $c_n = (c_{1n}, \ldots, c_{nn})$. Thus, in contrast to the original problem, in (14) we vary over all functions u_n with $u_n = 0$ on ∂G. Yet we add a purely technical condition. So that the integral in (14) will exist, we require that all the basic functions $w_1, \ldots, w_n$ are continuous and piecewise continuously differentiable on $\bar{G}$.

Corollary 18.7. *If u_n is a solution of* (14), *then*

$$\int_G \left(\sum_{i=1}^{N} D_i u_n D_i w_j - f w_j \right) dx = 0, \qquad j = 1, \ldots, n. \tag{15}$$

Obviously, this relation is equivalent to the so-called *Ritz equations*:

$$\sum_{k=1}^{n} c_{kn} \int_G \sum_{i=1}^{N} D_i w_k D_i w_j \, dx = \int_G f w_j \, dx, \qquad j = 1, \ldots, n. \tag{16}$$

This is a linear system of equations for determining the coefficients $c_{1n}, \ldots, c_{nn}$ of u_n. A comparison with (11) yields the following crucial observation:

The Ritz equations (15) *are obtained formally from the generalized problem* (11) *by replacing u by u_n and requiring that* (11) *holds for all* $v \in \operatorname{span}\{w_1, \ldots, w_n\}$.

PROOF. We write (14) as

$$f(c_n) = \min!, \qquad c_n \in \mathbb{R}^n.$$

For a solution c_n, we obtain

$$\frac{\partial f(c_n)}{\partial c_{jn}} = 0, \qquad j = 1, \ldots, n.$$

This immediately implies (16) and hence (15). □

A concrete example for the Ritz method will be considered in Problem 18.1.

18.3c. Discussion

Our considerations above give rise to the following questions:

(i) In what sense do equations (10), (11), and (12) have a solution u?
(ii) In which sense does (u_n) converge to u? How are $w_1, w_2, \ldots$ to be chosen so that convergence occurs?
(iii) How can the error estimates be obtained?

We give the answers to these questions in Sections 22.1 through 22.3. Here we content ourselves with a brief explanation. Let $g \equiv 0$.

Ad(i). In the general case, the original variational problem (10) possesses *no* solution $u \in C^1(\bar{G})$, but a unique solution u in the

$$\text{Sobolev space } \mathring{W}_2^1(G).$$

In order for the integral in (10) to exist, it suffices that u has generalized first derivatives $D_i u$ that are square integrable. These functions are in $W_2^1(G)$. One can motivate the choice of $\mathring{W}_2^1(G)$ by the boundary condition

$$u = 0 \quad \text{on } \partial G.$$

In this connection, note the following:

(a) the inclusion $\mathring{W}_2^1(G) \subseteq W_2^1(G)$ is valid; and
(b) the functions from $M = \{u \in C^1(\bar{G}): u = 0 \text{ on } \partial G\}$ are *dense* in $\mathring{W}_2^1(G)$.

In fact, the functions in the Sobolev space $\mathring{W}_2^1(G)$ have even boundary values in a certain generalized sense, i.e.,

$$u \in \mathring{W}_2^1(G) \qquad \text{implies} \qquad u = 0 \quad \text{on } \partial G$$

in a certain generalized sense (see Section 21.3).

Furthermore, we mention the fact that we can allow $f \in L_2(G)$ in (10), i.e., $\int_G f^2\,dx < \infty$. Thus, the function f may have discontinuities.

The final generalized boundary value problem (11) reads as follows:

For given $f \in L_2(G)$, a function $u \in \mathring{W}_2^1(G)$ is sought so that (11) *is valid for all $v \in \mathring{W}_2^1(G)$, i.e.,*

$$\text{(C*)} \qquad \int_G \left(\sum_{i=1}^N D_i u D_i v - uf \right) dx = 0 \qquad \text{for all} \quad v \in \mathring{W}_2^1(G).$$

This problem is called the generalized problem to the classical problem (12), i.e.,

$$\text{(C)} \qquad -\Delta u = f \quad \text{on } G, \qquad u = 0 \quad \text{on } \partial G.$$

According to Proposition 18.6, each classical solution of (C) is also a generalized solution, i.e., it is a solution of (C*). However, the converse is *not* true. Only with smooth data f and ∂G, can one verify that the generalized solutions of (C*) are also classical solutions of (C).

From the *physical point of view*, the generalized solutions are very natural, namely:

We determine solutions u for which the energy integral (10) *of the principle of minimal potential energy exists at least in a generalized sense.*

In this connection, note the following. In the original variational problem (10) and in the generalized boundary value problem (11), we only need *first-*

order derivatives. Thus, it is natural to work with the Sobolve space $\mathring{W}_2^1(G)$ in (C*), which contains functions having generalized first-order derivatives. In contrast to this, the classical boundary value problem (C) contains second-order derivatives.

Ad(ii). In Theorem 22.A we shall see that $\{w_1, w_2, \ldots\}$ must form a basis in $\mathring{W}_2^1(G)$ or, more generally, a so-called *Galerkin scheme*. By $A_2(57)$, this occurs, e.g., with regard to the functions $\{gp_1, gp_2, gp_3, \ldots\}$, where the p_j are polynomials and g, with certain regularity properties, vanishes on the boundary ∂G. Moreover, by $A_2(59)$, one can use piecewise linear functions, i.e., so-called finite elements. Trigonometric polynomials can also serve this purpose (cf. Problem 18.1).

Ad(iii). In order to obtain error estimates, one makes use of *duality theory*, i.e., one considers a dual maximum problem besides the initial minimum problem. We discuss a simple special case in Section 22.4. The general duality theory will be treated in Part III, Chapter 50, within the context of convex analysis.

Remark 18.8 (Reduction Principle). In conclusion, we describe a simple method by means of which one can reduce inhomogeneous boundary value conditions to homogeneous ones. To this end, we consider the *inhomogeneous* boundary value problem

$$\begin{aligned} -\Delta u &= f \quad \text{on } G,\\ u &= g \quad \text{on } \partial G. \end{aligned}$$

We choose a function $\bar{g}\colon \bar{G} \to \mathbb{R}$ with

$$\bar{g} = g \quad \text{on } \partial G.$$

If we now set $v = u - \bar{g}$, then we obtain the *homogeneous* boundary value problem

$$\begin{aligned} -\Delta v &= f + \Delta\bar{g} \quad \text{on } G,\\ v &= 0 \quad \text{on } \partial G. \end{aligned}$$

Analogously, the variational problem

$$\int_G \left(\frac{1}{2}\sum_{i=1}^N (D_i u)^2 - uf\right) dx = \min!, \qquad u = g \quad \text{on } \partial G,$$

is reduced to

$$\int_G \left(\frac{1}{2}\sum_{i=1}^N (D_i(v + \bar{g}))^2 - (v + \bar{g})f\right) dx = \min!, \qquad v = 0 \quad \text{on } \partial G.$$

In order to transform an inhomogeneous differential equation into a homogeneous one, we proceed similarly. For example, we consider the *inhomogeneous* equation

$$\begin{aligned} -\Delta u &= f \quad \text{on } G,\\ u &= g \quad \text{on } \partial G. \end{aligned}$$

If one knows a special solution $\bar{u}$ with

$$-\Delta\bar{u} = f \quad \text{on } G,$$

then we set $v = u - \bar{u}$, and hence we obtain the homogeneous equation

$$\begin{aligned} -\Delta v &= 0 \quad \text{on } G, \\ v &= g - \bar{u} \quad \text{on } \partial G. \end{aligned}$$

18.4. The Second and Third Boundary Value Problems and the Ritz Method

In this section we first of all encounter the task of explaining the concept of *natural boundary conditions*. These are boundary conditions that do *not* appear at all in the original variational problem, and first arise as necessary solvability conditions. Moreover, we shall discuss a method which allows us to force uniqueness for problems that are not uniquely solvable, by going over to a subspace (cf. Corollary 18.10). We consider the following problems:

(A) *Variational problem*

$$\int_G \left(\frac{1}{2}\sum_{i=1}^N (D_i u)^2 - uf\right)dx + \int_{\partial G}\left(\frac{1}{2}u^2 h - ug\right)dO = \min!, \qquad u \in C^1(\bar{G}). \tag{17}$$

(B) *Generalized boundary value problem*

$$\int_G \left(\sum_{i=1}^N D_i u D_i v - fv\right)dx + \int_{\partial G}(hu - g)v\,dx = 0 \quad \text{for all} \quad v \in C^1(\bar{G}). \tag{18}$$

(C) *Boundary value problem with natural boundary condition*

$$\begin{aligned} -\Delta u &= f \quad \text{on } G, \\ \frac{\partial u}{\partial n} + hu &= g \quad \text{on } \partial G. \end{aligned} \tag{19}$$

For $h \equiv 0$ (resp. $h \not\equiv 0$), (19) is called the second (resp. third) boundary value problem for the Poisson equation.

Proposition 18.9. *Let $f \in C(\bar{G})$ and $g, h \in C(\partial G)$. Suppose G satisfies* (7). *Then the following are valid:*

(a) *for $u \in C^2(\bar{G})$, problems* (17), (18), *and* (19) *are mutually equivalent;*
(b) *for $u \in C^1(\bar{G})$, problems* (17) *and* (18) *are equivalent;*
(c) *if $u \in C^1(\bar{G})$ is a solution of* (17) *with $h \equiv 0$, then the solvability condition*

$$\int_G f\,dx + \int_{\partial G} g\,dO = 0 \tag{20}$$

is valid.

In order to associate a physical picture with these assertions, we again, as in Figure 18.8 above, interpret $u(x)$ as the vertical displacement of a membrane at the point x over the region G in $\mathbb{R}^2$ under the influence of the external force f (e.g., the force of gravity) and the external boundary force g. The membrane is to be elastically joined to the boundary ∂G. Problem (17) contains the requirement for a minimal potential energy. The term $2^{-1}\int_{\partial G} u^2 h\, dO$ describes the potential energy of the elastic coupling at ∂G. Whereas *no* boundary condition appears in (17), there occurs in (19) the so-called *natural boundary condition*

$$\frac{\partial u}{\partial n} + hu = g \quad \text{on } \partial G$$

as a necessary solvability condition. This natural boundary condition is a necessary condition for the boundary forces g and the behavior of the displacement u on the boundary under equilibrium, whereas the solvability condition (20) represents an equilibrium condition for the external forces f and g.

In contrast to (19), the first boundary condition $u = g$ on ∂G in (12) prescribes the displacement of the membrane at the boundary points.

PROOF. We use the same strategy as in the proof of Proposition 18.6. To this end, we set

$$F(u) = \int_G \left(\frac{1}{2}\sum_{i=1}^N (D_i u)^2 - fu\right) dx + \int_G \left(\frac{1}{2}u^2 h - ug\right) dO.$$

Moreover, we define the real function

$$\varphi(t) = F(u + tv) \qquad \text{for all} \quad t \in \mathbb{R}$$

and fixed $v \in C^1(\bar{G})$.

Ad(a). Let $u \in C^2(\bar{G})$.

(I) (17) $\Rightarrow$ (18). If u is a solution of (17), then φ has a minimum at $t = 0$, i.e.,

$$\varphi'(0) = 0.$$

This implies (18).

(II) (18) $\Rightarrow$ (19). Integrating (18) by parts yields

$$\int_G (-\Delta u - f)v\, dx + \int_{\partial G} \left(hu - g + \frac{\partial u}{\partial n}\right) v\, dO = 0 \qquad \text{for all} \quad v \in C^1(\bar{G}).$$

It first of all follows from this that

$$\int_G (-\Delta u - f)v\, dx = 0 \qquad \text{for all} \quad v \in C_0^\infty(G),$$

therefore $-\Delta u - f = 0$ on G according to Proposition 18.2. Second, it follows from

$$\int_{\partial G} \left(hu - g + \frac{\partial u}{\partial n}\right) v\, dO = 0 \qquad \text{for all} \quad v \in C^1(G)$$

that $hu - g + \partial u/\partial n = 0$ on ∂G, since the boundary values of the functions $v \in C^1(\bar{G})$ are dense in $L_2(\partial G)$.

(III) (19) $\Rightarrow$ (18). Multiplication of $-\Delta u = f$ by $v \in C^1(\bar{G})$ and integration by parts yield (18).

(IV) (18) $\Rightarrow$ (17). One deduces as in the proof of Proposition 18.6(IV).

Ad(b). For (17) $\Leftrightarrow$ (18) only $u \in C^1(\bar{G})$ was used.

Ad(c). Choose $v \equiv 1$ in (18). □

One clearly recognizes in the proof that the natural boundary condition arises by means of integration by parts.

The second boundary value problem (19) with $h \equiv 0$ is *not* uniquely solvable since $u \equiv 1$ is a trivial solution. It is well known from the classical theory that the *uniqueness* is forced by means of the normalization condition

$$\int_G u\,dx = 0. \tag{21}$$

We shall see in Section 22.2f that (21) also suffices for the uniqueness of the corresponding generalized problem. As motivation for Definition 22.13 we prove the following.

Corollary 18.10. *Let* $h \equiv 0$ *and suppose that the solvability condition* (20) *is satisfied. Then the equivalence statements in Proposition* 18.9(a), (b) *remain correct in case one places the additional normalization condition* (21) *on* u *and* v *in* (17), (18), *and* (19).

Proof. Let $M = \{w \in C^1(\bar{G}): \int_G w\,dx = 0\}$. It suffices to prove that for $h \equiv 0$ condition (18) is equivalent to

$$\int_G \left(\sum_{i=1}^N D_i u D_i w - fw\right) dx + \int_{\partial G} (hu - g)w\,dx = 0 \qquad \text{for all} \quad w \in M. \tag{18*}$$

Each $v \in C^1(\bar{G})$ can be represented as $v = c + w$ with $w \in M$ and a constant c. Because of (21) and $h \equiv 0$,

$$\int_G \left(\sum_{i=1}^N D_i u D_i c - fc\right) dx + \int_{\partial G} (hu - g)c\,dO = 0$$

holds trivially. Therefore, (18) follows from (18*). The converse is obvious. □

18.4a. Generalized Problems in Sobolev Spaces

Motivated by (18*), in Section 22.2f we will formulate the final generalized problem to the *second boundary value problem* (19) with $h \equiv 0$ as follows. Let

$$M_1 = \left\{w \in W_2^1(G): \int_G w\,dx = 0\right\}.$$

For given $g \in L_2(\partial G)$ *and* $f \in L_2(G)$ *with* (20), *we are looking for a function*

$$u \in M_1$$

such that (18*) *holds for all* $w \in M_1$.

The choice of the Sobolev space $W_2^1(G)$ results from the fact that first-order partial derivatives appear in (18*).

Motivated by (18), in Section 22.2g we will formulate the final generalized problem to the *third boundary value problem* (19) as follows.

For given $g \in L_2(\partial G)$ *and* $h \in C(\partial G)$ *with* $h \geq h_0 > 0$ *on* ∂G *and* $f \in L_2(G)$ *with* (20), *we are looking for a function*

$$u \in W_2^1(G)$$

such that (18) *holds for all* $w \in W_2^1(G)$.

The condition "$h \geq h_0 > 0$ on ∂G" ensures that the elastic potential energy $\frac{1}{2}\int_{\partial G} hu^2\, dO$ is positive.

18.4b. The Ritz Method

In order to formulate the Ritz method for the original variational problem (17), we proceed from the trial

$$u_n = \sum_{k=1}^{n} c_{kn} w_k, \tag{22}$$

where the basis functions $w_1, \ldots, w_n$ are given. We determine the unknown real coefficients $c_{1n}, \ldots, c_{nn}$ in a way parallel to (17) from the variational problem

$$\int_G \left(\frac{1}{2}\sum_{i=1}^{N} (D_i u_n)^2 - u_n f\right) dx + \int_{\partial G} \left(\tfrac{1}{2}u_n^2 h - u_n g\right) dO = \min!, \qquad c_n \in \mathbb{R}^n, \tag{23}$$

where $c_n = (c_{1n}, \ldots, c_{nn})$. We assume that all $w_1, \ldots, w_n$ are continuous and piecewise differentiable on $\bar{G}$. In case $h \equiv 0$, we force the normalization condition

$$\int_G u_n\, dx = 0$$

by means of the requirement $\int_G w_k\, dx = 0$ for $k = 1, \ldots, n$.

Corollary 18.11. *If* u_n *is a solution of* (23), *then*

$$\int_G \left(\sum_{i=1}^{N} D_i u_n D_i w_j - f w_j\right) dx + \int_{\partial G} (h u_n - g) w_j\, dO = 0, \qquad j = 1, \ldots, n. \tag{24}$$

Because of (22), *this is a linear system of equations for determining* $c_{1n}, \ldots, c_{nn}$.

The proof runs analogous to that of Corollary 18.7. Equation (24) results formally from the generalized boundary value problem (18) if one replaces u there by u_n and requires that (18) is valid for all $v \in \operatorname{span}\{w_1, \ldots, w_n\}$.

Assertions concerning existence, uniqueness, and convergence of the Ritz method can be found in Sections 22.2f and 22.2g.

18.5. Eigenvalue Problems and the Ritz Method

In this section it will be shown that eigenvalue problems result from appropriate variational problems with *side conditions.* The method of Lagrange multipliers is typical for the investigation of variational problems with side conditions. We shall occupy ourselves therewith in detail in Part III. Here we use a direct method analogous to Sections 18.3 and 18.4. We consider the following problems. Let $M = \{u \in C^1(\bar{G}): u = 0 \text{ on } \partial G\}$.

(A) *Variational problem with side conditions*

$$\int_G \sum_{i=1}^N (D_i u)^2 \, dx = \min!, \qquad u \in M, \tag{25a}$$

$$\int_G u^2 \, dx = 1, \tag{25b}$$

$$\int_G u u_r \, dx = 0, \qquad r = 1, \ldots, m-1. \tag{25c}$$

This problem is to be understood as follows: We determine u_1 from (25a) and (25b). Then u_2 results from (25a)–(25c) with $m = 2$, and so forth.

(B) *Generalized boundary eigenvalue problem*

$$\begin{gathered} \int_G \left(\sum_{i=1}^N D_i u_m D_i v - \mu_m u_m v \right) dx = 0 \qquad \text{for all} \quad v \in M, \\ u_m = 0 \quad \text{on } \partial G. \end{gathered} \tag{26}$$

(C) *Boundary eigenvalue problem*

$$\begin{aligned} -\Delta u_m &= \mu_m u_m \quad \text{on } G, \\ u_m &= 0 \quad \text{on } \partial G. \end{aligned} \tag{27}$$

(D) *Ritz method*

$$\int_G \left(\sum_{i=1}^N D_i w D_i w_j - \mu w w_j \right) dx = 0, \qquad j = 1, \ldots, n. \tag{28}$$

The basis functions $w_1, \ldots, w_n$ are given. We seek $\mu \in \mathbb{R}$ and the real coefficients $c_1, \ldots, c_n$ of

$$w = \sum_{k=1}^n c_k w_k.$$

Problem (28) is an eigenvalue problem of the type

$$Ac = \mu Bc, \tag{28*}$$

where A and B are symmetric $n \times n$ matrices and the column matrix c contains the elements c_1 to c_n.

Parallel to Sections 18.3 and 18.4 one obtains (28) from the generalized boundary eigenvalue problem (26) if one only simply requires that (26) holds for all functions v in $\operatorname{span}\{w_1, \ldots, w_n\}$. So that w in (28) satisfies the boundary condition

$$w = 0 \quad \text{on } \partial G$$

we require that

$$w_k = 0 \quad \text{on } \partial G \qquad \text{for} \quad k = 1, \ldots, n.$$

So that the integral in (28) is meaningful, we require, say, that all w_k are continuous and piecewise continuously differentiable on $\bar{G}$. A precise analysis in Section 22.11 shows that the eigensolutions and eigenvalues of (28) approximate the eigenfunctions u_m and eigenvalues μ_m of (27) in an appropriate sense.

A numerical example for the Ritz method will be considered in Problem 18.1b.

Proposition 18.12. *Suppose that the region G satisfies* (7). *Then the following are valid*:

(a) *Every solution* $u_m \in C^2(\bar{G})$ *of* (25) *also satisfies* (26) *and* (27) *for* $m = 1, 2, \ldots$ *with suitable* $\mu_m \in \mathbb{R}$.
(b) *Every solution* $u_m \in C^1(\bar{G})$ *of* (25) *also satisfies* (26).
(c) *For* $u_m \in C^2(\bar{G})$, *the boundary eigenvalue problem* (27) *and the generalized problem* (26) *are equivalent.*

Proof. Ad(a), (c). Let $u_m \in C^2(\bar{G})$. We set

$$F(u) = \frac{\displaystyle\int_G \sum_{i=1}^N (D_i u)^2 \, dx}{\displaystyle\int_G u^2 \, dx}.$$

(I) (25) $\Rightarrow$ (26). First let $m = 1$. Recall that

$$M = \{u \in C^1(\bar{G}) \colon u = 0 \text{ on } \partial G\}.$$

If u_1 is a solution of (25), then it is also a solution of

$$F(u) = \min!, \qquad u \in M, \quad u \not\equiv 0. \tag{29}$$

We define the real function

$$\varphi(t) = F(u_1 + tv) \qquad \text{for all} \quad t \in \mathbb{R}$$

and fixed $v \in M$. Since $u_1 + tv \in M$ for all t, the real function φ has a minimum at $t = 0$, i.e.,

$$\varphi'(0) = 0.$$

This implies

$$\int_G \left(\sum_{i=1}^{N} D_i u_1 D_i v - \mu_1 u_1 v \right) dx = 0 \qquad \text{for all} \quad v \in M,$$

with $\mu_1 = F(u_1)$, after a simple calculation. This is (26) with $m = 1$.

Now let $m = 2$. We set

$$M_1 = \left\{ u \in M \colon \int_G u u_1 \, dx = 0 \right\}.$$

Let u_2 be a solution of (25). Then u_2 is also a solution of

$$F(u) = \min!, \qquad u \in M_1, \quad u \not\equiv 0.$$

We set

$$\psi(t) = F(u_2 + tw) \qquad \text{for all} \quad t \in \mathbb{R}$$

and fixed $w \in M_1$. Since $u_2 + tw \in M_1$ for all t, the real function ψ has a minimum at $t = 0$, i.e.,

$$\psi'(0) = 0.$$

This yields

$$\int_G \left(\sum_i D_i u_2 D_i w - \mu_2 u_2 w \right) dx = 0 \qquad \text{for all} \quad w \in M_1. \tag{30}$$

Furthermore, it follows from (26) with $m = 1$, $v = u_2$, that

$$\int_G \left(\sum_i D_i u_2 D_i u_1 - \mu_1 u_2 u_1 \right) dx = 0. \tag{31}$$

Each $v \in M$ can be represented as

$$v = cu_1 + w, \qquad w \in M_1, \quad c \in \mathbb{R}.$$

From $\int_G u_1 u_2 \, dx = 0$ it easily follows that (30) is also valid for all $w \in M$. This is (26) with $m = 2$.

One deduces the result for $m > 2$ analogously.

(II) (26) $\Rightarrow$ (27). Integrating (26) by parts yields

$$\int_G (-\Delta u_m - \mu_m u_m) v \, dx = 0 \qquad \text{for all} \quad v \in M.$$

Proposition 18.2 yields $-\Delta u_m - \mu_m u_m = 0$ on G, because $C_0^\infty(G) \subseteq M$.

(III) (27) $\Rightarrow$ (26). Multiplication of $-\Delta u_m = \mu_m u_m$ by $v \in M$ and integration by parts yields (26).

Ad(b). In the proof of (a) we used only that $u \in C^1(\bar{G})$. □

The final generalized problem in Section 22.13 reads as follows.

We are looking for a function $u_m \in \mathring{W}_2^1(G)$ *and a real number* μ_m *such that relation* (26) *holds for all* $v \in \mathring{W}_2^1(G)$.

As in Section 18.3, this is motivated by the fact that the set M is dense in the Sobolev space $\mathring{W}_2^1(G)$. The existence of solutions of the generalized problem and the convergence of the Ritz method will be proved in Section 22.13.

18.6. The Hölder Inequality and its Applications

The most important inequality in modern analysis is the following *Hölder inequality*:

$$\int_G |fg|\,dx \le \left(\int_G |f|^p\,dx\right)^{1/p} \left(\int_G |g|^q\,dx\right)^{1/q} \tag{32}$$

with

$$\frac{1}{p} + \frac{1}{q} = 1, \qquad 1 < p < \infty. \tag{33*}$$

If we set

$$\|f\|_p = \left(\int_G |f|^p\,dx\right)^{1/p},$$

then we may write (32) in the form

$$\left|\int_G fg\,dx\right| \le \int_G |fg|\,dx \le \|f\|_p \|g\|_q. \tag{33}$$

Proposition 18.13 (The Hölder Inequality). *Let G be a nonempty measurable set in* $\mathbb{R}^N$ *with* $N \ge 1$ *(e.g., G is open or closed). Then it follows from*

$$f \in L_p(G), \qquad g \in L_q(G)$$

with (33*) *that* (33) *holds.*

If $f \in L_1(G)$ *and* $g \in L_\infty(G)$, *then* (33) *holds with* $p = 1$ *and* $q = \infty$.

For the convenience of the reader, the basic material of modern analysis is summarized in the Appendix (e.g., measurable sets and measurable functions, the Lebesgue integral, the Lebesgue spaces $L_p(G)$, $1 \le p \le \infty$, etc.). For example, we have

$$f \in L_p(G), \qquad 1 \le p < \infty$$

iff $f\colon G \to \mathbb{R}$ is measurable and

$$\|f\|_p < \infty.$$

Furthermore, we have $f \in L_\infty(G)$ iff $f: G \to \mathbb{R}$ is measurable and there exists a real number C such that

$$|f(x)| \leq C \qquad \text{for almost all} \quad x \in G.$$

Then, $\|f\|_\infty$ is equal to the infimum of all possible C. We write

$$\|f\|_\infty = \operatorname*{ess\,sup}_{x \in G} |f(x)|.$$

For $1 \leq p \leq \infty$, $L_p(G)$ is a real B-space with the norm $\|\cdot\|_p$, provided we identify any two functions which differ only on a set of N-dimensional Lebesgue measure zero. That is, we identify f with g iff $f(x) = g(x)$ for almost all $x \in G$.

Note that only *pathological* sets and functions are *not* measurable. Nevertheless, in each case, one has to check that the functions and sets under consideration are measurable. To this end one can use the tools summarized in $A_2(1)$–$A_2(12)$. Furthermore, note that in each case, one has to check the existence of the integrals under consideration. In this connection, the most important tool is the majorant criterion of Lebesgue $A_2(17)$.

Roughly speaking, Proposition 18.13 tells us the following:

If the right-hand side in (33) *exists, then the remaining integrals also exist and the inequalities* (33) *hold.*

The Hölder inequality was first proved by Otto Hölder (1859–1937) in 1889 for series. The extension to integrals is due to Fryges Riesz. The following proof of Proposition 18.13 is based on the *Young inequality*:

$$ab \leq \frac{a^p}{p} + \frac{b^q}{q} \qquad \text{for all} \quad a, b \in \mathbb{R}_+ \tag{34}$$

and all p, q with (33*).

In Part III we shall show that (34) is the germ cell of the general duality theory.

Proof of Proposition 18.13.

(I) Proof of (34). If $p = q = 2$, then (34) follows from $(a - b)^2 \geq 0$. In the general case, we consider the function

$$F(a) = \frac{a^p}{p} + \frac{b^q}{q} - ab$$

for fixed $b > 0$. We have

$$F(0) > 0, \qquad F(b^{q/p}) = 0, \qquad \lim_{a \to +\infty} F(a) = +\infty.$$

Hence F has a minimum on $\mathbb{R}_+$. Thus, there exists a number $a_0 > 0$ such that

$$F(a) \geq F(a_0) \qquad \text{for all} \quad a \in \mathbb{R}_+.$$

From $F'(a_0) = 0$ it follows that $a_0 = b^{q/p}$, and hence $F(a_0) = 0$. This is (34).

(II) Proof of (33) for $1 < p < \infty$. We may assume that $\|f\|_p = \|g\|_q = 1$. Otherwise we use a division. Integration of

$$|fg| \le \frac{|f|^p}{p} + \frac{|g|^q}{q}$$

over G yields $\int_G |fg|\,dx \le 1$. This is (33).

(III) In the case $p = \infty$, we obtain

$$\left|\int_G fg\,dx\right| \le \int_G |f||g|\,dx \le \|f\|_\infty \int_G |g|\,dx. \qquad \square$$

We now consider some important applications of the Hölder inequality which will be used frequently.

18.6a. The 1-Trick

The 1-trick means that we apply the Hölder inequality (33) with $g = 1$. This is only possible for bounded measurable sets G. Otherwise $\|g\|_q = \infty$. This way we obtain

$$\int_G |f|\,dx \le (\text{meas } G)^{1/q} \left(\int_G |f|^p\,dx\right)^{1/p} \tag{35}$$

for all $f \in L_p(G)$ with $1 < p < \infty$.

EXAMPLE 18.14. Let G be a nonempty bounded measurable set in $\mathbb{R}^N$, $N \ge 1$. Then the embedding

$$L_r(G) \subseteq L_s(G), \qquad 1 \le s \le r \le \infty$$

is continuous, i.e., we have

$$\|h\|_s \le \text{const}\,\|h\|_r \qquad \text{for all} \quad h \in L_r(G). \tag{36}$$

PROOF.

(I) Let $1 \le s < r < \infty$ and let $h \in L_r(G)$. Then (36) follows from (35) with $f = |h|^s$ and $p = r/s$.

(II) Let $1 \le s < r = \infty$ and $h \in L_\infty(G)$. Then $|h(x)| \le C$ for almost all $x \in G$. Hence

$$\left(\int_G |h|^s\,dx\right)^{1/s} \le C \,\text{meas}\, G.$$

This implies (35), since $\|h\|_\infty$ is the infimum of all C. $\square$

EXAMPLE 18.15. Let G and H be two nonempty bounded measurable sets in $\mathbb{R}^N$ with $N \ge 1$ and $G \subseteq H$. Suppose that $f \in L_p(H)$ with $1 < p \le \infty$.

Then

$$\int_G |f|\,dx \le (\text{meas } G)^{1/q} \|f\|_{L_p(H)},$$

where $q^{-1} + p^{-1} = 1$ for $1 < p < \infty$ and $q = 1$ for $p = \infty$.
This follows from (35).

Remark 18.16. If $1 \le p \le \infty$ and $f \in L_p(H)$, then

$$\left|\int_G f\,dx\right| \to 0 \qquad \text{as} \quad \text{meas } G \to 0,$$

where H and G are given as in Example 18.15. This follows from the continuity of the embedding $L_p(H) \subseteq L_1(H)$ and the absolute continuity of the integral (cf. $A_2(20)$). In the case $1 < p \le \infty$, Example 18.15 contains precise information about this limiting process.

18.6b. The Sum Trick

The prototype of the sum trick is the inequality

$$2ab \le \varepsilon a^2 + \varepsilon^{-1} b^2 \qquad \text{for all} \quad a, b \in \mathbb{R}_+ \tag{37}$$

and all $\varepsilon > 0$, which follows from $(\sqrt{\varepsilon} a - b/\sqrt{\varepsilon}) \ge 0$. This inequality tells us that it is possible to estimate a product against a *sum* of squares, where one of the summands can be made arbitrarily small. From the Young inequality (34) there follows a generalization of (37), namely,

$$ab \le \frac{\varepsilon^p a^p}{p} + \frac{b^q}{q\varepsilon^q} \qquad \text{for all} \quad a, b \in \mathbb{R}_+ \tag{38}$$

and all $\varepsilon > 0$. Here, $1 < p < \infty$ and $p^{-1} + q^{-1} = 1$.

EXAMPLE 18.17. Let G be a nonempty measurable set in $\mathbb{R}^N$, $N \ge 1$. Suppose that $f \in L_p(G)$ and $g \in L_q(G)$ with $1 < p < \infty$ and $q^{-1} + p^{-1} = 1$. Then

$$\int_G |fg|\,dx \le \frac{\varepsilon^p}{p} \int_G |f|^p\,dx + \frac{1}{q\varepsilon^q} \int_G |g|^q\,dx \tag{39}$$

holds for all $\varepsilon > 0$. In particular, one can choose $p = q = 2$.
This follows from the Hölder inequality and from (38).

18.6c. The Nemyckii Trick

For the investigation of nonlinear differential and integral equations, the Nemyckii operator plays a fundamental role. This operator will be considered in Section 26.3. In this connection, the following result is crucial. Our goal is to show that the estimate

$$|w(x)| \le |u(x)| + |v(x)|^{p/q} \quad \text{on } G, \qquad 1 \le p, q < \infty, \tag{40}$$

implies

$$\|w\|_q \le \text{const}(\|u\|_q + \|v\|_p^{p/q}). \tag{41}$$

EXAMPLE 18.18. Suppose that:

(i) G is a nonempty measurable set in $\mathbb{R}^N$, $N \ge 1$.
(ii) $u \in L_q(G)$ and $v \in L_p(G)$.
(iii) The function $w: G \to \mathbb{R}$ is measurable (e.g., continuous) and (40) holds.

Then we have $w \in L_q(G)$, and (41) holds.

PROOF. Let $r > 0$. We use the inequality

$$(a + b)^r \le C(a^r + b^r) \qquad \text{for all} \quad a, b \in \mathbb{R}_+, \tag{41*}$$

where the constant C depends on r. From (40) we obtain

$$|w|^q \le C(|u|^q + |v|^p).$$

Integration over G yields

$$\int_G |w|^q \, dx \le C\left(\int_G |u|^q \, dx + \int_G |v|^p \, dx\right).$$

Now the assertion (41) follows from (41*) with $r = 1/q$. □

There are two completely different methods for the investigation of nonlinear differential and integral equations:

(i) One uses $L_p(G)$-functions.
(ii) One uses $C^k(G)$-functions or $C^{k,\alpha}(G)$-functions.

Note the following peculiarities. In the case (i), Example 18.18 above is responsible for the fact that one needs restrictive *growth conditions* for the nonlinearities. But one has the advantage that the $L_p(G)$-spaces are *reflexive* in case $1 < p < \infty$. In the case (ii), no growth conditions are needed, but the C^k-spaces and $C^{k,\alpha}$-spaces are *not* reflexive. Unfortunately, many existence theorems are only valid in reflexive B-spaces. The point is that in reflexive B-spaces, each bounded sequence has a weakly convergent subsequence.

In the Appendix, $A_2(29)$, we will consider the Hölder inequality and the Young inequality for n factors. If we combine these two inequalities, then we obtain the following *Hölder–Young inequality*:

$$\begin{aligned}\left|\int_G \prod_{i=1}^n f_i \, dx\right| &\le \prod_{i=1}^n \left(\int_G |f_i|^{p_i} \, dx\right)^{1/p_i} \\ &\le \sum_{i=1}^n \frac{1}{p_i} \int_G |f_i|^{p_i} \, dx.\end{aligned} \tag{42}$$

This inequality holds for all

$$f_i \in L_{p_i}(G), \qquad i = 1, \ldots, n,$$

where the real numbers $1 < p_1, \ldots, p_n < \infty$ satisfy the relation

$$\sum_{i=1}^{n} \frac{1}{p_i} = 1,$$

and G is a nonempty measurable subset of $\mathbb{R}^N$, $N \geq 1$.

18.7. The History of the Dirichlet Principle and Monotone Operators

In order to determine integrals of algebraic functions and their inverse functions (elliptic and Abelian functions), one can apply a principle which was used by Dirichlet in his lectures. This principle allows us to solve boundary problems for the Laplace equation.

Bernhard Riemann (1857)

By the *Dirichlet principle* we understand a method for solving boundary value problems via minimum problems for variational integrals. This principle goes back to Gauss (1839), Lord Kelvin, and Dirichlet. Weierstrass (1870) was the first to underline the shortcomings of this principle. He showed that there are variational problems which do not have any solution. In 1900 I showed that it is possible to justify rigorously the Dirichlet principle.

David Hilbert (1901)

Riemann (1826–1866) was a man of outstanding intuition. By his ingenuity he outshone all his contemporaries. Whereever his interest was awoken, he began anew, without allowing himself to be misled by tradition and without recognizing any compulsion of systematology.

Weierstrass (1815–1897) was first and foremost a logician. He proceeded slowly, systematically, gradually. Whereever he worked, he strove after the final form.

Certainly mathematics itself pronounces the verdict, even if it does so without rigorous proof. It will, however, remain for ever the secret of vigorous productivity to find new questions and to discover new theorems which lead to valuable results and new interrelations. Without new ideas and without the formulation of new objectives, mathematics would soon amount to nothing more than the rigidity of its proofs, and would begin to stagnate, its fuel running out.

So, in a certain sense, mathematics has been furthered most by those who distinguished themselves more through their intuition than through strong deduction.

Felix Klein (1926)

The well-known *bon mot* that "European philosophy" is only a footnote to Plato is perhaps true, but I would venture a much truer one: modern mathematics is only a footnote to Riemann.

Krysztof Maurin (1982)

18.7a. The Calculus of Variations

As the Dirichlet principle is very closely connected with the calculus of variations, we will next take a short glance at the history of this fundamental mathematical discipline.

In Parts III through V we will see that many basic equations of physics result from an important variational principle, namely from the principle of critical or stationary action, which is often incorrectly called the principle of least action. The general theory of relativity and the gauge field theories in modern elementary particle physics belong here.

It seems that the local symmetries of variational integrals, which lead to the gauge field theories, provide the key to understanding the four fundamental interactions in nature: the strong, weak, electromagnetic, and gravitational interactions. This will be discussed in detail in Part V.

About three hundred years ago, at the beginning of a fascinating development in mathematics and physics, there was the following prize competition.

> Johann Bernoulli, professor of mathematics, greets the most sophisticated mathematicians in the world.[1] Experience shows that noble intellectuals are driven to work for the pursuit of knowledge by nothing more than being confronted with difficult and useful problems.
>
> Six month ago, in the June edition of the *Leipzig Acta Euriditorium*, I presented such a problem. The allotted six-month deadline has now gone by, but no trace of a solution has appeared. Only the famous Leibniz informed me that he had unraveled the knot of this brilliant and outstanding problem, and he kindly asked me to extend the deadline until next Easter. I agreed to this honourable request .... I will repeat the problem here once more.
>
> Two points, at different distances from the ground and not in a vertical line, should be connected by such a curve so that a body under the influence of gravitational forces passes in the shortest possible time from the upper to the lower point.
>
> Johann Bernoulli (January, 1697)

> This paper solves my brother's problem, to whom I will set other problems in return.
>
> Jacob Bernoulli (May, 1697)

> How does one apply the methods of maxima and minima in the determination of unknown curves (solution to variational problems)?
>
> Leonhard Euler (1744)

> The famous Euler succeeded in tracing back to a general method all investigations on variational problems. But however sophisticated and fruitful his method may be, one has to admit that it is *not* simple. Here one now finds a method which only uses simple principles of calculus.
>
> Joseph Louis Lagrange (1762)

[1] Leibniz (1646–1716), Johann Bernoulli (1667–1748), Jacob Bernoulli (1654–1705) (brother of Johann Bernoulli), Daniel Bernoulli (1700–1782) (son of Johann Bernoulli), Euler (1707–1783), Lagrange (1736–1813), Legendre (1752–1833), Gauss (1777–1855), Jacobi (1804–1851), Dirichlet (1805–1859), Riemann (1826–1866), Weierstrass (1815–1897), and Hilbert (1862–1943).

The necessary conditions for the solutions of variational problems, found by Euler and Lagrange, will be considered in Section 18.18. We have already met the prototype in Proposition 18.6.

Euler used a difference method. Lagrange's simple method, based on integration by parts, is still used today. For example, we used Lagrange's method to prove Proposition 18.6. In contrast to Euler, Lagrange could also deal with variational problems for multi-dimensional variational integrals, which lead to partial differential equations.

> In most of the problems of the calculus of variations it follows from the nature of the problem as to whether or not a minimum or maximum is possible. There are, however, cases in which this distinction is not so easy. I therefore thought that the formulation of a criterion, which allows for a distinction between maxima and minima, would be useful.
>
> Adrien-Marie Legendre (1786)

This so-called Legendre condition will be considered in Section 18.17. It deals with a necessary condition for maxima and minima. Legendre believed that he had found a sufficient condition. As Jacobi, however, pointed out, Legendre's observations contained a serious mistake.

> I have succeeded in closing an essential gap in the calculus of variations. Until now *no* general sufficient condition was known by which to recognize whether a maximum or minimum existed. For example, the integral, considered in the principle of least action, can never have a maximum, as Lagrange mistakenly believed; in *no* way, however, will it always have a minimum.
>
> Carl Gustav Jacob Jacobi (1837)

In Section 29.12d we will consider a general functional analytic formulation of the classical Jacobi theory based on so-called accessory variational problems.

The decisive idea consists of studying a linear eigenvalue problem, which is closely connected with the quadratic form of the second variation. Therefore, our results on linear eigenvalue problems in Chapter 22 are of great interest for general variational problems.

> From this "independence theorem" arise not only the known sufficient conditions of Jacobi and Weierstrass for the appearance of minima, but also all the essential conclusions of the Hamilton–Jacobi theory.
>
> David Hilbert (1906)

Hilbert refers here to his invariant path integral, which we will consider in Part III within the framework of field theory in the calculus of variations (see Proposition 40.17). There we will also elucidate the connection with the famous Weierstrass E-function.

The Hamilton–Jacobi theory describes mathematically the fundamental propagation of action in nature and the structure of optimal processes in nature, technology, and economics. This will be studied in detail in Parts III through V.

18.7b. The Beginnings of Potential Theory

In 1782 Laplace remarked, in a study on the shape of the planets,[1] that the potential

$$u(y) = \frac{1}{4\pi}\int_G \frac{\rho(x)}{|x-y|}\,dx \tag{43}$$

of a mass distribution of density ρ on the region G satisfies the following partial differential equation

$$\Delta u = 0 \quad \text{outside } \bar{G}, \tag{44}$$

which is called the *Laplace equation.*

Interestingly, he first found this equation in the complex form of spherical coordinates. In 1787 he discovered the much simpler expression for Δu in Cartesian coordinates, namely,

$$\Delta u = u_{\xi\xi} + u_{\eta\eta} + u_{\zeta\zeta},$$

where $x = (\xi, \eta, \zeta)$. In 1813 Poisson found that for a ball G the following equation is valid in the case of a constant density ρ:

$$-\Delta u = \rho \quad \text{on } G. \tag{45}$$

This equation is called the *Poisson equation.*

Gauss proved the validity of this equation for more general cases in 1839 in his fundamental paper: "General Theorems about the Forces of Attraction and Repulsion which are Inversely Proportional to the Square of the Distance." Such forces are, for example, Newton's gravitational force and the electrostatical Coulomb force. Gauss also used such forces in order to explain magnetic phenomena. The above-mentioned work of Gauss is the beginning of the mathematical foundations of potential theory. In this work, for example, Gauss proved the fundamental mean value property of potential functions (solutions of the Laplace equation $\Delta u = 0$). He had already realized in 1813 that volume integrals can be transformed into surface integrals (Gauss' integral theorem). The formula of integration by parts, which is at the forefront of the modern theory of partial differential equations, is equivalent to the integral theorem of Gauss. We have already discussed the fundamental importance of this theorem for topology (mapping degree, cohomology) in the introduction to Chapter 12.

In 1807 Gauss became Professor of Astronomy and Director of the Observatory in Göttingen, where he worked until his death in 1855.

[1] Laplace (1749–1827), Poisson (1781–1840), Gauss (1777–1855), and Green (1793–1841).

For valuable contributions to potential theory we also have George Green to thank, who, in 1829, discovered the Green integral formulas and who introduced the important Green function, which allows us to represent the solutions of elliptic boundary value problems by integral expressions.

18.7c. Riemann and the Dirichlet Principle

From 1846 until 1847 Riemann[1] studied under Gauss in Göttingen. In the years 1847–1849 he was a student in Berlin and attended lectures by Dirichlet,[1] Jacobi, Eisenstein, and Steiner. He learnt about the *Dirichlet principle* in Dirichlet's lectures on potential theory. During this time, as a student in Berlin, the idea matured within Riemann to give a foundation to the theory of complex analytic functions by means of the theory of partial differential equations for real functions. To explain this let

$$f(z) = u + iv$$

be an analytic function. Then the real functions u and v satisfy the so-called *Cauchy–Riemann differential* equations

$$u_\xi = v_\eta, \qquad u_\eta = -v_\xi,$$

where $z = \xi + i\eta$.

By differentiation, we obtain the Laplace equation

$$u_{\xi\xi} + u_{\eta\eta} = 0, \tag{46}$$

and $v_{\xi\xi} + v_{\eta\eta} = 0$, i.e., $\Delta u = \Delta v = 0$. The corresponding variational problem to (46) is given by

$$\begin{gathered}\int_G (u_\xi^2 + u_\eta^2)\,dx = \min!,\\ u = g \quad \text{on } \partial G.\end{gathered} \tag{47}$$

In this way Riemann was able to give a *physical* interpretation, in terms of potential theory, to the mathematical problems of the theory of analytic functions. During his lifetime Riemann, like his teacher Gauss, tried to find *deep connections* between mathematics and physics.

In 1849 Riemann returned to Gauss in Göttingen, and in 1851 handed over his dissertation which was much praised by Gauss. In this fundamental paper of a mere fourty pages, Riemann laid the foundations for the complex function theory and proved his famous theorem on conformal mappings. In this paper, and in his fundamental paper on Abelian integrals on Riemannian surfaces written in 1857, he placed a general existence principle at the head of the theory, which he called the *Dirichlet principle*. In today's language the principle

[1] Riemann (1826–1866), Dirichlet (1805–1859), Jacobi (1804–1851), Eisenstein (1823–1852), Steiner (1796–1863), Weierstrass (1815–1897).

is as follows:

For given sufficiently smooth functions v and g, the minimum problem

$$\text{(D)} \qquad \int_G (u_\xi - v_\eta)^2 + (u_\xi + v_\eta)^2 \, dx = \min!, \qquad u = g \quad \text{on } \partial G$$

always has a solution in the set of piecewise continuously differentiable functions. Here, G is a bounded region in $\mathbb{R}^2$.

In the special case $v = 0$, problem (D) corresponds to (47). Like Dirichlet, Riemann made no attempt to prove this existence principle. Both considered it to be obvious that problem (D) possesses a solution. In fact, the time was not yet ripe for an existence proof for (D).

18.7d. The Theory of Abelian Integrals

The theory of Abelian integrals and their inverse functions was at the center of Riemann's and Weierstrass' research on complex function theory, which was carried out totally independently from one another.

In his obituary for Weierstrass, Hilbert writes:

> Karl Weierstrass studied law in Bonn. In 1839 he decided to study mathematics and therefore went to Münster, where he attended a mathematical lecture by Gudermann on elliptic functions. That was the *only* mathematical lecture which he ever heard in his life. After that Weierstrass worked from 1840 until 1855 as a schoolteacher, totally isolated, in small towns. During this time he published a series of fundamental papers on the most difficult problems of complex function theory.
>
> In his academic inaugural speech in 1857 as a newly elected member of the Berlin Academy, he described how the theory of elliptic functions had tremendously attracted him, even before his studies, and that he regarded it as one of the main tasks of mathematics to create a general theory on periodic functions of several complex variables (Abelian functions).
>
> To prepare himself for this difficult task, he first studied the available tools and occupied himself with less difficult problems. As a result of these studies he published, during the years 1841–1843, papers on the theory of power series and on the definition of analytic functions by means of algebraic differential equations.
>
> In 1854 Weierstrass first succeeded in finding the solution to the famous "inverse problem" posed by Jacobi in 1832. The generalization of this result to general Abelian integrals, which was later obtained by Riemann and then by Weierstrass, is one of the greatest achievements in analysis.
>
> Whereas Riemann based his theory of Abelian functions on the imperfect nonrigorous Dirichlet principle, Weierstrass' theory is based on purely algebraic arguments.

We would briefly like to explain the mathematical background to the reader. By an Abelian integral we understand an integral of the form

$$A = \int R(z, w(z))\,dz,$$

where $w = w(z)$ is an algebraic function and $R = R(z, w)$ is a rational function, i.e., the function $w = w(z)$ satisfies an algebraic equation of the form

$$a_n(z)w^n + \cdots + a_1(z)w + a_0(z) = 0,$$

where $a_0, \ldots, a_n$ are polynomials with respect to z. For example, let p_n be a polynomial of degree n and let

$$w^2 = p_n(z).$$

Then, $w = \sqrt{p_n(z)}$. If $n = 3, 4$ (resp. $n = 5, 6$), then the integral $\int R(z, w(z))\,dz$ is called elliptic (resp. hyperelliptic). In the special case $w^2 = 1 - z^2$, we obtain the Abelian integral

$$t = \int_0^z \frac{dz}{\sqrt{1 - z^2}}.$$

Here, the inverse function is given by $z = \sin t$.

In 1832 Jacobi formulated the following hypothesis:

Consider the two functions $u = u(a, b)$ and $v = v(a, b)$ given by the following system of hyperelliptic integrals:

$$\int_{u_0}^{u} \frac{dz}{\sqrt{p_6(z)}} + \int_{v_0}^{v} \frac{dz}{\sqrt{p_6(z)}} = a,$$

$$\int_{u_1}^{u} \frac{z\,dz}{\sqrt{p_6(z)}} + \int_{v_1}^{v} \frac{z\,dz}{\sqrt{p_6(z)}} = b.$$

Then the two functions $u + v$ and uv are univalent and they possess four different periods.

Riemann and Weierstrass proved a generalization of this assertion for systems of n equations with n general Abelian integrals. Whereas Weierstrass used algebraic methods, Riemann worked with the topological idea of the so-called *Riemannian surface*. To explain this, we shall consider the elliptic integral

$$\int R(z, w(z))\,dz$$

with $w(z) = \sqrt{p(z)}$, where p is a polynomial of degree $n = 3, 4$ with pairwise different zeros. Riemann's idea was to consider this integral on the Riemannian surface T of $w = w(z)$. Here T is homeomorphic to a two-dimensional torus. The point is that on the torus, two different paths exist, C_1 and C_2, as pictured

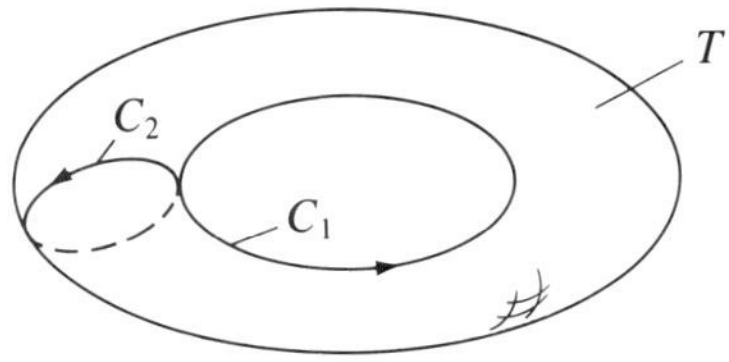

Figure 18.9

in Figure 18.9. Set

$$\omega_i = \int_{C_i} R(z, w(z))\,dz, \qquad i = 1, 2.$$

Each path on T can be deformed into

$$nC_1 + mC_2,$$

where n and m are integers. It follows from the Cauchy theorem that the deformation of paths on T does not change the corresponding integrals. Consequently, we obtain, without any calculations, that the elliptic integral $\int R(z, w(z))\,dz$ has the two additive periods ω_1 and ω_2. Therefore, the corresponding inverse functions, i.e., elliptic functions, have two periods.

Riemann generalized this argument to general Abelian integrals. He showed that the corresponding Riemannian surfaces are homeomorphic to spheres with p handles, and hence the corresponding Abelian integrals have $2p$ additive periods (Fig. 18.10(a)). The number p is called the genus of the Riemannian surface. This is the only topological invariant of Riemannian surfaces which correspond to algebraic functions. In the special case of a two-dimensional torus we have $p = 1$ (Fig. 18.10(b)).

Ingeniously, Riemann recognized that there are close and deep connections between analysis, topology, and algebra. With this recognition he decisively influenced mathematics in the twentieth century.

At the end of the last century Hilbert discovered that a new powerful instrument for number theory can be obtained by combining the idea of the Riemannian surface with the Galois theory. Thus he gave number theory a

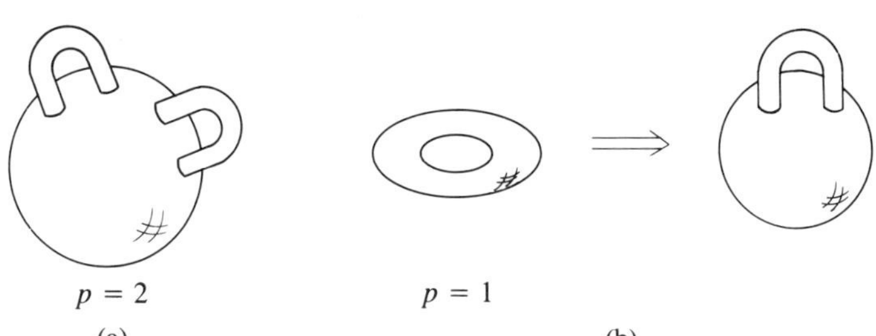

Figure 18.10

decisive new impulse. This is discussed in detail in the famous book by Felix Klein (1926): *The History of Mathematics in the Nineteenth Century.*

It is quite interesting that today Riemannian surfaces play a fundamental role in the so-called string theory of physicists (see Section 76.7) and in the theory of solitons (see Problem 30.7).

18.7e. The Criticism of the Dirichlet Principle by Weierstrass

In studying Riemann's work, Weierstrass found the Dirichlet principle unsatisfactory. In 1870 he constructed a counterexample which showed that the existence of solutions to variational problems is no trivial matter. He considered the following minimum problem:

$$J(f) \stackrel{\text{def}}{=} \int_{-1}^{1} (xf'(x))^2\,dx = \min!, \qquad f(-1) = a, \qquad f(1) = b \tag{48}$$

with $a \neq b$. The sequence of functions

$$f_n(x) = \frac{a+b}{2} - \frac{a-b}{2} \cdot \frac{\arctan nx}{\arctan n}, \qquad n = 1, 2, \ldots$$

satisfies the boundary conditions $f_n(-1) = a$ and $f_n(1) = b$, and we obtain

$$J(f_n) \to 0 \qquad \text{as} \quad n \to \infty.$$

Consequently, the infimum of the integral in (48) is equal to zero. However, there is no C^1-function which solves the minimum problem (48). In fact, if f is such a function, then it follows from (48) that $xf'(x) = 0$ on $]-1, 1[$. This implies $f = \text{constant}$. But this contradicts the boundary condition $f(-1) \neq f(1)$.

Because of this, Weierstrass' essential result is as follows:

(a) *Each variational problem which is bounded below possesses an infimum and hence minimal sequences exist.*
(b) *However, those minimal sequences do not necessarily converge to a solution of the variational problem.*

As the Dirichlet principle could not at first be justified, one looked for other methods to save Riemann's theory. To this end, Schwarz (1843–1921), C. Neumann (1832–1925), and Poincaré (1854–1912) developed special methods to solve the Laplace equation without using the Dirichlet principle.

18.7f. Hilbert's Justification of the Dirichlet Principle

In his Paris lecture in 1900 Hilbert formulated his famous twenty-three problems, which have substantially influenced mathematics in this century.

On this subject we recommend two anthologies by Aleksandrov (1971) and Browder (1976). In connection with the twentieth problem Hilbert said:

> The sophisticated methods of Schwarz, C. Neumann, and Poincaré essentially solved the boundary value problem for the Laplace equation. However, these methods cannot be directly extended to more general cases.... I am convinced that it will be possible to get these existence proofs by a *general basic idea*, towards the Dirichlet principle points. Perhaps it will then also be possible to answer the question of whether or not every regular variational problem possesses a solution if, with regard to boundary conditions, certain assumptions are fulfilled and if, when necessary, one sensibly *generalizes* the concept of *solution*.

At this time Hilbert had already found a possible means of saving the Dirichlet principle. He first presented a rough idea in a lecture in 1900, and then in greater detail in a later paper in 1901 (cf. Hilbert (1932), Vol. 3, pp. 10–37). Hilbert's idea was as follows:

(i) one constructs a minimal sequence; and
(ii) one shows that this minimal sequence contains a converging subsequence whose limit provides a solution to the original variational problem.

This is a typical compactness argument. In the second volume of the standard work of mathematical physics by Courant and Hilbert (1937), one finds the idea (i), (ii) worked out.

In today's terminology one works in Courant and Hilbert (1937) with *dense* subsets of the Sobolev spaces

$$W_2^1(G) \qquad \text{and} \qquad \mathring{W}_2^1(G),$$

where these subsets consist of piecewise smooth functions. The decisive compactness argument is based on a compactness theorem of Rellich (1930), which, in today's terminology, implies that the embedding

$$W_2^1(G) \subseteq L_2(G)$$

is compact.

In their fundamental paper on difference methods, Courant, Friedrichs, and Lewy (1928) had already emphasized the advantages of using generalized solutions of partial differential equations. The concept of generalized solutions is also stressed in Courant and Hilbert (1937).

In Part III we will formulate two general functional analytic existence principles for extremal problems, namely:

(a) a compactness principle which generalizes (i) and (ii) above (Chapter 38); and
(b) a convexity principle which is based on the Hahn–Banach theorem, i.e., on the separation of convex sets in B-spaces by hyperplanes (Chapter 39).

Our functional analytic justification of the Dirichlet principle in Section 18.10 is closely related to (b), i.e., in order to give a very elementary existence proof, we avoid compactness arguments.

The Ritz method was developed by Walter Ritz in his post-doctoral thesis, which he wrote in 1908 in Göttingen. Ritz applied his method, for example, to obtain numerical results for the clamped plate.

A few more words about Hilbert. The great mathematical tradition which Gauss founded in Göttingen was continued there by Riemann, Klein, and Hilbert. On Felix Klein's initiative, Hilbert, at the age of thirty-three, was appointed to a professorship in Göttingen, where he remained until his death. In the 1920's Göttingen was the leading center in mathematics and quantum physics. The Fascist seizure of power in 1933 also had the worst possible consequences for Göttingen. The world famous mathematicians Emmy Noether, Emil Artin, Paul Bernays, and Richard Courant were driven away, as well as the Nobel Prize winner for physics, Max Born. The number theory specialist Edmund Landau and the Nobel Prize winner for physics James Franck lost their positions. Hilbert's close personal friend Otto Blumenthal emigrated to the Netherlands and was murdered in the concentration camp Theresienstadt in 1944. When the Nazi minister for education asked Hilbert if mathematics in Göttingen was flourishing, Hilbert replied:

"There is no mathematics in Göttingen anymore."

Hilbert died at the age of ninety-one, in 1943. Scarcely a dozen friends accompanied the most important mathematician of this century to his final resting place. The words on his gravestone comprise his epistemological credo:

We must know.
We will know.

18.7g. The Work of Tonelli

The ideas of Hilbert for justifying the Dirichlet principle were substantially developed by the Italian mathematician Leonida Tonelli (1885–1946). The general compactness principle (a) mentioned in Section 18.7f above is based on the work of Tonelli.

> In 1884 Volterra had proposed the concept of functionals in the real field, and he considered the integrals of the calculus of variations as typical examples of functionals. The ideas of Volterra[1] were actively cultivated by Arzelà in Bologna (Italy). It was Arzelà who tried unsuccessfully to obtain a direct proof of the existence of the minimum for the Dirichlet integral. Hilbert in 1900 obtained such a direct proof, but his proof relied on *specific* properties of harmonic functions. The ideas were followed with enthusiasm by Beppo Levi, Lebesgue, and Fubini.
>
> In 1905 Baire had introduced his concept of semicontinuity for real functions showing that, in a compact domain, lower semicontinuous functions had

[1] Volterra (1860–1940), Arzelà (1847–1912), Ascoli (1887–1957), Beppo Levi (1875–1961), Fubini (1879–1943), Lebesgue (1875–1941), and Baire (1874–1932).

an absolute minimum, and upper semicontinuous functions had an absolute maximum.

Tonelli became acutely aware that a theory was needed for the direct existence proof of maxima and minima of general problems of the calculus of variations. Tonelli realized that in the Ascoli–Arzelà *compactness* theorem and Baire *semicontinuity* transferred from real functions to the functionals of the calculus of variations, he had the perfect tool for a "direct method in the calculus of variations" based on semicontinuity for an existence theory of maxima and minima. By this it is meant, as it is meant today, that to prove the existence of the absolute minimum of a functional F in a class C:

(i) one has to prove first that the functional $u \mapsto F(u)$ has a finite infimum on C (i.e., F is bounded below on C);
(ii) that the functional F is lower semicontinuous with respect to some type of convergence; and
(iii) that there is some minimizing sequence which converges in the stated type of convergence toward an element u_0 in C, and which then realizes the absolute minimum of the functional F in C.

The type of convergence is not assigned, but at the time, and for many more years, the natural convergence was the uniform convergence of the trajectories.

Indeed, when very young, Tonelli proved in 1911 the existence of the absolute minimum for general positive regular integrals of the calculus of variations over parametric continuous rectifiable curves in $\mathbb{R}^2$. The regularity was only a strong form of *convexity*, actually a strong form of the Legendre and Weierstrass conditions, and this assumption implies the *lower semicontinuity* of the integrals with respect to uniform convergence. The realization was reached that the classical Legendre and Weierstrass conditions are *not* so much conditions for a minimum or a maximum, but conditions for the lower or upper semicontinuity of the integral. For parametric integrals the conditions, required by Tonelli at the time, are expressed today by requesting only continuity and convexity.

In 1914–15 Tonelli proved the existence of the absolute minimum for ordinary curves in $\mathbb{R}^2$ (i.e., nonparametric), namely, absolute continuous functions, and under conditions at the time, which are expressed today by requesting only continuity, convexity, and coercivity.

After World War I, Tonelli collected the new ideas, and the impressive amount of work he had accumulated in so few years, into a complete theory, which appeared as his main opus, the *Fondamenti di Calcolo delle Variazioni*, a complex of 1150 pages, two volumes, in 1921–23.

Lamberdo Cesari (1987)

18.7h. Further Development

Since the 1930's one has systematically worked on the construction of a general theory of elliptic differential equations. The *Sobolev spaces* go back to Friedrichs (1934), who worked with the space $\mathring{W}_2^1(G)$ and other energetic spaces in connection with elliptic equations, and to Sobolev (1936), (1937), (1938), who studied systematically the spaces $W_p^m(G)$ together with his associate Kondračov and who applied his results to hyperbolic equations.

Friedrichs recognized the fundamental importance of estimates of the form

$$(C) \qquad (Au|u) \geq c\|u\|^2 \quad \text{for all} \quad u \in D(A)$$

for the functional analytic existence theory for elliptic differential equations. In the case $c > 0$, condition (C) means the strong *monotonicity* of the linear symmetric operator $A: D(A) \subseteq X \to X$ on the real H-space X. The theory of the *Friedrichs extension* will be studied in detail in Chapter 19.

The most general formulation of the notion of a generalized solution of partial differential equations can be obtained from within the framework of the theory of *distributions*, which was created by Laurent Schwartz (1950).

In the 1950's and 1960's the theory of linear elliptic differential equations was formally completed. By using *interpolation theory*, one was able to explicitly characterize the function spaces, which have the crucial property that linear elliptic differential operators represent *homeomorphisms* between those spaces. For example, let us consider the boundary value problem

$$\begin{aligned} -\Delta u &= f \quad \text{on } G, \\ u &= g \quad \text{on } \partial G, \end{aligned} \tag{49}$$

where G is a bounded region in $\mathbb{R}^3$ with a smooth boundary ∂G. If we write (49) in the form

$$Lu = (f, g),$$

then the operator

$$L: W_2^2(G) \to L_2(G) \times W_2^{3/2}(\partial G)$$

is a linear homeomorphism. In particular, this means that for the given functions

$$f \in L_2(G) \qquad \text{and} \qquad g \in W_2^{3/2}(\partial G),$$

the boundary value problem (49) has a unique solution u in the Sobolev space $W_2^2(G)$. Roughly speaking, the functions in the space $W_2^2(G)$ have square-integrable generalized second derivatives. Note that this final result is based on the use of Sobolev spaces $W_2^m(\partial G)$ of the fractional order $m = 3/2$. The precise definitions will be given later on. A detailed study of interpolation theory and its applications to partial differential equations can be found in the monographs by Lions and Magenes (1968), and Triebel (1978).

The development of the theory of partial differential equations in this century can be characterized by the following headline:

(P1) *From integral operators to differential operators.*

We want to explain this. The theory of integral equations, formulated by Fredholm and then by Hilbert around 1900, had its origin in potential theory. The basic idea was as follows:

(i) In order to solve the boundary value problem (49) in $\mathbb{R}^3$ with $f = 0$, we look for the function u in the form of a *double layer potential*

$$u(x) = \int_{\partial G} D(x, y)\rho(y)\, dO$$

with charge density ρ and the kernel

$$D(x, y) = -\frac{1}{4\pi} \frac{\partial}{\partial n_y} \frac{1}{|y - x|},$$

where $\partial/\partial n_y$ denotes the outer normal derivative on the boundary ∂G with respect to y (cf. Chapter 83 on electromagnetism).

(ii) To determine the unknown charge density ρ, we use the *jump conditions* for the double layer potential at boundary points. This way the boundary value problem (49) is transformed into the following integral equation for ρ:

$$g(x) = \tfrac{1}{2}\rho(x) + \int_{\partial G} D(x, y)\rho(y)\, dO.$$

Right up to the 1930's one frequently tried to solve differential equations

$$Au = f \tag{D}$$

by a passage to the integral representation

$$u = A^{-1}f, \tag{I}$$

where the kernel of the integral operator A^{-1} corresponds to the Green function K of (D), i.e., equation (I) has the form

$$u(x) = \int_G K(x, y)f(y)\, dy.$$

For example, if equation (D) corresponds to the boundary value problem (49) with $g = 0$, then

$$K(x, y) = \frac{1}{4\pi|y - x|} + H(x, y),$$

where, for each $y \in G$, the function $x \mapsto H(x, y)$ is uniquely determined by the following boundary value problem:

$$\begin{aligned} \Delta_x K(x, y) &= 0 \quad \text{for all} \quad x \in G \quad \text{with} \quad x \neq y, \\ K(x, y) &= 0 \quad \text{for all} \quad x \in \partial G. \end{aligned} \tag{K}$$

Note that problem (K) is independent of the function f.

This method led to many difficulties in more complicated elliptic problems than (49). Therefore, one began, in the 1930's, to study differential operators

$$A\colon D(A) \subseteq X \to X$$

as independent objects in Hilbert spaces. Kurt Otto Friedrichs (1934) was a pioneer in this field.

The theory of *pseudodifferential operators* and Fourier integral operators emerged in the 1960's and 1970's; it combines the theory of linear integral operators and differential operators. In this connection, one uses the Fourier transform and the theory of distributions. We refer to the four-volume exposition of Hörmander (1983).

18.7i. Monotone Operators

An operator $A: D(A) \subseteq X \to X$ on the real H-space X is called *monotone* iff

(M) $$(Au - Av|u - v) \geq 0 \quad \text{for all} \quad u, v \in D(A).$$

Furthermore, the operator A is called *strictly monotone* iff

$$(Au - Av|u - v) > 0 \quad \text{for all} \quad u, v \in D(A) \quad \text{with} \quad u \neq v.$$

Finally, the operator A is called *strongly monotone* iff there is a constant $c > 0$ such that

$$(Au - Av|u - v) \geq c\|u - v\|^2 \quad \text{for all} \quad u, v \in D(A).$$

If the operator A is linear, then the monotonicity condition (M) above is equivalent to the simpler positivity condition

$$(Au|u) \geq 0 \quad \text{for all} \quad u \in D(A).$$

Let X be a real B-space. Then the operator

$$A: D(A) \subseteq X \to X^*$$

is called monotone iff

$$\langle Au - Av, u - v\rangle \geq 0 \quad \text{for all} \quad u, v \in D(A).$$

Recall that $\langle f, u\rangle$ means the value $f(u)$ of the linear functional $f \in X^*$ at the point u. Furthermore, A is called strictly monotone iff

$$\langle Au - Av, u - v\rangle > 0 \quad \text{for all} \quad u, v \in D(A) \quad \text{with} \quad u \neq v,$$

and A is called strongly monotone iff there is a $c > 0$ such that

$$\langle Au - Av, u - v\rangle \geq c\|u - v\|^2 \quad \text{for all} \quad u, v \in D(A).$$

If X is a complex H-space (resp. complex B-space), then, in the monotonicity definitions above, we have to replace

$$(\cdot|\cdot) \quad \text{with} \quad \operatorname{Re}(\cdot|\cdot)$$

(resp. $\langle\cdot,\cdot\rangle$ with $\operatorname{Re}\langle\cdot,\cdot\rangle$). For example, the operator $A: D(A) \subseteq X \to X$ on the complex H-space X is called monotone iff

$$\operatorname{Re}(Au - Av|u - v) \geq 0 \quad \text{for all} \quad u, v \in D(A).$$

Similarly, the operator $A: D(A) \subseteq X \to X^*$ on the complex B-space X is called monotone iff

$$\operatorname{Re}\langle Au - Av, u - v\rangle \geq 0 \quad \text{for all} \quad u, v \in D(A).$$

Even in the 1930's, the notion of the monotone operator was implicitly hanging over the theory of linear *elliptic* differential equations. One also encountered this notion in the 1950's when investigating linear *parabolic* and *hyperbolic* differential equations by means of the theory of *semigroups*.

A further source for the theory of monotone operators is the classical *calculus of variations*. In Section 18.17 we shall show that the classical Legendre condition for minima is closely related to the convexity of the variational problem and hence to the monotonicity of the corresponding Euler–Lagrange equations. In order to explain the abstract basic idea, let

$$F: X \to \mathbb{R}$$

be a functional on the real B-space X. Then the minimum problem

$$\text{(P)} \qquad F(u) = \min!, \qquad u \in X,$$

corresponds to a classical variational problem. If u is a solution of (P), then u satisfies the operator equation

$$\text{(E)} \qquad F'(u) = 0,$$

which corresponds to the classical Euler–Lagrange equation. The point is that the *convexity* of the functional F implies the *monotonicity* of the operator

$$F': X \to X^*.$$

However, note the following very important fact:

The theory of monotone operators is also applicable to equations which are not related to a variational problem.

In the 1960's one recognized that the concept of monotone operators allows us to deal uniformly with many *apparently different* problems in a very elegant manner. A list of such problems is given in the Preface.

In connection with the general principle (P1) in Section 18.7h above, we mention the following principle:

(P2) *A priori estimates and compactness yield existence.*

This principle was used by the Russian mathematician Bernstein[1] at the beginning of this century and by Leray and Schauder in the 1930's, in order to obtain existence proofs for linear and nonlinear elliptic differential equations. This was discussed in the introduction to Chapter 6. The Leray–Schauder principle (Theorem 6.A) represents the most fruitful functional analytic formulation of (P2).

The theory of monotone operators allows us to free ourselves from the restrictive compactness assumption in (P2). It turns out that (P2) can be replaced by the following principle:

(P3) *Coercivity and monotonicity yield existence.*

In this connection the necessary *a priori* estimates follow easily from the coercivity condition.

[1] Bernstein (1880–1968), Schauder (1896–1943), and Leray (born 1906).

The interrelations between the calculus of variations, the development of the 19th and 20th problem of Hilbert, and the theory of monotone operators will be discussed in Section 18.19.

18.8. The Main Theorem on Quadratic Minimum Problems

That's what it was really all about.
Faust

We consider the minimum problem

$$\tfrac{1}{2}a(u, u) - b(u) = \min!, \qquad u \in u_0 + X_0. \tag{50}$$

In Section 18.10 below we shall show that the Dirichlet problem is a special case of (50). We also consider the equation

$$a(u, v) = b(v) \qquad \text{for all} \quad v \in X_0. \tag{50*}$$

Here we seek $u \in u_0 + X_0$. We make the following assumptions:

(H1) X_0 is a closed linear subspace of the real H-space X.

(H2) The map $a: X \times X \to \mathbb{R}$ is bilinear and symmetric, i.e., for all $u, v, w \in X$ and $\alpha, \beta \in \mathbb{R}$,

$$a(\alpha u + \beta v, w) = \alpha a(u, w) + \beta a(v, w),$$

$$a(u, v) = a(v, u).$$

(H3) There is a constant $d > 0$ such that

$$|a(u, v)| \leq d\|u\|\,\|v\| \qquad \text{for all} \quad u, v \in X.$$

(H4) There is a constant $c > 0$ such that

$$c\|u\|^2 \leq a(u, u) \qquad \text{for all} \quad u \in X_0.$$

(H5) The map $b: X \to \mathbb{R}$ is linear and continuous, i.e., for all $u, v \in X$ and $\alpha, \beta \in \mathbb{R}$,

$$b(\alpha u + \beta v) = \alpha b(u) + \beta b(v),$$

$$|b(v)| \leq \text{const}\,\|v\|.$$

Theorem 18.A (Main Theorem on Quadratic Minimum Problems). *Assume* (H1) *through* (H5). *Let $u_0 \in X$ be given. Then the original problem* (50) *has a unique solution. Moreover, problems* (50) *and* (50*) *are equivalent.*

The key to the proof of Theorem 18.A is the so-called *parallelogram identity*

$$\|x + y\|^2 + \|x - y\|^2 = 2\|x\|^2 + 2\|y\|^2 \tag{51}$$

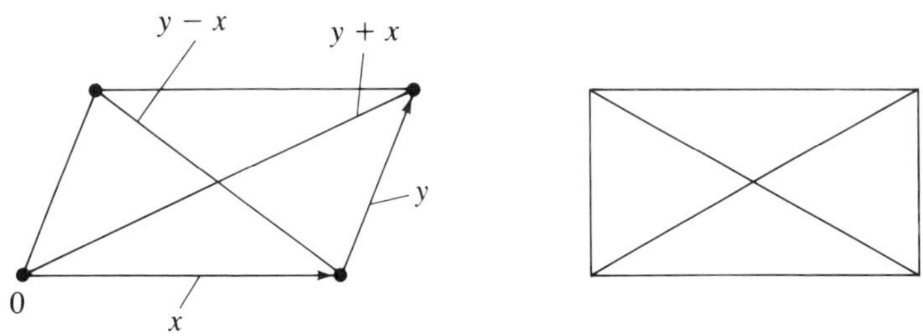

(a) (b)

Figure 18.11

for all $x, y \in Y$, where Y is a real or complex H-space. This follows from

$$\|x \pm y\|^2 = \|x\|^2 \pm (x|y) \pm (y|x) + \|y\|^2.$$

Figure 18.11(a) shows the geometrical meaning of this identity. By Figure 18.11(b) it is obvious that (51) generalizes the Pythagorean theorem.

Proof of Theorem 18.A.

(I) Equivalent equation. We show that (50) is equivalent to (50*). To this end, we set

$$F(u) = \tfrac{1}{2}a(u,u) - b(u) \qquad \text{for all} \quad u \in X.$$

Moreover, for fixed $u \in X$ and fixed $v \in X_0$, we set

$$\varphi(t) = F(u + tv) \qquad \text{for all} \quad t \in \mathbb{R}.$$

Then

$$\varphi(t) = \tfrac{1}{2}t^2 a(v,v) + t(a(u,v) - b(v)) + \tfrac{1}{2}a(u,u) - b(u).$$

Note that $a(v,v) > 0$ for all $v \in X_0$ with $v \neq 0$. Thus, the original problem

$$F(u) = \min!, \qquad u \in u_0 + X_0$$

has a solution u iff φ has a minimum at $t = 0$ for all $v \in X_0$, i.e.,

$$0 = \varphi'(0) = a(u,v) - b(v) \qquad \text{for all} \quad v \in X_0.$$

(II) Uniqueness. Let u and $\bar{u}$ be solutions of the original problem $F(u) = \min!$, $u \in u_0 + X_0$. By (I),

$$a(u,v) = b(v) \qquad \text{for all } v \in X_0,$$

$$a(\bar{u},v) = b(v).$$

Letting $v = u - \bar{u}$, we get

$$c\|u - \bar{u}\|^2 \leq a(u - \bar{u}, u - \bar{u}) = 0.$$

Hence $u = \bar{u}$.

(III) Energetic space. We set

$$(u|v)_E = a(u, v),$$
$$\|u\|_E = (u|u)_E^{1/2} \qquad \text{for all} \quad u, v \in X_0.$$

By (H3) and (H4),

$$c\|u\|^2 \le \|u\|_E^2 \le d\|u\|^2 \qquad \text{for all} \quad u \in X_0. \tag{52}$$

By definition, *the energetic space* X_E is equal to the set X_0 equipped with the so-called energetic scalar product $(\cdot|\cdot)_E$. Since X_0 is a real H-space with respect to the original scalar product $(\cdot|\cdot)$, it follows from (52) that X_E is a real H-space with respect to $(\cdot|\cdot)_E$.

In fact, let (u_n) be a Cauchy sequence in X_E. By (52), (u_n) is also a Cauchy sequence in X_0. Since X_0 is an H-space, there is a $u \in X_0$ such that $\|u_n - u\| \to 0$ as $n \to \infty$. By (52), we obtain $\|u_n - u\|_E \to 0$ as $n \to \infty$. Thus, X_E is an H-space.

(IV) Existence proof. Letting $w = u - u_0$, the original problem (50) is equivalent to the problem

$$G(w) = \min!, \qquad w \in X_0, \tag{53}$$

where

$$G(w) = \tfrac{1}{2}(w|w)_E - b_1(w) \qquad \text{for all} \quad w \in X_0,$$

and $b_1(w) = b(w) - a(u_0, w)$. By (52), there is a constant B such that

$$|b_1(w)| \le \|b\|\,\|w\| + d\|u_0\|\,\|w\|$$
$$\le B\|w\|_E \qquad \text{for all} \quad w \in X_0.$$

(IV-1) Minimal sequence (w_n). Set

$$\alpha = \inf_{w \in X_0} G(w).$$

Then there is a sequence (w_n) in X_0 such that

$$G(w_n) \to \alpha \qquad \text{as} \quad n \to \infty.$$

For all $w \in X_0$,

$$G(w) \ge \tfrac{1}{2}\|w\|_E^2 - B\|w\|_E.$$

This implies $G(w) \to +\infty$ as $\|w\|_E \to \infty$. Hence $\alpha > -\infty$.

(IV-2) Cauchy sequence. Adding $-b_1(w_n) - b_1(w_m)$ to the parallelogram identity (51) in X_E,

$$\tfrac{1}{2}\|w_n\|_E^2 + \tfrac{1}{2}\|w_m\|_E^2 = \tfrac{1}{4}\|w_n - w_m\|_E^2 + \tfrac{1}{4}\|w_n + w_m\|_E^2,$$

we obtain the *key* relation

$$G(w_n) + G(w_m) = \tfrac{1}{4}\|w_n - w_m\|_E^2 + 2G\left(\frac{w_n + w_m}{2}\right)$$
$$\ge \tfrac{1}{4}\|w_n - w_m\|_E^2 + 2\alpha.$$

Since $G(w_k) \to \alpha$ as $k \to \infty$, (w_n) is a Cauchy sequence in the H-space X_E. Thus, there is a $w \in X_E$ such that

$$w_n \to w \quad \text{in } X_E \qquad \text{as} \quad n \to \infty.$$

This implies

$$G(w_n) \to G(w) \qquad \text{as} \quad n \to \infty.$$

Hence $G(w) = \alpha$, i.e., w is a solution of (53). □

18.9. The Inequality of Poincaré–Friedrichs

In order to check assumption (H4) above in connection with the justification of the Dirichlet principle in the next section, we will use the following inequality of Poincaré–Friedrichs:

$$C \int_G u^2\,dx \le \int_G \sum_{i=1}^N (D_i u)^2\,dx \qquad \text{for all} \quad u \in C_0^\infty(G), \tag{54}$$

where $C > 0$.

Proposition 18.19. *Let G be a bounded region in $\mathbb{R}^N$, $N \ge 1$. Then* (54) *holds, where the constant C depends on G.*

Proof. We choose $N = 2$. The general case proceeds analogously. Let $u \in C_0^\infty(G)$. As in Figure 18.12, consider a rectangle $R = [a,b] \times [c,d]$ with $\bar{G} \subset \operatorname{int} R$. Note that u vanishes outside G. Then

$$u(\xi,\eta) = \int_c^\eta u_\eta(\xi,y)\,dy \qquad \text{for all} \quad (\xi,\eta) \in R.$$

The Hölder inequality with 1-trick yields

$$\begin{aligned}|u(\xi,\eta)|^2 &\le \left(\int_c^\eta dy\right)\left(\int_c^\eta u_\eta^2(\xi,y)\,dy\right)\\ &\le (d-c)\int_c^d u_\eta^2(\xi,y)\,dy.\end{aligned}$$

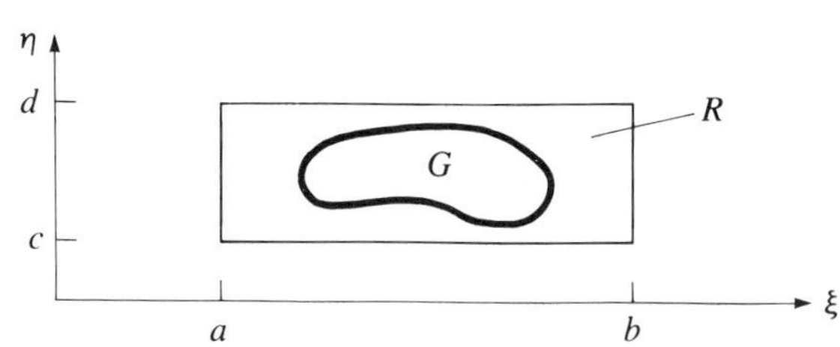

Figure 18.12

By integration over R,

$$\int_R u^2\,dx \le (d-c)^2 \int_R u_\eta^2\,dx. \qquad \square$$

Inequalities of the type (54) were considered by Poincaré (1890) in a famous paper on eigenvalue problems for the Laplace equation. Friedrichs (1934) recognized that such inequalities are the *key* to a functional analytic existence theory for linear elliptic differential equations.

18.10. The Functional Analytic Justification of the Dirichlet Principle

18.10a. The Dirichlet Problem

Let G be a bounded region in $\mathbb{R}^2$. We consider the Dirichlet problem

$$\begin{aligned} \frac{1}{2}\int_G (u_\xi^2 + u_\eta^2)\,dx &= \min!, \\ u = g \quad &\text{on } \partial G. \end{aligned} \tag{55}$$

If we regard $u(\xi, \eta)$ as the vertical displacement of a membrane at the point (ξ, η), then (55) corresponds to the principle of minimal potential energy.

Below we will use the so-called *energetic scalar product*

$$(u|v)_E = \int_G (u_\xi v_\xi + u_\eta v_\eta)\,dx.$$

Note that $2^{-1}(u|u)_E$ corresponds to the potential energy of the membrane. By Proposition 18.6, the corresponding Euler equation for (55) reads as follows:

$$\begin{aligned} \Delta u = 0 \qquad &\text{on } G, \\ u = g \qquad &\text{on } \partial G. \end{aligned} \tag{56}$$

Furthermore, the *generalized* problem for (56) reads as follows:

$$\begin{aligned} \int_G (u_\xi v_\xi + u_\eta v_\eta)\,dx = 0 \qquad &\text{for all} \quad v \in C_0^\infty(G), \\ u = g \quad &\text{on } \partial G. \end{aligned} \tag{57}$$

If u is sufficiently smooth, then (55), (56), and (57) are mutually equivalent by Proposition 18.6. Formally, one obtains (57) by multiplication of (56) by v and subsequent integration by parts. Moreover, integration by parts in (57) yields the so-called *weak* problem:

$$\int_G u\Delta v \, dx = 0 \qquad \text{for all} \quad v \in C_0^\infty(G),$$
$$u = g \quad \text{on } \partial G. \tag{58}$$

Relation (58) says that the differential equation $\Delta u = 0$ holds on G in the sense of the theory of distributions (cf. $A_2(62)$). Note that in (58) we do not assume that u is differentiable.

18.10b. Generalized Derivatives

The point of departure for the definition of generalized derivatives is the classical integration by parts formula:

$$\int_G uD_i v \, dx = -\int_G (D_i u)v \, dx \qquad \text{for all} \quad v \in C_0^\infty(G), \tag{59}$$

where $u \in C^1(G)$. Here, $D_1 = \partial/\partial\xi$ and $D_2 = \partial/\partial\eta$. The simple trick is to set

$$w = D_i u.$$

Then we obtain

$$\int_G uD_i v \, dx = -\int_G wv \, dx \qquad \text{for all} \quad v \in C_0^\infty(G). \tag{60}$$

The point is that this formula also remains valid for certain nonsmooth functions u and w.

Definition 18.20. Let $u, w \in L_2(G)$ and suppose that (60) holds. Then w is called a *generalized derivative* of u on G of type D_i.

As in the classical case we write $w = D_i u$.

Proposition 18.21. *The generalized derivative $w = D_i u$ is uniquely determined up to the values on a set of measure zero.*

PROOF. Suppose that (60) holds for $w, \overline{w} \in L_2(G)$. This yields

$$\int_G (w - \overline{w})v \, dx = 0 \qquad \text{for all } v \in C_0^\infty(G),$$

and the variational lemma (Proposition 18.2) implies that $w(x) = \overline{w}(x)$ for almost all $x \in G$. □

18.10c. The Sobolev Space $W_2^1(G)$

Definition 18.22. The space $W_2^1(G)$ consists exactly of all the functions $u \in L_2(G)$ which have generalized derivatives $D_i u \in L_2(G)$ for all i. Furthermore,

we set

$$(u|v)_{1,2} = \int_G (uv + u_\xi v_\xi + u_\eta v_\eta)\,dx \tag{61}$$

and $\|u\|_{1,2} = (u|u)_{1,2}^{1/2}$.

Proposition 18.23. *The space $W_2^1(G)$ together with $(\cdot|\cdot)_{1,2}$ becomes a real H-space in case we identify two functions whose values differ only on a set of measure zero.*

Proof. From $(u|u)_{1,2} = 0$ it follows that $u = 0$. Hence $(\cdot|\cdot)_{1,2}$ is a scalar product. In order to prove the completeness of $W_2^1(G)$, let (u_n) be a Cauchy sequence in $W_2^1(G)$, i.e.,

$$\|u_n - u_m\|_{1,2} < \varepsilon \qquad \text{for all} \quad n, m \geq n_0(\varepsilon). \tag{62}$$

Hence (u_n) and $(D_i u_n)$ are Cauchy sequences in $L_2(G)$. This implies

$$u_n \to u \quad \text{in } L_2(G) \qquad \text{and} \qquad D_i u_n \to w_i \quad \text{in } L_2(G) \qquad \text{as} \quad n \to \infty.$$

From

$$\int_G u_n D_i v\,dx = -\int_G (D_i u_n) v\,dx \qquad \text{for all} \quad v \in C_0^\infty(G)$$

we obtain

$$\int_G u D_i v\,dx = -\int_G w_i v\,dx \qquad \text{for all} \quad v \in C_0^\infty(G).$$

Hence $w_i = D_i u$. This implies $u \in W_2^1(G)$ and

$$\lim_{n\to\infty} \|u_n - u\|_{1,2} = 0.$$

In this connection, note that

$$\|u_n - u\|_{1,2}^2 = \|u_n - u\|^2 + \sum_i \|D_i u_n - D_i u\|^2,$$

where $\|\cdot\|$ denotes the norm on $L_2(G)$. □

Definition 18.24. Let $\mathring{W}_2^1(G)$ denote the closure of $C_0^\infty(G)$ in $W_2^1(G)$.

Consequently, we have $u \in \mathring{W}_2^1(G)$ iff $u \in W_2^1(G)$ and there exists a sequence (u_n) in $C_0^\infty(G)$ with

$$\|u_n - u\|_{1,2} \to 0 \qquad \text{as} \quad n \to \infty,$$

i.e., $u_n \to u$ in $L_2(G)$ and $D_i u_n \to D_i u$ *in* $L_2(G)$ as $n \to \infty$ for all i. This is equivalent to the following two conditions:

(i) $u \in L_2(G)$;
(ii) there exists a sequence (u_n) in $C_0^\infty(G)$, which is a Cauchy sequence with respect to $\|\cdot\|_{1,2}$, and we have $u_n \to u$ in $L_2(G)$ as $n \to \infty$.

Hence it is meaningful to say that the functions u in $\mathring{W}_2^1(G)$ satisfy the boundary condition

$$u = 0 \qquad \text{on } \partial G$$

in a generalized sense. This will be formulated more precisely in Theorem 21.A.

The space $\mathring{W}_2^1(G)$ is an H-space, since it is a closed linear subspace of the H-space $W_2^1(G)$.

18.10d. The Existence Theorem

We set

$$a(u, v) = \int_G (u_\xi v_\xi + u_\eta v_\eta)\, dx.$$

The following problem

$$a(u, u) = \min!, \qquad u - g \in \mathring{W}_2^1(G) \tag{63}$$

is called the *generalized* Dirichlet problem.

Obviously, problem (63) generalizes the classical Dirichlet problem (55), since the condition $u - g \in \mathring{W}_2^1(G)$ can be regarded as a generalization of the classical boundary condition $u - g = 0$ on ∂G.

Theorem 18.B. *For each given function $g \in W_2^1(G)$, problem* (63) *has a unique solution $u \in W_2^1(G)$.*

Corollary 18.25. *For all $v \in C_0^\infty(G)$,*

$$a(u, v) = 0, \tag{64}$$

$$\int_G u \Delta v\, dx = 0. \tag{65}$$

PROOF. We want to apply Theorem 18.A. To this end, we set

$$X = W_2^1(G), \qquad X_0 = \mathring{W}_2^1(G), \qquad u_0 = g.$$

(I) By the Hölder inequality, for all $u, v \in X$,

$$\begin{aligned} |a(u, v)| &\le \sum_{i=1}^{2} \int_G |D_i u||D_i v|\, dx \\ &\le \sum_{i=1}^{2} \left(\int_G (D_i u)^2\, dx\right)^{1/2} \left(\int_G (D_i v)^2\, dx\right)^{1/2} \\ &\le \|u\|_{1,2} \|v\|_{1,2}. \end{aligned}$$

(II) By the Poincaré–Friedrichs inequality, for all $v \in C_0^\infty(G)$,

$$C \int_G v^2\,dx \le \int_G \sum_i (D_i v)^2\,dx.$$

Adding $C \int_G \sum_i (D_i v)^2\,dx$, we obtain

$$c \int_G \left(v^2 + \sum_i (D_i v)^2\right) dx \le \int_G \sum_i (D_i v)^2\,dx$$

for all $v \in C_0^\infty(G)$, where $c = C/(1 + C)$, i.e.,

$$c\,\|v\|_{1,2}^2 \le a(v, v) \qquad \text{for all} \quad v \in C_0^\infty(G).$$

(III) Density argument. Let $u \in X_0$. Then there is a sequence (v_n) in $C_0^\infty(G)$ such that $\|u - v_n\|_{1,2} \to 0$ as $n \to \infty$. Hence

$$v_n \to u \quad \text{in } L_2(G) \qquad \text{and} \qquad D_i v_n \to D_i u \quad \text{in } L_2(G) \qquad \text{as} \quad n \to \infty.$$

From $c\,\|v_n\|_{1,2}^2 \le a(v_n, v_n)$ we get, as $n \to \infty$, the *key* inequality

$$c\,\|u\|_{1,2}^2 \le a(u, u) \qquad \text{for all} \quad u \in X_0.$$

Now Theorem 18.A yields Theorem 18.B and Corollary 18.25. □

In the proof of Theorem 18.A we introduced the energetic scalar product

$$(u|v)_E = a(u, v) = \int_G \sum_i D_i u D_i v\,dx$$

for all $u, v \in \mathring{W}_2^1(G)$. The corresponding *energetic space* X_E is equal to the set $\mathring{W}_2^1(G)$ equipped with $(\cdot|\cdot)_E$. Recall that X_E is an H-space.

18.10e. Regularity of the Generalized Solution

Corollary 18.26. *After changing the values of the solution u in Theorem* 18.B *on a set of measure zero, if necessary, u belongs to* $C^\infty(G)$ *and we have* $\Delta u = 0$ *on G in the classical sense.*

This follows from a more general result which we will prove in Section 18.15 (Lemma of Weyl). The smoothness of u on $\bar{G}$, i.e., the smoothness up to the boundary, will be considered in Section 22.21.

18.11. The Perpendicular Principle, the Riesz Theorem, and the Main Theorem on Linear Monotone Operators

The following results will be used frequently in this volume. Our goal is to underline the importance of the idea of orthogonality.

18.11a. Orthogonal Projection

We consider the minimum problem

$$\|u - v\| = \min!, \qquad v \in M, \tag{66}$$

and make the following assumptions:

(H) Let M be a closed linear subspace of the H-space X over $\mathbb{K} = \mathbb{R}, \mathbb{C}$, and let $u \in X$ be given.

Figure 18.13 shows the geometrical meaning of (66), i.e., we seek the foot v of a perpendicular from the point u to the plane M. Let $M^\perp$ denote the *orthogonal complement* to M, that is, by definition,

$$M^\perp = \{w \in X \colon (w|v) = 0 \text{ for all } v \in M\}.$$

Theorem 18.C (Perpendicular Principle). *Assume* (H). *Then the minimum problem* (66) *has a unique solution* v, *and* $u - v \in M^\perp$.

Corollary 18.27 (Decomposition Theorem). *If* (H) *holds, then there exists a unique decomposition of* u *of the form*

$$u = v + w, \qquad v \in M, \quad w \in M^\perp.$$

Proof of Theorem 18.C. Letting $w = u - v$, problem (66) is equivalent to

$$(w|w) = \min!, \qquad w \in u + M. \tag{66*}$$

(I) Let $\mathbb{K} = \mathbb{R}$. By Theorem 18.A, there exists a unique solution w of (66*), and $(w|v) = 0$ for all $v \in M$.

(II) Let $\mathbb{K} = \mathbb{C}$. Problem (66*) is identical to

$$a(w, w) = \min!, \qquad w \in u + M,$$

where $a(v, w) = \operatorname{Re}(v|w)$ for all $v, w \in X$. We regard X as a real H-space with respect to the scalar product $(u, v) \mapsto \operatorname{Re}(u|v)$. By Theorem 18.A, there exists a unique solution w of (66*). This implies that

$$\begin{aligned}(w|w) &\le (w - \lambda v | w - \lambda v)\\ &= (w|w) - \lambda(w|v) - \bar{\lambda}(v|w) + |\lambda|^2 \|v\|^2\end{aligned}$$

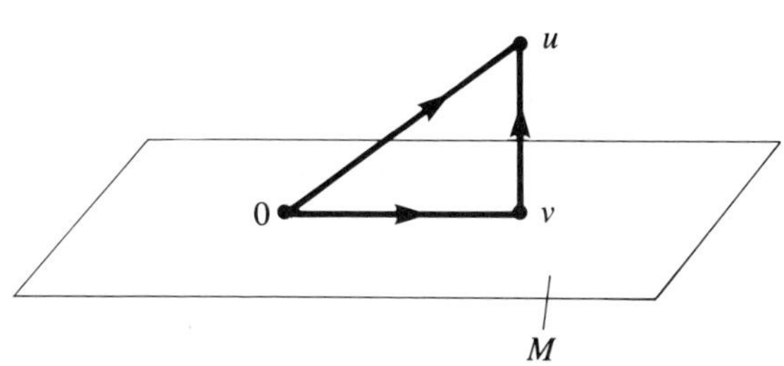

Figure 18.13

for all $v \in M$ and $\lambda \in \mathbb{C}$. Letting $\lambda = (v|w)/\|v\|^2$, we obtain $|(w|v)|^2 \leq 0$, i.e.,

$$(w|v) = 0 \qquad \text{for all} \quad v \in M. \qquad \square$$

Proof of Corollary 18.27. The existence of such a decomposition follows from Theorem 18.C. To prove the uniqueness of the decomposition, let $u = \bar{v} + \bar{w}$, $\bar{v} \in M$, $\bar{w} \in M^\perp$ be a second decomposition of u. Then

$$0 = (v - \bar{v}) + (w - \bar{w}),$$

$v - \bar{v} \in M$, $w - \bar{w} \in M^\perp$. Hence $(v - \bar{v}|v - \bar{v}) = 0$, i.e., $v = \bar{v}$. $\square$

18.11b. The Riesz Theorem

The geometrical meaning of the following theorem of Riesz is that in an H-space, there is a one-to-one correspondence between nonzero linear continuous functionals and closed linear subspaces of codimension one.

Theorem 18.D (Riesz (1934)). *Let $f: X \to \mathbb{K}$ be a linear continuous functional on the* H-*space X over $\mathbb{K} = \mathbb{R}, \mathbb{C}$. Then there exists a uniquely determined element $v \in X$ such that*

$$f(u) = (v|u) \qquad \text{for all} \quad u \in X. \tag{67}$$

Corollary 18.28. *Conversely, for each element $v \in X$, equation* (67) *defines a linear continuous functional f on X with $\|f\| = \|v\|$.*

In the case of the special space $X = L_2(G)$, this fundamental theorem of linear functional analysis was proved independently by the Hungarian mathematician Fryges Riesz (1880–1956) and the French mathematician René Maurice Fréchet (1878–1956) in the same issue of Volume 144 of *Comptes Rendus* (1907). In 1934, Riesz proved this theorem for abstract H-spaces.

Proof of Theorem 18.D.

(I) Uniqueness. It follows from

$$(v|u) = (\bar{v}|u) \qquad \text{for all} \quad u \in X$$

that $(v - \bar{v}|u) = 0$ and hence $(v - \bar{v}|v - \bar{v}) = 0$, i.e., $v = \bar{v}$.

(II) Existence.

(II-1) Let $\mathbb{K} = \mathbb{R}$. By Theorem 18.A, the minimum problem

$$\tfrac{1}{2}(v|v) - f(v) = \min!, \qquad v \in X,$$

has a unique solution v, and $(v|u) = f(u)$ for all $u \in X$. This is the assertion.

(II-2) Let $\mathbb{K} = \mathbb{C}$. Suppose that $f \neq 0$. The null space

$$N = \{u \in X: f(u) = 0\}$$

is a closed linear subspace of X. By the perpendicular principle (Corollary 18.27), there exists an element $u_0 \in N^\perp$ with $u_0 \neq 0$, since $N \neq X$. We may assume that $f(u_0) = 1$. For each $u \in X$,

$$f(u - f(u)u_0) = 0,$$

i.e., $u - f(u)u_0 \in N$. Hence we obtain the orthogonal decomposition

$$\text{(D)} \qquad u = w + f(u)u_0, \qquad w \in N, \quad u_0 \in N^\perp.$$

Scalar multiplication by u_0 yields

$$(u_0|u) = f(u)(u_0|u_0).$$

Letting $v = u_0/\|u_0\|^2$, we obtain (67). □

Relation (D) of this proof shows that the null space N of f is a closed plane in X with $\dim N^\perp = 1$, i.e.,

$$\operatorname{codim} N = 1.$$

Proof of Corollary 18.28. The Schwarz inequality yields

$$|(v|u)| \leq \|v\|\,\|u\| \qquad \text{for all} \quad u \in X.$$

Hence $\|f\| \leq \|v\|$. Because $f(v) = \|v\|^2$, we obtain $\|f\| = \|v\|$. □

18.11c. The Duality Map

Let X^* be the dual space of X, i.e., X^* consists of all linear continuous functionals $f: X \to \mathbb{K}$, where X is an H-space over $\mathbb{K} = \mathbb{R}, \mathbb{C}$. We define the so-called *duality map*

$$J: X \to X^*$$

by $J(v) = f$ with

$$f(u) = (v|u) \qquad \text{for all} \quad u \in X.$$

By Section 18.11b, the map J is bijective and isometric, i.e.,

$$\|J(v)\| = \|v\| \qquad \text{for all} \quad v \in X.$$

Moreover, J is antilinear, i.e., we have

$$J(\alpha v + \beta w) = \bar{\alpha} J(v) + \bar{\beta} J(w)$$

for all $v, w \in X$ and $\alpha, \beta \in \mathbb{K}$. Here, the bar denotes the conjugate complex number.

Let $\overline{X}^*$ be the space of all antilinear continuous functionals $g: X \to \mathbb{K}$. Set $\bar{f}(v) = \overline{f(v)}$. If $f \in X^*$, then $\bar{f} \in \overline{X}^*$. Conversely, if $g \in \overline{X}^*$, then $\bar{g} \in X^*$. We define the so-called *antiduality map*

$$\bar{J}: X \to \overline{X}^*$$

by $\bar{J}(v) = g$ with

$$g(u) = (u|v) \qquad \text{for all} \quad u \in X.$$

Obviously, $\bar{J}(v) = \overline{J(v)}$. The map $\bar{J}$ is linear, continuous, and isometric.

In the case of a real H-space X, we have $\bar{X}^* = X^*$ and $J = \bar{J}$, i.e., the duality map J is linear.

18.11d. Equivalent Existence Principles

The following three existence principles are mutually equivalent:

(i) *The existence principle for quadratic minimum problems* (*Theorem* 18.A).
(ii) *The perpendicular principle* (*Theorem* 18.C).
(iii) *The Riesz theorem* (*Theorem* 18.D).

These three principles represent the *idea of orthogonality* in H-spaces. Above we proved that

$$\text{(i)} \quad \Rightarrow \quad \text{(ii)} \quad \Rightarrow \quad \text{(iii)}.$$

Thus, it remains to show that (iii) $\Rightarrow$ (i).

To this end, we consider the minimum problem

$$\tfrac{1}{2}a(u, u) - b(u) = \min!, \qquad u \in u_0 + X_0,$$

as in Section 18.8. By the proof of Theorem 18.A, this problem is equivalent to the problem

$$\tfrac{1}{2}(v|v)_E - b_1(v) = \min!, \qquad v \in X_E,$$

where $b_1: X_E \to \mathbb{R}$ is linear and continuous. By the Riesz theorem there is an element $w \in X_E$ such that

$$b_1(v) = (w|v)_E \qquad \text{for all} \quad v \in X_E.$$

Thus we obtain the equivalent problem

$$\tfrac{1}{2}(v - w|v - w)_E = \min!, \qquad v \in X_E,$$

which has the unique solution $v = w$.

18.11e. Application to the Main Theorem on Linear Monotone Operators

Theorem 18.E. *Let $B: X \to X$ be a linear continuous operator on the* H-*space X over $\mathbb{K} = \mathbb{R}, \mathbb{C}$. Suppose that there is a $c > 0$ such that*

$$\text{(C)} \qquad |(Bu|u)| \geq c\|u\|^2 \qquad \textit{for all} \quad u \in X.$$

Then, for each given $f \in X$, the operator equation

$$Bu = f, \qquad u \in X$$

has a unique solution.

For example, condition (C) is satisfied if the operator B is *strongly monotone*, i.e., there is a $c > 0$ such that

$$\operatorname{Re}(Bu|u) \geq c\|u\|^2 \qquad \text{for all} \quad u \in X.$$

The following proof will be based on the perpendicular principle.

Theorem 18.E was proved by Lax and Milgram (1954). It is called the Lax–Milgram lemma. Applications to elliptic boundary value problems will be considered in Section 22.6.

PROOF.

(I) Uniqueness. Suppose that $Bu_1 = Bu_2 = f$. Set $u = u_1 - u_2$. Then $Bu = 0$ implies $c\|u\|^2 \leq |(Bu|u)| = 0$, and hence $u = 0$.

(II) Existence.

(II-1) Closedness of the range $R(B)$ of B. By the Schwarz inequality, for all $u \in X$,

$$c\|u\|^2 \leq |(Bu|u)| \leq \|Bu\| \, \|u\|.$$

Thus, we obtain the crucial *a priori* estimate

$$c\|u\| \leq \|Bu\| \qquad \text{for all} \quad u \in X.$$

This implies the closedness of $R(B)$. In fact, let

$$Bu_n \to v \qquad \text{as} \quad n \to \infty.$$

Thus, (Bu_n) is a Cauchy sequence. Now it follows from

$$\|u_n - u_m\| \leq c^{-1}\|Bu_n - Bu_m\|$$

that (u_n) is also a Cauchy sequence. Hence there is a u such that

$$u_n \to u \qquad \text{as} \quad n \to \infty.$$

The continuity of B implies that $Bu = v$, i.e., $v \in R(B)$. Consequently, $R(B)$ is closed.

(II-2) We show that $R(B)^\perp = \{0\}$. To this end, let $v \in R(B)^\perp$. Then

$$(Bu|v) = 0 \qquad \text{for all} \quad u \in X.$$

Choosing $u = v$, we obtain $c\|v\|^2 \leq |(Bv|v)| = 0$, i.e., $v = 0$.

(II-3) By the perpendicular principle (Corollary 18.27), the closedness of $R(B)$ and $R(B)^\perp = \{0\}$ imply $R(B) = X$. □

Theorem 18.F (Main Theorem on Linear Monotone Operators). *Let $A: X \to X^*$ be a linear continuous operator on the real* H*-space X. Suppose that A is strongly monotone, i.e., there is a $c > 0$ such that*

$$\langle Au, u \rangle \geq c\|u\|^2 \qquad \textit{for all} \quad u \in X.$$

Then, for each given $b \in X^*$, the operator equation

$$Au = b, \qquad u \in X$$

has a unique solution.

PROOF. Using the Riesz theorem, we will reduce Theorem 18.F to Theorem 18.E. Recall that X^* denotes the space of all linear continuous functionals on X, and $\langle Au, v\rangle$ denotes the value of the functional $Au \in X^*$ at the point v.

(I) Operator B. Fix $u \in X$. By the Riesz theorem (Theorem 18.D), there is an element $w \in X$ such that

$$\langle Au, v\rangle = (w|v) \qquad \text{for all} \quad v \in X,$$

and $\|w\| = \|Au\|$. We set $Bu = w$. Then

$$\|Bu\| = \|Au\| \leq \|A\|\,\|u\|.$$

Now consider Bu for all $u \in X$. Then the operator $B\colon X \to X$ is linear and continuous. Moreover, we have

$$(Bu|u) = \langle Au, u\rangle \geq c\|u\|^2 \qquad \text{for all} \quad u \in X.$$

Note that $(Bu|u)$ is real, since X is a real H-space.

(II) Again by the Riesz theorem, there exists an element $f \in X$ such that

$$\langle b, v\rangle = (f|v) \qquad \text{for all} \quad v \in X.$$

(III) Equivalent operator equation. The original operator equation $Au = b$, $u \in X$, is equivalent to

$$\langle Au, v\rangle = \langle b, v\rangle \qquad \text{for all} \quad v \in X,$$

that is,

$$(Bu|v) = (f|v) \qquad \text{for all} \quad v \in X.$$

This is equivalent to the operator equation

$$Bu = f, \qquad u \in X,$$

which has a unique solution according to Theorem 18.E. □

18.12. The Extension Principle and the Completion Principle

The classical idea of approximation corresponds to dense sets in functional analysis.

In order to simplify proofs, the following extension principle will be used frequently. We make the following assumptions:

(H1) Let X and Y be B-spaces over $\mathbb{K} = \mathbb{R}, \mathbb{C}$. The linear operator $A\colon D(A) \subseteq X \to Y$ satisfies the inequality

$$\|Au\| \leq C\|u\| \qquad \text{for all} \quad u \in D(A), \tag{68}$$

where C is a constant, i.e., A is *bounded*.

(H2) The set $D(A)$ is a linear subspace of X which is *dense* in X, i.e., $\overline{D(A)} = X$.

Explicitly, the density of $D(A)$ in X means that, for each $u \in X$ and each $\varepsilon > 0$, there exists a point $v \in D(A)$ with

$$\|u - v\| < \varepsilon.$$

Proposition 18.29. *Suppose that* (H1), (H2) *hold. Then*:

(a) *The operator A can be uniquely extended to a linear continuous operator $A: X \to Y$ with* (68) *for all $u \in X$.*
(b) *If $A: D(A) \subseteq X \to Y$ is compact, then so is $A: X \to Y$.*

PROOF. Ad(a).

(I) Existence. Let $u \in X$ and $u \notin D(A)$. Then there exists a sequence (u_n) in $D(A)$ with $u_n \to u$ as $n \to \infty$. In particular, (u_n) is a Cauchy sequence. By (68),

$$\|Au_n - Au_m\| \le C\|u_n - u_m\|.$$

Hence (Au_n) is also a Cauchy sequence, i.e., (Au_n) converges. We define

$$Au = \lim_{n\to\infty} Au_n. \tag{69}$$

We have to show that this definition of Au is independent of the choice of (u_n). To this end, let (v_n) be another sequence in $D(A)$ with $v_n \to u$ as $n \to \infty$. Then

$$\|Au_n - Av_n\| \le C\|u_n - v_n\| \to 0 \quad \text{as} \quad n \to \infty.$$

Hence $Av_n \to Au$ as $n \to \infty$.

A passage to the limit shows that (68) holds for all $u \in X$.

(II) Uniqueness. Each linear continuous extension $A: X \to Y$ of $A: D(A) \to Y$ satisfies (69). Hence the extension is unique.

Ad(b). Let (u_n) be a bounded sequence in X. Since $\overline{D(A)} = X$, there exists a bounded sequence (v_n) in $D(A)$ with $u_n - v_n \to 0$ as $n \to \infty$. Since $A: D(A) \to Y$ is compact, there exists a convergent subsequence $(Av_{n'})$. Hence $(Au_{n'})$ is convergent. □

In the context of Sobolev embedding theorems, the advantage of the extension principle is that it is sufficient to prove inequalities of type (68) for smooth functions.

The completion principle is closely related to the *extension principle.* This principle ensures that pre-Hilbert spaces (resp. normed spaces) can be completed to Hilbert spaces (resp. Banach spaces). This will be considered in Problem 18.3.

18.13. Proper Subregions

Recall that $A \subseteq B$ and $A \subset B$ means inclusion and proper inclusion, respectively.

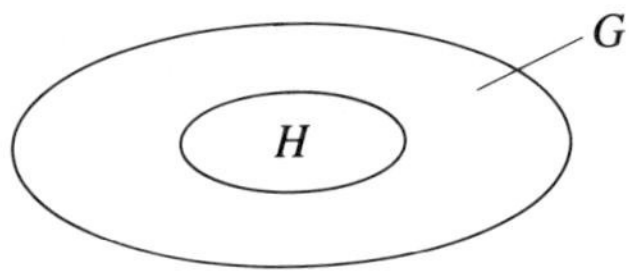

Figure 18.14

Definition 18.30. Let H and G be regions in $\mathbb{R}^N$. Then H is called a *proper* subregion of G iff

$$\bar{H} \subset G.$$

We write $H \subset\subset G$.

If G is bounded, then this definition means intuitively that a boundary strip of G is outside H (Fig. 18.14). In the same way, the symbol $H \subset\subset G$ will also be used for arbitrary sets in $\mathbb{R}^N$. In particular, if G is open and H is compact, then $H \subset G$ implies $H \subset\subset G$.

Definition 18.31. Let $f: D(f) \subseteq \mathbb{R}^N \to \mathbb{R}$ be a function. Then the closure of the set

$$\{x: f(x) \neq 0\}$$

in $\mathbb{R}^N$ is called the *support*, $\operatorname{supp} f$, of f.

In other words, $\operatorname{supp} f$ is the smallest closed set S in $\mathbb{R}^N$ with $f = 0$ outside S.

Definition 18.32. Let G be a region in $\mathbb{R}^N$. A function $u: G \to \mathbb{R}$ belongs to the space $L_{p,\text{loc}}(G)$ iff

$$u \in L_p(H)$$

holds for all compact subsets H of G.

Obviously, we have

$$L_p(G) \subseteq L_{p,\text{loc}}(G), \qquad 1 \leq p \leq \infty,$$

and the integral $\int_G uv\,dx$ exists for all

$$u \in L_{p,\text{loc}}(G), \quad 1 \leq p \leq \infty, \qquad \text{and all} \quad v \in C_0^\infty(G).$$

18.14. The Smoothing Principle

It is typical for modern analysis to work with nonsmooth functions (e.g., functions in Lebesgue spaces or Sobolev spaces). In order to simplify many investigations, nonsmooth functions u are frequently approximated by smooth

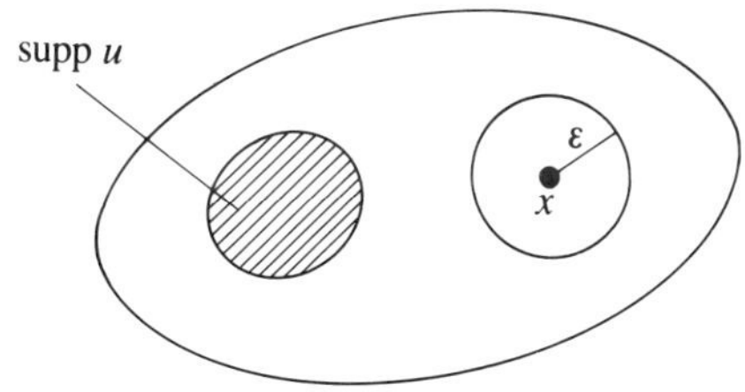

Figure 18.15

functions u_ε, where we set

$$u_\varepsilon = S_\varepsilon u. \tag{70}$$

In this section we want to study the properties of the so-called smoothing operator S_ε. As a special result, we shall obtain that $C_0^\infty(G)$ is dense in $L_p(G)$ for $1 \le p < \infty$. Let $\varepsilon > 0$.

The point of departure is the integral formula

$$u_\varepsilon(x) = \int_{\mathbb{R}^N} \varphi_\varepsilon(x-y)u(y)\,dy, \tag{71}$$

where $\varphi_\varepsilon(x) = \varepsilon^{-N}\varphi(x/\varepsilon)$, and φ has the following properties:

(i) $\varphi \in C_0^\infty(\mathbb{R}^N)$ and $\varphi(x) = 0$ for $|x| \ge 1$.
(ii) $\varphi \ge 0$ on $\mathbb{R}^N$.
(iii) $\int_{\mathbb{R}^N} \varphi(x)\,dx = 1$.

In particular, we obtain

$$\varphi_\varepsilon(x) = 0 \qquad \text{for} \quad |x| \ge \varepsilon.$$

This implies:

The values of $u_\varepsilon(x)$ depend only on the values of u on a ball around x of radius ε. (72)

If $\operatorname{dist}(x, \operatorname{supp} u) > \varepsilon$, then $u_\varepsilon(x) = 0$ (Fig. 18.15). (73)

A special choice is given by

$$\varphi(x) = \begin{cases} c\exp[(|x|^2 - 1)^{-1}] & \text{for} \quad |x| \le 1, \\ 0 & \text{for} \quad |x| > 1, \end{cases}$$

where the positive constant c is chosen in such a way that the normalization condition (iii) above is satisfied. Fig. 18.16 shows φ_ε in case $N = 1$.

Proposition 18.33 (Smoothing Operator). *Let G be a nonempty open set in $\mathbb{R}^N$ with $N \ge 1$ and let*

$$u \in L_p(G), \qquad 1 \le p < \infty.$$

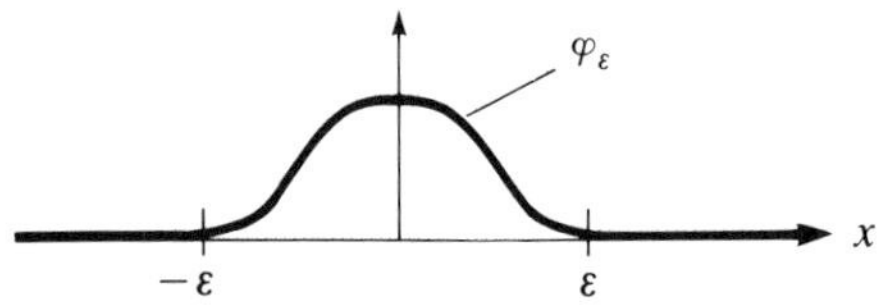

Figure 18.16

We set $u(x) = 0$ outside G. Then:

(a) *$S_\varepsilon u \in C^\infty(\mathbb{R}^N)$ and $\|S_\varepsilon u\|_p \le \|u\|_p$ for all $\varepsilon > 0$.*
(b) *$S_\varepsilon u \to u$ in $L_p(G)$ as $\varepsilon \to +0$.*
(c) *If* supp u *is a compact subset of G, then*

$$S_\varepsilon u \in C_0^\infty(G) \qquad \textit{for all} \quad 0 < \varepsilon < \varepsilon_0,$$

where ε_0 is the distance between the boundary of G and supp u.
(d) *Let the function u be continuous on an open neighborhood of the compact subset C of G. Then we obtain the uniform convergence*

$$S_\varepsilon u \rightrightarrows u \quad \textit{on } C \qquad \textit{as} \quad \varepsilon \to +0.$$

(e) *Let $1 \le p \le \infty$. Then statement* (a) *remains true. Moreover,*

$$(S_\varepsilon u)(x) \to u(x) \qquad \textit{as} \quad \varepsilon \to +0 \qquad \textit{for almost all} \quad x \in G.$$

PROOF. Ad(a). Let $u \in L_p(G)$, $1 \le p \le \infty$. Since φ is C^∞, the function u_ε in (71) has derivatives of arbitrary order. In this connection note that the function u is integrable on bounded open subsets of G, and hence we have majorants at hand (cf. $A_2(25)$). This implies $u_\varepsilon \in C^\infty(\mathbb{R}^N)$.

We set $\int = \int_{\mathbb{R}^N}$. From (71) we get

$$u_\varepsilon(x) = \int \varphi(z) u(x - \varepsilon z)\, dz.$$

If $p = \infty$, then

$$\|u_\varepsilon\|_\infty \le \|u\|_\infty \int \varphi(z)\, dz = \|u\|_\infty.$$

Now let $1 \le p < \infty$ and $p^{-1} + q^{-1} = 1$. Use $\varphi u = \varphi^{1/q}(\varphi^{1/p} u)$. By the Hölder inequality,

$$|u_\varepsilon(x)|^p \le \int |u(x - \varepsilon z)|^p \varphi(z)\, dz,$$

noting that $(\int \varphi(z)\, dz)^{p/q} = 1$. By the Fubini theorem $A_2(23)$,

$$\int |u(x - \varepsilon z)|^p \varphi(z)\, dz\, dx = \int \varphi(z) \left(\int |u(x - \varepsilon z)|^p\, dx \right) dz$$

$$= \int \varphi(z) \|u\|_p^p\, dz = \|u\|_p^p.$$

Hence $\|u_\varepsilon\|_p \le \|u\|_p$.

Ad(b). Let $B = \{y \in \mathbb{R}^N : |y| < 1\}$. It follows from (71) that

$$u_\varepsilon(x) = \int_B u(x - \varepsilon y)\varphi(y)\,dy$$

and hence

$$u_\varepsilon(x) - u(x) = \int_B (u(x - \varepsilon y) - u(x))\varphi(y)\,dy. \tag{74}$$

The Hölder inequality yields

$$|u_\varepsilon(x) - u(x)|^p \le d \int_B |u(x - \varepsilon y) - u(x)|^p\,dy,$$

where d is a constant. By the Fubini theorem $A_2(23)$,

$$\int_G |u_\varepsilon(x) - u(x)|^p\,dx \le d \int_B \left(\int_G |u(x - \varepsilon y) - u(x)|^p\,dx \right) dy.$$

The right-hand side goes to zero as $\varepsilon \to +0$ according to the p-mean continuity of the function $u \in L_p(G)$ (cf. $A_2(32)$).

Ad(c). This follows from (73).

Ad(d). This follows from (74) and the uniform continuity of u on an open neighborhood of C.

Ad(e). The validity of (a) for $p = \infty$ has been proved in (a) above. Let $B_\varepsilon = \{y \in \mathbb{R}^N : |y - x| < \varepsilon\}$. Then

$$\begin{aligned} |u_\varepsilon(x) - u(x)| &= \left| \varepsilon^{-N} \int_{B_\varepsilon} \varphi\left(\frac{x - y}{\varepsilon}\right)(u(y) - u(x))\,dy \right| \\ &\le \varepsilon^{-N} c \int_{B_\varepsilon} |u(y) - u(x)|\,dy \to 0 \end{aligned}$$

as $\varepsilon \to 0$ for almost all $x \in G$, by the Lebesgue theorem $A_2(25e)$. □

Corollary 18.34. *Let $1 \le p < \infty$ and let G be a nonempty open set in $\mathbb{R}^N$, $N \ge 1$. Then $C_0^\infty(G)$ is dense in $L_p(G)$.*

PROOF.

(I) Let G be bounded and $H \subset\subset G$. Let $u \in L_p(G)$. We set

$$v(x) = \begin{cases} u(x) & \text{on } H, \\ 0 & \text{on } G - H. \end{cases}$$

Then

$$\int_G |u - v|^p\,dx = \int_{G-H} |u|^p\,dx.$$

By the absolute continuity of integrals $A_2(20)$, the right integral is arbitrarily small provided the measure of $G - H$ is sufficiently small. Thus, we can choose the set H in such a way that

$$\|u - v\|_p < \varepsilon.$$

By Proposition 18.33(b), (c) there is a $w \in C_0^\infty(G)$ with $\|v - w\|_p < \varepsilon$. Hence $\|u - w\|_p < 2\varepsilon$.

(II) Let G be unbounded. Then, for each $\varepsilon > 0$, there exists a sufficiently large open ball B such that

$$\int_{G-H} |u|^p \, dx < \varepsilon,$$

where $H = G \cap B$. Now we can use the same argument as in (I). □

We now consider the derivatives of u_ε. In order to collect all the important properties of the smoothing operator in one place, we also investigate the case that u has generalized derivatives. The definition of $D^\alpha u$ can be found in Section 21.1.

Proposition 18.35 (Derivatives). *Let G be a nonempty bounded open set in $\mathbb{R}^N$, $N \geq 1$. Let $u \in L_{p,\text{loc}}(G)$, $1 \leq p \leq \infty$. Suppose that the generalized derivative $D^\alpha u$ exists with $D^\alpha u \in L_{p,\text{loc}}(G)$. Then:*

(a) *For all $\varepsilon > 0$ and all $x \in G$ with $\operatorname{dist}(x, \partial G) > \varepsilon$, we have*

$$D^\alpha(S_\varepsilon u)(x) = (S_\varepsilon D^\alpha u)(x). \tag{75}$$

(b) *If $H \subset\subset G$ and H is measurable, then*

$$\lim_{\varepsilon \to +0} \|D^\alpha(S_\varepsilon u) - D^\alpha u\|_{L_p(H)} = 0. \tag{76}$$

(c) *Let C be a compact subset of G. If $D^\alpha u$ is continuous on G, then we have the uniform convergence*

$$D^\alpha(S_\varepsilon u) \rightrightarrows D^\alpha u \quad \text{on } C \qquad \text{as} \quad \varepsilon \to +0.$$

(d) *If $G = \mathbb{R}^N$, then (75) holds for all $x \in \mathbb{R}^N$ and all $\varepsilon > 0$.*

PROOF. Ad(a). Note that

$$\frac{\partial}{\partial x}\varphi_\varepsilon(x - y) = -\frac{\partial}{\partial y}\varphi_\varepsilon(x - y).$$

By differentiation and integration by parts, it follows from (71) that

$$\begin{aligned}(D^\alpha S_\varepsilon u)(x) &= \int_G (-1)^{|\alpha|} D_y^\alpha \varphi_\varepsilon(x - y) u(y) \, dy \\ &= \int_G \varphi_\varepsilon(x - y) D^\alpha u(y) \, dy.\end{aligned}$$

This is (75). Note that, by the definition of generalized derivatives, the integration by parts formula remains true for generalized derivatives.

Ad(b), (c), (d). These follow from Proposition 18.33. □

Proposition 18.36 (Generalized Variational Lemma). *Let G be a nonempty open set in $\mathbb{R}^N$ with $N \geq 1$ and let $u \in L_{1,\mathrm{loc}}(G)$. Then it follows from*

$$\int_G uv\,dx = 0 \qquad \text{for all} \quad v \in C_0^\infty(G) \tag{77}$$

that $u(x) = 0$ for almost all $x \in G$.

Corollary 18.37. *If* (77) *holds for all $v \in C_0^\infty(G)$ with $\int_G v\,dx = 0$, then $u(x) =$ constant for almost all $x \in G$.*

PROOF. Since each function $v \in C_0^\infty(G)$ vanishes outside a bounded set, we can assume that G is bounded.

(I) Let $H \subset\subset G$. We set

$$w(x) = \begin{cases} \operatorname{sgn}|u(x)| & \text{on } H, \\ 0 & \text{on } G - H. \end{cases}$$

We shall show that

$$\int_G uw\,dx = 0. \tag{78}$$

This implies $\int_H |u|\,dx = 0$, i.e., $u(x) = 0$ for almost all $x \in H$. Since H is arbitrary, we obtain $u(x) = 0$ for almost all $x \in G$.

(II) We prove (78). Let $w_n = S_{1/n}w$ for all sufficiently large $n \geq n_0$. From (71) it follows that

$$\sup_{x \in G} |w_n(x)| \leq \sup_{x \in G} |w(x)| \leq 1. \tag{79}$$

By Proposition 18.33, we have $w_n \in C_0^\infty(G)$ for large n and $w_n \to w$ in $L_2(G)$ as $n \to \infty$. Thus, there exists a subsequence, again denoted by (w_n), such that $w_n(x) \to w(x)$ as $n \to \infty$ for almost all $x \in G$ (cf. $A_2(36)$). By (77),

$$\int_G uw\,dx = \lim_{n \to \infty} \int_G uw_n\,dx = 0.$$

In this connection, use (79) and the principle of majorized convergence $A_2(19)$. □

PROOF OF COROLLARY 18.37. Fix $h \in C_0^\infty(G)$ with $\int_G h\,dx = 1$. For $w \in C_0^\infty(G)$, we set

$$v(x) = w(x) - h(x)\int_G w\,dy.$$

Then $\int_G v\,dx = 0$. By assumption, $\int_G uv\,dx = 0$, i.e.,

$$\int_G (u - C)w\,dx = 0 \qquad \text{for all} \quad w \in C_0^\infty(G),$$

where $C = \int_G uh\,dx$. Proposition 18.36 yields $u = C$. □

18.15. The Idea of the Regularity of Generalized Solutions and the Lemma of Weyl

We consider the Laplace equation

$$\Delta u = 0 \quad \text{on } G. \tag{80}$$

Multiplication of (80) by $v \in C_0^\infty(G)$ and integration by parts yield

$$\int u \Delta v \, dx = 0 \qquad \text{for all} \quad v \in C_0^\infty(G). \tag{81}$$

Each function $u \in L_{1,\mathrm{loc}}(G)$ with (81) is called a *weak solution* of (80). Obviously, each classical solution $u \in C^2(G)$ of (80) is also a weak solution. The following famous so-called Lemma of Weyl tells us that the converse is true. This is the prototype of regularization theorems. More general results in this direction will be considered in Chapter 22.

Theorem 18.G (Weyl (1940)). *Let G be a bounded region in $\mathbb{R}^N$, $N \geq 1$. Suppose that $u \in L_{1,\mathrm{loc}}(G)$ is a weak solution of the Laplace equation* (80).

Then, after changing the values of u on a set of measure zero, if necessary, u is a classical solution of (80) *with $u \in C^\infty(G)$.*

Proof. We shall show that the weak solution u has the mean value property, i.e.,

$$u(x) = \frac{1}{V} \int_{B_r(x)} u(y) \, dy \tag{82}$$

for all $x \in G$ and all sufficiently small $r > 0$, where $B_r(x)$ is a ball around x of radius r, and V is the volume of $B_r(x)$. By a well-known classical result, it follows from (82) that u is a classical solution of (80) and $u \in C^\infty(G)$ (cf. Courant and Hilbert (1953, M), Vol. 2).

Step 1: The special case $u \in C(G)$, i.e., suppose first that u is continuous. Let $N \geq 2$. The simpler case $N = 1$ proceeds analogously.

(I) Choose a fixed ball $B_R(x)$. For $0 < \varepsilon < R$, set

$$\varphi(r) = \begin{cases} \text{constant} & \text{for} \quad r \leq \varepsilon, \\ 0 & \text{for} \quad r > R, \end{cases}$$

and extend φ to a function in $C^\infty(\mathbb{R}_+)$. Furthermore, let

$$\begin{aligned} v(y) &= \varphi(|x - y|), \\ w(r) &= r^{1-N} \int_{\partial B_r(x)} u \, dO. \end{aligned} \tag{83}$$

(II) We show that $w =$ constant on $]0, R[$. Let

$$\psi(r) = \frac{d}{dr}(r^{N-1}\varphi'(r)).$$

By a suitable choice of φ above, we may obtain each function $\psi \in C_0^\infty(\varepsilon, R)$ which satisfies the normalization condition

$$\int_\varepsilon^R \psi(r)\, dr = 0.$$

Using polar coordinates, it follows from (81) and (83) that

$$\int_\varepsilon^R w(r)\psi(r)\, dr = 0.$$

By Corollary 18.37, $w =$ constant on $[\varepsilon, R]$. This is true for each sufficiently small $\varepsilon > 0$. Thus, there exists a constant $C(x)$ depending on the point x such that

$$w(r) = \sigma_{N-1} C(x) \qquad \text{for all} \quad r \in]0, R[,$$

where σ_{N-1} denotes the surface measure of the $(N-1)$-dimensional unit sphere. In (83*) below, note that $V = \sigma_{N-1} r^N/N$.

(III) We now prove (82). Multiplication of w by r^{N-1} in (83) and subsequent integration over $[0, r]$ yield

$$C(x) = \frac{1}{V}\int_{B_r(x)} u(y)\, dy. \tag{83*}$$

Letting $r \to +0$, we obtain $C(x) = u(x)$.

Step 2: The general case $u \in L_{1,\text{loc}}(G)$.

We use the same argument as in Step 1. In this connection note the following peculiarities. By the Fubini theorem $A_2(23)$, the integral corresponding to $w(r)$ in (83) is defined for almost all r. If we choose a fixed r, then the absolute continuity of integrals $A_2(20)$ implies that $x \mapsto C(x)$ is continuous in (83*). Finally, by the Lebesgue theorem $A_2(25e)$, for almost all x, the right-hand side of (83*) goes to $u(x)$ as $r \to 0$. □

18.16. The Localization Principle

In modern analysis, global results are frequently proved by localization. To this end, one uses the following proposition.

Proposition 18.38 (Partition of Unity). *Let C be a compact set in $\mathbb{R}^N$, $N \geq 1$. Suppose that C is covered by a system $\{U_i\}$ of finitely many nonempty open sets*

U_i. Then there are functions $\varphi_i: \mathbb{R}^N \to \mathbb{R}$ with the following two properties:

(i) *$\varphi_i \in C_0^\infty(U_i)$ for all i.*
(ii) *$\sum_i \varphi_i(x) = 1$ for all $x \in C$.*

Motivated by (ii), $\{\varphi_i\}$ is called a *partition of unity*. For given $u: \mathbb{R}^N \to \mathbb{R}$, the function

$$u_i = \varphi_i u$$

is called a localization of u. By (i), u_i vanishes outside a compact subset of U_i, and by (ii),

$$u = \sum_i u_i.$$

PROOF. The main idea of proof is contained in (85) below. We may suppose that the sets U_i are bounded.

(I) We show that if V_i is open with $V_i \subset\subset U_i$, then there exists a function $\psi_i \in C_0^\infty(U_i)$ with

$$\psi_i = 1 \quad \text{on } V_i, \qquad 0 \le \psi_i \le 1 \quad \text{on } U_i.$$

To this end, we choose an open set W with $V_i \subset\subset W \subset\subset U_i$ (Fig. 18.17). Let

$$\chi(x) = \begin{cases} 1 & \text{on } W, \\ 0 & \text{otherwise.} \end{cases}$$

By Proposition 18.33, for sufficiently small ε, the function $\psi_i = S_\varepsilon \chi$ has the desired properties.

(II) For sufficiently small $\delta > 0$, the sets

$$V_i(\delta) = \{x \in U_i: \operatorname{dist}(x, \partial U_i) > \delta\}$$

cover C. Otherwise, there exists a sequence (x_n) in C with

$$x_n \notin \bigcup_i V_i(n^{-1}) \qquad \text{for all } n. \tag{84}$$

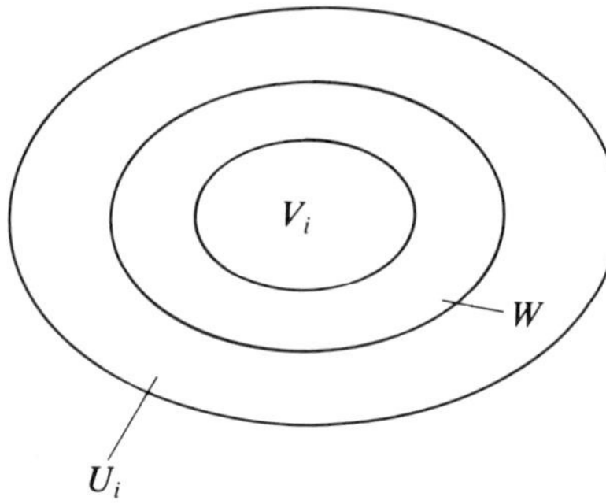

Figure 18.17

The set C is compact. Hence there exists a subsequence, again denoted by (x_n), such that $x_n \to x$ as $n \to \infty$, and $x \in C$. Thus, there is a j with $x \in U_j$, i.e., $x \in V_j(n^{-1})$ for sufficiently large n. This contradicts (84).

(III) In (I) we choose $V_i = V_i(\delta)$ for sufficiently small fixed $\delta > 0$. Then there exists an open set Z with

$$C \subset Z \subset \bigcup_i V_i.$$

By (I), $\sum_i \psi_i \geq 1$ on Z. As in (I), we construct a function $\psi \in C_0^\infty(Z)$ with $\psi = 1$ on C. Then the desired functions φ_i are given by

$$\varphi_i = \begin{cases} \psi\psi_i(\sum_i \psi_i)^{-1} & \text{on } Z, \\ 0 & \text{otherwise.} \end{cases} \tag{85}$$

□

18.17. Convex Variational Problems, Elliptic Differential Equations, and Monotonicity

In this section we want to explain a number of important interrelations contained in Figure 18.18. In particular, we want to show that convex variational problems lead to elliptic differential equations. The definition of regularly elliptic equations will be given below. According to Hilbert, a variational problem is called *regular* iff the corresponding Euler–Lagrange equation is regularly elliptic. Then the Legendre condition is fulfilled.

Note that there are elliptic differential equations which are *not* Euler–Lagrange equations to variational problems. In Chapters 26 and 27 we shall use the theory of monotone and pseudomonotone operators in order to treat elliptic equations which are not necessarily related to variational problems. This is one of the decisive *advantages* of the theory of monotone operators.

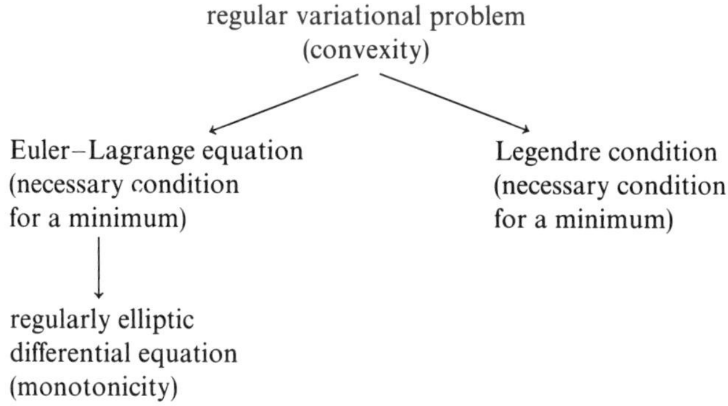

Figure 18.18

In order not to obscure the simple basic ideas by many indices, we begin with the case of $\mathbb{R}^2$.

18.17a. Variational Problems in $\mathbb{R}^2$

We consider the variational problem

$$\int_G L(x, u, p, q)\,dx = \min!, \qquad u \in C^2(\bar{G}), \qquad u = g \quad \text{on } \partial G. \tag{86}$$

Here, let $x = (\xi, \eta)$, $p = u_\xi$, $q = u_\eta$. Furthermore, let

$$P = (x, u, p, q)$$

and $P(x) = (x, u(x), p(x), q(x))$. We make the following assumptions:

(H1) G is a bounded region in $\mathbb{R}^2$.
(H2) The Lagrange function $L: \bar{G} \times \mathbb{R}^3 \to \mathbb{R}$ is C^2.

Important necessary conditions for (86) are given by the following *Euler–Lagrange equation*

$$\frac{\partial}{\partial \xi} L_p(P(x)) + \frac{\partial}{\partial \eta} L_q(P(x)) - L_u(P(x)) = 0 \quad \text{on } \bar{G} \tag{87}$$

and by the *Legendre condition*

$$L_{pp}(P(x))a^2 + 2L_{pq}(P(x))ab + L_{qq}(P(x))b^2 \geq 0 \quad \text{on } \bar{G} \tag{88}$$

for all real numbers a and b. Note the interesting fact that (88) contains only second partial derivatives of L with respect to p and q.

Proposition 18.39. *Under the assumptions* (H1) *and* (H2), *if* u *is a solution of* (86), *then* (87) *and* (88) *are satisfied.*

Remark 18.40 (Convexity). Suppose that L in (86) is convex with respect to (p, q). Then the Legendre condition is fulfilled automatically. To show this let

$$\psi(t) = L(x, u, p + ta, q + tb), \qquad t \in \mathbb{R}.$$

Then the real function ψ is convex, i.e.,

$$\psi''(t) \geq 0 \qquad \text{for all} \quad t \in \mathbb{R}.$$

This implies

$$L_{pp}(P)a^2 + 2L_{pq}(P)ab + L_{qq}(P)b^2 \geq 0$$

for all $P \in \bar{G} \times \mathbb{R}^3$ and all real a and b.

Remark 18.41 (Regularly Elliptic Equations). The partial differential equation

$$\frac{\partial}{\partial \xi} A(P(x)) + \frac{\partial}{\partial \eta} B(P(x)) + C(P(x)) = 0 \tag{89}$$

is called *regularly elliptic* iff

$$A_p(P)a^2 + (A_q(P) + B_p(P))ab + B_q(P)b^2 > 0$$

holds for all real a and b with $a^2 + b^2 \neq 0$ and all $P \in \bar{G} \times \mathbb{R}^3$. Note that the Euler–Lagrange equation (87) is a special case of (89) with

$$A = L_p, \qquad B = L_q, \qquad C = -L_u.$$

The variational problem (86) is called *regular* iff the Euler–Lagrange equation (87) is regularly elliptic. Then the Legendre condition (88) is fulfilled automatically. Explicity, the regularity of (86) means that

$$\begin{vmatrix} L_{pp}(P) & L_{pq}(P) \\ L_{pq}(P) & L_{qq}(P) \end{vmatrix} > 0, \qquad L_{pp}(P) > 0 \qquad \text{for all} \quad P \in \bar{G} \times \mathbb{R}^3.$$

The regularity of the variational problem implies $\psi''(t) > 0$ for all t in Remark 18.40, i.e., the Lagrange function L is strictly convex with respect to p and q.

Equation (89) is quasi-linear, i.e., it is linear with respect to the highest derivatives of u. In fact, (89) can be written in the form

$$A_p u_{\xi\xi} + (A_q + B_p)u_{\xi\eta} + B_q u_{\eta\eta} + \text{lower order derivatives} = 0.$$

The regular ellipticity of (89) means that the corresponding quadratic form is positive definite.

PROOF OF PROPOSITION 18.39. Let u be a solution of the minimum problem (86).

(I) Euler–Lagrange equation. We use the same simple idea as in Section 18.3. To this end, we set

$$w = u + th, \qquad h \in C_0^\infty(G), \qquad t \in \mathbb{R}.$$

Since $h = 0$ on ∂G, we get $w = g$ on ∂G. Thus, the real function

$$\varphi(t) = \int_G L(x, w, w_\xi, w_\eta)\, dx$$

has a minimum at $t = 0$, i.e., $\varphi'(0) = 0$. Hence

$$\varphi'(0) = \int_G (L_u h + L_p h_\xi + L_q h_\eta)\, dx = 0,$$

where L_u stands for $L_u(P(x))$, etc. Integration by parts yields

$$\int_G [L_u - (L_p)_\xi - (L_q)_\eta] h\, dx = 0 \qquad \text{for all} \quad h \in C_0^\infty(G).$$

By the variational lemma (Proposition 18.2), we get $[\dots] = 0$, i.e., (87) is satisfied.

(II) Legendre condition. Since φ has a minimum at $t = 0$, $\varphi''(0) \geq 0$, i.e.,

$$\varphi''(0) = \int_G (L_{pp}h_\xi^2 + 2L_{pq}h_\xi h_\eta + L_{qq}h_\eta^2 + F)\, dx \geq 0 \tag{90}$$

with the "unimportant" term

$$F = L_{uu}h^2 + 2L_{up}hh_\xi + 2L_{uq}hh_\eta.$$

Relation (90) holds for all $h \in C_0^\infty(G)$. Our goal is to obtain the Legendre condition from (90) by special choices of h.

First, let the point x_0 be fixed in the region G, and let B denote the unit ball. For each given $g \in C_0^\infty(B)$ and sufficiently small $\rho > 0$, the function

$$h(x) = \rho g\left(\frac{x - x_0}{\rho}\right)$$

lives in the space $C_0^\infty(G)$. Introducing the new variable $y = (x - x_0)/\rho$ and letting $\rho \to 0$, it follows immediately from (90) that

$$\int_B (\mathring{L}_{pp}g_\xi^2 + 2\mathring{L}_{pq}g_\xi g_\eta + \mathring{L}_{qq}g_\eta^2)\,dy \geq 0 \tag{90*}$$

for all $g \in C_0^\infty(B)$. Here, $\mathring{L}_{pp}$ denotes L_{pp} at the point x_0. Observe that $x = x_0 + \rho y$.

We now choose $g = g_1$ and $g = g_2$, where

$$g_1(x) = \psi(x)\cos t(a\xi + b\eta), \qquad g_2(x) = \psi(x)\sin t(a\xi + b\eta),$$

for fixed $a, b \in \mathbb{R}$, $t > 0$, and $\psi \in C_0^\infty(B)$. Adding the two expressions corresponding to (90*), we get

$$t^2(\mathring{L}_{pp}a^2 + 2\mathring{L}_{pq}ab + \mathring{L}_{pp}b^2)\int_B \psi^2\,dx + (\ldots) \geq 0,$$

for all $\psi \in C_0^\infty(B)$, where $(\ldots)$ is uniformly bounded with respect to all $t > 0$. Dividing by $t > 0$ and letting $t \to +\infty$, we obtain the desired Legendre condition

$$\mathring{L}_{pp}a^2 + 2\mathring{L}_{pq}ab + \mathring{L}_{pp}b^2 \geq 0 \qquad \text{for all} \quad a, b \in \mathbb{R}. \qquad \square$$

18.17b. Variational Problems in $\mathbb{R}^N$ for Several Unknown Functions

The preceding results can be generalized immediately to the following variational problem:

$$\int_G L(x, u, p)\,dx = \min!, \qquad u \in C^2(\bar{G})^K,$$
$$u = g \quad \text{on } \partial G. \tag{91}$$

Here let

$$x = (\xi_1, \ldots, \xi_N), \qquad u = (u^1, \ldots, u^K).$$

Furthermore, we set

$$p_i^k = D_i u^k,$$

and denote by p the tuple of all the p_i^k. The *Euler–Lagrange equations* are given by

$$\frac{\partial}{\partial \xi_i} L_{p_i^k} - L_{u^k} = 0, \qquad k = 1, \ldots, K, \tag{92}$$

and the *Legendre condition* is given by

$$L_{p_i^k p_j^m} a_i a_j c_k c_m \geq 0 \tag{93}$$

for all $a \in \mathbb{R}^N$ and $c \in \mathbb{R}^K$. In this connection, the argument of L in (92) and (93) is $(x, u(x), p(x))$. We sum over two equal indices where $i, j = 1, \ldots, N$ and $k, m = 1, \ldots, K$. Condition (93) is also called the Legendre–Hadamard condition.

Proposition 18.42. *Let G be a bounded region in $\mathbb{R}^N$, $N \geq 1$, and suppose that the Lagrange function*

$$L: \bar{G} \times \mathbb{R}^{K+KN} \to \mathbb{R}$$

is C^2. Then each solution u of (91) *satisfies* (92) *and* (93).

The proof is the same as the proof of Proposition 18.39.

18.17c. Regularly Elliptic Systems

The system

$$\frac{\partial}{\partial \xi_i} A_i^k(x, u(x), p(x)) + C^k(x, u(x), p(x)) = 0, \qquad k = 1, \ldots, K \tag{94}$$

is called *regularly elliptic* iff

$$\frac{\partial A_i^k}{\partial p_j^m}(x, u, p) a_i^k a_j^m > 0 \qquad \text{for all} \quad a \neq 0 \tag{95}$$

and for all possible arguments, i.e., $x \in \bar{G}$, $u \in \mathbb{R}^K$, $p, a \in \mathbb{R}^{KN}$. We sum over two equal indices where $i, j = 1, \ldots, N$ and $k, m = 1, \ldots, K$. The variational problem (91) is called *regular* iff the Euler–Lagrange equations (92) are regularly elliptic. Then the Legendre condition (93) is fulfilled automatically.

18.18. The General Euler–Lagrange Equations

We consider the variational problem

$$\begin{gathered} \int_G L(x, u, Du)\, dx = \min!, \qquad u \in C^{2m}(\bar{G})^K, \\ D^\alpha u = g_\alpha \quad \text{on } \partial G \qquad \text{for all} \quad \alpha: |\alpha| \leq m - 1. \end{gathered} \tag{96}$$

In this connection we set $u = (u_1, \ldots, u_K)$. The Lagrange function L depends on u and all partial derivatives of u up to order m. Furthermore, $D^\alpha u$ denotes an arbitrary partial derivative of order $|\alpha|$. Let $Du = (D^\alpha u)_{1 \le |\alpha| \le m}$.

Theorem 18.H. *Let G be a bounded region in $\mathbb{R}^N$ with $N \ge 1$, and suppose that the Lagrange function $L: \bar{G} \times \mathbb{R}^{K+M} \to \mathbb{R}$ is C^2. Then each solution u of the variational problem* (96) *is also a solution of the Euler–Lagrange equations*:

$$\sum_{|\alpha| \le m} (-1)^{|\alpha|} D^\alpha \left[\frac{\partial L}{\partial D^\alpha u_k}(x, u(x), Du(x)) \right] = 0, \qquad k = 1, \ldots, K \tag{97}$$

for all $x \in \bar{G}$.

Moreover, u is also a solution of the generalized Euler–Lagrange equations:

$$\sum_{|\alpha| \le m} \int_G \frac{\partial L}{\partial D^\alpha u_k}(x, u(x), Du(x)) D^\alpha h_k(x)\, dx = 0$$

for all $h_k \in C_0^\infty(G)$ and all $k = 1, \ldots, K$.

The proof, which proceeds analogously to the proof of Proposition 18.39, will be given in Problem 18.2.

The calculus of variations was founded by Euler (1744) who obtained (97) for $N = 1$ by means of a difference method. Lagrange (1762) invented those elegant methods which we use today. In contrast to Euler, Lagrange obtained (97) for $N \ge 1$. For brevity in the following, the Euler–Lagrange equations will be simply called the Euler equations.

18.19. The Historical Development of the 19th and 20th Problems of Hilbert and Monotone Operators

18.19a. The 20th Problem

> The solution of the 20th Hilbert problem was closed in some sense by the study of monotone operators.
>
> Jindřic Nečas (1982)

We consider the regular variational problem

$$\begin{gathered} \int_G L(x, u(x), Du(x))\, dx = \min!, \\ u = g \quad \text{on } \partial G. \end{gathered} \tag{98}$$

Here, Du denotes the tupel of the first-order partial derivatives of the real function u. In the 20th problem of his Paris lecture, Hilbert asked the following question.

Is it true that (98) *has a solution u in case we generalize the notion of solution in an appropriate sense?*

Today we know a positive answer. There exists a solution u in the Sobolev space $W_p^1(G)$, $1 < p < \infty$, in the case where the following assumptions are satisfied:

(i) G is a bounded region in $\mathbb{R}^N$, $N \geq 1$.
(ii) The functions L and g are sufficiently smooth, and L is convex with respect to Du. More precisely, let $g \in W_p^1(G)$ and $L: \bar{G} \times \mathbb{R}^{N+1} \to \mathbb{R}$ is C^1.
(iii) There are positive constants a, b, c such that L satisfies the *growth condition*

$$|L(x, u, Du)| \leq a(1 + |u|^p + |Du|^p) \tag{99}$$

and the *coerciveness condition*

$$L(x, u, Du) \geq b|Du|^p - c. \tag{100}$$

This result and its generalizations will be considered in Chapter 42. The theory of monotone operators can be regarded as a natural development of the 20th problem. Roughly speaking, the main idea is the following:

(i) If the function L is *convex* with respect to u and Du, then the integral in (98) corresponds to a convex functional $J = J(u)$ where the derivative J' is a *monotone operator*. The Euler equation corresponds to the operator equation $J'(u) = 0$.
(ii) If L is only convex with respect to Du, then the operator J' is *pseudomonotone*.
(iii) The existence theorems for monotone and pseudomonotone operator equations allow us to prove the existence of solutions for nonlinear elliptic differential equations which are not necessarily the Euler equations to variational problems.

These results are also valid for problems which contain derivatives up to order n. In this case the function L in (i) must be convex with respect to u and all derivatives. In (ii) we only need the convexity of L with respect to the nth derivatives.

Note that the coerciveness condition (100) is not satisfied for the problem of minimal surfaces, i.e., for

$$L = \sqrt{1 + u_\xi^2 + u_\eta^2}.$$

In this case one has to use special sophisticated geometrical methods. In particular, if the region G is not convex, then there need not exist solutions of (98). This will be considered in Chapter 52.

Višik (1961) was the first to apply monotonicity methods to quasi-linear elliptic differential equations.

18.19b. The 19th Problem

In the 19th problem Hilbert asked the following question: Is it true that the analyticity of the functon L implies the analyticity of all the solutions of the regular variational problem (98)? A generalization of this problem can be formulated as follows. Consider the regular elliptic second-order system

$$\frac{\partial}{\partial \xi_i} A_i^k(x, u(x), Du(x)) + C^k(x, u(x), Du(x)) = 0, \qquad k = 1, \dots, K, \quad (101)$$

on the bounded region G in $\mathbb{R}^N$ where $u \in \mathbb{R}^K$. We now ask the following question:

(Q) *Are the solutions u of* (101) *sufficiently smooth if the functions A_i^k, C^k are sufficiently smooth*?

For $N = 2$, a positive answer to this question was given by Morrey (1940). For $N > 2$ and $K = 1$ (the case of one equation), a positive answer was given by De Giorgi (1957) and Nash (1958). By generalizing the classical Harnack inequality, Moser (1961) simplified the proofs of De Giorgi and Nash.

In very rough terms, the main idea of the regularity proofs for (101) is the following. We consider (101) with $K = 1$. For the sake of simplicity, we assume that $C^k(\cdot)$ does not depend on Du. Put

$$v = \frac{\partial u}{\partial \xi_m} \qquad \text{for fixed } m.$$

Then differentiation of (101) with respect to ξ_m yields the key equation

$$\frac{\partial}{\partial \xi_i}\left(a_{ij} \frac{\partial v}{\partial \xi_j}\right) = f \quad \text{on } G, \qquad (101^*)$$

where a_{ij} and f depend on u.

Step 1: Existence of generalized second-order derivatives. Let $u \in W_2^1(G)$ be a generalized solution of (101), i.e., the function u has first-order generalized derivatives. We show that $u \in W_2^2(G)$, i.e., u has second-order generalized derivatives. Hence $v \in W_2^1(G)$. Furthermore, we show that v satisfies (101*).

Step 2: Boundedness of the first-order derivatives. We show that $u \in W_\infty^1(G)$, i.e.,

$$\sup_{x \in G} |Du(x)| < \infty.$$

Step 3: Reduction to the regularity theory for *linear* elliptic equations. For *fixed* u, equation (101*) represents a linear second-order elliptic equation for v, where we assume that there is a $c > 0$ such that

$$a_{ij}(x) b_i b_j \geq c|b|^2 \qquad \text{for all} \quad x \in \bar{G}, \quad b \in \mathbb{R}^N.$$

Essentially, this condition follows from the regular ellipticity of (101). We now use the following two basic results for linear elliptic equations:

(i) Theorem of De Giorgi and Nash. *Let G be a bounded region in $\mathbb{R}^N$, $N \geq 1$, and let $a_{ij}, f \in L_\infty(G)$ for all i, j. Then, each generalized solution v of equation* (101*) *with*

$$v \in W_2^1(G)$$

is locally Hölder continuous in G, i.e., there exists an $\alpha \in]0, 1[$ such that

$$v \in C^\alpha_{\text{loc}}(G).$$

This statement remains true if $f \in L_p(G)$, where $p > N/2$.

(ii) *The smoother the coefficients a_{ij} and the right member f are, the smoother is the solution v of* (101*).

This way we obtain the regularity of the generalized solution u of (101) step by step. Note that (i) is a very strong result since the functions a_{ij} and f are "bad," i.e., they are only measurable and bounded. For example, it follows from (i) that $u \in C^{1,\alpha}$ and hence $a_{ij}, f \in C^\alpha$ for all i, j. Then the regularity theory for linear elliptic equations tells us that the weak solution $v \in W_2^1$ of (101*) belongs to $C^{1,\alpha}$ and hence $u \in C^{2,\alpha}$. This process can be continued if A_i^k, C^k in (101) are sufficiently smooth.

In this connection, note the following. By definition, the function v is a *generalized* solution of equation (101*) iff

$$-\int_G a_{ij} \frac{\partial v}{\partial \xi_j} \frac{\partial \varphi}{\partial \xi_i} dx = \int_G f\varphi \, dx$$

for all $\varphi \in C_0^\infty(G)$. This identity follows formally from (101*) by multiplying (101*) with φ and integrating by parts.

Similarly, the function u is a *generalized* solution of equation (101) iff

$$\int_G A_i^k(x, u(x), Du(x)) \frac{\partial \varphi(x)}{\partial \xi_i} dx = \int_G C^k(x, u(x), Du(x)) \varphi(x) \, dx$$

for all $\varphi \in C_0^\infty(G)$ and $k = 1, \ldots, K$.

The regularity theory for linear elliptic equations can be found in Ladyženskaja and Uralceva (1973, M) and in Gilbarg and Trudinger (1983, M). We also recommend Giaquinta (1983, M) where the following techniques are used: the Caccioppoli inequality, Campanato spaces, and the freezing of coefficients (cf. also Problem 18.7).

We now consider the case $K > 1$ where (101) represents a regularly elliptic *system*. Unfortunately, the answer to Hilbert's question (Q) above is *negative*. From the physical point of view this is not completely surprising since the deformation of elastic bodies can be described by nonlinear elliptic systems where rupture may occur. Influenced by the counterexamples of De Giorgi (1968) and Giusti and Miranda (1968), Nečas (1977) constructed an elliptic system of the form (101) with *analytic* functions A_i^k, C^k which has a generalized solution being merely *Lipschitz continuous*. For systems one has two options:

(a) One proves *partial regularity*, i.e., one shows that the generalized solution is regular up to a set of small Hausdorff dimension. Therefore, the set of singularities is "thin."
(b) One proves the full regularity for special systems of so-called Liouville type.

In this connection, we recommend Giaquinta (1983, M), Nečas (1983, M), and Kohn (1984, S). In Problem 72.11 we shall discuss an important partial regularity theorem for the Navier–Stokes equations which describe viscous flows. This theorem allows the interpretation that the possible domain of irregular behavior (turbulence) is "thin."

As an introduction to the regularity theory for nonlinear problems we recommend Nečas (1982, S), (1983, M), Giaquinta (1983, M), and Giaquinta and Hildebrandt (1989, M). Furthermore, we recommend the standard works by Ladyženskaja and Uralceva (1964, M), (1973, M), Morrey (1966, M), and Gilbarg and Trudinger (1983, M).

18.19c. Two Basic Strategies in Regularity Theory

The main idea of regularity theory is to use the integral identity for the generalized solution u in order to obtain integral inequalities for appropriate quantities. From these inequalities one obtains information about u. There are the following important possibilities:

(a) Integral inequalities for difference quotients ensure the existence of generalized derivatives (cf. Section 22.18).
(b) Integral inequalities for "mean values" yield the Hölder continuity of functions (see the Morrey lemma and the theory of Campanato spaces below).
(c) Let G be a nonempty measurable set in $\mathbb{R}^N$ with $N \geq 1$ and $\operatorname{meas}(G) < \infty$. Let $1 \leq p_1 \leq p_2 \leq \cdots$ such that $p_n \to \infty$ as $n \to \infty$ and

$$u \in \bigcap_{n=1}^{\infty} L_{p_n}(G) \qquad \text{with} \qquad \sup_n \|u\|_{p_n} < \infty.$$

Then $u \in L_\infty(G)$ and

$$\|u\|_\infty = \lim_{n \to \infty} \|u\|_{p_n}.$$

Statement (c) can be used to prove the boundedness of the function u via L_p-estimates for u. A simple proof of the well-known result (c) can be found in Kufner, John, and Fučik (1977, M), p. 85.

First strategy: The difference quotient method.

This is related to (a) above. In Section 22.18ff we will use this method in order to obtain general regularity results for linear elliptic equations of order $2m$. The same method can be used in Step 1 from Section 18.19b. The idea is to use the integral identity for the generalized solution and the ellipticity condition in order to obtain integral inequalities for appropriate difference quotients. These estimates yield the existence of additional derivatives.

Second strategy: Campanato spaces.

This is related to (b) above. The main idea is contained in the following lemma.

Fundamental Lemma 18.43. *Let G be a nonempty open set in $\mathbb{R}^N$, $N \geq 1$. Set $B_r(y) = \{x \in \mathbb{R}^N : |x - y| < r\}$ and define the mean value with respect to the ball $B_r(y)$ by*

$$u_{y,r} = \frac{1}{\text{meas } B_r(y)} \int_{B_r(y)} u(x)\, dx.$$

Let $0 < \alpha \leq 1$, $1 \leq p < \infty$, and $R > 0$ be given. Then the inequality

$$\int_{B_r(y)} |u(x) - u_{y,r}|^p \, dx \leq \text{const}\, r^{N+p\alpha}$$

for all $r < \min(R, \text{dist}(y, \partial G))$ implies $u \in C^\alpha_{\text{loc}}(G)$, i.e., u is locally Hölder continuous on G.

Proof. Cf. Giaquinta (1983, M), p. 72. □

We want to show that this result generalizes the famous classical Morrey lemma.

Corollary 18.44. *Let $u \in W^1_p(G)$. Then the inequality*

$$\int_{B_r(y)} |Du|^p \, dx \leq \text{const}\, r^{N-p+p\alpha}$$

for all $r < \text{dist}(y, \partial G)$ and $y \in G$ implies $u \in C^\alpha_{\text{loc}}(G)$.

Proof. Set $B = B_1(y)$. By A_2(53a), the norm

$$\left(\int_B |Dv|^p \, dx + \left| \int_B v \, dx \right|^p \right)^{1/p}$$

is an equivalent norm on $W^1_p(B)$ (inequality of Poincaré). Let $w \in W^1_p(B)$. We set $v = w - w_{y,1}$. Then $\int_B v \, dx = 0$, and hence

$$\int_B |v|^p \, dx \leq \text{const} \int_B |Dv|^p \, dx.$$

Let $u \in W^1_p(G)$. Using the transformation $x \mapsto rx$ we get

$$\int_{B_r(y)} |u(x) - u_{y,r}|^p \, dx \leq \text{const}\, r^p \int_{B_r(y)} |Du|^p \, dx.$$

The assertion now follows from Lemma 18.43. □

Finally, we want to formulate a global variant of the Fundamental Lemma 18.43.

Definition 18.45 (Campanato Space $\mathscr{L}^{p,\lambda}(G)$). Let $1 \le p < \infty$ and $0 \le \lambda < \infty$. Let G be a bounded region in $\mathbb{R}^N$, $N \ge 1$. We set $G_r(y) = \{x \in G: |x - y| < r\}$ and

$$[u]_{p,\lambda} = \sup_{y \in G, r>0} r^{-\lambda} \int_{G_r(y)} |u(x) - u_{y,r}|^p \, dx,$$

where

$$u_{y,r} = \frac{1}{\operatorname{meas} G_r(y)} \int_{G_r(y)} u(x)\, dx.$$

The *Campanato space* $\mathscr{L}^{p,\lambda}(G)$ consists of all $u \in L_p(G)$ with $[u]_{p,\lambda} < \infty$.

$\mathscr{L}^{p,\lambda}(G)$ is a B-space with the norm $\|u\|_{L_p(G)} + [u]_{p,\lambda}$.

Let X and Y be B-spaces over $\mathbb{K}$. Then X and Y are called *equivalent* iff there exists a linear homeomorphism $h: X \to Y$. In particular, there are constants $c, d > 0$ such that

$$\|h(u)\|_Y \le c\|u\|_X \qquad \text{and} \qquad \|u\|_X \le d\|h(u)\|_Y$$

for all $u \in X$. Usually, one identifies $h(u)$ with u.

If X and Y are function spaces, then, by definition, the equivalence between X and Y includes the fact that u and $h(u)$ denote the same function.

The two B-spaces X and Y are called *normisomorphic* iff they are equivalent and we have

$$\|h(u)\|_Y = \|u\|_X \qquad \text{for all} \quad u \in X.$$

Proposition 18.46. *Let $N < \lambda \le N + p$ and $\alpha = (\lambda - N)/p$. Let G be a bounded region in $\mathbb{R}^N$ with $\partial G \in C^{0,1}$, $N \ge 1$. Then the Campanato space $\mathscr{L}^{p,\lambda}(G)$ is equivalent to the Hölder space $C^\alpha(\overline{G})$.*

PROOF. Cf. Giaquinta (1983, M), p. 70. □

In particular, this important result tells us that it follows from $u \in L_p(G)$ and $[u]_{p,\lambda} < \infty$ that $u \in C^\alpha(\overline{G})$, i.e., u is Hölder continuous and

$$\|u\|_{C^\alpha(\overline{G})} \le \text{const}(\|u\|_{L_p(G)} + [u]_{p,\lambda}).$$

Definition 18.47 (Morrey Spaces $L^{p,\lambda}(G)$). We use the same notation as in Definition 18.45. Set

$$[u]^*_{p,\lambda} = \sup_{y \in G, r>0} r^{-\lambda} \left(\int_{G_r(y)} |u|^p \, dx \right)^{1/p}.$$

The *Morrey space* $L^{p,\lambda}(G)$ consists of all $u \in L_p(G)$ with $[u]^*_{p,\lambda} < \infty$.

$L^{p,\lambda}(G)$ is a B-space with the norm $[u]^*_{p,\lambda}$.

Proposition 18.48. *For each $1 \leq p < \infty$, the Morrey space $L^{p,N}(G)$ is equivalent to $L_\infty(G)$ where $N = \dim G$.*

Let $\partial G \in C^{0,1}$. Then, for each $1 \leq p < \infty$ and $0 \leq \lambda < N$, the Morrey space $L^{p,\lambda}(G)$ is equivalent to the Campanato space $\mathscr{L}^{p,\lambda}(G)$.

PROOF. Cf. Nečas (1983, M), p. 35. □

In particular, it follows from $u \in L_p(G)$ and $[u]^*_{p,N} < \infty$ with $1 \leq p < \infty$ that $u \in L_\infty(G)$, i.e., u is *bounded*.

18.20. Sufficient Conditions for Local and Global Minima and Locally Monotone Operators

The Euler–Lagrange equations from Section 18.18 are *necessary* conditions for local minima. In order to obtain sufficient conditions for minima, one has the following four important possibilities:

(i) *Monotone operators in the convex case.* In Chapter 25 we show that in the case of the convex minimum problem

$$\text{(M)} \qquad F(u) = \min!, \qquad u \in X,$$

on the real B-space X, the Euler equation

$$\text{(E)} \qquad F'(u) = 0, \qquad u \in X,$$

is a necessary and sufficient condition for *global* minima of (M). Here, the operator $F': X \to X^*$ is *monotone.*

If F' is strictly monotone, then the unique solution u of (E) is the unique minimal point of (M).

(ii) *Definiteness of the second variation, accessory quadratic problems, eigenvalue criteria and locally regularly monotone operators.* This will be studied in detail in Chapter 29.

In this connection, we will use the theory of *eigenvalue problems* for bilinear symmetric functionals in H-spaces and the theory of strongly elliptic differential equations. These tools will be investigated in Chapter 22.

(iii) *Field theory.* See Chapter 40.

(iv) *Duality theory.* See Chapter 52.

PROBLEMS

18.1. *Examples for the Ritz method.*

18.1a. Solve the boundary value problem

$$u''(x) + \cos x = 0, \qquad 0 < x < \pi,$$
$$u(0) = u(\pi) = 0$$

with the aid of the Ritz method.

Solution: As basis functions we choose

$$w_k = \sin kx, \qquad k = 1, 2, \ldots.$$

These functions satisfy the boundary condition. The corresponding variational problem reads as follows:

$$\int_0^\pi (2^{-1}u'^2 - u\cos x)\,dx = \min!,$$
$$u(0) = u(\pi) = 0.$$

The Ritz method yields an approximate solution

$$u_n = \sum_{k=1}^n c_k w_k,$$

where the unknown coefficients c_k are determined by the minimum problem

$$F(c) \stackrel{\text{def}}{=} \int_0^\pi (2^{-1}u_n'^2 - u_n \cos x)\,dx = \min!.$$

From

$$\frac{\partial F}{\partial c_k} = 0, \qquad k = 1, \ldots, n$$

we obtain c_k. Explicitly,

$$u_{2n} = \frac{2}{\pi} \sum_{r=1}^n \frac{\sin 2rx}{r(4r^2 - 1)}.$$

As $n \to \infty$, this series converges uniformly on $[0, \pi]$ to the exact solution $u(x) = \cos x - 2\pi^{-1}x - 1$.

18.1b. Solve approximately the boundary eigenvalue problem

$$-u''(x) = \mu u(x), \qquad -1 < x < 1,$$
$$u(-1) = u(1) = 0$$

with the normalization condition

$$\int_{-1}^1 u^2\,dx = 1$$

by means of the Ritz method. This example was considered by Ritz (1909) in his basic paper.

Solution: In order to compute approximately the even eigenfunctions, we proceed from the trial

$$v = (1 - x^2)(c_0 + c_1 x^2 + \cdots + c_n x^{2n}),$$

which satisfies the boundary condition. Let

$$F(c) = \int_{-1}^1 (v'^2 - \mu v^2)\,dx.$$

Then the Ritz equation (28) reads as follows:

$$\frac{\partial F(c)}{\partial c_k} = 0, \qquad k = 0, \ldots, n.$$

(I) Let $n = 1$. We obtain the eigenvalue problem

$$(35 - 14\mu)c_0 + (7 - 2\mu)c_1 = 0,$$
$$(21 - 6\mu)c_0 + (33 - 2\mu)c_1 = 0.$$

The determinant of this system must vanish, i.e.,

$$\mu^2 - 28\mu + 63 = 0.$$

The solutions are

$$\mu = 2.467\,440, \qquad \mu = 25.6.$$

The exact solutions of the problem are the eigenfunctions

$$u = \cos \pi x/2, \qquad \sin \pi x, \qquad \cos 3\pi x/2, \ldots$$

with the eigenvalues

$$\mu = \pi^2/4 = 2.467\,401, \qquad \mu = \pi^2, \qquad \mu = 9\pi^2/4 = 22.207, \ldots$$

Thus, our approximate eigenvalue $\mu = 2.46\,744$ is already a good approximation of the lowest eigenvalue.

(II) Let $n = 2$. Then we obtain

$$\mu = 2.467\,401,$$

and $v = (1 - x^2)(1 - 0.233\,430x^2 + 0.018\,962x^4)$ which is a very good approximation of the lowest eigensolution.

The odd eigenfunctions are obtained through the trial

$$w = (1 - x^2)(c_0 x + c_1 x^3 + \cdots + c_n x^{2n+1}).$$

18.2. *Proof of Theorem* 18.H.

Solution: We set

$$w = u + th, \qquad h \in C_0^\infty(G), \qquad t \in \mathbb{R}.$$

Then the real function

$$\varphi(t) = \int_G L(x, w, Dw)\,dx$$

has a minimum at $t = 0$, i.e.,

$$\varphi'(0) = \int_G \sum_{\alpha,k} L_{D^\alpha u_k} D^\alpha h_k\,dx = 0.$$

Integration by parts yields

$$\int_G \sum_k \left[\sum_\alpha (-1)^{|\alpha|} D^\alpha L_{D^\alpha u_k}\right] h_k\,dx = 0$$

for all $h_k \in C_0^\infty(G)$. The variational lemma (Proposition 18.2) implies $[\ldots] = 0$.

18.3. *The completion principle for* B-*spaces. Let D* be a *normed space* over $\mathbb{K} = \mathbb{R}, \mathbb{C}$, i.e., D is a linear space over $\mathbb{K}$ and there exists a function $u \mapsto \|u\|$ on D with the following properties:

(i) $\|u\| \geq 0$ for all $u \in D$, and $\|u\| = 0$ iff $u = 0$;

(ii) $\|u + v\| \leq \|u\| + \|v\|$ for all $u, v \in D$;
(iii) $\|\alpha u\| = |\alpha| \, \|u\|$ for all $u \in D$, $\alpha \in \mathbb{K}$.

The B-space X over $\mathbb{K}$ is called a *completion* of D iff D is dense in X and the X-norm and the D-norm coincide on D.

Two B-spaces X and Y over $\mathbb{K}$ are called *normisomorphic* iff there exists a linear bijective map $j: X \to Y$ with

$$\|j(u)\| = \|u\| \qquad \text{for all} \quad u \in X.$$

18.3a. *Uniqueness of completions.* Show that two completions X and Y of D are normisomorphic.

Solution: Let $D \subseteq X$ and $D \subseteq Y$. We define the operator $j: D \subseteq X \to Y$ by $j(u) = u$. Then we have $\|j(u)\| = \|u\|$ on D. By the extension principle in Section 18.12, there exists a unique extension $j: X \to Y$ with $\|j(u)\| = \|u\|$ on X.

In order to prove that $j(X) = Y$, let (u_n) be a sequence in D with $u_n \to v$ in Y as $n \to \infty$. Then (u_n) is a Cauchy sequence in X and Y. Thus, we obtain $u_n \to u$ in X as $n \to \infty$, i.e., $j(u) = v$.

18.3b. *Existence of a completion.* Show that there exists a B-space X which is a completion of D.

Solution: Two Cauchy sequences (u_n) and (v_n) in D are called equivalent iff

$$\|u_n - v_n\| \to 0 \qquad \text{as} \quad n \to \infty.$$

Let X be the set of the corresponding equivalence classes $u = [(u_n)]$. We define operations by

$$[(u_n)] + [(v_n)] = [(u_n + v_n)],$$

$$\alpha[(u_n)] = [(\alpha u_n)].$$

It is easy to show that these operations are independent of the choice of the representatives. This way X becomes a linear space.

Furthermore, we define

$$\|u\| = \lim_{n \to \infty} \|u_n\|. \tag{102}$$

Since $|\|u_n\| - \|u_m\|| \leq \|u_n - u_m\|$, this limit exists. Moreover, it follows from

$$|\|u_n\| - \|v_n\|| \leq \|u_n - v_n\|$$

that the limit in (102) is independent of the choice of the representative (u_n) of u. It follows easily from (102) that X is a normed space.

Let $w \in D$. Then the constant sequence (w) is a Cauchy sequence and hence $[(w)]$ lies in X. We identify w with the equivalence class $[(w)]$. This way D becomes a subset of X, i.e., $D \subseteq X$.

Each Cauchy sequence (u_n) in D converges to $u = [(u_n)]$ in X. This follows from $u - u_m = [(u_n - u_m)]$ for fixed m. Hence D is *dense* in X.

Finally, we show that X is a B-space. To this end let (w_n) be a Cauchy sequence in X. We choose a sequence (u_n) in D with $\|u_n - w_n\| < 1/n$ for all n. Hence (u_n) is a Cauchy sequence and $u_n \to u$ as $n \to \infty$. This implies $w_n \to u$ as $n \to \infty$.

18.3c. *Example.* If D is equal to the set of rational numbers, then the completion of D is $X = \mathbb{R}$. Our method of proof in Problem 18.3b corresponds exactly to the well-known construction of real numbers due to Cantor (1845–1918). In Section 19.10 we shall use this completion procedure in order to construct the

Friedrichs extension. This way we shall obtain existence theorems for partial differential equations.

18.4. *The completion principle for* H-*spaces.* Let D be a pre-Hilbert space over $\mathbb{K} = \mathbb{R}$, $\mathbb{C}$. Then the Hilbert space X over $\mathbb{K}$ is called a *completion* of D iff D is dense in X and the X-scalar product and the D-scalar product coincide on D.

Two H-spaces X and Y over $\mathbb{K}$ are called H-*isomorphic* iff there exists a linear bijective map $j: X \to Y$ with

$$(j(u)|j(v)) = (u|v) \qquad \text{for all} \quad u, v \in X.$$

Such operators j are called *unitary* operators.

18.4a. *Existence and uniqueness of completions.* Show that there exists an H-space X which is a completion of D and that two such completions of D are H-isomorph.

Solution: Let $\mathbb{K} = \mathbb{C}$. For scalar products, one has the identity

$$4(u|v) = \|u + v\|^2 - \|u - v\|^2 - i\|u + iv\|^2 + i\|u - iv\|^2. \qquad (103)$$

Hence each normisomorphism between H-spaces is also an H-isomorphism.

According to Problem 18.3b, let X be a B-space which is a completion of D. Then, by (103), X is also an H-space, where the scalar product on X is given by the limit

$$(u|v) = \lim_{n\to\infty} (u_n|v_n).$$

Here, (u_n) and (v_n) are sequences in D with $u_n \to u$ and $v_n \to v$ as $n \to \infty$. The existence of this limit follows from (103).

If $\mathbb{K} = \mathbb{R}$, then use (103) with $i = 0$.

18.4b. *Example.* Let G be a bounded region in $\mathbb{R}^N$, $N \geq 1$. We equip the real linear space $C_0^\infty(G)$ with the scalar product

$$(u|v)_{1,2,0} = \int_G \sum_{i=1}^N D_i u D_i v \, dx.$$

Show that the Sobolev space $\mathring{W}_2^1(G)$ is the completion of $C_0^\infty(G)$ with respect to $(\cdot|\cdot)_{1,2,0}$.

Solution: By Section 18.10, the space $\mathring{W}_2^1(G)$ is the closure of $C_0^\infty(G)$ with respect to $(\cdot|\cdot)_{1,2}$. Hence $\mathring{W}_2^1(G)$ is the completion of $C_0^\infty(G)$ with respect to $(\cdot|\cdot)_{1,2}$. By the inequality of Poincaré–Friedrichs (54), we have

$$C(u|u)_{1,2} \leq (u|u)_{1,2,0} \leq (u|u)_{1,2}$$

for all $u \in \mathring{W}_2^1(G)$, i.e., the convergences with respect to $(\cdot|\cdot)_{1,2}$ and $(\cdot|\cdot)_{1,2,0}$ coincide on $\mathring{W}_2^1(G)$.

18.5. *A density criterion in* H-*spaces.* Let S be a set in an H-space X over $\mathbb{K}$. Show that if the set S is dense in X, then

$$(u|v) = 0 \qquad \text{for all} \quad v \in S \qquad \text{implies} \qquad u = 0. \qquad (104)$$

Conversely, it follows from (104) that span S is dense in X.

Solution: Let S be dense in X. Then $(u|v) = 0$ for all $v \in S$ implies $(u|v) = 0$ for all $v \in X$, i.e., in particular, $(u|u) = 0$. Hence $u = 0$.

Conversely, suppose that (104) holds. If span S is not dense, then the closure of span S in X is a proper linear subspace of X. By the *perpendicular principle* in Section 18.11, there exists a $u \neq 0$ in X which is orthogonal to span S. This contradicts (104).

18.6. *Density and duality*. Let X and Y be B-spaces over $\mathbb{K} = \mathbb{R}, \mathbb{C}$, such that the embedding

$$X \subseteq Y \tag{105}$$

is continuous, and X is *dense* in Y. Show the following:

(i) The embedding $Y^* \subseteq X^*$ is continuous.
(ii) If X is reflexive, then Y^* is dense in X^*.

Solution: Ad(i). It follows from (105) that

$$\|x\|_Y \leq \text{const}\|x\|_X \qquad \text{for all} \quad x \in X.$$

Let $y \in Y^*$ be given. Then, for all $x \in X$,

$$\begin{aligned} |\langle y^*, x\rangle_Y| &\leq \|y^*\|_{Y^*}\|x\|_Y \\ &\leq \text{const}\|y^*\|_{Y^*}\|x\|_X. \end{aligned} \tag{106}$$

Let $\bar{y}^*\colon X \to \mathbb{K}$ denote the restriction of the functional $y^*\colon Y \to \mathbb{K}$ to the subset X of Y. By (106), we get $\bar{y}^* \in X^*$, where

$$\langle \bar{y}^*, x\rangle_X = \langle y^*, x\rangle_Y \qquad \text{for all} \quad x \in X, \tag{107}$$

and

$$\|\bar{y}^*\|_{X^*} \leq \text{const}\|y^*\|_{Y^*} \qquad \text{for all} \quad y^* \in Y^*. \tag{108}$$

We want to show that $\bar{y}^* = 0$ implies $y^* = 0$. In fact, let $\bar{y}^* = 0$. Since X is *dense* in Y, it follows from (107) that $\langle y^*, x\rangle_Y = 0$ for all $x \in Y$, and hence $y^* = 0$.

Set $j(y^*) = \bar{y}^*$. Then, it follows from our considerations above that the operator

$$j\colon Y^* \to X^*$$

is *injective* and continuous. Therefore, the functional $y^* \in Y^*$ can be identified with the functional $\bar{y}^* \in X^*$. In this sense, we have $Y^* \subseteq X^*$. Moreover, it follows from (107) and (108) that

$$\langle y^*, x\rangle_X = \langle y^*, x\rangle_Y \qquad \text{for all} \quad y^* \in Y^*, \quad x \in X, \tag{109}$$

and

$$\|y^*\|_{X^*} \leq \text{const}\|y^*\|_{Y^*} \qquad \text{for all} \quad y^* \in Y^*.$$

Ad(ii). If the assertion is not true, then the closure of Y^* in the B-space X^* is a *proper* closed linear subspace of X^*. By the *Hahn–Banach theorem*, there exists a functional $f \in (X^*)^*$ such that

$$f(x^*) = 0 \qquad \text{for all} \quad x^* \in Y^* \tag{110}$$

and $f \neq 0$. Since X is *reflexive*, there exists an $x \in X$ such that

$$f(x^*) = \langle x^*, x\rangle_X \qquad \text{for all} \quad x^* \in X^*.$$

By (109) and (110),

$$\langle x^*, x\rangle_Y = 0 \qquad \text{for all} \quad x^* \in Y^*.$$

Since $X \subseteq Y$, and hence $x \in Y$, we get $x = 0$. This implies $f = 0$, which contradicts $f \neq 0$.

18.7. *Elementary approach to the regularity theory for linear and nonlinear elliptic partial differential equations of second order.* Such an approach, including the theorem of De Giorgi and Nash, can be found in Giusti (1984, M), Appendix C (approach via the Harnack inequality), and in Kinderlehrer and Stampacchia (1980, M), Appendix to Chapter 2 (the De Giorgi technique).

More general results can be found in Gilbarg and Trudinger (1983, M) (Chapter 8), Ladyženskaja and Uralceva (1973, M), (1986, S), and Krylov (1985, M). An elegant modern approach to the regularity theory can be found in Giaquinta (1983, M) and in Giaquinta and Hildebrandt (1989, M).

References to the Literature

Classical works: Euler (1744) and Lagrange (1762) (foundation of the calculus of variations), Laplace (1782) (Laplace equation), Gauss (1839) (potential theory), Riemann (1851), (1857) (foundation of complex function theory), Weierstrass (1870) (criticism of the Dirichlet principle), Hilbert (1901) (justification of the Dirichlet principle), Friedrichs (1934), Sobolev (1936), Courant and Hilbert (1937, M), Schwartz (1950) (distributions).

Collected works: Euler (1911), Vols. 1–72, Lagrange (1867), Vols. 1–14, Laplace (1878), Vols. 1–14, Gauss (1863), Vols. 1–12, Riemann (1892), Vols. 1–2, Weierstrass (1894), Vols. 1–7, Poincaré (1916), Vols. 1–11, Hilbert (1932), Vols. 1–3.

History of mathematics of the nineteenth century: Klein (1926, M).

History of functional analysis: Dieudonné (1981, M).

History of calculus from Euler to Riemann and Weierstrass: Bottazzini (1986, M).

Hilbert's problems, their solution, and their influence on the mathematics of the twentieth century: Aleksandrov (1971, S), Browder (1976, S).

Biographies of Hilbert: Blumenthal (1932), Reid (1970).

History of potential theory and of the theory of partial differential equations: Burkhardt and Meyer (1900, S), Sommerfeld (1900, S), and Lichtenstein (1921, S) (three articles in the encyclopedia of mathematics), Dieudonné (1981, M).

Collection of important classical papers in analysis: Birkhoff (1973).

Potential theory, integral equations, and boundary value problems: Courant and Hilbert (1953, M), Günter (1957, M).

Classical mathematical physics: Courant and Hilbert (1953, M), Vols. 1, 2.

Modern mathematical physics: Reed and Simon (1971, M), Vols. 1–4.

Interpolation theory and linear partial differential equations: Lions and Magenes (1968, M), Vols. 1–3, Triebel (1978, M), (1983, M).

General theory of linear differential operators, pseudodifferential operators, and Fourier integral operators: Hörmander (1983, M), Vols. 1–4.

Variational methods: Courant and Hilbert (1953, M) (recommended as an introduction), Morrey (1966, M) (standard work), Klötzler (1971, M), Velte (1976, M) (elementary introduction), Blanchard and Brüning (1982, M), Giaquinta and Hildebrandt (1989, M) (modern standard work).

Survey on minimal surfaces: Osserman (1986, M).

Minimal surfaces, modern geometric measure theory, and functions of bounded variations: Giusti (1984, M).

Harmonic mappings: Jost (1984, L), (1985, L), (1988, L), Hildebrandt (1985, L), Hildebrandt and Leis (1989, S).

Essays on mathematics and optimal form: Hildebrandt and Tromba (1985, M).

Regularity theory: Nečas (1982, S), (1983, M) and Giaquinta (1983, M) (introduction), Ladyženskaja and Uralceva (1964, M), (1973, M), Morrey (1966, M), Friedman (1969, M) (linear equations), Gilbarg and Trudinger (1983, M), Koshelev (1985, L), Giaquinta and Hildebrandt (1989, M) (cf. also Problem 18.7).

Classical papers on the Ritz method: Ritz (1909), Courant (1943).

Numerical realization of the Ritz method: Velte (1976, M), Stoer and Bulirsch (1976, M) (elementary introduction), Michlin (1962, M), (1969, M), (1985, M), Rektorys (1977, M), Glowinski (1984, M), Hackbusch (1986, M).

The Ritz method and the method of finite elements: Ciarlet (1977, M). Cf. also the References to the Literature to the Appendix.

Multigrid methods: Hackbusch (1985, M).

Software system ELLPACK for solving elliptic equations on computers: Rice and Boisvert (1984, M) (Ritz method and finite elements, finite differences, SFT, multigrid methods, ect.).

Software system for general classes of mathematical problems: IMSL (1987).

Handbook of numerical analysis: Ciarlet and Lions (1988, M), Vols. 1ff (finite element method, difference method, etc.).

Supercomputing: Murman (1985, P), Lichnewsky and Saguez (1987, S), Martin (1988, S).

Numerical recipies—the art of scientific computing: Press (1986, M).

CHAPTER 19

The Galerkin Method for Differential and Integral Equations, the Friedrichs Extension, and the Idea of Self-Adjointness

In mathematics, it's the idea that counts.

Sofia Vassilevna Kovalevskaja (1850–1891)

At the very beginning the given elliptic differential operator is only defined on a space of twice continuously differentiable functions. Then this operator is extented to an abstractly defined operator by using a formal closure. The main task is to show that the extended operator is self-adjoint. In this case it is possible to apply the methods of John von Neumann (1929).

Kurt Otto Friedrichs (1934)

In 1971, I read the beautiful paper by Fujita and Kato (1964) on the quasilinear Navier–Stokes equations and was delighted to find that, properly viewed, it looked like an *ordinary* differential equation, and the analysis proceeded in ways familiar for ordinary differential equations. This is perhaps no surprise to people in partial differential equations, but my training was in ordinary and functional differential equations, and my attempts to read partial differential equations usually became bogged down in technalities.

Many partial differential equation problems can be written as ordinary differential equations in Banach spaces, involving unbounded operators. Rewritten as Volterra integral equations via semigroups, unbounded operators no longer appear in parabolic problems and the analysis is entirely analogous to the case of ordinary differential equations. One works exclusively with generalized (mild) solutions, which often turn out to be classical solutions. The major technical differences with the case of ordinary differential equations are that we work in *two* spaces (or more) and the Gronwall inequality must be modified.

Dan Henry (1981)

The Basic Ideas of the Galerkin Method

In the preceding chapter we have seen how generalized problems and the Ritz method result, in a natural way, in a connection with variational problems.

In this chapter we shall show how this discussion can be carried over to a large class of problems which are *not* related to variational problems. In this connection, we make use of the following heuristic method:

(i) Multiplication of the differential equation for u by the functions $v \in K$ and integration over G with subsequent integration by parts yield the *generalized problem*.

For integral equations the integration by parts naturally drops out.

(ii) In this connection, it is decisive that the function class K is so chosen that a sufficiently smooth solution u of the generalized problem is also a solution of the original classical problem in the case where the data are sufficiently smooth (boundary, coefficients, boundary and initial values, etc.).

(iii) Restriction of u and v in the generalized problem to appropriate finite-dimensional subspaces yields the *Galerkin method* as a generalization of the Ritz method.

It is important for the reader that he learns, by a study of the given examples, to carry over a given problem: first, formally to a generalized form and then to obtain therefrom the Galerkin method in a simple way. In this connection, one must always pay attention, especially to (ii).

For variational problems, the Ritz method and the Galerkin method coincide. The generalized problems to be defined later in Chapters 22 through 24 result from the generalized problems given in this chapter, if one, roughly speaking, *replaces* the classical derivatives by generalized derivatives. For the practical realization of the Galerkin method one needs the *basis functions* $w_1, \ldots, w_n$. We give a number of possibilities in the Appendix, $A_2(56)$–(60) (polynomials and so-called finite elements, i.e., piecewise polynomial functions). Table 19.1 shows those problems that the Galerkin method leads to. The Galerkin method distinguishes itself by its great *universality*. It can be applied to linear as well as to nonlinear problems. However, this method is not only of great practical significance but also possesses, for nonlinear problems within the context of the theory of monotone operators, central theoretical meaning for obtaining existence statements, as we shall explain later in detail. In Section 19.4 we justify the designation projection method for the Galerkin method.

Table 19.1

Original problem		Approximate problem	
Elliptic differential equation and integral equation	linear	System of equations in $\mathbb{R}^N$	linear
	nonlinear		nonlinear
Differential equation	parabolic	System of ordinary differential equations	first order
	hyperbolic		second order

The Basic Ideas of the Friedrichs Extension

In the middle of this chapter, there stand existence theorems based on the notion of self-adjoint operators and the method of the Friedrichs extension. Figures 19.1 and 19.2 show important interrelations and applications to be considered in this chapter. We want to show that:

The Friedrichs extension and the more general energetic extension of symmetric operators are the functional analytic core of mathematical physics and the germ cell of the theory of monotone operators.

Note that such an elegant functional analytic formulation of mathematical physics is only possible because we use the notion of generalized solutions.

We first discuss *stationary problems* which are related to elliptic differential equations and variational problems. Instead of the original operator equation

(A) $$Au = f,$$

where the linear operator $A: D(A) \subseteq X \to X$ is symmetric and strongly monotone on the real H-space X, we consider the following two operator equations

(B) $$A_F u = f$$

and

(C) $$A_E u = f$$

with $A \subseteq A_F \subseteq A_E$, i.e., A_F is an extension of A and, in turn, A_E is an extension of A_F. Here the *Friedrichs extension*

$$A_F: D(A_F) \subseteq X \to X$$

is linear bijective, self-adjoint, and generally *unbounded*, whereas the *energetic extension*

$$A_E: X_E \to X_E^*$$

is a linear, strongly monotone homeomorphism on the so-called energetic space X_E. In particular, A_E is a *bounded* operator in contrast to A. To be precise, we have the following situation:

(a) The *energetic space* X_E is the completion of the set $D(A)$ with respect to the so-called energetic scalar product defined by

$$(u|v)_E = (Au|v) \qquad \text{for all} \quad u, v \in D(A).$$

(b) The operator A_E is the *duality map* of X_E which is closely related to the theorem of Riesz by Section 18.11.

(c) The Friedrichs extension A_F of the original operator A is defined through the relation

$$A_F^{-1} f = A_E^{-1} f \qquad \text{for all} \quad f \in X.$$

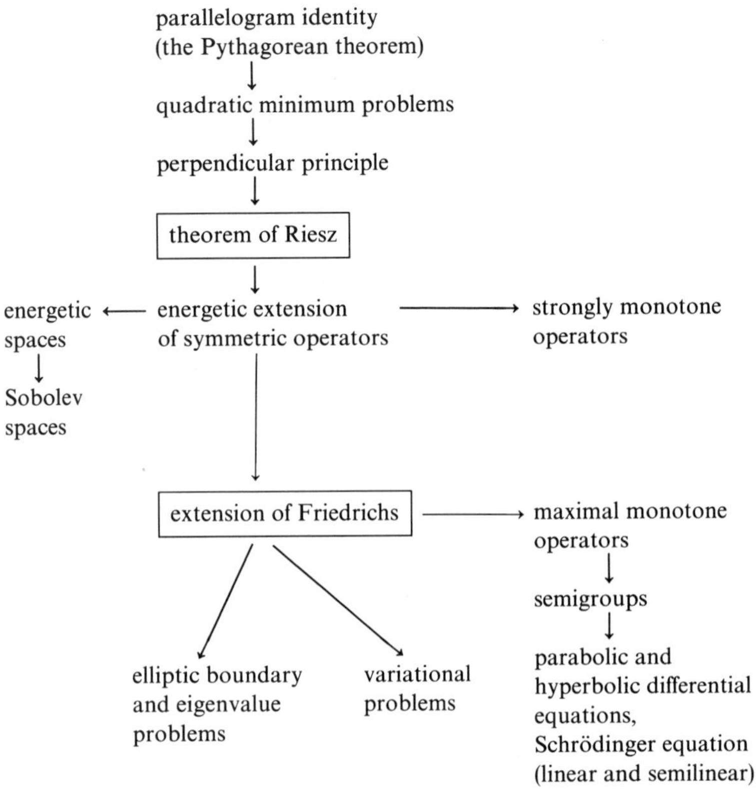

Figure 19.1

Note that the range X_E^* of the operator A_E leaves the original space X, since

$$D(A) \subseteq D(A_F) \subseteq X_E \subseteq X \subseteq X_E^*.$$

In applications to elasticity, the elastic energy of the body is given by

$$E_{\text{elast}} = 2^{-1}(u|u)_E,$$

where u corresponds to the deformation. This motivates the designation "energetic scalar product." Now to the point.

In contrast to the original equation (A) *above, equations* (B) *and* (C) *have a solution u for each given $f \in X$ and $f \in X_E^*$, respectively.*

These solutions are called *generalized solutions* to the original equation (A). Each solution of (A) is also a solution of (B) and (C). However, the converse is not always true.

For given $f \in X$, equation (B) is equivalent to the variational problem

(V) $$\tfrac{1}{2}(u|u)_E - (f|u)_X = \min!, \qquad u \in X_E,$$

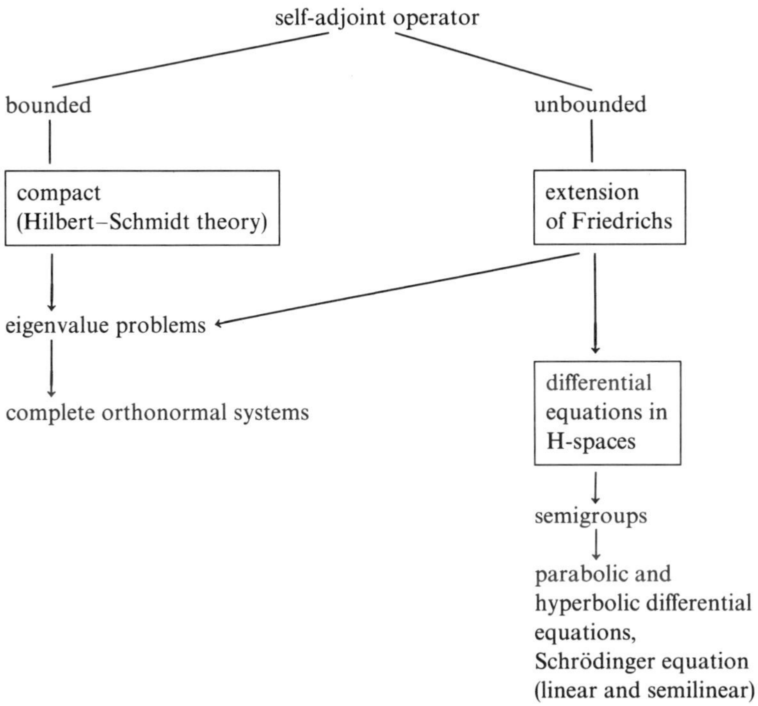

Figure 19.2

which corresponds to the principle of minimal potential energy in elasticity. In this connection, f is related to the outer forces.

In the special case of the Dirichlet problem, equation (A) corresponds to the classical problem

$$\text{(D)} \qquad \begin{aligned} -\Delta u &= f \quad \text{on } G, \\ u &= 0 \quad \text{on } \partial G, \end{aligned}$$

with $D(A) = C_0^\infty(G)$ and $X = L_2(G)$. In Section 19.11, we will obtain that

$$X_E = \mathring{W}_2^1(G), \qquad D(A_F) = W_2^2(G) \cap X_E.$$

The variational problem (V) above corresponds to

$$\int_G \left(\frac{1}{2} \sum_{i=1}^N (D_i u)^2 - fu \right) dx = \min!, \qquad u \in X_E,$$

which generalizes the classical variational problem to (D).

Therefore, we have the following two options:

(a) We consider equation (B) and work with *unbounded* operators in an H-space X. This will be done in this chapter.
(b) We consider equation (C) and work with *bounded* operators between the *two* H-spaces X_E and X_E^*. This will be done in Chapter 22.

Later on, the approaches (a) and (b) will be generalized to nonlinear maximal monotone operators and to nonlinear strongly monotone operators, respectively.

The theory of *monotone operators* can be regarded as a *generalization* of the Friedrichs extension and the energetic extension to linear *nonsymmetric* operators and to nonlinear operators. The theory of linear monotone operators has the following two important *sources*:

(i) A general existence principle for the equation $Au = b$, where A is a linear continuous strongly monotone operator on an H-space (Theorem 18.F).
(ii) The generation of linear nonexpansive semigroups by maximal dissipative operators B, i.e., $-B$ is maximal monotone (Theorem 19.E).

In (i) and (ii) we use linear monotone operators which are *not* necessarily symmetric. Theorem 19.E shows that the theory of semigroups leads in a natural way to the notion of monotone operators. The importance of the theory of semigroups for the mathematical description of evolution processes in nature will be investigated in detail in Section 19.17.

We now discuss briefly *nonstationary problems* which are related to parabolic and hyperbolic differential equations. The classical ordinary differential equation

$$x'(t) = Bx(t)$$

has the solution

$$x(t) = e^{Bt}x(0), \tag{I}$$

and

$$x''(t) + B^2x(t) = 0$$

has the solution

$$x(t) = (\cos tB)x(0) + B^{-1}(\sin Bt)x'(0). \tag{II}$$

In Sections 19.15 and 19.16, we show that the same formulas (I) and (II) describe the solutions of the heat equation and the wave equation, respectively. To this end, we only need an appropriate definition of $f(B)$, where f is a real function and B is a self-adjoint operator. This shows the *beauty* and *elegance* of functional analysis.

In particular, if the operator B has a complete orthonormal system of eigenvectors (u_i) with the corresponding eigenvalues (λ_i), then

$$Bu = \sum_i \lambda_i(u_i|u)u_i \qquad \text{for all} \quad u \in D(B),$$

and we simply define

$$f(B)u = \sum_i f(\lambda_i)(u_i|u)u_i$$

for all u for which this series converges.

The Basic Ideas for Semilinear Problems

Along with linear equations we also consider semilinear equations in this chapter. The basic idea is to transform the semilinear operator equation

$$Au = f(u) \tag{E}$$

into the equivalent equation

$$u = A^{-1}f(u) \tag{E*}$$

and to solve (E*) by using the *Banach fixed-point theorem*. Here, A is a linear operator and f is a nonlinear operator. In the case of the operator differential equation

$$u' = Bu + f(u) \tag{D}$$

for $u = u(t)$, which corresponds to parabolic and hyperbolic partial differential equations, we pass to the integral equation

$$u(t) = e^{tB}u(0) + \int_0^t e^{(t-s)B}f(u(s))\,ds \tag{D*}$$

and solve (D*) by using the Banach fixed-point theorem. In order to define the operator e^{tB}, we need the theory of *semigroups*. The solutions of (D*) are called generalized or *mild* solutions of (D).

In connection with (E*) and (D*) we encounter the following typical difficulty:

In order to handle nonlinear problems, it is necessary to choose the appropriate spaces.

We want to discuss this.

(a) We first consider (D*). In order to be able to apply the Banach fixed-point theorem to (D*), we need a space in which e^{tB} and f behave in such a way that the right-hand side of (D*) has the contractivity property. In Section 19.26 we shall use spaces defined by *fractional powers* B^α of B, which are closely related to Sobolev spaces and to interpolation theory.

(b) We now consider (E*). Let G be a bounded region in $\mathbb{R}^3$ and let $X = L_2(G)$. Set

$$f(u) = u^3.$$

Then f is *not* an operator from X to X, since $u \in L_2(G)$ does not always imply $u^3 \in L_2(G)$. However, the operator

$$f: \mathring{W}_2^1(G) \to L_2(G)$$

is locally Lipschitz continuous. This follows from the continuous embedding

$\mathring{W}_2^1(G) \subseteq L_6(G)$ which is a special case of the general Sobolev embedding theorems. This property of f will be used in Section 19.24 in order to solve the semilinear wave equation

$$u_{tt} - \Delta u + m^2 u = -\lambda u^3$$

which is related to mesons of mass m.

Generally, one uses the Sobolev embedding theorems, the Gagliardo–Nirenberg inequalities, and the abstract methods of interpolation theory in order to investigate the properties of nonlinear operators. As for (D*), it is frequently necessary to use more than one space.

As we shall see in this section, Sobolev spaces are closely related to:

(i) the energetic spaces of Friedrichs; and to
(ii) the spaces defined via fractional powers of operators.

In the following Sections 19.1 through 19.4 we use *convention* (7) in Chapter 18. Furthermore, we restrict ourselves to homogeneous boundary conditions. Inhomogeneous boundary conditions can be reduced to homogeneous boundary conditions by using Remark 18.8.

19.1. Elliptic Differential Equations and the Galerkin Method

EXAMPLE 19.1. We consider the boundary value problem

$$\begin{aligned} Lu &= f \quad \text{on } G, \\ u &= 0 \quad \text{on } \partial G, \end{aligned} \tag{1}$$

with

$$Lu = -\Delta u + \sum_{i=1}^{N} a_i D_i u + a_0 u$$

and $a_i = a_i(x)$. Multiplication by $v \in C_0^\infty(G)$ and integration yield

$$\int_G (Lu - f)v\,dx = 0 \qquad \text{for all} \quad v \in C_0^\infty(G). \tag{2}$$

After integration by parts we obtain the *generalized problem* corresponding to (1):

$$\begin{gathered} \int_G \left(\sum_{i=1}^{N} D_i u D_i v + (a_i D_i u)v + a_0 uv - fv \right) dx = 0 \qquad \text{for all} \quad v \in C_0^\infty(G), \\ u = 0 \quad \text{on } \partial G. \end{gathered} \tag{3}$$

The designation "generalized problem" is justified by the fact that, for

$u \in C^2(\bar{G})$ and $a_0, \ldots, a_N, f \in C(\bar{G})$, equation (3) implies (2) after integration by parts, and, in turn, equation (2) implies (1) by Proposition 18.2.

Remark 19.2. The choice of the generalized problem is *not* unique. It would suffice that one requires (3) for all $v \in M$ with $M = \{v \in C^1(\bar{G}): v = 0 \text{ on } \partial G\}$. The final generalized problem in Section 22.6, for which we prove existence statements there, reads as follows:

We seek a function $u \in \mathring{W}_2^1(G)$ *so that* (3) *holds for all* $v \in \mathring{W}_2^1(G)$.

Note that $C_0^\infty(G)$ as well as M are dense in $\mathring{W}_2^1(G)$.

Remark 19.3. In the general case, problem (1) does *not* arise from a variational problem. In Example 41.7 we shall show that for this it is necessary and sufficient that the integral on the left-hand side of (3) is symmetric in u and v for $f \equiv 0$. This means that $a_1 = \cdots = a_N \equiv 0$.

In order to construct an approximate solution of (1) with the aid of the *Galerkin method*, we choose functions $w_1, \ldots, w_n$ with

$$w_k = 0 \quad \text{on } \partial G \qquad \text{for all } k$$

and make the trial

$$u_n = \sum_{k=1}^n c_{kn} w_k.$$

It is thus guaranteed that $u_n = 0$ on ∂G. If we now replace u in (3) by u_n and require that (3) holds only for all

$$v \in \operatorname{span}\{w_1, \ldots, w_n\},$$

then we obtain the so-called *Galerkin equations*

$$\int_G \left(\sum_{i=1}^N D_i u_n D_i w_j + (a_i D_i u_n) w_j + a_0 u_n w_j - f w_j \right) dx = 0, \qquad j = 1, \ldots, n. \quad (3^*)$$

This is a linear system of equations for the real numbers $c_{1n}, \ldots, c_{nn}$. For $a_0 = \cdots = a_N \equiv 0$, equation (3*) is identical to the Ritz equations (16) in Corollary 18.7.

The following example concerns natural boundary conditions.

EXAMPLE 19.4. We study the second boundary value problem

$$\begin{aligned} Lu &= f \quad \text{on } G, \\ \frac{\partial u}{\partial n} &= 0 \quad \text{on } \partial G, \end{aligned} \tag{4}$$

where L has the same meaning as in Example 19.1. Multiplication by $v \in C^\infty(\bar{G})$ and integration by parts yield the *generalized problem* corresponding to (4):

$$\int_G \left(\sum_{i=1}^N D_i u D_i v + (a_i D_i u) v + a_0 u v - f v \right) dx = 0 \qquad \text{for all} \quad v \in C^\infty(\bar{G}). \quad (5)$$

For smooth functions we obtain backwards equation (4) from (5). For example, let $u \in C^2(\bar{G})$ and $a_0, \ldots, a_N, f \in C(\bar{G})$. Then (5) implies

$$\int_G (Lu - f)v\,dx + \int_{\partial G} \frac{\partial u}{\partial n} v\,dO = 0 \qquad \text{for all} \quad v \in C^\infty(\bar{G}), \tag{6}$$

after integration by parts. This yields (4) as in the proof of Proposition 18.9.

In contrast to Example 19.1, here we multiply by $v \in C^\infty(\bar{G})$ and not by $v \in C_0^\infty(G)$. In the latter case we would not obtain the equivalence of (4) and (5) for smooth functions. It is remarkable that the boundary condition "$\partial u/\partial n = 0$ on ∂G" in (4) does *not* appear in the generalized problem (5). It is a so-called *natural boundary condition.*

In order to formulate the *Galerkin method* for (4), we choose functions $w_1, \ldots, w_n$ and proceed from the trial

$$u_n = \sum_{k=1}^n c_{kn} w_k.$$

If we replace u in (5) by u_n and require that (5) holds for all

$$v \in \operatorname{span}\{w_1, \ldots, w_n\},$$

then we obtain for u_n the linear system of equations (3*) to determine the real numbers $c_{1n}, \ldots, c_{nn}$. The only difference from Example 19.1 now consists of the fact that we place no boundary condition on w_k.

As our next example, we consider a *nonlinear* problem.

EXAMPLE 19.5. Suppose, for given $p > 2$, the boundary value problem is

$$\begin{aligned} -\sum_{i=1}^N D_i(|D_i u|^{p-2} D_i u) + a_0 u &= f \quad \text{on } G, \\ u &= 0 \quad \text{on } \partial G. \end{aligned} \tag{7}$$

Multiplication by $v \in C_0^\infty(G)$ and integration by parts yield the *generalized problem* corresponding to (7):

$$\begin{gathered} \int_G \left(\sum_{i=1}^N |D_i u|^{p-2} D_i u D_i v + a_0 uv - fv \right) dx = 0 \qquad \text{for all} \quad v \in C_0^\infty(G), \\ u = 0 \quad \text{on } \partial G. \end{gathered} \tag{8}$$

For $u \in C^2(\bar{G})$, equation (7) easily follows from (8) analogously to Example 19.1. To formulate the *Galerkin method* we choose functions $w_1, \ldots, w_n$ with

$$w_k = 0 \quad \text{on } \partial G \qquad \text{for} \quad k = 1, \ldots, n$$

and proceed from the trial

$$u_n = \sum_{k=1}^n c_{nk} w_k.$$

Then we have $u_n = 0$ on ∂G. If, as usual, we replace u in (8) by u_n and require that (8) holds for all $v \in \operatorname{span}\{w_1, \ldots, w_n\}$, then we obtain the Galerkin equations for $j = 1, \ldots, n$:

$$\int_G \left(\sum_{i=1}^{N} |D_i u_n|^{p-2} D_i u_n D_i w_j + a_0 u_n w_j - f w_j \right) dx = 0. \tag{8*}$$

This is a *nonlinear* system of equations for determining the real numbers $c_{1n}, \ldots, c_{nn}$ of the form

$$g_j(c_{1n}, \ldots, c_{nn}) = 0, \qquad j = 1, \ldots, n.$$

We occupy ourselves with the interative solution of such equations in Sections 26.4 and 26.5. We investigate a difference method for (8) in Chapter 35.

19.2. Parabolic Differential Equations and the Galerkin Method

EXAMPLE 19.6. Let $Q_T = G \times]0, T[$. We consider the parabolic initial–boundary value problem

$$u_t(x,t) - \Delta u(x,t) = f(x,t) \quad \text{on } Q_T, \tag{9a}$$

$$u(x,t) = 0 \quad \text{on } \partial G \times [0,T] \qquad \text{(boundary condition)}, \tag{9b}$$

$$u(x,0) = u_0(x) \quad \text{on } G \qquad \text{(initial condition)} \tag{9c}$$

(Fig. 19.3). If we interpret $u(x,t)$ as the temperature at the point $x \in G$ at time t, then (9c) means that the temperature at the initial time $t = 0$ is prescribed. Condition (9b) means that the temperature on the boundary ∂G is equal to zero for all times $t \in [0, T]$. The function f describes an external heat source. We are given f and u_0. We seek u. A physical motivation for (9) can be found in Section 69.2.

Multiplication of (9a) by $v \in C_0^\infty(G)$ and subsequent integration over G yield

$$\int_G (u_t - \Delta u - f) v \, dx = 0 \qquad \text{for all} \quad v \in C_0^\infty(G). \tag{10}$$

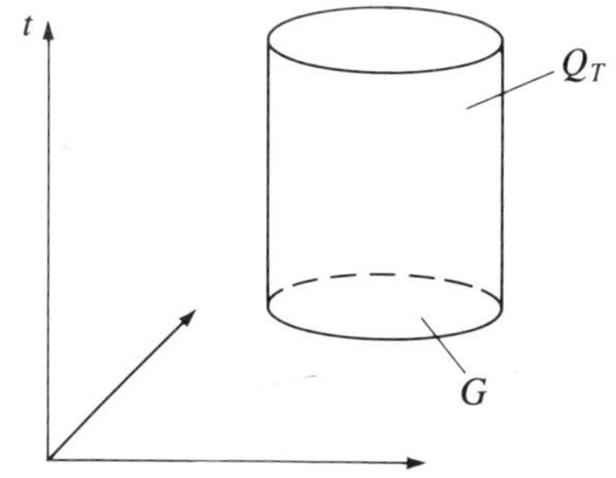

Figure 19.3

Integration by parts yields the *generalized problem* corresponding to (9):

$$\frac{d}{dt}\int_G u(x,t)v(x)\,dx + \int_G\left(\sum_{i=1}^N D_iuD_iv - fv\right)dx = 0 \qquad \text{for all} \quad v \in C_0^\infty(G), \tag{11a}$$

$$u(x,t) = 0 \quad \text{on } \partial G \times [0,T], \tag{11b}$$

$$u(x,0) = u_0(x) \quad \text{on } G. \tag{11c}$$

We note that the function v depends only on the space variable x, but not on time t.

Conversely, for sufficiently smooth u and f, from the generalized problem (11a) one first obtains equation (10) by means of integration by parts and (10) implies the original classical problem (9), according to Proposition 18.2.

In order to construct an approximate solution u_n with the aid of the *Galerkin method*, we make the trial

$$u_n(x,t) = \sum_{k=1}^n c_{kn}(t)w_k(x).$$

In contrast to Section 19.1, the unknown real coefficients c_{kn} now depend on time t. If we replace u in the generalized problem (11a) by u_n and require that (11a) holds for all $v \in \operatorname{span}\{w_1, \ldots, w_n\}$, then we obtain the Galerkin equations for $j = 1, \ldots, n$:

$$\sum_{k=1}^n c_{kn}'(t)\int_G w_kw_j\,dx + \sum_{k=1}^n c_{kn}\int_G\sum_{i=1}^N D_iw_kD_iw_j\,dx = \int_G fw_j\,dx. \tag{12a}$$

This is a linear system of first-order ordinary differential equations for the real functions $c_{1n}, \ldots, c_{nn}$. We must still add an *initial condition*

$$c_{jn}(0) = \alpha_{jn}, \qquad j = 1, \ldots, n. \tag{12b}$$

This must correspond approximately to the initial condition

$$u(x,0) = u_0(x) \quad \text{on } G$$

of the original problem. For this we choose a sequence (u_{n0}) with

$$u_{n0}(x) = \sum_{k=1}^n \alpha_{kn}w_k(x)$$

so that (u_{n0}) converges to u_0 as $n \to \infty$, and indeed does so in the norm of $L_2(G)$, i.e.,

$$\int_G (u_{n0} - u_0)^2\,dx \to 0 \qquad \text{as} \quad n \to \infty.$$

Then the equation

$$u_n(x,0) = u_{n0}(x) \quad \text{on } G$$

corresponds to (12b). An especially favorable choice of u_{n0} results from the requirement

$$\int_G (u_0(x) - u_{n0}(x))^2\,dx = \min!, \tag{13}$$

which corresponds to the least squares method. If we write (13) in the form

$$g(\alpha_{1n}, \ldots, \alpha_{nn}) = \min!,$$

then from the necessary extremal condition

$$\frac{\partial g}{\partial \alpha_{jn}} = 0, \qquad j = 1, \ldots, n,$$

we obtain a linear system of equations to determine $\alpha_{1n}, \ldots, \alpha_{nn}$. In order to satisfy the *boundary condition* (11b) also, i.e.,

$$u_n(x, t) = 0 \quad \text{on } \partial G \times [0, T],$$

we choose w_k so that

$$w_k(x) = 0 \quad \text{on } \partial G \qquad \text{for} \quad k = 1, \ldots, n.$$

Remark 19.7. If one replaces Δu in (9) by a *nonlinear* differential operator, then there arises in place of the Galerkin equation (12a) a nonlinear system of ordinary differential equations, but for this one finds in the standard literature well worked-out approximation methods for numerical computations.

The exact treatment of Example 19.6 will be considered in Section 23.8.

19.3. Hyperbolic Differcential Equations and the Galerkin Method

EXAMPLE 19.8. Parallel to Example 19.6 we consider the hyperbolic initial–boundary value problem

$$u_{tt}(x, t) - \Delta u(x, t) = f(x, t) \quad \text{on } Q_T, \tag{14a}$$

$$u(x, t) = 0 \quad \text{on } \partial G \times [0, T] \qquad \text{(boundary condition)}, \tag{14b}$$

$$u(x, 0) = u_0(x) \quad \text{on } G \qquad \text{(initial conditions)}, \tag{14c}$$

$$u_t(x, 0) = u_1(x) \quad \text{on } G.$$

The functions f, u_0, u_1 are given. We seek u.

If we interpret $u(x, t)$ in the case where $G \subset \mathbb{R}^2$ as the vertical displacement of a membrane at the point $x \in G$ at time t (Fig. 19.4), then (14c) means the prescription of the displacement $u(x, 0)$ and the displacement velocity $u_t(x, 0)$ at the initial time $t = 0$. Condition (14b) means that the membrane remains tightly coupled with the boundary for all times $t \in [0, T]$.

Multiplication of equation (14a) by $v \in C_0^\infty(G)$ and integration by parts yield

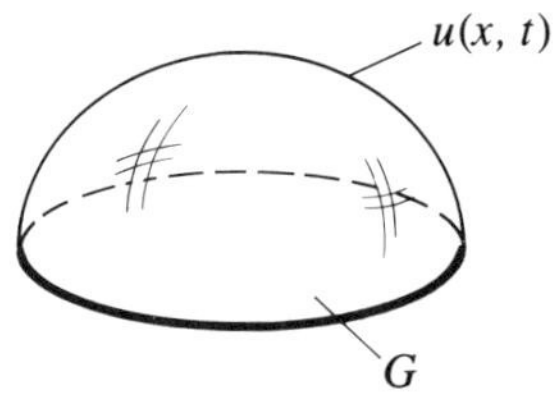

Figure 19.4

the *generalized problem* corresponding to (14):

$$\frac{d^2}{dt^2}\int_G u(x,t)v(x)\,dx + \int_G\left(\sum_{i=1}^{N} D_i u D_i v - fv\right)dx = 0 \qquad \text{for all} \quad v \in C_0^\infty(G), \tag{15a}$$

$$u(x,t) = 0 \quad \text{on } \partial G \times [0,T], \tag{15b}$$

$$\begin{aligned} u(x,0) &= u_0(x) \quad \text{on } G, \\ u_t(x,0) &= u_1(x) \quad \text{on } G. \end{aligned} \tag{15c}$$

One again obtains (14), for sufficiently smooth u and f, backward from (15), by means of integration by parts and an application of Proposition 18.2.

In order to construct an approximate solution with the aid of the *Galerkin method*, we again proceed from the trial

$$u_n(x,t) = \sum_{k=1}^{n} c_{kn}(t) w_k(x).$$

If we replace u in the generalized problem (15a) by u_n and require that (15a) holds for all $v \in \operatorname{span}\{w_1, \ldots, w_n\}$, then we obtain the Galerkin equations for $j = 1, \ldots, n$:

$$\sum_{k=1}^{n} c''_{kn}(t)\int_G w_k w_j\,dx + \sum_{k=1}^{n} c_{kn}(t)\int_G \sum_{i=1}^{N} D_i w_k D_i w_j\,dx = \int_G f w_j\,dx. \tag{16a}$$

This is a linear system of second-order ordinary differential equations for the real functions $c_{1n}, \ldots, c_{nn}$. To this we add the *initial conditions*

$$c_{jn}(0) = \alpha_{jn}, \qquad c'_{jn}(0) = \beta_{jn}, \qquad j = 1, \ldots, n. \tag{16b}$$

These are to realize (15c) approximately. To this end, we set

$$u_{n0} = \sum_{k=1}^{n} \alpha_{kn} w_k, \qquad u_{n1} = \sum_{k=1}^{n} \beta_{kn} w_k.$$

Then (16b) means that

$$u_n(x,0) = u_{n0}(x), \qquad \frac{\partial u_n}{\partial t}(x,0) = u_{n1}(x).$$

As in Example 19.6, we choose (u_{n0}) and (u_{n1}) so that

$$u_{n0} \to u_0 \quad \text{and} \quad u_{n1} \to u_1 \quad \text{in } L_2(G) \quad \text{as} \quad n \to \infty.$$

In particular, one can choose u_{n0} and u_{n1} as solutions of the corresponding minimum problem (13). In order to fulfill the *boundary condition* (15b), i.e.,

$$u_n(x, t) = 0 \quad \text{on } \partial G \times [0, T],$$

we choose w_k so that

$$w_k(x) = 0 \quad \text{on } \partial G \qquad \text{for} \quad k = 1, \ldots, n.$$

The exact treatment of this problem follows in Section 24.2.

19.4. Integral Equations and the Galerkin Method

EXAMPLE 19.9. We consider the integral equation

$$u(x) - \int_G k(x, y)u(y)\,dy = f(x) \qquad \text{for all} \quad x \in G. \tag{17}$$

In this connection, let $k \in L_2(G \times G)$ and $f \in L_2(G)$, i.e., the integrals

$$\int_{G \times G} k(x, y)^2\,dx\,dy \qquad \text{and} \qquad \int_G f(x)^2\,dx$$

exist. Multiplication of (17) by $v \in C_0^\infty(G)$ and integration over G yield the following problem that is equivalent to (17):

$$\int_G uv\,dx - \int_{G \times G} k(x, y)u(y)v(x)\,dy\,dx = \int_G fv\,dx \qquad \text{for all} \quad v \in C_0^\infty(G). \tag{18}$$

Here we take the Fubini theorem $A_2(23)$ into account. We designate (18) as the *generalized problem* corresponding to (17).

In order to construct an approximate solution with the aid of the *Galerkin method* we make the trial

$$u_n = \sum_{k=1}^{n} c_{kn} w_k.$$

If we replace u in the generalized problem (18) by u_n and require that (18) is valid only for all $v \in \operatorname{span}\{w_1, \ldots, w_n\}$, then we obtain the Galerkin equations for $j = 1, \ldots, n$:

$$\int_G u_n w_j\,dx - \int_{G \times G} k(x, y)u_n(y)w_j(x)\,dy\,dx = \int_G fw_j\,dx. \tag{18*}$$

This is a linear system of equations for the real numbers $c_{1n}, \ldots, c_{nn}$.

In conclusion, having this example at hand, we shall explain why the Galerkin method is also designated as a *projection method*. To this end we set

$X = L_2(G)$ and write (17) in the form

$$u - Ku = f, \qquad u \in X. \tag{19}$$

Let $w_1, \ldots, w_n \in X$ and let

$$X_n = \operatorname{span}\{w_1, \ldots, w_n\}.$$

Then, by Section 21.12 below, there exists an orthogonal projection operator

$$P_n\colon X \to X_n$$

onto X_n. The operator P_n is self-adjoint, i.e.,

$$(P_n u|v) = (u|P_n v) \qquad \text{for all} \quad u, v \in X,$$

and $P_n u = u$ holds for all $u \in X_n$. Instead of the original equation (19), we consider the approximate equation

$$P_n(u_n - Ku_n) = P_n f, \qquad u_n \in X_n. \tag{19a}$$

We shall show that (19a) is equivalent to

$$(u_n - Ku_n|w_j) = (f|w_j), \qquad u_n \in X_n, \qquad j = 1, \ldots, n. \tag{19b}$$

These are exactly the Galerkin equations (18*) above. In fact, first of all (19a) is equivalent to

$$(P_n(u_n - Ku_n)|v) = (P_n f|v) \qquad \text{for all} \quad v \in X.$$

This is equivalent to

$$(u_n - Ku_n|P_n v) = (f|P_n v) \qquad \text{for all} \quad v \in X,$$

and thus equivalent to

$$(u_n - Ku_n|v) = (f|v) \qquad \text{for all} \quad v \in X_n,$$

which is equivalent to (19b).

19.5. Complete Orthonormal Systems and Abstract Fourier Series

The following results on orthonormal systems are generalizations of the classical Fourier expansion and other expansions with respect to important special functions of mathematical physics. The following propositions will be used frequently in connection with eigenfunction expansions, functional calculus, convergence of the Ritz and Galerkin methods, etc. We make the following assumption:

(H) Let X be an H-space over $\mathbb{K} = \mathbb{R}, \mathbb{C}$, and let $\{u_n\}$ be an at most countable orthonormal system in X, i.e.,

$$(u_i|u_j) = \delta_{ij} \qquad \text{for all } i, j.$$

We consider the partial sum of the Fourier series

$$s_m = \sum_{n=1}^{m} (u_n|u)u_n.$$

Our goal is to study the convergence of the Fourier series

$$u = \sum_{n=1}^{\infty} (u_n|u)u_n. \tag{20}$$

Definition 19.10. Assume (H). The countable orthonormal system $\{u_n\}$ is called *complete* in X iff (20) holds for all $u \in X$. The finite orthonormal system $\{u_n\}$ is called *complete* in X iff it is a basis in X.

We want to motivate (20). In order to find a good approximation of u by a linear combination $c_1 u_1 + \cdots + c_m u_m$, we set

$$f(c_1, \ldots, c_m) = \left\| u - \sum_{n=1}^{m} c_n u_n \right\|^2$$

and consider the minimum problem

$$f(c_1, \ldots, c_m) = \min!, \qquad c_1, \ldots, c_m \in \mathbb{K}, \tag{21}$$

according to the least squares method of Gauss.

Proposition 19.11. *The unique solution of* (21) *is given by the Fourier coefficients* $c_n = (u_n|u)$.

PROOF. A short computation shows that

$$f(c) = \|u\|^2 - \sum_{n=1}^{m} |(u_n|u)|^2 + \sum_{n=1}^{m} |(u_n|u) - c_n|^2. \qquad \square$$

In particular, this implies

$$\|u - s_m\|^2 \leq f(c) \qquad \text{for all} \quad c \in \mathbb{K}^m \tag{22}$$

and the Bessel equation

$$\|u - s_m\|^2 = \|u\|^2 - \sum_{n=1}^{m} |(u_n|u)|^2 \qquad \text{for all} \quad u \in X. \tag{23}$$

Hence we obtain the important *Bessel inequality*

$$\sum_{n=1}^{m} |(u_n|u)|^2 \leq \|u\|^2 \qquad \text{for all} \quad u \in X. \tag{24}$$

Proposition 19.12 (Convergence Criterion). *Assume* (H). *Then the series*

$$\sum_n c_n u_n, \qquad c_n \in \mathbb{K}$$

is convergent iff $\sum_n |c_n|^2$ *is convergent.*

PROOF. This follows from

$$\left\| \sum_{n=m}^{m+k} c_n u_n \right\|^2 = \sum_{n=m}^{m+k} |c_n|^2. \qquad \square$$

Thus, by (24), the *Fourier series* on the right-hand side of (20) *is always convergent*. The following theorem contains conditions which ensure that the limit of the Fourier series is equal to u.

Theorem 19.A (Completeness Criterion). *Assume* (H). *Then the following four conditions are mutually equivalent*:

(i) *The orthonormal system* $\{u_n\}$ *is complete.*
(ii) *The linear hull of* $\{u_n\}$ *is dense in* X.
(iii) *For all* $u \in X$, *there holds the Parseval equation*

$$\|u\|^2 = \sum_n |(u_n|u)|^2.$$

(iv) *For a given dense set* D *in* X *it follows, from* $v \in D$ *and* $(u_n|v) = 0$ *for all* n, *that* $v = 0$.

PROOF. (i) $\Leftrightarrow$ (ii) $\Leftrightarrow$ (iii). This follows from (22) and (23).

(i) $\Rightarrow$ (iv). Use (iii).

(iv) $\Rightarrow$ (ii). Let $S = \operatorname{span}\{u_1, u_2, \ldots\}$. Denote by $S^\perp$ the orthogonal complement to S. By (iv), $S^\perp \cap D = \{0\}$. Since D is dense in X, the set $S^\perp \cap D$ is dense in $S^\perp$. Hence $S^\perp = \{0\}$. Now it follows from $X = \bar{S} \oplus S^\perp$ that $\bar{S} = X$, i.e., S is dense in X. $\square$

As a typical example we consider the classical Fourier expansion. To this end, let $X = L_2(0, 2\pi)$ with the scalar product

$$(u|v) = \int_0^{2\pi} uv\,dx.$$

Moreover, we set $u_0 = 1/\sqrt{2\pi}$ and

$$u_{2k} = \frac{1}{\sqrt{\pi}}\cos kx, \qquad u_{2k+1} = \frac{1}{\sqrt{\pi}}\sin kx, \qquad k = 1, 2, \ldots.$$

Then

$$(u_m|u_n) = \delta_{mn} \qquad \text{for all } m, n,$$

i.e., $\{u_m\}$ is an orthonormal system in X.

Let $C_p[0, 2\pi]$ denote the set of all the continuous functions $u\colon [0, 2\pi] \to \mathbb{R}$ with $u(0) = u(2\pi)$. We equip $C_p[0, 2\pi]$ with the usual norm $\|u\| = \max_{0 \le x \le 2\pi} |u(x)|$. By the classical theorem of Fejér, for each $u \in C_p[0, 2\pi]$, we have the uniform convergence

$$\frac{s_0 + \cdots + s_{m-1}}{m} \rightrightarrows u \quad \text{on } [0, 2\pi],$$

where

$$s_m = \sum_{n=0}^{m} (u_n|u)u_n.$$

Hence the linear hull of $\{u_m\}$ is dense in $C_p[0, 2\pi]$. Since $C_p[0, 2\pi]$ is dense in $X = L_2(0, 2\pi)$, the linear hull of $\{u_m\}$ is also dense in X. By Theorem 19.A, $\{u_m\}$ forms a complete orthonormal system in X, i.e.,

$$\lim_{m\to\infty} \int_0^{2\pi} |u - s_m|^2\, dx = 0 \qquad \text{for all} \quad u \in L_2(0, 2\pi).$$

This means that, for each $u \in L_2(0, 2\pi)$, the classical Fourier series converges to u in the sense of the space $L_2(0, 2\pi)$.

19.6. Eigenvalues of Compact Symmetric Operators (Hilbert–Schmidt Theory)

We consider the eigenvalue problem

$$Bu = \lambda u, \qquad u \in X, \quad \lambda \in \mathbb{K}. \tag{25}$$

We want to prove the existence of eigensolutions $\{u_n, \lambda_n\}$ and the convergence of the Fourier series

$$u = \sum_{n=1}^{\infty} (u_n|u)u_n \qquad \text{for all} \quad u \in X. \tag{26}$$

Definition 19.13. A linear operator $B\colon D(B) \subseteq X \to X$ on the H-space X is called *symmetric* iff $\overline{D(B)} = X$ and

$$(Bu|v) = (u|Bv) \qquad \text{for all } u, v \in D(B).$$

The operator B is called *skew-symmetric* iff $\overline{D(B)} = X$ and

$$(Bu|v) = -(u|Bv) \qquad \text{for all} \quad u, v \in D(B).$$

We assume:

(H) The operator $B\colon X \to X$ is linear, symmetric, and *compact* on the H-space X over $\mathbb{K} = \mathbb{R}, \mathbb{C}$. Let $B \neq 0$.

Theorem 19.B. *Suppose that* (H) *holds with* $\dim X = \infty$ *and* $Bu = 0$ *implies* $u = 0$, *i.e., zero is not an eigenvalue of B. Then*:

(a) *The operator B has a countable system* $\{u_n, \lambda_n\}$ *of eigensolutions containing all the eigenvalues of B.*
(b) *All the eigenvalues* λ_n *are real and of finite multiplicity, and* $\lambda_n \to 0$ *as* $n \to \infty$.
(c) *The eigenvectors* $\{u_n\}$ *form a complete orthonormal system in X, i.e.,* (26) *holds.*

This theorem will be proved in the next section together with the following more general result.

Proposition 19.14. *Suppose that* (H) *holds. Then*:

(a) *The operator B has an at most countable system $\{u_n, \lambda_n\}$ of eigensolutions containing all the nonzero eigenvalues of B.*
(b) *All the eigenvalues λ_n are real. There exists at least one nonzero eigenvalue of B and all the nonzero eigenvalues of B have finite multiplicity.*
(c) *The eigenvectors $\{u_n\}$ form an orthonormal system in X and*

$$Bu = \sum_n \lambda_n (u_n|u)u_n \qquad \text{for all} \quad u \in X. \tag{27}$$

If $\lambda_m \neq 0$, then all the u_n corresponding to λ_m form a basis in the eigenspace to λ_m.
(d) *If $Bu = 0$ implies $u = 0$, then* (26) *holds.*

Corollary 19.15. *If* (H) *holds, where X is separable, then the operator B has an at most countable complete orthonormal system of eigenvectors in X.*

Proof of Corollary 19.15. We will use the fact that, in each separable nonzero H-space, there exists an at most countable complete orthonormal system. This will be proved in Problem 19.16.

Let $N = \{u \in X: Bu = 0\}$ and suppose that $N \neq \{0\}$. Then the linear subspace N is closed, and hence N is an H-space. We consider the orthogonal decomposition

$$X = N \oplus N^\perp.$$

Let $\{v_m\}$ be a complete orthonormal system in N. Then $Bv_m = 0$ for all m, *i.e.*, v_m is an eigenvector with the corresponding eigenvalue $\lambda = 0$. Furthermore, $B(N^\perp) \subseteq N^\perp$. In fact, if $u \in N^\perp$, then

$$(Bu|v) = (u|Bv) = 0 \qquad \text{for all} \quad v \in N,$$

i.e., $Bu \in N^\perp$. By Proposition 19.14(d), B has a complete orthonormal system $\{u_n\}$ of eigenvectors in the H-space $N^\perp$. Thus, all the u_n, v_m form a complete orthonormal system of eigenvectors of the operator B in X. □

In order to obtain another important consequence of Proposition 19.14, we consider the equation

$$\lambda u - Bu = f, \qquad u \in X. \tag{28}$$

Proposition 19.16 (The Fredholm Alternative). *Let $\lambda \neq 0$ and assume* (H). *Then*:

(a) *If λ is not an eigenvalue of B, then for each $f \in X$, equation* (28) *has a unique solution.*

(b) *If λ is an eigenvalue of B, then equation* (28) *has a solution iff*

$$(f|u) = 0$$

for all eigenvectors u of B corresponding to λ.

Proof. Let $Bu_n = \lambda_n u_n$. Then

$$(u_n|Bu) = (Bu_n|u) = \lambda_n(u_n|u).$$

From (28) we obtain the key formula

$$(\lambda - \lambda_n)(u_n|u) = (u_n|f). \tag{29}$$

Ad(a). Let $\lambda \neq \lambda_n$ for all n. Let u be a solution of (28). From (27) and (29) we obtain

$$u = \frac{1}{\lambda}\left(f + \sum_n \frac{\lambda_n}{\lambda - \lambda_n}(u_n|f)u_n\right). \tag{30}$$

Conversely, from $\sum_n |(u_n|f)|^2 \leq \|f\|^2$ and $\lambda_n \to 0$ as $n \to \infty$ we obtain

$$\sum_n \left|\frac{\lambda_n}{\lambda - \lambda_n}\right|^2 |(u_n|f)|^2 < \infty. \tag{31}$$

Hence the right-hand side of (30) converges and represents a solution of (28).

Ad(b). Let $\lambda = \lambda_m$ for fixed m. Let u be a solution of (28). It follows from (29) that $(u_n|f) = 0$ for all u_n corresponding to λ_m.

Conversely, suppose that $(u_n|f) = 0$ for all u_n corresponding to λ_m. Then (30) represents a solution of (28) in case we do not sum over all those n which satisfy $\lambda_n = \lambda_m$. □

19.7. Proof of Theorem 19.B

The basic idea is to use a variational principle in order to obtain the largest eigenvalue of B.

Lemma 19.17. *Let $B: D(B) \subseteq X \to X$ be a linear symmetric operator on the* H-*space X over $\mathbb{K} = \mathbb{R}, \mathbb{C}$. Then:*

(a) *$(Bu|u)$ is real for all $u \in D(B)$ and all the eigenvalues of B are real.*
(b) *Eigenvectors of B with different eigenvalues are orthogonal.*
(c) *If $\{u_n\}$ is a complete orthonormal system of eigenvectors of B, then the sequence $\{\lambda_n\}$ of the corresponding eigenvalues contains all the eigenvalues of B.*
(d) *If $D(B) = X$ and B is continuous, where $X \neq \{0\}$, then*

$$\sup_{\|u\|=1} |(Bu|u)| = \|B\|. \tag{32}$$

The proof will be given in Problem 19.3. We now consider Proposition 19.14 in the case where $\dim X = \infty$. The simpler case $\dim X < \infty$ can be treated analogously.

Step 1: Variational problem. We consider the maximum problem

$$|(Bu|u)| = \max!, \qquad \|u\| = 1. \tag{33}$$

By (32), the maximal value is equal to $\|B\|$. Recall $B \neq 0$. Let (v_n) be a sequence with $\|v_n\| = 1$ for all n and

$$|(Bv_n|v_n)| \to \|B\| \qquad \text{as} \quad n \to \infty.$$

By passing to a subsequence, if necessary, we obtain

$$(Bv_n|v_n) \to \lambda_1 \qquad \text{as} \quad n \to \infty.$$

Here,

$$|\lambda_1| = \|B\|$$

and $\|Bv_n\| \leq |\lambda_1|$ for all n. Hence

$$\|Bv_n - \lambda_1 v_n\|^2 = \|Bv_n\|^2 - 2\lambda_1(Bv_n|v_n) + \lambda_1^2 \to 0 \qquad \text{as} \quad n \to \infty.$$

The operator B is compact. Thus there exists a subsequence, again denoted by (Bv_n), such that (Bv_n) converges. Since $Bv_n - \lambda_1 v_n \to 0$ and $\lambda_1 \neq 0$, the sequence (v_n) also converges to a certain u_1. This implies

$$Bu_1 = \lambda_1 u_1, \qquad \|u_1\| = 1,$$

and u_1 is a solution of (33).

Step 2: Induction. Let

$$Y = \{u \in X : (u|u_1) = 0\}.$$

Then Y is a linear closed subspace of X. The key to our induction argument is the relation

$$B(Y) \subseteq Y,$$

i.e., Y is invariant with respect to B. In fact, it follows from $u \in Y$ that

$$(Bu|u_1) = (u|Bu_1) = \lambda_1(u|u_1) = 0$$

and hence $Bu \in Y$. We assume that $B \neq 0$ on Y. Otherwise, we finish our procedure.

We now apply Step 1 to the restricted operator $B: Y \to Y$. This way we obtain

$$Bu_2 = \lambda_2 u_2, \qquad u_2 \in Y,$$

where $|\lambda_2|$ is the norm of B on Y. This implies

$$\|Bv\| \leq |\lambda_2| \, \|v\| \qquad \text{for all} \quad v \in Y \tag{34}$$

and $|\lambda_2| \leq \|B\| = |\lambda_1|$.

We now set $Z = \{u \in Y: (u|u_2) = 0\}$ and continue this procedure. This way we obtain the eigensolutions

$$Bu_n = \lambda_n u_n, \quad n = 1, 2, \ldots, \qquad |\lambda_1| \geq |\lambda_2| \geq \cdots > 0,$$

and $\{u_n\}$ is an orthonormal system.

Step 3: We show that $\lambda_n \to 0$ as $n \to \infty$. Otherwise, the sequence $(\lambda_n^{-1} u_n)$ is bounded. The operator B is compact. Since $B(\lambda_n^{-1} u_n) = u_n$, the sequence (u_n) contains a convergent subsequence. But this is impossible, since $(u_n|u_m) = 0$ and

$$\|u_n - u_m\|^2 = \|u_n\|^2 + \|u_m\|^2 = 2 \qquad \text{for} \quad n \neq m.$$

Step 4: We show that

$$Bu = \sum_{n=1}^{\infty} \lambda_n (u_n|u) u_n. \tag{35}$$

To this end, let

$$w_m = u - \sum_{n=1}^{m} (u_n|u) u_n.$$

We define $V = \{u \in X: (u|u_i) = 0, i = 1, \ldots, m\}$. By Step 2, $B(V) \subseteq V$ and

$$\|Bv\| \leq |\lambda_{m+1}| \|v\| \qquad \text{for all} \quad v \in V.$$

Obviously, $w_m \in V$, and hence

$$\|Bw_m\| \leq |\lambda_{m+1}| \, \|w_m\|.$$

The Bessel equation (23) yields

$$\|w_m\| \leq \|u\|.$$

This implies $Bw_m \to 0$ as $m \to \infty$, and hence we obtain (35).

By Proposition 19.12, each Fourier series is convergent. Hence

$$w_m \to w \qquad \text{as} \quad m \to \infty$$

and $Bw = 0$. If $Bw = 0$ implies $w = 0$, then $w_m \to 0$ as $m \to \infty$, i.e., the orthonormal system $\{u_m\}$ is complete.

Step 5: We show that the sequence (λ_n) contains all the eigenvalues $\lambda \neq 0$ of B. In fact, it follows from

$$Bu = \lambda u, \qquad u \neq 0, \qquad \lambda \neq 0,$$

and $\lambda \neq \lambda_n$ for all n, that $(u_n|u) = 0$ for all n, according to Lemma 19.17(b). By (35), $Bu = 0$. This is a contradiction.

An analogous argument shows that all the eigenvectors corresponding to $\lambda_m \neq 0$ form a basis in the eigenspace to the eigenvalue λ_m. Since $\lambda_m \to 0$ as $m \to \infty$, each $\lambda_m \neq 0$ has finite multiplicity.

The proof of Proposition 19.14 and Theorem 19.B is complete.

19.8. Self-Adjoint Operators

While working with operators $A: D(A) \subseteq X \to X$, which are not defined on the total space X, observe carefully the specific form of the domain of definition $D(A)$ of A.

Definition 19.18. Let $A: D(A) \subseteq X \to X$ be a linear operator on the H-space X with $\overline{D(A)} = X$. The adjoint operator $A^{*'}: D(A^{*'}) \subseteq X \to X$ is defined by

$$(u|Av) = (A^{*'}u|v).$$

To be precise, we first define $u \in D(A^{*'})$ iff $u \in X$ and there exists an $f \in X$ with

$$(u|Av) = (f|v) \qquad \text{for all} \quad v \in D(A), \tag{36}$$

and, for $u \in D(A^{*'})$, we then set $A^{*'}u = f$.

The operator A is called *self-adjoint* iff $A = A^{*'}$, and A is called *skew-adjoint* iff $A = -A^{*'}$.

The definition of $A^{*'}$ makes sense. In fact, if (36) also holds for $\bar{f}$, then $(f - \bar{f}|v) = 0$ for all $v \in D(A)$. Since $D(A)$ is dense in X, we obtain $f = \bar{f}$.

In the following we agree to speak about the existence of the adjoint operator $A^{*'}$ only in the case where A is *densely* defined. We use the notation $A^{*'}$ in order to emphasize the difference between the adjoint operator $A^{*'}$ on X and the dual operator A^* on the dual space X^*. However, for simplicity of notation, and if it is stated expressively, the adjoint operator to A is denoted by A^*. In particular, this *convention* will be used throughout this chapter.

We write

$$A \subseteq B$$

iff the operator B is an *extension* of the operator A, i.e., $D(A) \subseteq D(B)$ and $Au = Bu$ on $D(A)$. Using Definition 19.13, it follows easily from (36) that the following hold:

(i) *A is symmetric* iff $A \subseteq A^*$.
(ii) *A is skew-symmetric* iff $A \subseteq -A^*$.

This immediately implies the following:

(iii) *If A is self-adjoint, i.e., $A = A^*$, then A is symmetric.*
(iv) *If A is skew-adjoint, i.e., $A = -A^*$, then A is skew-symmetric.*

Generally, the converse of (iii) and (iv) is *not* true. However, it follows from (i) that each symmetric operator $A: D(A) \subseteq X \to X$ with $D(A) = X$ is self-adjoint. In this special case the operator A is also continuous, by Problem 19.5e. But note that, in the general case $D(A) \neq X$, the self-adjointness of A is a much stronger property than the symmetry of A. This difference plays a fundamental role in the theory of partial differential equations. In particular, self-adjoint operators are *not* necessarily continuous. However, we have the following weaker result:

(v) *Self-adjoint and skew-adjoint operators are graph closed.*

This follows from the fact that adjoint operators are always graph closed, by Problem 19.5a. Finally, we mention the following result:

(vi) *If $A: X \to X$ is a linear continuous operator, then $D(A^*) = X$ and $A^*: X \to X$ is also linear and continuous.*

To prove this we use the Schwarz inequality

$$\text{(S)} \qquad |(u|Av)| \le \|u\| \, \|Av\| \le \|u\| \, \|A\| \, \|v\| \qquad \text{for all} \quad u, v \in X.$$

Thus, for each fixed $u \in X$, the mapping $v \mapsto (u|Av)$ is a linear continuous functional on X. By the Riesz theorem in Section 18.11, there exists an $f \in X$ with

$$(u|Av) = (f|v) \qquad \text{for all} \quad v \in X$$

and $\|f\| \le \|A\| \, \|u\|$, by (S). This implies $A^*u = f$ and $\|A^*u\| \le \|A\| \, \|u\|$ for all $u \in X$.

Let X and Y be two H-spaces over $\mathbb{K} = \mathbb{R}, \mathbb{C}$. Then the operator $U: X \to Y$ is called *unitary* (or an H-isomorphism) iff U is linear, bijective, and

$$(Uu|Uv) = (u|v) \qquad \text{for all} \quad u, v \in X.$$

For a linear operator $U: X \to Y$, the following two conditions are equivalent:

(i) *U is unitary.*
(ii) *U is a normisomorphism, i.e., U is surjective and $\|Uu\| = \|u\|$ for all $u \in X$.*

The proof was given in Problem 18.4a.

For a linear operator $U: X \to X$, the following two conditions are equivalent:

(a) *U is unitary.*
(b) *$U^*U = UU^* = I$.*

To prove this first suppose that (a) holds. Then $(Uu|Uv) = (u|v)$ implies

$$(U^*Uu|v) = (u|v) \qquad \text{for all} \quad u, v \in X,$$

i.e., $U^*U = I$. Furthermore, we have $UU^*u = UU^*UU^{-1}u = u$, i.e., $UU^* = I$.

Conversely, suppose that (b) holds. From $UU^* = I$ it follows that $D(U^*) = X$ and U is surjective. Moreover, $U^*U = I$ implies that U is injective (i.e., $Uu = 0$ implies $u = 0$) and that $(Uu|Uv) = (u|v)$.

19.9. The Friedrichs Extension of Symmetric Operators

Let $A: D(A) \subseteq X \to X$ be a linear symmetric operator. We consider the equation

$$Au = f, \qquad u \in D(A). \tag{37a}$$

If this equation corresponds to a classical boundary value problem for a linear elliptic equation, then, as a rule, problem (37a) is *not* solvable for each $f \in X$. Therefore, along with (37a), we study the following two *generalized* problems:

$$A_F u = f, \qquad u \in D(A_F), \tag{37b}$$

$$(u|Av) = (f|v) \qquad \text{for fixed} \quad u \in X_E \quad \text{and all} \quad v \in D(A). \tag{37c}$$

Here $A \subseteq A_F$. Indeed, we shall show that (37b) and (37c) are equivalent and that, for each $f \in X$, both equations have a unique solution u. This solution can also be obtained by the following variational problem:

$$2^{-1}(u|u)_E - (f|u) = \min!, \qquad u \in X_E. \tag{38}$$

We make the following assumption:

(H) The linear operator $A: D(A) \subseteq X \to X$ is symmetric on the real H-space X with $\overline{D(A)} = X$ and A is *strongly monotone*, i.e.,

$$(Au|u) \geq c\|u\|^2 \qquad \text{for all} \quad u \in D(A) \quad \text{and fixed} \quad c > 0. \tag{39}$$

Our goal is to construct the extensions

$$A \subseteq A_F \subseteq A_E,$$

where A_F is called the *Friedrichs extension* and A_E is called the *energetic extension* of A. Note that the range X_E^* of the operator $A_E: X_E \to X_E^*$ leaves the original space X. The so-called *energetic space* X_E is the completion of $D(A)$ with respect to the so-called energetic scalar product $(u|v)_E = (Au|v)$. This space will be studied in greater detail in Section 19.10.

Theorem 19.C (Friedrichs (1934)). *Suppose that* (H) *holds. Then*:

(a) *There exists a self-adjoint extension $A_F: D(A_F) \subseteq X \to X$ of the operator A with $D(A_F) \subseteq X_E \subseteq X$ and*

$$(A_F u|u) \geq c\|u\|^2 \qquad \textit{for all} \quad u \in D(A_F).$$

(b) *The inverse operator $A_F^{-1}: X \to X$ exists and is linear, continuous, and self-adjoint. Consequently, for each $f \in X$, equation* (37b) *has a unique solution.*

(c) *The operator $A_F^{-1}: X \to X_E$ is linear and continuous.*

(d) *The embeddings $X_E \subseteq X \subseteq X_E^*$ are continuous.*

(e) *The operator A_F has the extension*

$$A_E: X_E \to X_E^*,$$

where A_E *is the duality map of* X_E, *i.e.,* A_E *a linear homeomorphism with*

$$\langle A_E u, u\rangle = \|u\|_E^2 \qquad \textit{for all} \quad u \in X_E. \tag{40}$$

Moreover,

$$A_F^{-1} f = A_E^{-1} f \qquad \textit{for all} \quad f \in X.$$

(f) *If the embedding* $X_E \subseteq X$ *is compact, then the operator* $A_F^{-1}\colon X \to X$ *is compact.*

Corollary 19.19. *For given* $f \in X$, *problems* (37b), (37c), *and* (38) *are mutually equivalent. Each solution of the original problem* (37a) *is also a solution of* (37b), (37c), *and* (38).

The proof of Theorem 19.C will be given in the next section.

PROOF OF COROLLARY 19.19. From $A \subseteq A_F$ we obtain (37a) $\Rightarrow$ (37b). Moreover, it follows from

$$(A_F u|v) = (u|A_F v) = (u|Av) \qquad \text{for all} \quad v \in D(A)$$

that (37b) $\Rightarrow$ (37c). In the next section we shall prove that (37c) $\Rightarrow$ (37b) and (37b) $\Leftrightarrow$ (38). $\square$

We now consider the *eigenvalue problem*

$$Au = \mu u + f, \qquad u \in D(A), \quad \mu \in \mathbb{R}, \tag{41a}$$

along with the two generalized problems

$$A_F u = \mu u + f, \qquad u \in D(A_F), \tag{41b}$$

$$(u|Av) = \mu(u|v) + (f|v) \qquad \text{for fixed} \quad u \in X_E \quad \text{and all} \quad v \in D(A). \tag{41c}$$

By Corollary 19.19, problems (41b) and (41c) are equivalent, and each solution of (41a) is also a solution of (41b) and (41c).

Corollary 19.20 (Eigenvalue Problem). *Suppose that* (H) *holds with* $\dim X = \infty$, *and that the embedding* $X_E \subseteq X$ *is compact. Then*:

(a) *The operator* $A_F\colon D(A_F) \subseteq X \to X$ *has a countable system of eigensolutions* $\{u_n, \mu_n\}$ *which contains all the eigenvalues of* A_F.
(b) *All the eigenvalues of* A_F *have finite multiplicity. Furthermore, we have* $\mu_n \geq c$ *for all* n *and* $\mu_n \to +\infty$ *as* $n \to \infty$.
(c) *The eigenvectors* $\{u_n\}$ *form a complete orthonormal system in* X.

PROOF. If $A_F u = \mu u$ with $u \neq 0$, then $(A_F u|u) \geq c\|u\|^2$ implies $\mu \geq c$.

We set $B = A_F^{-1}$. By Theorem 19.C, the operator $B\colon X \to X$ is compact and $Bu = 0$ implies $u = 0$. In the case where $\mu \neq 0$, equation (41b) is equivalent to equation

$$\lambda u = Bu + \lambda Bf \tag{42}$$

with $\lambda = \mu^{-1}$. Theorem 19.B yields the assertion. $\square$

Corollary 19.21 (The Fredholm Alternative). *Suppose that* (H) *holds with* $\dim X = \infty$, *and that the embedding* $X_E \subseteq X$ *is compact. Let* $\mu \neq 0$. *Then*:

(a) *If* μ *is not an eigenvalue of* A_F, *then, for each* $f \in X$, *equation* (41b) *has a unique solution.*
(b) *If* μ *is an eigenvalue of* A_F, *then, for given* $f \in X$, *equation* (41b) *has a solution iff*

$$(f|u) = 0$$

holds for all eigenvectors u *of* A_F *corresponding to* μ.

PROOF. This follows immediately from (42) and Proposition 19.16. In (b), note that $A_F u = \mu u$ implies $u = \mu B u$ and

$$(Bf|u) = (f|Bu) = \mu^{-1}(f|u). \tag{43}$$

Hence $(\lambda Bf|u) = 0$ iff $(f|u) = 0$. □

Concrete examples will be considered in Sections 19.11 and 19.12.

Discussion

Roughly speaking, we have the following situation.

(i) If the original equation (37a) above corresponds to a boundary value problem for a linear elliptic differential equation, then the variational problem (38) corresponds to the classical variational problem related to the boundary value problem (37a). The inverse operator A_F^{-1} corresponds to an integral operator whose kernel is an appropriate Green's function.

(ii) The point is that the classical problem (37a) has only solutions for sufficiently smooth right-hand sides f. The energetic space X_E is constructed in such a way that the variational problem (38) has a solution in X_E for each $f \in X$.

(iii) In classical mathematical physics, one studied eigenvalue problems for differential operators by reducing them to integral equations. This corresponds to the passage from

$$A_F u = \mu u + f$$

to

$$u = \mu A_F^{-1} u + A_F^{-1} f,$$

which we have used in the proof of Corollaries 19.20 and 19.21.

(iv) The energetic space X_E corresponds to Sobolev spaces, and the compact embedding $X_E \subseteq X$ corresponds to Sobolev embedding theorems which have their roots in the compactness theorem of Rellich (1930).

(v) In Sections 19.15 and 19.16, we shall show that the Friedrichs extension is also important for solving linear parabolic and hyperbolic equations.

(vi) In terms of the theory of monotone operators, the energetic extension A_E: $X_E \to X_E^*$ is a *strongly monotone* operator. In Section 31.4 we shall show that the Friedrichs extension A_F: $D(A_F) \subseteq X \to X$ from Theorem 19.C is a *maximal monotone* extension of A in X. The theory of monotone operators can be regarded as a generalization of the Friedrichs extension to linear operators (without any symmetry properties) and to nonlinear operators.

(vii) The embeddings $X_E \subseteq X \subseteq X_E^*$ are the prototype of *evolution triples* which will play a fundamental role in the investigation of nonlinear evolution equations.

(viii) If we treat elliptic equations with the aid of the unbounded Friedrichs extension A_F, then we work in terms of the theory of *unbounded* operators. This will be done in this chapter. In Chapter 22, we will use the bounded energetic extension A_E in order to study elliptic differential equations. This approach is based on the theory of *bounded* operators. However, note that in quantum theory, the use of unbounded operators is indispensable.

19.10. Proof of Theorem 19.C

The basic idea of the proof is the following:

(i) The energetic space X_E is the completion of $D(A)$ with respect to the energetic scalar product $(u|v)_E = (Au|v)$. It is important that X_E can be identified with a subset of X.

(ii) The energetic extension A_E: $X_E \to X_E^*$ is the duality map of X_E, i.e.,

$$\langle A_E u, v\rangle = (u|v)_E \qquad \text{for all} \quad u, v \in X_E.$$

(iii) The Friedrichs extension is constructed by $D(A_F) = A_E^{-1}(X)$ and $A_F u = A_E u$ on $D(A_F)$, i.e., $A_F^{-1} = A_E^{-1}$ on X.

Having this in mind, the proof proceeds straightforwardly by using the Riesz theorem (Theorem 18.D).

Step 1: The energetic space X_E. We define the energetic scalar product by

$$(u|v)_E = (Au|v) \qquad \text{for all} \quad u, v \in D(A), \tag{44}$$

and we set $\|u\|_E = (u|u)_E^{1/2}$. From (39) we obtain the *first key formula*:

$$\|u\| \le d\|u\|_E \qquad \text{for all} \quad u \in D(A), \tag{45}$$

where $d = 1/\sqrt{c}$, By definition, X_E is the completion of $D(A)$ with respect to $(\cdot|\cdot)_E$. In particular, $D(A)$ is dense in X_E (cf. Problem 18.4).

Step 2: There exists a linear continuous injective map j: $X_E \to X$. Thus if we identify u with $j(u)$, then $X_E \subseteq X$.

To begin with we construct j: $D(A) \subseteq X_E \to X$ by $j(u) = u$. According to

(45), and the extension principle in Section 18.12, there exists a uniquely determined continuous extension $j: X_E \to X$. Passing to the limit in (44), we get

$$(u|v)_E = (Au|j(v)) \qquad \text{for all} \quad u \in D(A), \quad v \in X_E. \tag{46}$$

Therefore, $j(v) = 0$ implies $v = 0$, i.e., j is injective.

Identifying $j(v)$ with v in (46), we obtain the *second key formula*

$$(u|v)_E = (Au|v) \qquad \text{for all} \quad u \in D(A), \quad v \in X_E. \tag{47}$$

Passing to the limit in (45), we get

$$\|u\| \le d\|u\|_E \qquad \text{for all} \quad u \in X_E. \tag{48}$$

Step 3: We show that there exists a linear continuous injective map $h: X \to X_E^*$. Thus if we identify u with $h(u)$, then $X \subseteq X_E^*$.

For fixed $f \in X$, we set

$$f^*(v) = (f|v) \qquad \text{for all} \quad v \in X_E.$$

Then $f^* \in X_E^*$, since

$$|f^*(v)| \le \|f\|\,\|v\| \le d\|f\|\,\|v\|_E \qquad \text{for all} \quad v \in X_E.$$

We set $h(f) = f^*$. Because $\|f^*\| \le d\|f\|$, the map $h: X \to X_E^*$ is continuous. Since X_E is dense in X, it follows from $f^* = 0$ that $f = 0$, i.e., h is injective.

In the sense of the identification $f = f^*$, we have

$$\langle f, v\rangle_{X_E} = (f|v) \qquad \text{for all} \quad f \in X, \quad v \in X_E. \tag{49}$$

Step 4: Construction of the energetic extension $A_E: X_E \to X_E^*$.

Let A_E be the duality map of X_E. Recall the following construction of A_E from Section 18.11. Let $u \in X_E$. We set

$$b(v) = (u|v)_E \qquad \text{for all} \quad v \in X_E. \tag{50}$$

According to the theorem of Riesz (Theorem 18.D) we obtain $b \in X_E^*$ and $\|b\| = \|u\|_E$. We define

$$A_E u = b.$$

Again by the theorem of Riesz, the operator $A_E: X_E \to X_E^*$ is bijective. From

$$\|A_E u\| = \|u\|_E \qquad \text{for all} \quad u \in X_E$$

it follows that A_E is isometric and hence A_E is a *linear homeomorphism*.

From (48) and (50) it follows that

$$\begin{aligned} \langle A_E u, v\rangle &= (u|v)_E \qquad \text{for all} \quad u, v \in X_E, \\ \langle A_E u, u\rangle &= \|u\|_E^2 \ge c\|u\|^2. \end{aligned} \tag{51}$$

Let $f \in X$. By (49) and (51), the equation

$$A_E u = f \tag{52}$$

is equivalent to

$$(u|v)_E = (f|v) \qquad \text{for all} \quad v \in X_E. \tag{52*}$$

The operator $A_E: X_E \to X_E^*$ is an extension of $A: D(A) \subseteq X \to X$. In fact, it follows from $Au = f$ that

$$(u|v)_E = (Au|v) = (f|v) \qquad \text{for all} \quad v \in X_E$$

and hence $A_E u = f$.

Step 5: Construction of the Friedrichs extension $A_F: D(A_F) \subseteq X \to X$. We set $D(A_F) = A_E^{-1}(X)$ and $A_F u = A_E u$ on $D(A_F)$. The last argument of Step 4 shows that $A \subseteq A_F \subseteq A_E$.

Step 6: The inverse operator $A_F^{-1}: X \to X$. We set $B = A_F^{-1}$. The operator B is the restriction of $A_E^{-1}: X_E^* \to X_E$ to X. Since the embeddings $X_E \subseteq X \subseteq X_E^*$ are continuous, the operators $B: X \to X_E$ and $B: X \to X$ are continuous. If the embedding $X_E \subseteq X$ is compact, then $B: X \to X$ is compact.

We will show that $B: X \to X$ is symmetric. Let $f, g \in X$. We set $u = Bf$ and $v = Bg$. By (52*),

$$(Bf|Bg)_E = (f|Bg).$$

Hence $(f|Bg) = (g|Bf)$, i.e., B is symmetric. Since $D(B) = X$, the operator $B: X \to X$ is self-adjoint.

As an inverse operator of the self-adjoint operator B, the operator A_F is also self-adjoint (cf. Problem 19.5d). By (52) and (52*),

$$(A_F u|u) = (u|u)_E \ge c\|u\|^2 \qquad \text{for all} \quad u \in D(A_F).$$

Step 7: For each $b \in X_E^*$, the variational problem

$$2^{-1}(u|u)_E - b(u) = \min!, \qquad u \in X_E, \tag{53}$$

has a unique solution given by $u = A_E^{-1}b$. In fact, if we set $a = A_E^{-1}b$, then

$$b(v) = (a|v)_E \qquad \text{for all} \quad v \in X_E,$$

by construction of A_E. Hence (53) is equivalent to

$$2^{-1}(u - a|u - a)_E = \min!, \qquad u \in X_E,$$

with the unique solution $u = a$.

In particular, if we choose $f \in X$ and $b(u) = (f|u)$ for all $u \in X_E$, then $u = A_E^{-1}b = A_F^{-1}f$.

Consequently, problems (38) and (37b) are equivalent.

Step 8: Let $f \in X$ be given and let u be a solution of (37c), i.e.,

$$(u|Av) = (f|v) \qquad \text{for fixed} \quad u \in X_E \quad \text{and all} \quad v \in D(A).$$

By (47), $(u|v)_E = (f|v)$ and hence $u = A_E^{-1}f = A_F^{-1}f$, i.e., u is a solution of (37b).

The proof of Theorem 19.C and Corollary 19.19 is complete.

19.11. Application to the Poisson Equation

We consider the classical boundary value problem

$$\begin{aligned} -\Delta u &= f \quad \text{on } G, \\ u &= 0 \quad \text{on } \partial G, \end{aligned} \tag{54}$$

where G is a bounded region in $\mathbb{R}^N$ with $N \geq 1$. Generally, this problem has *no* classical solution if f and the boundary ∂G are not sufficiently smooth. In order to apply the results of Section 19.9 to this problem, we set

$$X = L_2(G), \qquad D(A) = C_0^\infty(G) \qquad \text{and} \qquad Au = -\Delta u.$$

Then (54) corresponds to the operator equation

$$Au = f, \qquad u \in D(A), \qquad f \in X. \tag{55}$$

Integration by parts yields

$$(Au|v) = \int_G \sum_{i=1}^N D_i u D_i v \, dx \qquad \text{for all} \quad u, v \in D(A).$$

Hence A is symmetric. The inequality of *Poincaré–Friedrichs* from Section 18.9 yields the *key estimate*

$$(Au|u) \geq c\|u\|_X^2 \qquad \text{for all} \quad u \in D(A) \quad \text{and fixed} \quad c > 0.$$

According to Section 19.10, the energetic scalar product is given by

$$(u|v)_E = (Au|v) \qquad \text{for all} \quad u, v \in D(A).$$

The energetic space X_E is the completion of $C_0^\infty(G)$ with respect to $(\cdot|\cdot)_E$. By Problem 18.4, the Sobolev space $\mathring{W}_2^1(G)$ is the completion of $C_0^\infty(G)$ with respect to $(\cdot|\cdot)_E$. Hence

$$X_E = \mathring{W}_2^1(G).$$

By Theorem 19.C, the embeddings

$$X_E \subseteq X \subseteq X_E^* \tag{56}$$

are continuous. In Section 19.13 we shall prove the following:

$$\textit{The embedding } X_E \subseteq X \textit{ is compact.} \tag{57}$$

By Theorem 19.C, the operator A has the energetic extension

$$A_E\colon X_E \to X_E^*,$$

which is a linear homeomorphism, namely, A_E is equal to the duality map of X_E. For each $f \in X_E^*$, the equation

$$A_E u = f, \qquad u \in X_E$$

has a unique solution. For given arbitrary $f \in X$, exactly all the solutions u of the latter equation form the set $D(A_F)$. The Friedrichs extension A_F of the operator A is the restriction of the energetic extension A_E to the set $D(A_F)$.

For given fixed $f \in X$, the equation

$$A_F u = f$$

is equivalent to each of the following two problems:

$$(u|Av) = (f|v) \quad \text{for fixed} \quad u \in X_E \quad \text{and all} \quad v \in D(A) \tag{58}$$

and

$$\tfrac{1}{2}(u|u)_E - (f|u) = \min!, \qquad u \in X_E. \tag{59}$$

Explicitly, problem (58) reads as follows:

$$\int_G \sum_{i=1}^N D_i u D_i v \, dx = \int_G fv \, dx \tag{58*}$$

for fixed $u \in \mathring{W}_2^1(G)$ and all $v \in C_0^\infty(G)$. Furthermore, the equivalent variational problem (59) reads as follows:

$$\int_G \left(\frac{1}{2} \sum_{i=1}^N (D_i u)^2 - fu \right) dx = \min!, \qquad u \in \mathring{W}_2^1(G). \tag{59*}$$

Note that, by Section 18.3, (59*) is the variational problem which corresponds to the original problem (54), and (58*) is the generalized problem for (54).

Summarizing we obtain the following from Theorem 19.C and Corollary 19.19.

Proposition 19.22. *Problems* (58*) *and* (59*) *are equivalent and, for each given* $f \in L_2(G)$, *they have a unique solution* $u \in \mathring{W}_2^1(G)$ *which can be regarded as a generalized solution of the original classical boundary value problem* (54).

Frequently, one writes $A_E = -\Delta$. Then, by (56), the embeddings

$$\mathring{W}_2^1(G) \subseteq L_2(G) \subseteq \mathring{W}_2^1(G)^*$$

are continuous and

$$-\Delta\colon \mathring{W}_2^1(G) \to \mathring{W}_2^1(G)^*$$

is a linear homeomorphism, namely, the duality map of $\mathring{W}_2^1(G)$. In the literature, one also uses the notation

$$\mathring{H}^1(G) = \mathring{W}_2^1(G), \qquad H^0(G) = L_2(G), \qquad H^{-1}(G) = \mathring{W}_2^1(G)^*.$$

The Friedrichs extension

$$-\Delta\colon D(A_F) \to L_2(G)$$

is bijective and has a linear symmetric compact inverse operator

$$B\colon L_2(G) \to L_2(G),$$

i.e., $B = A_F^{-1}$. The equation

$$u = Bf$$

corresponds to the classical formula

$$u(x) = \int_G K(x, y)f(y)\,dy, \tag{60}$$

where $K(x, y) = K(y, x)$ for all x, $y \in G$, and K is the classical *Green function* to (54), i.e., (60) is the classical integral representation for the solutions u of the original classical problem (54).

In Section 22.21 we shall show the following: If $\partial G \in C^\infty$ and $f \in L_2(G)$, then the solution u of the generalized problem (58*) belongs to the Sobolev space $W_2^2(G)$. Moreover, we shall obtain an explicit characterization of the domain of definition $D(A_F)$ of the Friedrichs extension A_F of A, namely,

$$D(A_F) = \mathring{W}_2^1(G) \cap W_2^2(G).$$

19.12. Application to the Eigenvalue Problem for the Laplace Equation

We now consider the classical boundary eigenvalue problem

$$\begin{aligned} -\Delta u &= \mu u + f \quad \text{on } G, \\ u &= 0 \quad \text{on } \partial G. \end{aligned} \tag{61}$$

Here G is a bounded region in $\mathbb{R}^N$ with $N \geq 1$ and μ is a real number. We will use the same notation as in Section 19.11.

Instead of (61), we consider the operator equation

$$Au = \mu u + f, \qquad u \in D(A)$$

together with the two equivalent generalized problems

$$A_F u = \mu u + f, \qquad u \in D(A_F)$$

and

$$(u|Av) = \mu(u|v) + (f|v) \qquad \text{for fixed} \quad u \in X_E \quad \text{and all} \quad v \in D(A).$$

Explicitly, that means

$$\int_G \sum_{i=1}^N D_i u D_i v\,dx = \mu \int_G uv\,dx + \int_G fv\,dx \tag{62}$$

for fixed $u \in \mathring{W}_2^1(G)$ and all $v \in C_0^\infty(G)$. Note that we obtain (62) from (61) by multiplication by v and subsequent integration by parts.

Proposition 19.23 (Eigenvalue Problem). *Let $f = 0$. Then equation* (62) *has a countable system of eigensolutions $\{u_n, \mu_n\}$ which contains all the eigenvalues.*

We have

$$\mu_n > 0 \qquad \text{for all } n \qquad \text{and} \qquad \mu_n \to \infty \qquad \text{as} \quad n \to \infty.$$

The eigenfunctions u_n are in $\mathring{W}_2^1(G)$ and form a complete orthonormal system in $L_2(G)$.

PROOF. This follows from (57) and Corollary 19.20. □

Proposition 19.24 (The Fredholm Alternative). *Let $\mu \neq 0$. Then:*

(a) *If μ is not an eigenvalue of the homogeneous equation (62) with $f \equiv 0$, then, for each $f \in L_2(G)$, equation (62) has a unique solution $u \in \mathring{W}_2^1(G)$.*
(b) *If μ is an eigenvalue of the homogeneous equation (62), then, for fixed $f \in L_2(G)$, equation (62) has a solution $u \in \mathring{W}_2^1(G)$ iff*

$$\int_G fw\,dx = 0$$

for all eigenvectors w corresponding to μ.

PROOF. This follows from (57) and Corollary 19.21. □

19.13. The Inequality of Poincaré and the Compactness Theorem of Rellich

Proposition 19.25 (Rellich (1930)). *If G is a bounded region in $\mathbb{R}^N$, $N \geq 1$, then the embedding $\mathring{W}_2^1(G) \subseteq L_2(G)$ is compact.*

This proposition is equivalent to the following statement: Each bounded sequence (u_n) in $\mathring{W}_2^1(G)$ has a subsequence $(u_{n'})$ which converges in $L_2(G)$ to a certain function u, i.e., from

$$\int_G \left[u_n^2 + \sum_{i=1}^N (D_i u_n)^2 \right] dx \leq \text{constant} \qquad \text{for all } n, \tag{63}$$

it follows that

$$\int_G (u_{n'} - u)^2\,dx \to 0 \qquad \text{as} \quad n \to \infty. \tag{64}$$

Our proof of Proposition 19.25 will be based on the following special inequality of Poincaré:

$$\int_C u^2\,dx \leq 2R^2 N \int_C \sum_{i=1}^N (D_i u)^2\,dx + (2R)^{-N} \left| \int_C u\,dx \right|^2. \tag{65}$$

Lemma 19.26. *Let C be a closed cube in $\mathbb{R}^N$, $N \geq 1$, with edge length $2R$, then (65) holds for all $u \in C^1(C)$.*

The simple proof, based on the Hölder–Young inequality, will be given in Problem 19.11.

PROOF OF PROPOSITION 19.25. To simplify the notation, let $N = 2$ and $x = (\xi, \eta)$. For general N, the proof proceeds analogously.

Since $C_0^\infty(G)$ is dense in $\mathring{W}_2^1(G)$, it is sufficient to prove that the embedding operator

$$j\colon C_0^\infty(G) \subseteq \mathring{W}_2^1(G) \to L_2(G)$$

is compact, according to the extension principle in Proposition 18.29. Let $B = \{u \in C_0^\infty(G)\colon \|u\|_{1,2} \le 1\}$. Recall

$$\|u\|_{1,2} = \left(\int_G (u^2 + u_\xi^2 + u_\eta^2)\,dx\right)^{1/2}.$$

Since j is linear, it suffices to show that $j(B)$ is relatively compact in $L_2(G)$. By $A_1(21)$, we have to prove that for each $\varepsilon > 0$, the set $j(B)$ has a finite ε-net, i.e., there are elements $u_1, \ldots, u_r \in B$ with

$$\min_i \| j(u) - j(u_i)\|^2 \equiv \min_i \int_G (u - u_i)^2\,dx < \varepsilon \tag{66}$$

for all $u \in B$.

(I) Boundary strip. For each $\delta > 0$, there exists a proper subregion $H \subset\subset G$ such that

$$\int_{G-H} u^2\,dx < \delta \qquad \text{for all} \quad u \in B \tag{67}$$

(Fig. 19.5). This will be proved in (III) below.

(II) Inequality of Poincaré on H. We choose closed cubes $C_1, \ldots, C_s$ of edge length $2R$, which cover H with $C_k \subseteq G$ for all k. Let $D = \bigcup_k C_k$. By the special inequality of Poincaré (65) with $N = 2$,

$$\int_H u^2\,dx \le \int_D u^2\,dx$$

$$\le 4R^2 \int_D (u_\xi^2 + u_\eta^2)\,dx + (2R)^{-2} \sum_k \left|\int_{C_k} u\,dx\right|^2 \tag{68}$$

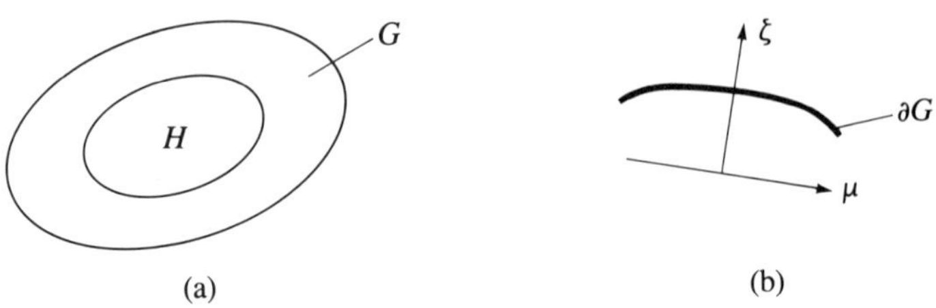

Figure 19.5

for all $u \in B$. We set

$$F(u) = (y_1, \ldots, y_s), \qquad y_k = \int_{C_k} u\,dx.$$

By (68), for all $u \in B$,

$$\int_H u^2\,dx \le 4R^2 \|u\|_{1,2}^2 + (2R)^{-2}|F(u)|^2, \tag{69}$$

where $|F(u)|^2 = \sum_k |y_k|^2$. The Hölder inequality yields

$$|y_k|^2 \le \text{const} \int_{C_k} u^2\,dx.$$

Thus, the set $F(B)$ is bounded in $\mathbb{R}^s$ and hence relatively compact. Consequently, for each $\eta > 0$, the set $F(B)$ has a finite η-net, i.e., there exist elements $u_1, \ldots, u_r \in B$ with

$$\min_i |F(u) - F(u_i)| < \eta \qquad \text{for all} \quad u \in B.$$

From (67) and (69) we obtain the *key formula*

$$\int_G (u - u_i)^2\,dx = \int_{G-H} (u - u_i)^2\,dx + \int_H (u - u_i)^2\,dx$$
$$\le 4\delta + 4R^2 \|u - u_i\|_{1,2}^2 + (2R)^{-2}|F(u) - F(u_i)|^2$$

for all $u \in B$. In this connection, note that $(u - u_i)^2 \le 2u^2 + 2u_i^2$. Now if we choose δ, R, and η sufficiently small, then we obtain the desired estimate (66).

(III) We prove (I) under the assumption $\partial G \in C^{0,1}$. To this end we choose a local (μ, ζ)-coordinate system as in Figure 19.5(b). More precisely, we assume that the boundary ∂G has a local representation of the form

$$\zeta = g(\mu), \qquad \mu \in J,$$

where $g: J \to \mathbb{R}$ is a $C^{0,1}$-function and $J =]-\alpha, \alpha[$ with $\alpha > 0$. Let

$$\Omega_\beta = \{(\mu, \zeta): \mu \in J, g(\mu) - \beta < J < g(\mu)\}.$$

For small $\beta > 0$, the boundary strip Ω_β belongs to G according to the definition of "$\partial G \in C^{0,1}$" in Section 6.2. Let $u \in B$. For all points (μ, ζ) and (μ, z) in Ω_β,

$$u(\mu, \zeta) = \int_z^\zeta u_\zeta(\mu, t)\,dt + u(\mu, z).$$

From $(a + b)^2 \le 2a^2 + 2b^2$ and the Hölder inequality with 1-trick it follows that

$$u(\mu, \zeta)^2 \le 2\beta \int_{g(\mu)-\beta}^{g(\mu)} u_\zeta(\mu, t)^2\,dt + 2u(\mu, z)^2.$$

Integration over z yields

$$\beta u(\mu, \zeta)^2 \leq \int_{g(\mu)-\beta}^{g(\mu)} (2\beta^2 u_\zeta(\mu, t)^2 + 2u(\mu, t)^2)\, dt,$$

and integration over Ω_ε yields

$$\beta \int_{\Omega_\varepsilon} u^2\, dx \leq \varepsilon \int_{\Omega_\beta} (2\beta^2 u_\zeta^2 + 2u^2)\, dx$$

$$\leq \varepsilon \operatorname{const} \|u\|_{1,2}^2.$$

Since a finite number of local boundary strips covers a boundary strip of G, we obtain (67) in the case where ε is sufficiently small.

(IV) The assertion (I) remains true if the assumption $\partial G \in C^{0,1}$ drops out. In this connection, note that the functions u in B vanish outside a boundary strip, and hence we can pass to a smaller region with a $C^{0,1}$-boundary in the case where G does not have a $C^{0,1}$-boundary. □

Let $\mathscr{W}_2^1(G)$ denote the closure of $C^\infty(\bar{G})$ in $W_2^1(G)$. Then the same proof as above shows that the embedding

$$\mathscr{W}_2^1(G) \subseteq L_2(G)$$

is compact in the case where $\partial G \in C^{0,1}$. We have only to replace $C_0^\infty(G)$ by $C^\infty(\bar{G})$. In Section 21.4 we shall prove that

$$\mathscr{W}_2^1(G) = W_2^1(G)$$

provided $\partial G \in C^{0,1}$. However, this fact will not be used before Chapter 21.

19.14. Functions of Self-Adjoint Operators

Our goal is to construct a simple functional calculus for an important class of self-adjoint operators. Applications to the heat equation and the wave equation will be considered in the following two sections. Our basic tool will be the *convergence criterion* for abstract Fourier series (Proposition 19.12). We make the following assumption:

(H) The linear operator $A: D(A) \subseteq X \to X$ is self-adjoint on the separable H-space X over $\mathbb{K} = \mathbb{R}, \mathbb{C}$, and A possesses a complete orthonormal system of eigenvectors $\{u_n\}$ in X, i.e., $Au_n = \lambda_n u_n$ for all n.

Proposition 19.27. *Assume* (H). *Then*

$$Au = \sum_n \lambda_n (u_n|u) u_n \qquad \textit{for all} \quad u \in D(A). \tag{70}$$

Furthermore, the following three conditions are mutually equivalent:

(i) $u \in D(A)$.
(ii) *The right-hand side of* (70) *converges.*
(iii) $\sum_n |\lambda_n|^2 |(u_n|u)|^2 < \infty$.

PROOF. For all $v \in X$,

$$v = \sum_n (u_n|v)u_n, \tag{71}$$

since $\{u_n\}$ is complete. If $u \in D(A)$, then

$$(u_n|Au) = (Au_n|u) = \lambda_n(u_n|u).$$

Hence (71) implies (70).

By Proposition 19.12, it follows that (ii) and (iii) are equivalent.

We now construct a linear operator $B: D(B) \subseteq X \to X$ by

$$Bu = \sum_n \lambda_n(u_n|u)u_n$$

for all $u \in X$ with (iii). From

$$(Bu|v) = \sum_n \lambda_n(u|u_n)(u_n|v) = (u|Bv)$$

it follows that B is symmetric. By our construction, B is an extension of A. Since A is self-adjoint, there does not exist a proper symmetric extension (cf. Problem 19.5c). Hence $A = B$. □

Let the function $f: \mathbb{R} \to \mathbb{K}$ be given. We define the operator

$$f(A): D(f(A)) \subseteq X \to X$$

by the natural formula

$$f(A)u = \sum_n f(\lambda_n)(u_n|u)u_n. \tag{72}$$

Moreover, by definition, u belongs to $D(f(A))$ iff the right-hand side in (72) converges. By Proposition 19.12, $u \in D(f(A))$ iff

$$\sum_n |f(\lambda_n)|^2 |(u_n|u)|^2 < \infty. \tag{73}$$

In particular, we have that $f(A)u_n = f(\lambda_n)u_n$ for all n.

Proposition 19.28. *Assume* (H). *Then the operator $f(A)$ is self-adjoint for each real function $f: \mathbb{R} \to \mathbb{R}$.*

PROOF. Let $B = f(A)$ and let B^* denote the adjoint operator. As in the proof of Proposition 19.27, we obtain that B is symmetric, i.e., $B \subseteq B^*$. In order to show that $B^* \subseteq B$, let $u \in B^*$. Since

$$(u_n|B^*u) = (Bu_n|u) = f(\lambda_n)(u_n|u),$$

we obtain

$$B^*u = \sum_n (u_n|B^*u)u_n = \sum_n f(\lambda_n)(u_n|u)u_n,$$

i.e., $u \in D(B)$. □

With a view to applications in the next two sections, we now consider functions

$$f(A, t) = \sum_n f(\lambda_n, t)(u_n|u)u_n$$

depending on a parameter t, which will play the role of time. We set

$$x(t) = f(A, t)u$$

for fixed u. In order to obtain the continuity and differentiability of $x(\cdot)$, we need the following two majorant conditions:

$$\sum_n |c_n(u_n|u)|^2 < \infty, \qquad c_n = \sup_{t \in J} |f(\lambda_n, t)|, \tag{74}$$

$$\sum_n |d_n(u_n|u)|^2 < \infty, \qquad d_n = \sup_{t \in J} |f_t(\lambda_n, t)|. \tag{75}$$

Proposition 19.29. *Suppose that* (H) *holds and let the function* $f\colon \mathbb{R} \times J \to \mathbb{R}$ *be given, where* J *is a real interval. Then:*

(a) *If* $t \mapsto f(\lambda, t)$ *is continuous on* J *for each* $\lambda \in \mathbb{R}$, *then* $t \mapsto x(t)$ *is continuous on* J *provided* (74) *holds.*
(b) *If* $t \mapsto f(\lambda, t)$ *is continuously differentiable on* J *for each* $\lambda \in \mathbb{R}$, *then* $t \mapsto x(t)$ *is continuously differentiable on* J *provided* (74) *and* (75) *hold. If we set* $h(\lambda, t) = f_t(\lambda, t)$, *then*

$$x'(t) = h(A, t)u \qquad \textit{for all} \quad t \in J.$$

PROOF. Ad(a). We set $f_n(t) = f(\lambda_n, t)$ and $a_n = (u_n|u)$. Then

$$\begin{aligned}\|f(A,t)u - f(A,s)u\|^2 &= \sum_n |(f_n(t) - f_n(s))a_n|^2 \\ &\leq \sum_{n \leq N} |(f_n(t) - f_n(s))a_n|^2 + 4 \sum_{n > N} |c_n a_n|^2 \\ &< \varepsilon + \varepsilon,\end{aligned}$$

in the case where N is sufficiently large and $|t - s|$ is sufficiently small.

Ad(b). Use the mean value theorem

$$\frac{f_n(t) - f_n(s)}{t - s} - f_n'(s) = f_n'(s + \vartheta(t - s)) - f_n'(s), \qquad 0 < \vartheta < 1,$$

and an analogous argument as in (a). □

This functional calculus is the prototype of the general functional calculus for self-adjoint operators which we summarize in the Appendix, $A_2(96)$.

19.15. Application to the Heat Equation

We consider the initial value problem

$$\begin{aligned} u'(t) + A_F u(t) &= 0, \qquad 0 \le t < \infty, \\ u(0) &= w. \end{aligned} \tag{76}$$

We make the following assumptions:

(H1) The operator $A: D(A) \subseteq X \to X$ is symmetric[1] and strongly monotone on the real, separable H-space X.

(H2) Let A_F denote the Friedrichs extension of A with the energetic space X_E, and suppose that the embedding $X_E \subseteq X$ is compact.

Proposition 19.30. *Assume* (H1), (H2). *Then for each* $w \in D(A_F)$, *equation* (76) *has a unique* C^1-solution $u: \mathbb{R}_+ \to X$. *This solution is given by*

$$u(t) = e^{-tA_F} w \qquad \textit{for all} \quad t \ge 0. \tag{77}$$

Corollary 19.31. *If* $w \in X$, *then the function* u *in* (77) *is continuous on* $\mathbb{R}_+$. *We call this function* $u(\cdot)$ *a generalized (mild) solution of* (76). *The family* $\{e^{-tA_F}\}$ *forms a semigroup, with generator* $-A_F$, *i.e., the operator* $e^{-tA_F}: X \to X$ *is linear and continuous for all* $t \ge 0$,

$$e^{-(t+s)A_F} = e^{-tA_F} e^{-sA_F} \qquad \textit{for all} \quad t, s \ge 0, \tag{78}$$

and the limit

$$\lim_{h \to +0} \left(\frac{e^{-hA_F} - I}{h} \right) w$$

exists exactly for all $w \in D(A_F)$ *and is equal to* $-A_F w$.

This corollary is a special case of a much more general result which will be considered in Section 19.18.

Example 19.32. Let G be a bounded region in $\mathbb{R}^N$, $N \ge 1$. We consider the initial–boundary value problem

$$\begin{aligned} u_t - \Delta u &= 0 \qquad \text{on } G \times \mathbb{R}_+, \\ u(x,t) &= 0 \qquad \text{on } \partial G \times \mathbb{R}_+, \\ u(x,0) &= w(x) \quad \text{on } G. \end{aligned} \tag{79}$$

We set $X = L_2(G)$ and $Au = -\Delta u$ with $D(A) = C_0^\infty(G)$. Then the conditions (H1), (H2) are fulfilled, by Section 19.11.

[1] By Definition 19.13, this implies $\overline{D(A)} \subseteq X$, i.e., $D(A)$ is dense in X.

We regard both the solutions of (76) and the generalized (mild) solutions of (76) as generalized solutions of the classical problem (79). By Corollary 19.31, for each initial value function $w \in L_2(G)$, we obtain a uniquely determined mild solution.

PROOF OF PROPOSITION 19.30 and COROLLARY 19.31.

(I) Uniqueness. Let $u: \mathbb{R}_+ \to X$ be a C^1-solution of (76) with $u(0) = 0$. Then it follows from

$$\frac{d}{dt}(u(t)|u(t)) = 2(u'(t)|u(t)) = -2(A_F u(t)|u(t)) \leq 0$$

that $u(t) \equiv 0$.

(II) Existence. We set $B = A_F$. By Corollary 19.20, the operator B has a complete orthonormal system (u_n) of eigenvectors with

$$Bu_n = \lambda_n u_n \qquad \text{and} \qquad \lambda_n \geq c > 0 \qquad \text{for all } n.$$

We now use the simple functional calculus from Section 19.14. Then

$$Bw = \sum_n \lambda_n (u_n|w) u_n,$$

and $w \in D(B)$ iff $\sum_n |\lambda_n (u_n|w)|^2 < \infty$. By definition,

$$e^{-tB} w = \sum_n e^{-t\lambda_n} (u_n|w) u_n,$$

and $w \in D(e^{-tB})$ iff $\sum_n |e^{-t\lambda_n}(u_n|w)|^2 < \infty$.

For all $w \in X$ and $t \geq 0$,

$$\|e^{-tB}w\|^2 = \sum_n |e^{-t\lambda_n}(u_n|w)|^2$$
$$\leq \sum_n |(u_n|w)|^2 = \|w\|^2.$$

Hence $D(e^{-tB}) = X$ and $\|e^{-tB}\| \leq 1$. From

$$e^{-(t+s)\lambda} = e^{-t\lambda} e^{-s\lambda}$$

we obtain the semigroup property (78).

We now set

$$u(t) = e^{-tB} w.$$

For $w \in X$ and all $t \geq 0$, we have the majorant condition

$$\sum_n |e^{-t\lambda_n}(u_n|w)|^2 \leq \sum_n |(u_n|w)|^2 < \infty.$$

By Proposition 19.29, the function $t \mapsto u(t)$ is continuous on $\mathbb{R}_+$. For $w \in D(B)$ and all $t \geq 0$, we have the majorant condition

$$\sum_n |\lambda_n e^{-t\lambda_n}(u_n|w)|^2 \leq \sum_n |\lambda_n (u_n|w)|^2 < \infty.$$

By Proposition 19.29, there exists the derivative

$$u'(t) = -\sum_n \lambda_n e^{-t\lambda_n}(u_n|w)u_n$$

for all $t \geq 0$, and u' is continuous on $\mathbb{R}_+$. Because of $e^{-tB}u_n = e^{-t\lambda_n}u_n$ and the self-adjointness of the operators B and e^{-tB}, we obtain

$$u'(t) = -\sum_n \lambda_n(u_n|e^{-tB}w)u_n = -Be^{-tB}w = -Bu(t)$$

for all $t \geq 0$, i.e., $u(\cdot)$ is a solution of (76).

Let us define the operator $C: D(C) \subseteq X \to X$ by

$$Cw = \lim_{h \to +0} \left(\frac{e^{-hB} - I}{h}\right) w,$$

where w belongs to $D(C)$ iff this limit exists. We want to show that $C = -B$. In fact, differentiation of the relation

$$(e^{-tB}u|v) = (u|e^{-tB}v) \qquad \text{for all} \quad u, v \in D(C)$$

with respect to t at $t = 0$ yields $(Cu|v) = (u|Cv)$ for all $u, v \in D(C)$, i.e., C is symmetric. Since $u'(0) = -Bw$, the operator C is a symmetric extension of the self-adjoint operator $-B$, and hence $C = -B$ (cf. Problem 19.5c). □

19.16. Application to the Wave Equation

We consider the initial value problem

$$\begin{aligned} u''(t) + A_F u(t) = 0, &\qquad -\infty < t < \infty, \\ u(0) = v, &\qquad u'(0) = w, \end{aligned} \tag{80}$$

and make the following assumptions:

(H1) The operator $A: D(A) \subseteq X \to X$ is symmetric[1] and strongly monotone on the real, separable H-space X.

(H2) Let A_F denote the Friedrichs extension of A with the energetic space X_E, and suppose that the embedding $X_E \subseteq X$ is compact.

In the trivial special case $X = \mathbb{R}$ and $A_F > 0$, equation (80) describes the motion of a harmonic oscillator with the classical solution

$$u(t) = (\cos tB)v + B^{-1}(\sin tB)w, \tag{81}$$

where $B = A_F^{1/2}$. The energy of the harmonic oscillator is given by

$$E(t) = \tfrac{1}{2}(|u'(t)|^2 + |Bu(t)|^2).$$

[1] By Definition 19.13, this implies $\overline{D(A)} \subseteq X$, i.e., $D(A)$ is dense in X.

By definition, a *classical solution* of (80) is a C^2-function $u: \mathbb{R} \to X$ such that (80) holds and $t \mapsto Bu(t)$ is C^1 from $\mathbb{R}$ into X.

Below we shall prove that $D(A_F) \subseteq D(B)$. This implies that $(A_F u|u) = (B^2 u|u) = (Bu|Bu)$ for all $u \in D(A_F)$, by Section 19.14. According to our definition above, a classical solution of (80) satisfies $u(t) \in D(A_F)$ for all t, and hence $u(t) \in D(B)$ for all t.

Generalizing the classical expression for the energy of the harmonic oscillator above, we define the *energy* of a classical solution of (80) at time t by

$$E(t) = \tfrac{1}{2}(\|u'(t)\|_X^2 + \|Bu(t)\|_X^2).$$

Since $\|Bu(t)\|_X^2 = (A_F u(t)|u(t))$, we obtain

$$E(t) = \tfrac{1}{2}[\|u'(t)\|_X^2 + (A_F u(t)|u(t))]. \tag{81*}$$

Proposition 19.33. *Assume* (H1), (H2). *Then for each $v \in D(A_F)$ and $w \in D(A_F^{1/2})$, equation* (80) *has exactly one classical solution. This solution is given by* (81) *with $E(t) = $ constant.*

EXAMPLE 19.34. We consider the initial–boundary value problem for the classical wave equation

$$\begin{aligned} u_{tt} - \Delta u &= 0 && \text{on } G \times \mathbb{R}, \\ u(x,t) &= 0 && \text{on } \partial G \times \mathbb{R}, \\ u(x,0) &= v(x) && \text{on } G, \\ u_t(x,0) &= w(x) && \text{on } G, \end{aligned} \tag{82}$$

where G is a bounded region in $\mathbb{R}^N$, $N \geq 1$. We set $X = L_2(G)$ and $Au = -\Delta u$ with $D(A) = C_0^\infty(G)$. Then, the conditions (H1), (H2) are fulfilled, by Section 19.11.

We regard the solutions of (80) as generalized solutions of the classical problem (82). By (81*), for sufficiently smooth functions u, we obtain the energy

$$E(t) = \frac{1}{2}\int_G (u_t^2 - (\Delta u)u)\,dx,$$

where $u = 0$ on ∂G. Integration by parts yields the well-known classical formula

$$E(t) = \frac{1}{2}\int_G \left(u_t^2 + \sum_{i=1}^N (D_i u)^2\right) dx.$$

The latter expression makes sense for a broad class of functions u.

PROOF OF PROPOSITION 19.33.

(I) Uniqueness. Let u be a classical solution of (80) with $v = w = 0$. As in the proof of Proposition 19.30, we obtain

$$A_F^\alpha w = \sum_n \lambda_n^\alpha (u_n|w)u_n \qquad \text{for each} \quad \alpha > 0,$$

where $w \in D(A_F^\alpha)$ iff $\sum_n \lambda_n^{2\alpha} |(u_n|w)|^2 < \infty$. Hence $D(A_F) \subseteq D(A_F^{1/2})$, since $\lambda_n \to +\infty$ as $n \to \infty$ in the case where $\dim X = \infty$. The operator $B = A_F^{1/2}$ is self-adjoint and hence graph closed; therefore

$$(Bu(t))' = Bu'(t)$$

(cf. Problem 19.4d). Differentiation of the energy function yields

$$\begin{aligned} E'(t) &= (u''(t)|u'(t)) + (Bu'(t)|Bu(t)) \\ &= -(A_F u(t)|u'(t)) + (u'(t)|A_F u(t)) = 0, \end{aligned}$$

and $E(0) = 0$. This implies $E(t) \equiv 0$, and hence $u(t) \equiv 0$.

(II) Existence. Using the same method as in the proof of Proposition 19.30, one easily verifies that (81) is a classical solution of (80). The conditions $v \in D(A_F)$ and $w \in D(A_F^{1/2})$ guarantee the convergence of simple majorants obtained in a natural way. □

In Section 19.23 we shall construct a one-parameter unitary group $\{S(t)\}$ which yields generalized (mild) solutions of equation (80) of the form

$$(u(t), u'(t)) = S(t)(v, w),$$

where $v \in X_E$ and $w \in X$ are given. In Example 19.34 above, this means $v \in \mathring{W}_2^1(G)$ and $w \in L_2(G)$.

19.17. Semigroups and Propagators, and Their Physical Relevance

The notion of a semigroup is the most important notion for describing time-dependent processes in nature in terms of functional analysis. The *key relations* are

$$S(t + s) = S(t)S(s) \qquad \text{for all} \quad t, s \in \mathbb{R}_+, \tag{83}$$

$$S(0) = I, \tag{84}$$

and (85) below.

Definition 19.35. A *semigroup* $\{S(t)\}$ on a B-space X consists of a family of operators $S(t)\colon X \to X$ for all $t \in \mathbb{R}_+$ with (83) and (84). The *generator* $B\colon D(B) \subseteq X \to X$ of the semigroup $\{S(t)\}$ is defined by

$$Bw = \lim_{t \to +0} \frac{S(t)w - w}{t}, \tag{85}$$

where w belongs to $D(B)$ iff the limit in (85) exists.

A *one-parameter* group $\{S(t)\}$ on the B-space X consists of a family of operators $S(t): X \to X$ for all $t \in \mathbb{R}$ with (83) for all $t, s \in \mathbb{R}$ and (84).

Sometimes generators are also called infinitesimal generators.

STANDARD EXAMPLE 19.36. We consider the differential equation

$$\begin{aligned} u'(t) &= Bu(t) \quad \text{on } \mathbb{R}_+, \\ u(0) &= w. \end{aligned} \tag{86}$$

Let $B: X \to X$ be a linear continuous operator on the B-space X over $\mathbb{K}$. We set $S(t) = e^{tB}$, where

$$e^{tB} = \sum_{n=0}^{\infty} \frac{t^n B^n}{n!}. \tag{87}$$

This series converges on the B-space $L(X, X)$ for all $t \in \mathbb{R}$, i.e., this series converges with respect to the operator norm. Then $\{S(t)\}$ is a one-parameter group and the unique solution of (86) is given by

$$u(t) = S(t)w. \tag{88}$$

The operator B is the generator of $\{S(t)\}$, since $u'(0) = Bu(0) = Bw$.

In Section 19.15 we have already encountered a semigroup in connection with the heat equation. In this case, the generator of the semigroup was an unbounded operator.

19.17a. Classification of Semigroups

Let $\mathscr{S} = \{S(t)\}$ be a semigroup on the B-space X.

(i) $\mathscr{S}$ is called *strongly continuous* iff $t \mapsto S(t)w$ is continuous on $\mathbb{R}_+$ for all $w \in X$, i.e.,

$$\lim_{t \to s} S(t)w = S(s)w \qquad \text{for all} \quad s \in \mathbb{R}_+.$$

At $s = 0$ we have to take the limit from the right.

Sometimes strongly continuous semigroups are also briefly called continuous.

(ii) $\mathscr{S}$ is called *uniformly continuous* iff all the operators $S(t): X \to X$ are linear and continuous, and $t \mapsto S(t)$ is continuous on $\mathbb{R}_+$ with respect to the operator norm, i.e.,

$$\lim_{t \to s} \|S(t) - S(s)\| = 0 \qquad \text{for all} \quad s \in \mathbb{R}_+.$$

At $s = 0$ we have to take the limit from the right.

(iii) $\mathscr{S}$ is called *nonexpansive* iff all the operators $S(t): X \to X$ are nonexpansive and

$$\lim_{t \to +0} S(t)w = w$$

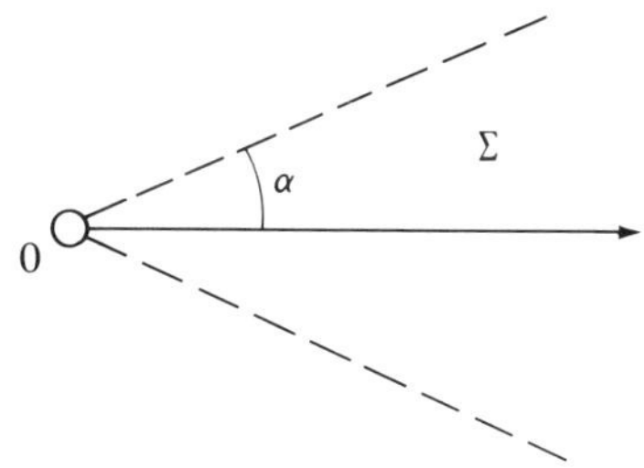

Figure 19.6

holds for all $w \in X$. In Problem 19.9 we will show that each nonexpansive semigroup is strongly continuous.

(iv) $\mathscr{S}$ is called a *linear* semigroup iff all the operators $S(t)\colon X \to X$ are linear and continuous.

If one replaces $\mathbb{R}_+$ with $\mathbb{R}$, then one obtains the analogous notions for one-parameter groups. By a *one-parameter unitary group* we understand a strongly continuous one-parameter group $\{S(t)\}$ of linear unitary operators $S(t)\colon X \to X$ on the H-space X.

(v) Let X be a complex B-space and let Σ be an open sector with the sector angle α as in Figure 19.6, i.e.,

$$\Sigma = \{z \in \mathbb{C}\colon -\alpha < \arg z < \alpha, z \neq 0\}.$$

By an analytic semigroup $\{S(t)\}$, we understand an open sector Σ and a family of linear continuous operators $S(t)\colon X \to X$ for all $t \in \Sigma$ with $S(0) = I$ and the following three properties:

(a) $t \mapsto S(t)$ is an analytic map from Σ into $L(X, X)$;
(b) $S(t + s) = S(t)S(s)$ for all $t, s \in \Sigma$;
(c) $\lim S(t)w = w$ as $t \to 0$ in Σ for all $w \in X$.

Therefore, an analytic semigroup is the analytic continuation of a strongly continuous semigroup on $\mathbb{R}_+$ to an open sector Σ. Note that, by definition, analytic semigroups are linear semigroups.

By a *bounded analytic* semigroup $\{S(t)\}$ we understand an analytic semigroup as defined above which has the additional property that, for each $\beta \in]0, \alpha[$,

$$\sup_{t \in \Sigma_\beta} \|S(t)\| < \infty,$$

where $\Sigma_\beta = \{z \in \mathbb{C}\colon -\beta < \arg z < \beta, z \neq 0\}$.

EXAMPLE 19.37. (Fundamental Situations).

(i) The one-parameter group $\{e^{tB}\}$ in Standard Example 19.36 is uniformly continuous and analytic on $\mathbb{C}$.

(ii) If $B\colon D(B) \subseteq X \to X$ is a linear self-adjoint operator on the H-space X with $(Bu|u) \leq 0$ on $D(B)$, then B is the generator of a linear nonexpansive semigroup. This is a special case of Theorem 19.E in Section 19.20. In terms of the general functional calculus for self-adjoint operators this semigroup is given by $\{e^{tB}\}$.

Such semigroups describe, in particular, heat conduction and diffusion processes.

(iii) If $H: D(H) \subseteq X \to X$ is a linear self-adjoint operator on the complex H-space X, then $-iH$ generates a one-parameter unitary group. This will be proved in Section 19.21. In terms of the functional calculus, this group is given by $\{e^{-itH}\}$. Such groups describe the dynamics of quantum systems as will be shown in Part V. The operator H corresponds to the energy of the quantum system and is called the Hamiltonian of the system.

(iv) If $C: D(C) \subseteq X \to X$ is a skew-adjoint operator on the real H-space X, then C generates a one-parameter unitary group. This will be shown in Section 19.21. Such groups describe, for example, the dynamics of wave processes.

19.17b. Inhomogeneous Linear Evolution Equations and Semilinear Evolution Equations

We consider the inhomogeneous equation

$$\begin{aligned} u'(t) &= Bu(t) + f(t) \quad \text{on } \mathbb{R}_+, \\ u(0) &= w. \end{aligned} \tag{89}$$

In the classical case, u is a real function and B is a real number. Then Lagrange's method of the variation of the constant yields the solution

$$u(t) = S(t)w + \int_0^t S(t-s)f(s)\,ds \tag{90}$$

with $S(t) = e^{tB}$. In the general case of an unbounded operator B, the situation is more complex. Formal differentiation of (90) yields (89). Unfortunately, this formal procedure can only be justified if f is sufficiently regular. For example, if f is merely continuous, then $u(\cdot)$ in (90) need not be a solution of (89). This is a *typical difficulty* of the theory of semigroups. However, it is quite natural to call the solution $u(\cdot)$ of (90) a *mild* (or generalized) solution of (89). This way the *semilinear* problem

$$\begin{aligned} u'(t) &= Bu(t) + f(t, u(t)) \quad \text{on } \mathbb{R}_+, \\ u(0) &= w, \end{aligned} \tag{91}$$

can be reduced to the Volterra integral equation

$$u(t) = S(t)w + \int_0^t S(t-s)f(s, u(s))\,ds. \tag{92}$$

The solutions of the *fixed-point problem* (92) are called mild solutions of (91). If f is sufficiently regular, then the mild solutions are also classical solutions of (91).

19.17c. Nonautonomous Equations and Propagators

We consider the initial value problem

$$\begin{aligned} u'(t) &= B(t)u(t) \quad \text{on } [s, \infty[, \\ u(s) &= w. \end{aligned} \tag{93}$$

Here the point is that the operator B depends on time t. Such equations correspond to processes in nature which are *not* homogeneous in time (e.g., time-dependent outer forces or the fundamental scattering processes of elementary particles in accelerators). Suppose that for each fixed time s, equation (93) has a unique solution $u(\cdot)$. We define the so-called *propagator* $\{P(t, s)\}$ by

$$P(t, s)w = u(t) \qquad \text{for all} \quad t \geq s. \tag{94}$$

The assumed uniqueness of the solution implies

$$P(t, r)(P(r, s)w) = P(t, s)w.$$

If B is independent of time t, and if B generates the semigroup $\{S(t)\}$, then $u(t) = S(t - s)w$, i.e.,

$$P(t, s) = S(t - s).$$

This shows that propagators represent generalizations of semigroups. Formal differentiation shows that the so-called series of Dyson

$$P(t, s) = I + \sum_{n=1}^{\infty} \int_s^t \int_s^{t_1} \cdots \int_s^{t_{n-1}} B(t_1) \ldots B(t_n)\, dt_n \ldots dt_1$$

yields a formal solution of (93) by means of (94). As we shall see in Part V, this is the key formula of quantum field theory.

Furthermore, formal differentiation of

$$u(t) = P(t, s)w + \int_s^t P(t, r) f(r, u(r))\, dr \tag{95}$$

with respect to t yields

$$u'(t) = B(t)u(t) + f(t, u(t)), \qquad u(s) = w. \tag{96}$$

The solutions of the fixed-point problem (95) are called *mild* solutions of (96).

19.17d. Quasi-Linear Equations

We consider the quasi-linear equation

$$u' = C(u, t)u + f(t, u), \qquad u(s) = w \tag{97}$$

for the unknown function $u = u(t)$. Such equations correspond to physical processes with interactions. Comparing (97) with (96), equation (97) can be reduced to a *fixed-point problem* for u of the form (95) by letting

$$B(t) = C(u(t), t).$$

Note that B depends on u, and hence P in (95) also depends on u.

19.17e. Abstract Semigroups and Abstract Propagators, and Their Physical Interpretation

Definition 19.38. Let X be an arbitrary set. By a *propagator* $\{P(t,s)\}$ on X, we understand a family of operators $P(t,s)\colon X \to X$ for all real $t \geq s$ with the following two properties:

$$P(t,t) = I \qquad \text{for all } t, \tag{98}$$

$$P(t,s) = P(t,r)P(r,s) \qquad \text{for all} \quad t \geq r \geq s. \tag{99}$$

By a *semigroup* $S(t)$ on X (resp. one-parameter group), we understand a family of operators $S(t)\colon X \to X$ for all real $t \geq 0$ (resp. all real t) with $S(0) = I$ and

$$S(t+s) = S(t)S(s) \qquad \text{for all} \quad t, s \geq 0 \tag{100}$$

(resp. for all real t, s).

Let us discuss the following four fundamental *principles* for the mathematical description of time-dependent processes in nature:

(i) *Propagators describe causal processes.*
(ii) *One-parameter groups describe causal processes which are homogeneous in time and reversible.*
(iii) *Proper semigroups describe causal processes which are homogeneous in time and irreversible (e.g., the growth in biology, heat conduction, or diffusion).*
(iv) *The description of time-dependent processes by propagators or semigroups is more natural than the description by differential equations.*

In order to have an intuitive picture at hand, let X be a manifold (e.g. the surface of the earth). Let $\{S(t)\}$ be a semigroup on X (or a one-parameter group). Then

$$u(t) = S(t)w$$

can be regarded as the motion of a fluid particle on X. Here, $u(t)$ is the position of the particle at time t and $u(0) = w$. Therefore, semigroups and one-parameter groups are also called *semiflows* and *flows*, respectively. Moreover, semigroups are also called abstract dynamical systems.

Propagators and Causality

We regard the elements of the set X as the possible states of a physical system. Suppose that the system is in the state w at time s. Then

$$u(t) = P(t,s)w \tag{101}$$

describes the time-evolution of the system, i.e., $u(t)$ is the state of the system at time t. The causality of the process is reflected by the propagator property (99). In fact, relation (99) tells us that

$$u(t) = P(t, r)u(r)$$

for all r with $s \leq r \leq t$, i.e., the final state $u(t)$ is uniquely determined by each of the intermediate states $u(r)$.

If $\{S(t)\}$ is a semigroup, then we obtain a propagator by letting

$$P(t, s) = S(t - s) \qquad \text{for all} \quad t \geq s. \tag{102}$$

In fact, the propagator property (99) follows from the semigroup property (100). Conversely, if the process is homogeneous in time, i.e.,

$$P(t + a, s + a) = P(t, s) \tag{102*}$$

for all real a, then we obtain a semigroup $\{S(t)\}$ by letting

$$S(t) = P(t, 0).$$

In this case relation (102*) implies (102).

Semigroups and Irreversibility

Let $\{S(t)\}$ be a semigroup and let

$$u(t) = S(t)w \tag{103}$$

be the time-evolution of the initial state w at time $t = 0$. Then, in general, $u(t)$ is only defined for $t \geq 0$ and the inverse operator $S(t)^{-1}$ does not exist, i.e., the past state w is not uniquely determined by the future state $u(t)$. The situation changes if $\{S(t)\}$ is a one-parameter group. In this case, it follows from the group property (100) that

$$S(t)S(-t) = S(-t)S(t) = I,$$

i.e., $S(t)\colon X \to X$ is surjective and injective and hence bijective with the inverse operator $S(t)^{-1} = S(-t)$. This means

$$u(-t) = S(-t)w = S(t)^{-1}w.$$

In particular, we obtain that if the process $t \mapsto u(t)$ is possible, then the inverse process $t \mapsto u(-t)$, which corresponds to a time reflection, is also possible. Such processes are called reversible. Many processes do not have this property, i.e., they are irreversible. For example, the growth of a human being is irreversible. In Part IV we shall show that the second law of thermodynamics is responsible for the fact that most processes in nature are irreversible.

Standard Example 19.36 shows that linear bounded operators $B\colon X \to X$ on a B-space X generate one-parameter groups. Therefore, irreversible processes in nature *cannot* be described by differential equations $u' = Bu$ where B is bounded. This underlines the *importance of unbounded operators.*

One-Parameter Groups and the Representation Theory of Groups

Let X be a set and let $B(X, X)$ denote the set of all bijective mappings $F: X \to X$. Then $B(X, X)$ forms a group. By a *representation* of a given group G in $B(X, X)$, we understand a mapping $H: G \to B(X, X)$ with

$$H(gh) = H(g)H(h) \qquad \text{for all} \quad g, h \in G,$$

i.e., a representation is a homomorphism from G into $B(X, X)$.

In Part V we shall show that the representation theory of groups plays an important role in modern physics. From the abstract mathematical point of view, modern physics is based on the following three fundamental principles:

(a) Elementary particles correspond to irreducible representations of groups.
(b) Conservation laws correspond to global symmetries (group invariance) of processes.
(c) The four fundamental interactions (i.e., gravitative, weak, electromagnetic, and strong interaction) correspond to the curvature of manifolds (fiber bundles with structural groups), and the interaction is related to local symmetries (gauge transformations).

For example, if processes are invariant with respect to *time translation* (resp. spatial translation or rotation), then there exists a mathematical conservation quantity which corresponds to *energy* (resp. momentum or angular momentum). In this connection, invariance with respect to time translation means that if the process $t \mapsto u(t)$ is possible, then all the processes $t \mapsto u(t + a)$ are also possible for arbitrary $a \in \mathbb{R}$. This result explains mathematically why conservation of energy, momentum, and angular momentum exist in all fields of physics (e.g., mechanics, electromagnetism, quantum theory, etc.).

Semigroups and Differential Equations

Suppose that a physical process is described by a classical partial differential equation of the structure

$$u'(t) = B_0 u(t), \qquad u(0) = w, \tag{104}$$

(e.g., the heat equation). In order to treat (104) by means of functional analysis, we have to extend B_0 to an operator B. Then, instead of (104), we consider the operator equation

$$u'(t) = Bu(t), \qquad u(0) = w, \tag{104*}$$

(e.g., B is the Friedrichs extension of B_0). From the point of view of the theory of semigroups, this extension is necessary since B_0 is generally *not* the generator of a semigroup. Let $B: D(B) \subseteq X \to X$ be the generator of the semigroup $\{S(t)\}$. In Section 19.18 we shall show that the solution of (104*) is given by

$$u(t) = S(t)w \tag{104**}$$

for all initial values $w \in D(B)$. However, the operator $S(t)$ is defined on the *entire* space X. Thus, the function $u = u(t)$ in (104**) is defined for *each* initial value $w \in X$. We call this function u in (104**) a generalized (mild) solution of the original problem (104).

Note that the derivation of the classical partial differential equations of mathematical physics is based on the (unrealistic) assumption that the situations are sufficiently smooth. Therefore, it is quite natural to regard the *semigroups* as *primary* and the differential equations (104) and (104*) as secondary.

The complexity of the general theory of nonlinear semigroups reflects the complexity of time-dependent processes in nature. For example, one observes:

(i) explosions (e.g., supernovae and quasars),
(ii) shock waves, and
(iii) turbulence.

Mathematically, this corresponds to blowing-up effects, instabilities and discontinuities, and strange attractors.

19.18. Main Theorem on Abstract Linear Parabolic Equations

We consider the initial value problem

$$\begin{aligned} u'(t) &= Bu(t) + f(t) \quad \text{on }]t_0, T[, \\ u(t_0) &= w, \end{aligned} \tag{105}$$

together with the formula

$$u(t) = S(t - t_0)w + \int_{t_0}^{t} S(t - s)f(s)\,ds. \tag{106}$$

We make the following assumptions:

(H1) Let $\{S(t)\}$ be a strongly continuous linear semigroup on the B-space X over $\mathbb{K} = \mathbb{R}, \mathbb{C}$ with the generator B, i.e., $\{S(t)\}$ is a semigroup of linear continuous operators $S(t)\colon X \to X$ for all $t \geq 0$, and $t \mapsto S(t)w$ is continuous on $\mathbb{R}_+$ for all $w \in X$.
(H2) The function $f\colon [t_0, T[\to X$ is continuous.

Definition 19.39. Let the function $u\colon [t_0, T[\to X$ be given.

(i) u is called a *classical solution* of (105) iff u is C^1 on $]t_0, T[$ and is continuous on $[t_0, T[$, and (105) holds.
(ii) u in (106) is called a *mild* solution of the original problem (105).

Theorem 19.D. *Assume* (H1), (H2). *Then:*

(a) *There exists at most one classical solution of* (105), *and each classical solution is also a mild solution.*

(b) *If f is C^1 and $w \in D(B)$, then the mild solution* (106) *is also a classical solution of* (105).

In particular, for each $w \in D(B)$, the homogeneous problem (105) *with $f \equiv 0$ has a unique classical solution given by $u(t) = S(t - t_0)w$.*

(c) *If the operator $B: X \to X$ is linear and continuous, then, for each $w \in X$ and each continuous f, the mild solution* (106) *is also a classical solution of* (105).

This theorem shows clearly the importance of semigroups for the solution of the initial value problem (105).

Corollary 19.40. *Assume* (H1). *Then*:

(i) *There exist constants $C \geq 1$ and $a \geq 0$ such that*

$$\|S(t)\| \leq Ce^{at} \qquad \text{for all} \quad t \geq 0. \tag{107}$$

(ii) *The generator $B: D(B) \subseteq X \to X$ of the semigroup $\{S(t)\}$ is a linear graph closed operator and $D(B)$ is dense in X.*

(iii) *The semigroup is uniquely determined by its generator.*

The proofs will be given in the next section.

In order to be able to apply Theorem 19.D to many concrete problems, we need to know properties of B which ensure assumption (H1).

STANDARD EXAMPLE 19.41. Let $B: D(B) \subseteq X \to X$ be a linear densely defined operator on the H-space X over $\mathbb{K} = \mathbb{R}, \mathbb{C}$. Then assumption (H1) above is satisfied if one of the following two conditions is valid:

(i) B is *maximal dissipative*, i.e.,

$$\operatorname{Re}(Bu|u) \leq 0 \qquad \text{for all} \quad u, v \in D(B)$$

and $R(I - \mu B) = X$ for some $\mu > 0$.

(ii) $-B$ is *sectorial* (cf. Definition 19.52).

We prove (i) in Section 19.20 (Theorem 19.E). The proof of (ii) can be found in Friedman (1969, M).

A process in physics is called dissipative iff the entropy is increasing (e.g., heat conduction or diffusion). Mathematically, such processes are frequently described by semigroups with maximal dissipative operators. This motivates the notion "dissipative operator." Maximal dissipative operators generate nonexpansive semigroups. Such semigroups are very important for the following reason. We replace u, w, f in (106) with u_i, w_i, f_i, respectively. Then $\|S(s)\| \leq 1$ for all $s \geq 0$ implies

$$\|u_1(t) - u_2(t)\| \leq \|w_1 - w_2\| + (t - t_0) \max_{t_0 \leq s \leq t} \|f_1(s) - f_2(s)\|,$$

for all $t \geq t_0$, i.e., if w_1 and f_1 are close to w_2 and f_2, respectively, then the solution u_1 is close to u_2.

EXAMPLE 19.42. Let $B: D(B) \subseteq X \to X$ be a densely defined linear operator on the H-space X over $\mathbb{K} = \mathbb{R}, \mathbb{C}$. Then assumption (H1) above is satisfied if one of the following four conditions is met:

(i) B is self-adjoint and $(Bu|u) \leq 0$ for all $u \in D(B)$.
(ii) $-B$ is the Friedrichs extension of a symmetric strongly monotone operator.
(iii) $B = -iH$, where H is self-adjoint and the H-space X is complex.
(iv) B is skew-adjoint.

In all these cases, B is maximal dissipative.

PROOF. Ad(i). All $\lambda > 0$ belong to the resolvent set of B (cf. Problem 19.7). Hence $R(\lambda I - B) = X$, i.e., B is maximal dissipative.

Ad(ii). This is a special case of (i).

Ad(iii). All $\lambda \in \mathbb{C}$ with $\operatorname{Im} \lambda \neq 0$ belong to the resolvent set of the self-adjoint operator H (cf. Problem 19.7). Thus, all real $\lambda \neq 0$ belong to the resolvent set of $B = -iH$. The operator B is skew-adjoint. Hence $\operatorname{Re}(Bu|u) = 0$, i.e., B is maximal dissipative. Note that $(Bu|u) = (u|-Bu)$.

Ad(iv). If X is complex, then iB is self-adjoint, i.e., (iv) is a special case of (iii). Note that $(iB)^* = -iB^* = iB$.

If X is real, then we extend B to the complexification of X. Here B remains skew-adjoint (cf. Problem 19.6). □

EXAMPLE 19.43. If assumption (H1) above is satisfied for the operator B, then it is also satisfied for all the operators $B + \lambda I$ with $\lambda \in \mathbb{R}$.

In particular, assumption (H1) holds for B in the case where $B - \lambda I$ is maximal dissipative for some $\lambda \in \mathbb{R}$.

PROOF. If B generates the semigroup $\{S(t)\}$, then $B + \lambda I$ generates the semigroup $\{e^{\lambda t} S(t)\}$. □

19.19. Proof of Theorem 19.D

After a time translation, if necessary, we can assume that $t_0 = 0$. In the following we will frequently make use of the following two simple results:

(i) Let X be a B-space. If (f_n) is a sequence of continuous functions $f_n: [a, b] \to X$ on the compact interval $[a, b]$ and (f_n) converges uniformly on $[a, b]$ to the function $f: [a, b] \to X$, then f is continuous and

$$\int_a^b f_n \, ds \to \int_a^b f \, ds \quad \text{as} \quad n \to \infty.$$

(ii) If $f: [a, b] \to X$ is continuous, then

$$\frac{d}{dt}\int_0^t f(s)\,ds = f(t) \qquad \text{for all} \quad t \in [a, b].$$

The proofs are the same as in the classical case $X = \mathbb{R}$.

Step 1: We show that

$$\sup_{0 \le t \le T} \|S(t)\| < \infty \qquad \text{for all} \quad T > 0. \tag{108}$$

By the semigroup property (83), it suffices to show that (108) is valid for a fixed T. If this is not true, then there exists a sequence $t_n \to +0$ as $n \to \infty$ with $\|S(t_n)\| \ge n$ for all n. Since $S(t_n)u \to u$ as $n \to \infty$ for each $u \in X$, we obtain $\sup_n \|S(t_n)\| < \infty$ by the *uniform boundedness theorem* $\text{A}_1(35)$. This is a contradiction.

Step 2: We prove (107). By (108), there is a constant C with $\|S(t)\| \le C$ on $[0, T]$. Since $S(0) = I$, $C \ge 1$. Let $a = T^{-1} \ln C$. Each $t \ge 0$ can be represented as $t = nT + s$ with $n = 0, 1, \ldots$ and $0 \le s < T$. Hence

$$\begin{aligned} \|S(t)\| &= \|S(nT)S(s)\| \le \|S(T)\|^n \|S(s)\| \\ &\le C^{n+1} \le CC^{t/T} = Ce^{at}. \end{aligned}$$

Step 3: Let $w \in D(B)$ and $t \ge 0$. We show that

$$\frac{d}{dt}S(t)w = S(t)Bw = BS(t)w. \tag{109}$$

This is the *key relation* of the theory of semigroups which generalizes the classical formula

$$\frac{d}{dt}e^{tB} = Be^{tB}$$

for real functions and for linear continuous operators $B: X \to X$ on a B-space X. In particular, it follows from (109) that $S(t)w \in D(B)$ for all $w \in D(B)$ and all $t \ge 0$.

(I) Letting $h \to +0$,

$$\frac{S(h) - I}{h}S(t)w = S(t)\frac{S(h) - I}{h}w \to S(t)Bw.$$

By definition of the generator B, we obtain that $S(t)w \in D(B)$ and $BS(t)w = S(t)Bw$ as well as

$$\frac{d^+}{dt}S(t)w = S(t)Bw.$$

Note that $S(t + h) = S(h)S(t)$.

(II) Let $t > 0$. In order to show that

$$\frac{d^-}{dt}S(t)w = S(t)Bw,$$

we use the identity

$$\frac{S(t)w - S(t-h)w}{h} - S(t)Bw$$
$$= S(t-h)\left(\frac{S(h)w - w}{h} - Bw\right) + S(t-h)Bw - S(t)Bw.$$

By (108) and by the definition of the generator B, the right-hand side goes to zero as $h \to +0$.

Step 4: Let $t \geq 0$. Integration of (109) yields

$$S(t)w - w = \int_0^t S(s)Bw\,ds = \int_0^t BS(s)w\,ds \tag{110}$$

for all $w \in D(B)$. Furthermore, for all $w \in X$,

$$S(t)w - w = B\int_0^t S(s)w\,ds. \tag{111}$$

This follows from the identity

$$\frac{S(h) - I}{h}\int_0^t S(s)w\,ds = \frac{1}{h}\int_0^t (S(s+h)w - S(s)w)\,ds$$
$$= \frac{1}{h}\int_t^{t+h} S(s)w\,ds - \frac{1}{h}\int_0^h S(s)w\,ds.$$

By letting $h \to +0$, the right-hand side goes to $S(t)w - S(0)w$ and $S(0) = I$.

Step 5: $D(B)$ is dense in X. Let $w \in X$ and $t > 0$. Then

$$w_t \stackrel{\text{def}}{=} \frac{1}{t}\int_0^t S(s)w\,ds \to w \qquad \text{as} \quad t \to +0,$$

and $w_t \in D(B)$ by (111).

Step 6: B is graph closed. Let $w_n \to w$ and $Bw_n \to z$ as $n \to \infty$. By (110),

$$S(t)w_n - w_n = \int_0^t S(s)Bw_n\,ds.$$

Letting $n \to \infty$ and noting (108) we get

$$S(t)w - w = \int_0^t S(s)z\,ds.$$

Hence $(d/dt)S(t)w = S(t)z$, i.e., $Bw = z$ by letting $t = 0$.

Step 7: Let $f\colon [0, T[\to X$ be continuous and let u be a classical solution of (105). We set

$$v(s) = S(t-s)u(s), \qquad 0 < s < t < T.$$

Differentiation yields

$$\begin{aligned} v'(s) &= -BS(t-s)u(s) + S(t-s)u'(s) \\ &= -BS(t-s)u(s) + S(t-s)(Bu(s) + f(s)) \\ &= S(t-s)f(s). \end{aligned}$$

Note that the product rule is valid because of (108). Integration yields

$$v(t-\varepsilon) - v(\varepsilon) = \int_{\varepsilon}^{t-\varepsilon} S(t-s)f(s)\,ds.$$

Letting $\varepsilon \to +0$,

$$v(t) - v(0) = \int_0^t S(t-s)f(s)\,ds.$$

Hence

$$u(t) = S(t)w + \int_0^t S(t-s)f(s)\,ds, \tag{112}$$

i.e., u is a mild solution of (105).

Consequently, the classical solution u of (105) is unique.

Step 8: Let B be the generator of a linear strongly continuous semigroup $\{T(t)\}$. The preceding arguments show that, for each $w \in D(B)$, the function $u(t) = T(t)w$ is a classical solution of the initial value problem (105) with $t_0 = 0$ and $f = 0$. Hence $T(t)w = S(t)w$ for all $t \geq 0$ and all $w \in D(B)$. Since $D(B)$ is dense in X, this implies $T(t) = S(t)$ for all $t \geq 0$, i.e., the generator B uniquely determines the corresponding semigroup.

Step 9: Let $f\colon [0, T[\to X$ be C^1. We show that u in (106) is a classical solution of (105) with $t_0 = 0$.

By (109) it suffices to show that the function

$$g(t) = \int_0^t S(t-s)f(s)\,ds$$

is a classical solution of (105) for $t_0 = 0$ and $w = 0$, i.e., $u(0) = 0$. Let $0 \leq t < T$. From the identity

$$\begin{aligned} \Delta \stackrel{\text{def}}{=} \frac{g(t+h) - g(t)}{h} &= \int_0^t S(s)\frac{f(t+h-s) - f(t-s)}{h}\,ds \\ &\quad + \frac{1}{h}\int_t^{t+h} S(s)f(t+h-s)\,ds \end{aligned}$$

and (108) we obtain that the derivative $g'(t)$ exists and that

$$g'(t) = \int_0^t S(s)f'(t-s)\,ds + S(t)f(0).$$

Furthermore, it follows from

$$\Delta = \frac{1}{h}\left(\int_0^{t+h} S(t+h-s)f(s)\,ds - \int_0^t S(t-s)f(s)\,ds\right)$$

that

$$\frac{S(h)-I}{h}\int_0^t S(t-s)f(s)\,ds = \Delta - \frac{1}{h}\int_t^{t+h} S(t+h-s)f(s)\,ds.$$

Letting $h \to +0$,

$$Bg(t) = B\int_0^t S(t-s)f(s)\,ds = g'(t) - f(t).$$

Step 10: Let $B: X \to X$ be linear and continuous and let $f: [0, T[\to X$ be continuous. By Standard Example 19.36, the operator B generates the semigroup $\{S(t)\}$ with $S(t) = e^{tB}$. Here the series

$$e^{tB} = \sum_{n=0}^{\infty} \frac{t^n}{n!} B^n$$

converges absolutely with respect to the operator norm for all $t \in \mathbb{R}$. As in the classical case $X = \mathbb{R}$, we obtain that $(d/dt)e^{tB} = Be^{tB}$ with respect to the operator norm and that u in (106) is a classical solution of (105).

The proof of Theorem 19.D and Corollary 19.40 is complete.

19.20. Monotone Operators and the Main Theorem on Linear Nonexpansive Semigroups

Recall that a linear nonexpansive semigroup on an H-space X is a semigroup $\{S(t)\}$ of linear continuous operators $S(t): X \to X$ with $\|S(t)\| \leq 1$ for all $t \geq 0$ and $t \mapsto S(t)w$ is continuous on $\mathbb{R}_+$ for each $w \in X$. If $\{S(t)\}$ is an arbitrary linear strongly continuous semigroup on X, then by Corollary 19.40, there are constants $C \geq 1$ and a so that $\{C^{-1}e^{-at}S(t)\}$ is a linear nonexpansive semigroup. Thus, in order to *characterize* the class of all linear strongly continuous semigroups, it is sufficient to characterize the class of all linear nonexpansive semigroups.

Definition 19.44. Let $B: D(B) \subseteq X \to X$ be a linear operator on the H-space X over $\mathbb{K} = \mathbb{R}, \mathbb{C}$.

(i) B is called *dissipative* iff

$$\operatorname{Re}(Bu|u) \leq 0 \qquad \text{for all} \quad u \in D(B).$$

(ii) B is called *maximal dissipative* iff it is dissipative and $R(I - \mu B) = X$ for some $\mu > 0$.
(iii) B is called *accretive* iff $(I + \mu B)$ is injective and $(I + \mu B)^{-1}$ is nonexpansive for all $\mu > 0$.
(iv) B is called *maximal accretive* iff it is accretive and $R(I + \mu B) = X$ for all $\mu > 0$.

Theorem 19.E (Hille (1948), Yosida (1948)). *Let $B: D(B) \subseteq X \to X$ be a linear operator on the* H-*space X over $\mathbb{K} = \mathbb{R}, \mathbb{C}$ with $\overline{D(B)} = X$. Then the following three conditions are mutually equivalent:*

(i) *B is the generator of a linear nonexpansive semigroup.*
(ii) *B is maximal dissipative.*
(iii) *$-B$ is maximal accretive.*

The proof will be given in Section 19.22. Let $\mathbb{K} = \mathbb{R}$. Then the linear operator $B: D(B) \subseteq X \to X$ is dissipative iff

$$(Bu - Bv | u - v) \leq 0 \qquad \text{for all} \quad u, v \in D(B),$$

i.e., $-B$ is monotone. In Section 31.2 we shall prove that B is maximal dissipative iff $-B$ is maximal monotone. This shows that there is a very *close connection* between the theory of *monotone operators* and the theory of *semigroups*. Later on we will generalize this connection to nonlinear operators.

Corollary 19.45. *The operator B is dissipative* iff *$-B$ is accretive.*

PROOF. For all $v \in D(B)$ and $\mu > 0$,

$$\|(I - \mu B)v\|^2 = \|v\|^2 - 2\,\mathrm{Re}(Bv|v) + \mu^2 \|Bv\|^2.$$

By definition, $-B$ is accretive iff $\|(I - \mu B)v\|^2 \geq \|v\|^2$ for all $v \in D(B)$, $\mu > 0$. □

19.21. The Main Theorem on One-Parameter Unitary Groups

Recall that a one-parameter unitary group $\{S(t)\}$ on the H-space X is a family of unitary operators $S(t): X \to X$ for all $t \in \mathbb{R}$ so that

$$S(t + s) = S(t)S(s) \qquad \text{for all} \quad t, s \in \mathbb{R}, \tag{113}$$

and $t \mapsto S(t)w$ is continuous on $\mathbb{R}$ for each $w \in X$.

Theorem 19.F. *Let $B: D(B) \subseteq X \to X$ be a linear operator on the* H-*space X over $\mathbb{K} = \mathbb{R}, \mathbb{C}$. Then the following three assertions are mutually equivalent:*

(i) *B is the generator of a one-parameter unitary group.*

(ii) *B and $-B$ are maximal dissipative and* $\overline{D(B)} = X$.
(iii) *B is skew-adjoint.*

Corollary 19.46 (Theorem of Stone (1932)). *The linear operator $B\colon D(B) \subseteq X \to X$ on the complex* H-*space X is the generator of a one-parameter unitary group* iff $B = -iH$, *where H is self-adjoint.*

Proof of Theorem 19.F. By Corollary 19.40, generators are always densely defined.

(i) $\Rightarrow$ (ii). Let B be the generator of the one-parameter unitary group $\{S(t)\}$. We set $S_\pm(t) = S(\pm t)$ for all $t \geq 0$. Then $\{S_+(t)\}$ and $\{S_-(t)\}$ is a nonexpansive semigroup with the generator B and $-B$, respectively. Now (ii) follows from Theorem 19.E.

(ii) $\Rightarrow$ (i). By Theorem 19.E, the operators B and $-B$ generate a nonexpansive semigroup $\{S_+(t)\}$ and $\{S_-(t)\}$, respectively. We set

$$S(t) = \begin{cases} S_+(t) & \text{for} \quad t \geq 0, \\ S_-(-t) & \text{for} \quad t \leq 0. \end{cases}$$

For all $w \in D(B)$ and $t \geq 0$,

$$\frac{d}{dt}(S_+(t)S_-(t)w - S_-(t)S_+(t)w)$$
$$= BS_-(t)w - S_+(t)Bw + BS_+(t)w - S_-(t)Bw = 0.$$

Hence

$$S_+(t)S_-(t) = S_-(t)S_+(t) = I \qquad \text{for all} \quad t \geq 0.$$

The semigroup property of $\{S_+(t)\}$ implies (113). From $\|S_\pm(t)\| \leq 1$ and $1 \leq \|S_+(t)\|\,\|S_-(t)\|$ it follows that $\|S_\pm(t)\| = 1$, i.e., $\{S(t)\}$ is a one-parameter unitary group. Obviously, B is the generator of $\{S(t)\}$.

(iii) $\Rightarrow$ (ii). Cf. Example 19.42.

(i) $\Rightarrow$ (iii). Suppose that B generates the unitary group $\{S(t)\}$. Differentiation of

$$(S(t)v|S(t)w) = (v|w) \qquad \text{for all} \quad v, w \in D(B)$$

yields $(Bv|w) + (v|Bw) = 0$, i.e., B is skew-symmetric; therefore, $-B \subseteq B^*$ where B^* denotes the adjoint operator.

We show that $B^* \subseteq -B$. Let $z \in D(B^*)$. According to (i) $\Rightarrow$ (ii), the operator $\pm B$ is maximal dissipative and hence $\mp B$ is maximal accretive by Theorem 19.E. In particular, $R(I \pm \mu B) = X$ for fixed $\mu > 0$. Thus there exists a $w \in D(B)$ such that

$$(I - \mu B)w = (I + \mu B^*)z.$$

Since $-Bw = B^*w$, $(I + \mu B^*)(z - w) = 0$. Thus, for all $y \in D(B)$,

$$((I + \mu B)y|z - w) = 0.$$

Hence $z - w = 0$. □

PROOF OF COROLLARY 19.46. From $(iB)^* = -iB^*$ it follows that B is skew-adjoint iff iB is self-adjoint. □

19.22. Proof of Theorem 19.E

Step 1: If B generates a linear nonexpansive semigroup $\{S(t)\}$, then $-B$ is maximal accretive.

To prove this let $\lambda > 0$. Then $\{e^{-\lambda t}S(t)\}$ is a nonexpansive semigroup whose generator is $B - \lambda I$. By (110) and (111),

$$e^{-\lambda t}S(t)w - w = \int_0^t e^{-\lambda s}S(s)(B - \lambda I)w\,ds \qquad \text{for all} \quad w \in D(B),$$

$$e^{-\lambda t}S(t)v - v = (B - \lambda I)\int_0^t e^{-\lambda s}S(s)v\,ds \qquad \text{for all} \quad v \in X.$$

The operator $B - \lambda I$ is graph closed, by Corollary 19.40. Thus, letting $t \to +\infty$, we find that

$$w = \int_0^\infty e^{-\lambda s}S(s)(\lambda I - B)w\,ds \qquad \text{for all} \quad w \in D(B),$$

$$v = (\lambda I - B)\int_0^\infty e^{-\lambda s}S(s)v\,ds \qquad \text{for all} \quad v \in X$$

(cf. Problem 19.4d).

Consequently, the operator $\lambda I - B$ is injective and surjective, and

$$\|(\lambda I - B)^{-1}v\| \le \|v\|\int_0^\infty e^{-\lambda s}\,ds = \lambda^{-1}\|v\|$$

for all $v \in X$, i.e., $-B$ is maximal accretive.

Step 2: If $-B$ is maximal accretive, then B generates a linear nonexpansive semigroup.

In order to prove this, we use the idea of the *Yosida approximation* B_μ. We set $R_\mu = (I - \mu B)^{-1}$. By assumption, the operator

$$R_\mu\colon X \to D(B)$$

is nonexpansive and bijective. We set

$$B_\mu = \mu^{-1}(R_\mu - I) \qquad \text{for all} \quad \mu > 0.$$

Then $B_\mu\colon X \to X$ is linear and continuous. Hence B_μ generates the semigroup

$$S_\mu(t) = e^{tB_\mu}.$$

Formally, $B_\mu \to B$ as $\mu \to +0$. Our *goal* is to prove that the following limit exists:

$$S(t)w \overset{\text{def}}{=} \lim_{\mu \to +0} S_\mu(t)w \qquad \text{for all} \quad w \in X, \quad t \ge 0. \tag{114}$$

(I) Since $R_\mu^{-1} B_\mu w = Bw$,

$$B_\mu w = R_\mu Bw \qquad \text{for all} \quad w \in D(B).$$

Because $(I - \mu B)R_\mu w = w$,

$$B_\mu w = BR_\mu w \qquad \text{for all} \quad w \in X.$$

Since $\|R_\mu\| \leq 1$,

$$\|B_\mu w\| \leq \|Bw\| \qquad \text{for all} \quad w \in D(B).$$

It follows from

$$\|R_\mu w - w\| = \|\mu B_\mu w\| \leq \mu \|Bw\|$$

that $R_\mu w \to w$ as $\mu \to +0$ for each $w \in D(B)$. Since $D(B)$ is dense in X and $\|R_\mu\| \leq 1$, we obtain $R_\mu w \to w$ as $\mu \to +0$ for each $w \in X$. Hence

$$B_\mu w = R_\mu Bw \to Bw \qquad \text{as} \quad \mu \to +0$$

for each $w \in D(B)$.

(II) Let $\alpha = \mu^{-1}$. For all $t \geq 0$,

$$\|S_\mu(t)\| = \|e^{-\alpha t} e^{\alpha R_\mu t}\| \leq e^{-\alpha t} e^{\|\alpha R_\mu\| t} \leq 1,$$

i.e., $\{S_\mu(t)\}$ is nonexpansive.

(III) For all $t \geq 0$ and all positive λ, μ, we get

$$\begin{aligned} S_\lambda(t) - S_\mu(t) &= \int_0^t \frac{d}{ds}(S_\mu(t-s)S_\lambda(s))\,ds \\ &= \int_0^t S_\mu(t-s)S_\lambda(s)(B_\lambda - B_\mu)\,ds, \end{aligned}$$

and hence, for all $w \in D(B)$,

$$\|S_\lambda(t)w - S_\mu(t)w\| \leq t\|B_\lambda w - B_\mu w\| \leq 2t\|Bw\|.$$

Consequently, for each $w \in D(B)$, the sequence $(S_\mu(t)w)$ converges as $\mu \to +0$ uniformly on compact t-intervals. Since $\|S_\mu(t)\| \leq 1$ and $D(B)$ is dense in X, the same is valid for each $w \in X$.

(IV) We now define the operator $S(t)\colon X \to X$ through the limit (114). Recall that $\{S_\mu(t)\}$ is a nonexpansive semigroup on X for each $\mu > 0$. Since

$$\|S_\mu(t)w\| \leq \|w\| \qquad \text{for all} \quad w \in X, \quad \mu > 0, \quad t \geq 0,$$

and since $(S_\mu(t)w)$ is uniformly convergent on compact t-intervals as $\mu \to +0$, the limit function $t \mapsto S(t)w$ is continuous on $\mathbb{R}_+$ for each $w \in X$, and $\{S(t)\}$ is a nonexpansive semigroup on X.

(V) Let C be the generator of $\{S(t)\}$. Let $w \in D(B)$. As $\mu \to +0$, the sequence $(S_\mu(t)B_\mu w)$ converges to $(S(t)Bw)$ uniformly on compact t-intervals. Thus, letting $\mu \to +0$, it follows from

$$S_\mu(t)w - w = \int_0^t S_\mu(s)B_\mu w\,ds,$$

by (110), that

$$S(t)w - w = \int_0^t S(s)Bw\,ds.$$

Differentiation at $t = 0$ shows that $w \in D(C)$ and $Cw = Bw$, i.e., $B \subseteq C$. Since $-B$ is maximal accretive, $I - B$ is surjective. By Step 1, $I - C$ is injective. Hence $B = C$.

Step 3: If $-B$ is maximal accretive, then B is maximal dissipative. This follows from Corollary 19.45.

Step 4: If B is maximal dissipative, then $-B$ is maximal accretive.

By Corollary 19.45, it suffices to prove that $R(I - \mu B) = X$ for all $\mu > 0$. By assumption, there is a $\mu_0 > 0$ such that $I - \mu_0 B\colon X \to X$ is bijective and

$$\|(I - \mu_0 B)^{-1}\| \leq 1.$$

By Problem 1.9, there exists the inverse operator $(I - \mu B)^{-1}$ on X for all μ with $|\mu - \mu_0| < 1$. A simple continuation argument shows that $(I - \mu B)^{-1}$ exists on X for all $\mu > 0$.

The proof of Theorem 19.E is complete.

19.23. Abstract Semilinear Hyperbolic Equations

We consider the initial value problem

$$\begin{gathered} u''(t) + Au(t) = f(u(t)), \qquad 0 < t < \infty, \\ u(0) = u_0, \qquad u'(0) = v_0. \end{gathered} \tag{115}$$

We make the following assumptions:

(H1) The linear operator $A\colon D(A) \subseteq X \to X$ is self-adjoint and strongly monotone on the H-space X over $\mathbb{K} = \mathbb{R}, \mathbb{C}$. Let X_E be the energetic space of A with the norm $\|\cdot\|_E$, i.e., X_E is the completion of $D(A)$ with respect to the energetic scalar product $(u|v)_E = (Au|v)$.

(H2) The operator $f\colon X_E \to X$ is locally Lipschitz continuous, i.e., for each $R > 0$ there is an L such that

$$\|f(u) - f(v)\| \leq L\|u - v\|_E$$

for all $u, v \in X_E$ with $\|u\|_E, \|v\|_E \leq R$.

We set $v = u'$. Instead of (115) we consider the first-order system

$$\begin{pmatrix} u' \\ v' \end{pmatrix} = \begin{pmatrix} 0 & I \\ -A & 0 \end{pmatrix} \begin{pmatrix} u \\ v \end{pmatrix} + \begin{pmatrix} 0 \\ f \end{pmatrix}. \tag{116}$$

Moreover, we set $z = (u, v)$ and write (116) in the form

$$\begin{aligned} z'(t) &= Cz(t) + F(z(t)), \qquad 0 < t < \infty, \\ z(0) &= z_0. \end{aligned} \tag{117}$$

Let $Z = X_E \times X$ and $D(C) = D(A) \times X_E$.

Proposition 19.47. *Assume* (H1). *Then the operator* $C: D(C) \subseteq Z \to Z$ *is skew-adjoint and generates a one-parameter unitary group* $\{S(t)\}$.

Corollary 19.48. *Assume* (H1) *and let* $f \equiv 0$, *i.e.,* $F \equiv 0$. *Then, for each* $z_0 \in D(C)$, *equation* (117) *has a unique classical solution, in the sense of Definition* 19.39, *given by*

$$u(t) = S(t)z_0.$$

For each $z_0 \in Z$, the function u is called a mild solution of equation (117) with $F \equiv 0$.

PROOF. We set $B = A^{1/2}$. By Problem 19.8b, $(u|\bar{u})_E = (Bu|B\bar{u})$ for all $u, \bar{u} \in X_E$. Hence, for all $z, \bar{z} \in Z$,

$$(z|\bar{z})_Z = (u|\bar{u})_E + (v|\bar{v}) = (Bu|B\bar{u}) + (v|\bar{v}),$$

where $(\cdot|\cdot)$ is the scalar product on X. Note that $Cz = (v, -Au)$. For all z, $\bar{z} \in D(C)$,

$$\begin{aligned} (z|C\bar{z})_Z &= (Bu|B\bar{v}) - (v|A\bar{u}) = (Au|\bar{v}) - (v|A\bar{u}), \\ (Cz|\bar{z})_Z &= (Bv|B\bar{u}) - (Au|\bar{v}) = (v|A\bar{u}) - (Au|\bar{v}). \end{aligned}$$

Hence $(z|C\bar{z})_Z = -(Cz|\bar{z})_Z$, i.e., C is skew-symmetric.

Furthermore, $R(I \pm C) = Z$. In fact, the equation $\bar{z} = z \pm Cz$ is equivalent to

$$\bar{u} = u \pm v, \qquad \bar{v} = v \mp Au,$$

and the latter equation has the solution $u = (I + A)^{-1}(\bar{u} \mp \bar{v})$, $v = \pm(\bar{u} - u)$. Consequently, both the operators C and $-C$ are maximal dissipative. Theorem 19.F yields the assertion.

In this connection note the following important fact. Since $(Au|u) \geq 0$ for all $u \in D(A)$, the resolvent $(I + A)^{-1}: X \to X$ exists, by Problem 19.7b. $\square$

Corollary 19.48 follows from Theorem 19.D.

In order to solve the original problem (115) with $f \not\equiv 0$ we consider the integral equation

$$z(t) = S(t)z_0 + \int_0^t S(t-s)F(z(s))\,ds. \tag{118}$$

The solutions of (118) are called *mild solutions* of (117) and of (115). Note that each sufficiently regular solution of (115) is a solution of (117) and hence a mild solution of (117), by Theorem 19.D. We set $Y = C([-T, T], Z)$ with the usual norm

$$\|z\|_Y = \max_{-T \le t \le T} \|z(t)\|_Z.$$

Theorem 19.G. *Suppose that* (H1), (H2) *hold. Then, for each* $z_0 = (u_0, v_0)$ *in* Z, *there are numbers* $T > 0$ *and* $r > 0$ *so that equation* (115) *has exactly one mild solution* $z \in Y$ *with* $\|z - z_0\|_Y \le r$.

PROOF. Let $\|\cdot\|$ denote the norm on Y and let

$$M = \{z \in Y: \|z - z_0\| \le r\}.$$

We write (118) in form of the following *fixed-point* equation:

$$z = Kz, \qquad z \in M. \tag{118*}$$

Noting (H2) and $\|S(t)\|_Z \le 1$ we obtain that

$$\begin{aligned} \|Kz - K\bar{z}\| &\le \max_{-T \le t \le T} \left| \int_0^t \|F(z(s)) - F(\bar{z}(s))\|_Z \, ds \right| \\ &\le TL(r)\|z - \bar{z}\| \qquad \text{for all} \quad z, \bar{z} \in M. \end{aligned}$$

This implies

$$\begin{aligned} \|Kz - z_0\| &\le \|Kz - Kz_0\| + \|Kz_0 - z_0\| \\ &\le TL(r)\|z - z_0\| + \|Kz_0 - z_0\| \qquad \text{for all} \quad z \in M. \end{aligned}$$

Hence we can choose r and T in such a way that $K(M) \subseteq M$ and $K: M \to M$ is k-contractive. Consequently, (118*) has a unique solution by the Banach fixed-point theorem (Theorem 1.A). □

19.24. Application to Semilinear Wave Equations

We consider the initial value problem

$$\begin{aligned} u_{tt} - \Delta u + m^2 u &= -\lambda u^3 && \text{on } G \times]0, T[, \\ u(x, 0) &= u_0(x) && \text{on } G, \\ u_t(x, 0) &= v_0(x) && \text{on } G, \\ u(x, t) &= 0 && \text{on } \partial G \times]0, T[. \end{aligned} \tag{119}$$

Here G is a bounded region in $\mathbb{R}^N$, $N \ge 1$. In the case where $\lambda = 0$, this is the so-called Klein–Gordon equation. Problem (119) describes the classical field u of a meson of mass m. The nonlinear term $-\lambda u^3$, with the so-called coupling constant $\lambda > 0$, describes the self-interaction of the field. This is only a classical model of a quantum field, since quantum fields correspond mathematically to distributions with values in an H-space (cf. Part V). In order to formulate (119) as an operator equation of the form

$$\begin{aligned} u'' + Au &= \lambda f(u) \quad \text{on }]0, T[, \\ u(0) &= u_0, \qquad u'(0) = v_0 \end{aligned} \tag{120}$$

with $f(u) = -u^3$, we set $X = L_2(G)$, $D(A_0) = C_0^\infty(G)$, and

$$A_0 u = (-\Delta + m^2)u.$$

For all $u \in D(A_0)$,

$$(A_0 u|u) = (-\Delta u|u) + m^2(u|u) \geq m^2(u|u).$$

Let $A\colon D(A) \subseteq X \to X$ denote the Friedrichs extension of A_0. Then the energetic spaces of A_0 and $-\Delta$ are the same up to an equivalent norm, i.e., the energetic space of A is equal to $X_E = \mathring{W}_2^1(G)$. In Problem 19.10b it will be proved that the operator $f\colon X_E \to X$ is locally Lipschitz continuous in the case where $N = 3$. Hence we obtain the following result.

Example 19.49. Let $m \geq 0$.

(a) Let $\lambda = 0$. Then, by Corollary 19.48, for each $u_0 \in D(A)$ and $v_0 \in X_E$, equation (120) has a unique classical solution. For each $u_0 \in X_E$, $v_0 \in X$, there exists a mild solution of (120).
(b) Let $\lambda > 0$ and $N = 3$. Then Theorem 19.G applies to (120), i.e., for each $u_0 \in X_E$ and $v_0 \in X$ there exists a unique local mild solution of (120).

The energy of (119) is defined by

$$E(t) = 2^{-1} \int_G \left[u_t^2 + m^2 u^2 + \sum_{i=1}^N (D_i u)^2 + 2^{-1}\lambda u^4 \right] dx.$$

In the case where $\lambda = m = 0$, this corresponds to the energy of the wave equation in Section 19.16. If u is a sufficiently smooth solution of (119), then integration by parts yields

$$E'(t) = \int_G u_t[u_{tt} - \Delta u + m^2 u + \lambda u^3]\, dx = 0,$$

i.e., $E(t) = \text{constant}$. This describes the conservation of energy.

19.25. The Semilinear Schrödinger Equation

We consider the Schrödinger equation

$$\begin{aligned} u'(t) &= -iHu(t) + F(u(t)) \quad \text{on }]-T, T[, \\ u(0) &= u_0. \end{aligned} \tag{121}$$

We make the following assumptions:

(H1) The linear operator $H: D(H) \subseteq X \to X$ is self-adjoint on the complex H-space X.
(H2) The operator $F: X \to X$ is locally Lipschitz continuous.

As we shall show in Part V, equation (121) with $F \equiv 0$ is the basic equation of quantum mechanics. The operator H corresponds to the energy of the quantum system.

Proposition 19.50. *Suppose that* (H1) *holds. Then $-iH$ is the generator of a one-parameter unitary group $\{S(t)\}$.*

Corollary 19.51. *Suppose that* (H1) *holds and let $F \equiv 0$. For each $u_0 \in D(H)$, equation* (121) *has a unique classical solution given by*

$$u(t) = S(t)u_0.$$

For each $u_0 \in X$, this is called a mild solution of (121).

PROOF. Cf. Theorems 19.D and 19.F. □

Instead of (121), we now consider the integral equation

$$u(t) = S(t)u_0 + \int_0^t S(t-s)F(u(s))\,ds. \tag{121*}$$

The solutions of (121*) are called *mild* solutions of (121). Let $Y = C([-T, T], X)$.

Theorem 19.H. *Suppose that* (H1), (H2) *hold. Then, for each $u_0 \in X$, there exist numbers $T > 0$ and $r > 0$ so that equation* (121) *has exactly one mild solution with $\|u - u_0\|_Y \leq r$.*

PROOF. Use exactly the same argument as in the proof of Theorem 19.G. □

19.26. Abstract Semilinear Parabolic Equations, Fractional Powers of Operators, and Abstract Sobolev Spaces

We consider the semilinear equation

$$\begin{aligned} u'(t) + Au(t) &= f(t, u(t)) \quad \text{on }]t_0, t_0 + T[, \\ u(t_0) &= w, \end{aligned} \tag{122}$$

and we make the following assumptions:

(H1) The linear operator $A: D(A) \subseteq X \to X$ is self-adjoint on the H-space X over $\mathbb{K} = \mathbb{R}, \mathbb{C}$, and A is semibouned, i.e., there is a real constant c such that

$$(Au|u) \geq c(u|u) \qquad \text{for all} \quad u \in D(A).$$

In the following, let $\|u\|$ denote the norm on X.

(H2) We choose a fixed real number a with $a + c > 0$.

Then the operator $B \stackrel{\text{def}}{=} A + aI$ is strongly monotone. Hence, in the sense of the general functional calculus for self-adjoint operators, the operator $B^\alpha: X \to X$ is self-adjoint for each $\alpha \geq 0$. We equip $X_\alpha = D(B^\alpha)$ with the graph norm

$$\|u\|_\alpha^* \stackrel{\text{def}}{=} \|u\| + \|B^\alpha u\|.$$

Then X_α becomes a B-space over $\mathbb{K}$ (cf. Problem 19.4a). In Lemma 19.55 below we show that

$$\|B^\alpha u\| \geq d\|u\| \qquad \text{for all} \quad u \in D(B^\alpha) \quad \text{and fixed} \quad d > 0.$$

Hence

$$\|u\|_\alpha \stackrel{\text{def}}{=} \|B^\alpha u\|$$

is an equivalent norm on X_α. For $\alpha = 0$, $X_\alpha = X$. The space X_α is called an *abstract Sobolev space*.

(H3) There is an $\alpha \in [0, 1[$ such that

$$\|f(t, u) - f(s, v)\| \leq L(|t - s|^\beta + \|u - v\|_\alpha)$$

for all (t, u), (s, v) in a neighborhood of the point (t_0, w) in $\mathbb{R} \times X_\alpha$. Here $L > 0$ and $0 < \beta \leq 1$ are constants and the initial value $w \in X_\alpha$ is given.

The decisive *advantage* of the spaces X_α is that we can treat more general nonlinearities than in X. In applications, the functions in X_α are smoother than those in X. A concrete simple example will be considered in the next section.

By Examples 19.42 and 19.43, the operator $-A$ generates a linear strongly continuous semigroup $\{S(t)\}$. Instead of the original problem (122) we consider the integral equation

$$u(t) = S(t - t_0)w + \int_{t_0}^{t} S(t - s)f(s, u(s))\, ds. \tag{123}$$

We set $Y = C([t_0, t_0 + T], X_\alpha)$ with the norm

$$\|u\|_Y = \max_{t_0 \leq t \leq t_0 + T} \|u\|_\alpha.$$

The solutions of (123) are called *mild solutions* of (122).

Theorem 19.I. *Suppose that* (H1) *through* (H3) *hold. Then there are positive numbers T and r such that the original problem* (122) *has exactly one mild solution $u \in Y$ with $\|u - w\|_Y \leq r$.*

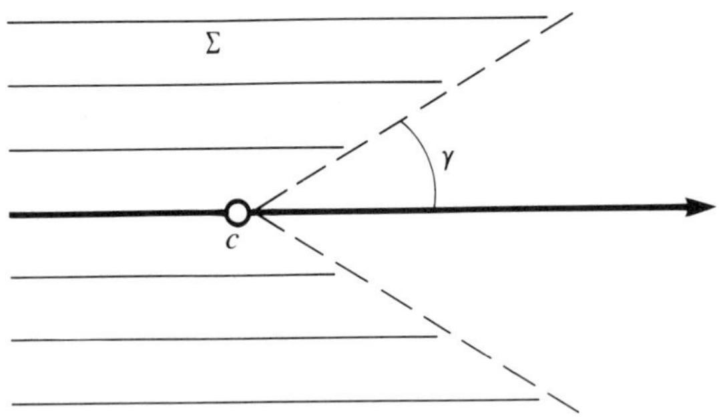

Figure 19.7

Definition 19.52. The operator $A: D(A) \subseteq X \to X$ on the complex B-space X is called *sectorial* iff the following are true:

(i) A is linear, graph closed, and $\overline{D(A)} = X$.
(ii) There are numbers $c \in \mathbb{R}$, $M \geq 1$, and $\gamma \in]0, \pi/2[$ such that the open sector

$$\Sigma = \{\lambda \in \mathbb{C}: \gamma < |\arg(\lambda - c)| \leq \pi, \lambda \neq c\}$$

is a subset of the resolvent set of A and

$$\|(\lambda I - A)^{-1}\| \leq M|\lambda - c|^{-1} \qquad \text{for all} \quad \lambda \in \Sigma$$

(cf. Fig. 19.7).

Corollary 19.53. *Theorem* 19.I *remains true if we replace* (H1) *by the more general assumption that the operator A is sectorial.*

The proof of Theorem 19.I will be given in Section 19.28. This proof is based on properties of the fractional powers B^α which we shall prove in Lemma 19.55. The proof of Corollary 19.53 proceeds completely analogously to the proof of Theorem 19.I by noting the following. If A is sectorial, then $-A$ generates an analytic semigroup and it is possible to define fractional powers B^α which have the same properties as in Lemma 19.55 below. The definition of B^α in the general case and its properties can be found in $A_2(110)$.

Remark 19.54 (Regularization). Using the properties of fractional powers, one can show that the mild solutions above are just classical solutions of (122). The proof for sectorial operators can be found in Henry (1981, L). The technique of proof is the same as in the theory of singular integral equations.

Theorem 19.I and Corollary 19.53 allow important applications to general classes of parabolic equations and parabolic systems, since such problems are related to sectorial operators. This will be discussed in $A_2(110c)$.

19.27. Application to Semilinear Parabolic Equations

We consider the initial–boundary value problem

$$\begin{aligned} u_t - \Delta u &= \lambda u u_x \quad \text{on } G \times]0, T[, \\ u(x, t) &= 0 \quad \text{on } \partial G \times [0, T[, \\ u(x, 0) &= w(x) \quad \text{on } G, \end{aligned} \tag{124}$$

with a constant λ. Here G is a bounded open interval in $\mathbb{R}$, i.e., $\Delta u = u_{xx}$. Let $X = L_2(G)$ and let A be the Friedrichs extension of $-\Delta$ with $D(-\Delta) = C_0^\infty(G)$. By Problem 19.8b,

$$X_{1/2} = X_E = \mathring{W}_2^1(G),$$

i.e., $X_{1/2}$ is the energetic space of $-\Delta$. We set $f(u) = \lambda u u_x$. By Problem 19.10d, the mapping

$$f: X_E \to X$$

is locally Lipschitz continuous.

Therefore, we can apply Theorem 19.I to (124). This way, for each $w \in \mathring{W}_2^1(G)$, we obtain a mild solution $u: [0, T] \to \mathring{W}_2^1(G)$ for (124) on a sufficiently small time interval, and this solution is unique in the sense of Theorem 19.I.

19.28. Proof of Theorem 19.I

We set $t_0 = 0$. Furthermore, we make the following additional assumption:

(H2*) We have $c > 0$ and $a = 0$, i.e., $B = A$ and $\|u\|_\alpha = \|A^\alpha u\|$.

We free ourselves from (H2*) in Step 3 below.

Step 1: Properties of fractional powers.

In what follows we will use the general functional calculus for self-adjoint operators summarized in the Appendix, $A_2(96)$. The reader should convince himself that the following proof becomes completely elementary if, in addition, we assume that the operator A possesses a complete orthonormal system of eigenvectors. In this case we can use the special functional calculus for self-adjoint operators introduced in Section 19.14. Finally, the reader should also convince himself that the following proof can be immediately translated to a proof of Corollary 19.53, if we assume that the operator A is sectorial. In this case we have to use the calculus of fractional powers for sectorial operators which we summarize in the Appendix, $A_2(110)$.

Lemma 19.55. *There is a constant* $C > 0$ depending on $\alpha \in [0, 1[$ *such that the following are true*:

(i) $\|u\| \leq C\|A^\alpha u\|$ *for all* $u \in D(A^\alpha)$, *i.e., the embedding* $X_\alpha \subseteq X$ *is continuous.*
(ii) $\|(e^{-tA} - I)u\| \leq Ct^\alpha \|A^\alpha u\|$ *for all* $u \in D(A^\alpha)$, $t > 0$.
(iii) $\|A^\alpha e^{-tA}\| \leq Ct^{-\alpha}$ *and* $e^{-tA}A^\alpha \subseteq A^\alpha e^{-tA}$ *for all* $t > 0$.

PROOF. Let $J = [c, \infty[$ and $t \geq 0$. Then

$$A^\alpha u = \int_J \lambda^\alpha \, dE_\lambda u,$$

where $u \in D(A^\alpha)$ iff $\int_J \lambda^{2\alpha} \, d\|E_\lambda u\|^2 < \infty$. Furthermore,

$$e^{-tA}u = \int_J e^{-\lambda t} \, dE_\lambda u,$$

where $u \in D(e^{-tA})$ iff $\int_J e^{-2\lambda t} \, d\|E_\lambda u\|^2 < \infty$. In particular,

$$\begin{aligned} \|e^{-tA}u\|^2 &\leq \int_J |e^{-\lambda t}|^2 \, d\|E_\lambda u\|^2 \\ &\leq \int_J d\|E_\lambda u\|^2 = \|u\|^2. \end{aligned}$$

It follows analogously to Section 19.15 that $\{e^{-tA}\}$ is a semigroup generated by $-A$ (cf. Problem 19.13).

Ad(i). For $u \in D(A^\alpha)$,

$$\begin{aligned} \|A^\alpha u\|^2 &= \int_J \lambda^{2\alpha} \, d\|E_\lambda u\|^2 \geq \int_J c^{2\alpha} \, d\|E_\lambda u\|^2 \\ &= c^{2\alpha}\|u\|^2. \end{aligned}$$

Ad(ii). The real function $x \mapsto (e^{-x} - 1)/x^\alpha$ is bounded on $]0, \infty[$. Thus, for all $u \in D(A^\alpha)$,

$$\begin{aligned} \|(e^{-tA} - I)u\|^2 &= \int_J |e^{-t\lambda} - 1|^2 \, d\|E_\lambda u\|^2 \\ &\leq \int_J Ct^{2\alpha}\lambda^{2\alpha} \, d\|E_\lambda u\|^2 = Ct^{2\alpha}\|A^\alpha u\|^2. \end{aligned}$$

Ad(iii). Let $\psi(\lambda) = \lambda^\alpha e^{-t\lambda}$ for $\lambda \geq 0$. The real function $x \mapsto x^\alpha e^{-x}$ is bounded on $\mathbb{R}_+$. Thus, for all $u \in X$, $t > 0$,

$$\begin{aligned} \|\psi(A)u\|^2 &= \int_J |\psi(\lambda)|^2 \, d\|E_\lambda u\|^2 \\ &\leq \int_J Ct^{-2\alpha} \, d\|E_\lambda u\|^2 = Ct^{-2\alpha}\|u\|^2. \end{aligned}$$

By A_2(96v), $\psi(A) = A^\alpha e^{-tA}$ and $e^{-tA}A^\alpha \subseteq \psi(A)$. □

Step 2: The fixed-point theorem of Banach.

Let $w \in D(A^\alpha)$ and $M = \{u \in Y: \|u - w\|_Y\} \le r$. We write the integral equation

$$u(t) = e^{-tA}w + \int_0^t e^{-(t-s)A} f(s, u(s))\, ds$$

in the form of the fixed-point equation

$$u = Ku, \qquad u \in M, \tag{125}$$

i.e.,

$$(Ku)(t) = e^{-tA}w + \int_0^t e^{-(t-s)A} f(s, u(s))\, ds.$$

By the Lipschitz condition (H3),

$$\|(Ku)(t) - (Kv)(t)\|_\alpha \le L \int_0^t \|A^\alpha e^{-(t-s)A}\|\, ds\, \|u - v\|_Y,$$

and, for all $t \in T$,

$$\begin{aligned}\|(Ku)(t) - w\|_\alpha &\le \|(Ku)(t) - Kw\|_\alpha + \|Kw - w\|_\alpha \\ &\le \|A^\alpha(e^{-tA} - I)w\| + \int_0^t \|A^\alpha e^{-(t-s)A}\|\, ds (L\|u - w\|_Y + F),\end{aligned}$$

where $\|f(t, w)\| \le F$ on $[0, T]$.

By Lemma 19.55, there exist numbers $r > 0$ and $T > 0$ such that $K(M) \subseteq M$ and the operator $K: M \to M$ is k-contractive. In this connection, note that

$$\int_0^t \|A^\alpha e^{-(t-s)A}\|\, ds \le \int_0^t \frac{C}{(t-s)^\alpha}\, ds = \frac{Ct^{1-\alpha}}{1-\alpha}.$$

Furthermore, note that $w \in D(A^\alpha)$ and hence

$$A^\alpha(e^{-tA} - I)w = e^{-tA}A^\alpha w - A^\alpha w \to 0$$

as $t \to +0$, since the semigroup $\{e^{-tA}\}$ is strongly continuous.

Now the Banach fixed-point theorem (Theorem 1.A) yields the existence of a unique solution of (125).

Step 3: We set $u(t) = e^{at}v(t)$. Then the original equation

$$u'(t) + Au(t) = f(t, u(t))$$

is transformed into

$$v'(t) + (A + aI)v(t) = e^{-at} f(t, e^{at}v(t)).$$

Condition (H2*) is satisfied for the latter equation. Hence, without loss of generality, we can assume that (H2*) holds.

The proof of Theorem 19.I is complete.

19.29. Five General Uniqueness Principles and Monotone Operators

We consider the equation

$$\text{(E)} \qquad Au = f, \qquad u \in X.$$

Theorem 19.J. *Let $A: D(A) \subseteq X \to X$ be a linear operator on the* H-*space X over $\mathbb{K} = \mathbb{R}, \mathbb{C}$. Then, for each $f \in X$, equation* (E) *has at most one solution in case one of the following five conditions is satisfied*:

(i) Strict monotonicity (positivity):

$$\operatorname{Re}(Au|u) > 0 \qquad \text{for all} \quad u \in D(A) \quad \text{with} \quad u \neq 0.$$

(ii) A priori estimate (stability):

$$c\|u\| \leq \|Au\| \qquad \text{for all} \quad u \in D(A) \quad \text{and fixed} \quad c > 0.$$

(iii) Contractivity of $A - I$:

$$\|Au - u\| < \|u\| \qquad \text{for all} \quad u \in D(A) \quad \text{with} \quad u \neq 0.$$

(iv) Monotone type: *With respect to an order cone on X,*

$$Au \leq Av \qquad \text{implies} \qquad u \leq v.$$

(v) Duality: *There exists an operator $B: D(B) \subseteq X \to X$ with $\overline{R(B)} = X$ and*

$$(Au|v) = (u|Bv) \qquad \text{for all} \quad u \in D(A), \quad v \in D(B).$$

Proof. We have to show that $Au = 0$ implies $u = 0$. This is obviously satisfied in (i), (ii), and (iii).

In (iv), $Au = 0$ implies that $Au \leq 0$ and $Au \geq 0$. Hence $u \leq 0$ and $u \geq 0$, i.e., $u = 0$.

In (v), $Au = 0$ implies that $(u|w) = 0$ for all w in the dense set $R(B)$. Hence $u = 0$. □

This theorem summarizes important uniqueness principles which allow many diversified applications to concrete problems.

The following result tells us in which sense condition (v) is necessary and sufficient for uniqueness. Here A^* denotes the adjoint operator.

Corollary 19.56. *Let $A: D(A) \subseteq X \to X$ be a linear operator on the* H-*space X over $\mathbb{K}$, where $D(A)$ and $D(A^*)$ are dense in X. Then the following two statements are equivalent*:

(i) *For each $f \in X$, equation* (E) *has at most one solution.*
(ii) *$R(A^*)$ is dense in X.*

Proof. (i) $\Rightarrow$ (ii). If $R(A^*)$ is not dense in X, then there exists a $u \neq 0$ in $\overline{R(A^*)}^{\perp}$ by the *perpendicular principle* in Section 18.11. Since $D(A)$ is dense in X, we may assume that $u \in D(A)$. Hence

$$(Au|v) = (u|A^*v) = 0 \qquad \text{for all} \quad v \in D(A^*).$$

Since $D(A^*)$ is dense in X, we obtain $Au = 0$. This contradicts $u \neq 0$.

(ii) $\Rightarrow$ (i). Use Theorem 19.J(v) with $B = A^*$. □

In the following we generalize Theorem 19.J to nonlinear operators.

Proposition 19.57 (Strict Monotonicity and Uniqueness). *Equation* (E) *has at most one solution in the case where one of the following two conditions is valid*:

(i) *Let X be an* H-*space over* $\mathbb{K}$. *The operator* $A: D(A) \subseteq X \to X$ *is strictly monotone, i.e.*,

$$\operatorname{Re}(Au - Av|u - v) > 0 \qquad \text{for all} \quad u, v \in D(A) \quad \textit{with} \quad u \neq v.$$

(ii) *Let X be a* B-*space over* $\mathbb{K}$. *The operator* $A: D(A) \subseteq X \to X^*$ *is strictly monotone, i.e.*,

$$\operatorname{Re}\langle Au - Av, u - v\rangle > 0 \qquad \textit{for all} \quad u, v \in D(A) \quad \textit{with} \quad u \neq v.$$

PROOF. $Au = Av$ implies $u = v$. □

Proposition 19.58. *Let $A: D(A) \subseteq X \to Y$ be an operator, where X and Y are* B-*spaces. Then, for each $f \in Y$, equation* (E) *has at most one solution in the case where one of the following three conditions is satisfied*:

(i) Stability:

$$c\|u - v\| \leq \|Au - Av\| \qquad \textit{for all} \quad u, v \in D(A) \quad \textit{and fixed} \quad c > 0.$$

(ii) Contractivity of $A - I$:

$$\|(Au - u) - (Av - v)\| < \|u - v\| \qquad \textit{for all} \quad u, v \in D(A) \quad \textit{with} \quad u \neq v.$$

(iii) Monotone type: *With respect to order cones on X and Y*,

$$Au \leq Av \qquad \textit{implies} \qquad u \leq v.$$

PROOF. Let $Au = Av$. Then $u = v$ by (i) or (ii).

In (iii), $Au = Av$ implies $Au \leq Av$ and $Au \geq Av$. Hence $u \leq v$ and $u \geq v$, i.e., $u = v$. □

19.30. A General Existence Principle and Linear Monotone Operators

The equation

$$\text{(E)} \qquad Au = f, \qquad u \in X$$

has a solution for each $f \in Y$ iff $R(A) = Y$, i.e., A is surjective.

Theorem 19.K. *Let $A: D(A) \subseteq X \to Y$ be a linear operator where X and Y are* B-*spaces over* $\mathbb{K} = \mathbb{R}, \mathbb{C}$. *Then, we have $R(A) = Y$* iff *the following two conditions are satisfied*:

(i) *$R(A)$ is closed.*
(ii) $R(A)^\perp = \{0\}$.

Corollary 19.59 (*A priori* Estimates and Existence). *Let $A: D(A) \subseteq X \to Y$ be a linear graph closed operator where X and Y are* B-*spaces over* $\mathbb{K}$. *Suppose that*:

(i) A priori estimate. *There is a $c > 0$ such that*

$$c\|u\| \leq \|Au\| \qquad \text{for all} \quad u \in D(A).$$

(ii) $R(A)^\perp = \{0\}$.

Then, for each $f \in Y$, equation (E) *has a unique solution.*

Let Y be a B-space (resp. H-space). Then, by definition, $R(A)^\perp = \{0\}$ means that

$$\langle v, Au \rangle = 0 \qquad \text{for all} \quad u \in D(A) \qquad \text{implies} \qquad v = 0$$

(resp. $(v|Au) = 0$ for all $u \in D(A)$ implies $v = 0$). If $\overline{D(A)} = Y$, then there exists the dual operator A^* (resp. adjoint operator A^*). In this case we have

$$R(A)^\perp = \{0\} \qquad \text{iff} \qquad N(A^*) = \{0\}.$$

Proof of Theorem 19.K. Obviously, $R(A) = Y$ implies (i), (ii). Conversely, suppose that (i) and (ii) hold and $R(A) \neq Y$. Then $R(A)$ is a linear closed subspace of Y.

(I) Let Y be an H-space. By the *perpendicular principle* in Section 18.11, there is a $v \neq 0$ in $R(A)^\perp$. This contradicts $R(A)^\perp = \{0\}$.
(II) Let Y be a B-space. Now we use the separation theorems for convex sets in Section 39.1 which follow from the Hahn–Banach theorem. To this end we choose an element $y \in Y$ with $y \notin R(A)$. Then $y \neq 0$. The compact set $\{y\}$ and the closed convex set $R(A)$ can be strictly separated by a closed hyperplane, i.e., there exists a $v \in Y^*$ with $\langle v, y \rangle > 0$ and $\langle v, Au \rangle \leq 0$ for all $u \in D(A)$. Hence $\langle v, Au \rangle = 0$ for all $u \in D(A)$. This contradicts $R(A)^\perp = \{0\}$. □

This proof makes use of the following general principle:

Passing from H-*spaces to* B-*spaces, replace the perpendicular principle by the separation of convex sets (the Hahn–Banach theorem).*

Recall that we have shown in Section 18.11 that the following three principles are mutually equivalent in H-spaces: the perpendicular principle, the Riesz theorem, and the existence principle for quadratic minimum problems. This underlines the importance of the Hahn–Banach theorem as a *general existence principle* in functional analysis.

PROOF OF COROLLARY 19.59. The *a priori* estimate yields the closedness of the range $R(A)$. Hence Corollary 19.59 is a special case of Theorem 19.K. In fact, let (f_n) be a sequence in $R(A)$ with $f_n \to f$ as $n \to \infty$. Then $f_n = Au_n$ and

$$c\|u_n - u_m\| \le \|Au_n - Au_m\| \qquad \text{for all } n, m.$$

Hence $u_n \to u$ as $n \to \infty$. Since A is graph closed, $Au = f$. □

Remark 19.60 (Application to Linear Monotone Operators). We want to show that Theorem 19.K generalizes our results on linear monotone operators in Section 18.11e. Let X be an H-space over $\mathbb{K} = \mathbb{R}, \mathbb{C}$.

(i) We first consider the equation

$$Bu = f, \qquad u \in X, \tag{126}$$

where the operator $B: X \to X$ is linear, continuous, and strongly monotone, i.e., there is a $c > 0$ such that

$$\operatorname{Re}(Bu|u) \ge c\|u\|^2 \qquad \text{for all} \quad u \in X.$$

This implies $c\|u\|^2 \le |(Bu|u)| \le \|Bu\|\,\|u\|$ and hence

$$c\|u\| \le \|Bu\| \qquad \text{for all} \quad u \in X.$$

Moreover, $R(B)^\perp = \{0\}$. Indeed, $(v|Bu) = 0$ for all $u \in X$ implies

$$c\|v\|^2 \le \operatorname{Re}(Bv|v) = 0,$$

i.e., $v = 0$.

By Corollary 19.59, equation (126) has a *unique solution* for each $f \in X$.

(ii) Next we consider the equation

$$Au = b, \qquad u \in X, \tag{126*}$$

where the operator $A: X \to X^*$ is linear, continuous, and strongly monotone, i.e., there is a $c > 0$ such that

$$\operatorname{Re}\langle Au, u\rangle \ge c\|u\|^2 \qquad \text{for all} \quad u \in X.$$

This implies $c\|u\|^2 \le |\langle Au, u\rangle| \le \|Au\|\,\|u\|$ and hence

$$c\|u\| \le \|Au\| \qquad \text{for all} \quad u \in X.$$

Furthermore, $R(A)^\perp = \{0\}$. Indeed, $\langle v, Au\rangle = 0$ for all $u \in X$ and fixed $v \in X^{**}$ implies

$$c\|v\|^2 \le \operatorname{Re}\langle Av, v\rangle = \operatorname{Re}\langle v, Av\rangle = 0,$$

i.e., $v = 0$. In this connection, note that each H-space X is reflexive. Thus, each element $v \in X^{**}$ can be regarded as an element in X with

$$\langle v, w \rangle = \langle w, v \rangle \qquad \text{for all} \quad v \in X, \quad w \in X^*.$$

In particular, $\langle Av, v \rangle = \langle v, Av \rangle$ (cf. Sections 21.5 and 21.6).

By Corollary 19.59, equation (126*) has a *unique solution* for each $b \in X^*$.

Problems

19.1. *Mixed boundary value problem.* Formulate the generalized problem for

$$\begin{aligned} -\Delta u &= f \quad \text{on } G, \\ u &= g \quad \text{on } \partial_1 G, \\ \frac{\partial u}{\partial n} + au &= h \quad \text{on } \partial_2 G, \end{aligned} \tag{127}$$

and find the corresponding Galerkin equations. In this connection, let $\partial_1 G$ and $\partial_2 G$ be disjoint sufficiently regular subsets of ∂G with $\partial G = \partial_1 G \cup \partial_2 G$.

Solution: The generalized problem reads as follows:

$$\int_G \left(\sum_{i=1}^N D_i u D_i v - fv \right) dx + \int_{\partial_2 G} (au - h) v \, dO = 0 \tag{128}$$

for all $v \in M$ and

$$u = g \quad \text{on } \partial_1 G,$$

where

$$M = \{ v \in C^1(\bar{G}) \colon v = 0 \text{ on } \partial_1 G \}.$$

(I) (127) $\Rightarrow$ (128). From (127), by multiplication by v and subsequent integration by parts, it follows that

$$\int_G \left(\sum_{i=1}^N D_i u D_i v - fv \right) dx - \int_{\partial G} \frac{\partial u}{\partial n} v \, dO = 0. \tag{129}$$

This yields (128), since $v = 0$ on $\partial_1 G$.

(II) (128) $\Rightarrow$ (127). Let $u \in C^2(\bar{G})$ be a solution of (128). Then integration by parts yields

$$\int_G (-\Delta u - f) v \, dx + \int_{\partial G} \left(au - h + \frac{\partial u}{\partial n} \right) v \, dO = 0$$

for all $v \in M$. This implies (127).

We suppose that we can extend the boundary function g to the set $\bar{G}$. Then the *Galerkin equations*, which follow from (128), read as follows:

$$\int_G \left(\sum_{i=1}^N D_i u D_i v_j - f v_j \right) dx + \int_{\partial_2 G} (au - h) v_j \, dO = 0, \qquad j = 1, \ldots, n,$$

where $u = \sum_{j=1}^n c_j v_j + g$ and $v_j \in M$ for all j.

19.2. *Variational problem.* Show that (127) is the Euler equation to a variational problem.

Solution: We consider

$$\int_G \left(\frac{1}{2} \sum_{i=1}^{N} (D_i u)^2 - fu \right) dx + \int_{\partial_2 G} (\tfrac{1}{2} au^2 - hu)\, dO = \min!,$$
$$u = g \quad \text{on } \partial_1 G. \tag{130}$$

Let $u \in C^2(\bar{G})$ be a solution of (130). Denote the left-hand side of (130) by $J(u)$ and set $\varphi(t) = J(u + tv)$ where $v \in M$. Then $\varphi'(0) = 0$. This is precisely (128), which implies (127).

19.3. *Symmetric operators.* Let $B: D(B) \subseteq X \to X$ be a linear symmetric operator on the H-space X over $\mathbb{K} = \mathbb{R}, \mathbb{C}$.

19.3a. Show that all eigenvalues of B are real.

Solution: From $Bu = \lambda u$ and $(u|u) = 1$ it follows that

$$\lambda = (u|Bu) = (Bu|u) = \bar{\lambda}.$$

19.3b. Show that two eigenvectors of B with different eigenvalues are orthogonal.

Solution: From $Bu = \lambda u$ and $Bv = \mu v$ with $\lambda \neq \mu$ it follows that

$$(\lambda - \mu)(u|v) = (Bu|v) - (u|Bv) = 0,$$

and hence $(u|v) = 0$.

19.3c. Let $\{u_n\}$ be a *complete* orthonormal system of eigenvectors of B. Show that the corresponding system $\{\lambda_n\}$ of eigenvalues contains *all* the eigenvalues of B.

Solution: Let $Bu = \lambda u$ with $u \neq 0$ and $\lambda \neq \lambda_n$ for all n. By (26),

$$u = \sum_n (u_n|u)u_n.$$

By Problem 19.3b, $(u_n|u) = 0$ for all n. Hence $u = 0$. This is a contradiction.

19.3d. Show that if $D(B) = X$ and B is continuous, then

$$\beta \stackrel{\text{def}}{=} \sup_{\|u\|=1} |(Bu|u)| = \|B\|.$$

Here assume $X \neq \{0\}$.

Solution: From $|(Bu|u)| \leq \|B\| \|u\|^2$ it follows that $\beta \leq \|B\|$. Conversely, we show that $\|B\| \leq \beta$. To this end, we set

$$v_\pm = \lambda u \pm \lambda^{-1} Bu, \qquad \lambda > 0.$$

It follows from $(B^2 u|u) = (Bu|Bu)$ that

$$\begin{aligned}\|Bu\|^2 &= 4^{-1}[(Bv_+|v_+) - (Bv_-|v_-)] \\ &\leq 4^{-1}\beta(\|v_+\|^2 + \|v_-\|^2) \\ &= 2^{-1}\beta(\lambda^2\|u\|^2 + \lambda^{-2}\|Bu\|^2).\end{aligned}$$

Letting $\lambda^2 = \|Bu\|$ and $\|u\| = 1$, we find that $\|Bu\|^2 \leq \beta\|Bu\|$. Hence $\|B\| \leq \beta$.

19.4. *Graph closed operators.* The operator $A: D(A) \subseteq X \to Y$, where X and Y are B-spaces, is called *graph closed* iff the graph

$$G(A) = \{(u, Au): u \in D(A)\}$$

is closed in $X \times Y$, i.e., for each sequence (u_n) in $D(A)$, it follows from

$$u_n \to u \quad \text{in } X \qquad \text{as} \quad n \to \infty,$$
$$Au_n \to v \quad \text{in } Y \qquad \text{as} \quad n \to \infty,$$

that $u \in D(A)$ and $Au = v$. This is an important generalization of the notion of continuity.

Let $A: D(A) \subseteq X \to X$ be a linear operator. We define the *graph norm*

$$\|u\|_A = \|u\| + \|Au\| \qquad \text{for all} \quad u \in D(A).$$

The linear operator $B: D(B) \subseteq X \to X$ with $A \subseteq B$ is called the closure of A iff

$$\overline{G(A)} = G(B).$$

We write $B = \bar{A}$. If the closure $\bar{A}$ exists, then it is unique. The linear operator A is called *essentially self-adjoint* iff the closure $\bar{A}$ is self-adjoint. This is a fundamental notion in quantum theory.

19.4a. Show that the linear operator $A: D(A) \subseteq X \to X$ on the B-space X is graph closed iff $D(A)$ is a B-space with respect to the graph norm.

Solution: Let A be graph closed and let (u_n) be a sequence in $D(A)$ which is a Cauchy sequence with respect to the graph norm $\|\cdot\|_A$. This implies the convergence $u_n \to u$ as $n \to \infty$ and $Au_n \to v$ as $n \to \infty$. Hence $Au = v$, i.e., $D(A)$ is a B-space with respect to $\|\cdot\|_A$.

The converse follows analogously.

19.4b. Let X be an H-space. We set

$$(u|v)_A = (u|v) + (Au|Av) \qquad \text{for all} \quad u, v \in D(A).$$

Show that the linear operator $A: D(A) \subseteq X \to X$ is graph closed iff $D(A)$ is an H-space with respect to the *graph scalar product* $(\cdot|\cdot)_A$.

Solution: This follows as in Problem 19.4a.

19.4c. Let $A: D(A) \subseteq X \to X$ a linear operator on the B-space X. Show that if A is injective and graph closed, then A^{-1} is also graph closed.

Solution: $G(A) = G(A^{-1})$.

19.4d. Our goals are the following two important formulas:

$$\frac{d}{dt}(Au(t)) = A\left(\frac{du(t)}{dt}\right), \tag{131}$$

$$A \int_a^b u(t)\,dt = \int_a^b Au(t)\,dt. \tag{132}$$

Let $A: D(A) \subseteq X \to X$ be a linear *graph closed* operator on the B-space X. Show that:

(i) If the functions $u: U(t) \subseteq \mathbb{R} \to D(A)$ and $s \mapsto Au(s)$ are differentiable at t, then (131) holds.

(ii) If the functions $u: [a, b] \to D(A)$ and $Au(\cdot): [a, b] \to X$ are continuous on the compact interval $[a, b]$, then (132) holds.

Solution: Let $w_h(t) = (u(t + h) - u(t))/h$. Then

$$Aw_h(t) = \frac{Au(t + h) - Au(t)}{h}.$$

Letting $h \to 0$, we have $w_h(t) \to u'(t)$ and the right-hand side goes to $(Au(t))'$. Since A is graph closed, we obtain (131).

Formula (132) follows analogously by using the fact that the integral is the limit of sums.

19.5. *Adjoint operators.* Let X be an H-space over $\mathbb{K} = \mathbb{R}, \mathbb{C}$. We consider two linear densely defined operators $A: D(A) \subseteq X \to X$ and $B: D(B) \subseteq X \to X$. Let A^* and B^* denote the corresponding adjoint operators.

19.5a. Show that A^* is graph closed.

Solution: Let $A^*u_n \to v$ and $u_n \to u$ as $n \to \infty$. It follows from

$$(w|A^*u_n) = (Aw|u_n) \qquad \text{for all} \quad w \in D(A)$$

that $(w|v) = (Aw|u)$ for all $w \in D(A)$, and hence $A^*u = v$.

19.5b. If $A \subseteq B$, then $B^* \subseteq A^*$. Furthermore, A is symmetric iff $A \subseteq A^*$ and A is skew-symmetric iff $A \subseteq -A^*$.

Solution: Use the corresponding definitions in Section 19.8.

In particular, we obtain that if A is symmetric, then so is A^{**}. In fact, $A \subseteq A^*$ implies $A^{**} \subseteq A^*$, and hence $A^{**} \subseteq (A^{**})^*$.

19.5c. Show that if A is self-adjoint, then A has *no* proper symmetric extension, i.e., A is maximal symmetric.

Solution: Let $A \subseteq B$ where B is symmetric, i.e., $B \subseteq B^*$. From $A = A^*$ we obtain that

$$A \subseteq B \subseteq B^* \subseteq A^* = A,$$

and hence $A = B$.

Analogously, we obtain that each skew-adjoint operator is maximal skew-symmetric.

19.5d. *Inverse operator.* Let $A: D(A) \subseteq X \to X$ be symmetric with $\overline{R(A)} = X$. Then

$$(A^{-1})^* = (A^*)^{-1},$$

where all the appearing inverse and adjoint operators exist.

In particular, if A is self-adjoint, then so is A^{-1}.

Solution: The operator A is injective. In fact, $Au = 0$ implies

$$(u|Av) = (Au|v) = 0 \qquad \text{for all} \quad v \in D(A).$$

Since $R(A)$ is dense, $u = 0$.

The same argument shows that A^* is injective. Since $\overline{R(A)} = X$, the adjoint operator $B = (A^{-1})^*$ exists. We have

$$(u|v) = (A^{-1}Au|v) = (Au|Bv) \qquad \text{for all} \quad u \in D(A), \quad v \in D(B),$$

$$(z|w) = (AA^{-1}z|w) = (A^{-1}z|A^*w) \qquad \text{for all} \quad w \in D(A^*), \quad z \in D(A^{-1}).$$

Hence $A^*Bv = v$ and $BA^*w = w$, i.e., $B = (A^*)^{-1}$.

19.5e. If $A: X \to X$ is linear and symmetric, then A is continuous.

Solution: Since $A = A^*$, the operator A is graph closed. By the closed graph theorem A_1(39e), A is continuous.

19.5f. *Continuous linear operators.* Let $A, B: X \to X$ be continuous linear operators on the H-space X over $\mathbb{K} = \mathbb{R}, \mathbb{C}$. Show that:

(i) $(A^*)^* = A$.
(ii) $A^*(\lambda u + \mu v) = \bar{\lambda} A^* u + \bar{\mu} A^* v$ for all $u, v \in X$ and all $\lambda, \mu \in \mathbb{K}$.
(iii) $(AB)^* = B^* A^*$.
(iv) $\|A^*\| = \|A\|$.

Solution: Ad(i)–(iii). Use the definition of A^* and B^*.

Ad(iv). From $(Au|v) = (u|A^*v)$ it follows that

$$\|Au\| = \sup_{\|v\| \le 1} |(Au|v)| \le \|u\| \, \|A^*\|,$$

and hence $\|A\| \le \|A^*\|$. By (i), $\|A^*\| \le \|A\|$.

19.5g.* *Positive self-adjoint operators.* Let $L_s(X, X)$ denote the set of all linear continuous self-adjoint operators $A: X \to X$ on the H-space X over $\mathbb{K}$. For A, $B \in L_s(X, X)$, we define

$$A \le B \quad \text{iff} \quad (Au|u) \le (Bu|u) \quad \text{for all} \quad u \in X.$$

For A, B, $C \in L_s(X, X)$, prove the following:
(i) If $A \ge 0$, then $A^{1/2} \ge 0$.
(ii) It follows from

$$A \le B$$

and $C \ge 0$, $AC = CA$, $BC = CB$ that $AC \le BC$.
(iii) It follows from

$$aI \le A \le bI$$

and $a, b > 0$ that A^{-1} belongs to $L_s(X, X)$ and

$$b^{-1} I \le A^{-1} \le a^{-1} I.$$

Hint: Cf. Riesz and Nagy (1956, M), Section 104. The statement (iii) will be used in the proof of Theorem 26.B.

19.6. *Complexification.* Let X be a real H-space. By definition, $X_{\mathbb{C}}$ is the set of all "complex" elements

$$u + iv \quad \text{with} \quad u, v \in X.$$

We set

$$(\alpha + i\beta)(u + iv) \stackrel{\text{def}}{=} (\alpha u - \beta v) + i(\beta u + \alpha v) \qquad \text{for all} \quad \alpha, \beta \in \mathbb{R}$$

and

$$(u + iv|x + iy) \stackrel{\text{def}}{=} i(u|y) - i(v|x) + (u|x) + (v|y).$$

This definition is motivated by a formal multiplication. A linear operator $A: D(A) \subseteq X \to X$ is extended to $X_{\mathbb{C}}$ by

$$A_{\mathbb{C}}(u + iv) \stackrel{\text{def}}{=} Au + iAv.$$

19.6a. Show that $X_{\mathbb{C}}$ becomes a complex H-space which is an extension of X.
Solution: Elementary computation.

19.6b. Show that $(A_{\mathbb{C}})^* = (A^*)_{\mathbb{C}}$.
Solution: By definition,

$$(w|A_{\mathbb{C}} z) = ((A_{\mathbb{C}})^* w|z)$$

with $w = u + iv$ and $z = x + iy$. Now set $x = 0$ or $y = 0$.

This result implies the following: If A is self-adjoint or skew-adjoint, then so is $A_{\mathbb{C}}$.

19.7. *Spectrum of self-adjoint operators.* Let $A: D(A) \subseteq X \to X$ be a self-adjoint operator on the real or complex H-space.

19.7a. Show that all $\lambda \in \mathbb{C}$ with $\operatorname{Im} \lambda \neq 0$ belong to the resolvent set of A.

Note that this statement refers to the complexification of A in case X is real.

Solution: We use the functional calculus for self-adjoint operators summarized in $A_2(96)$. Let X be complex. We fix a $\lambda \in \mathbb{C}$ with $\operatorname{Im} \lambda \neq 0$ and set $\psi(\mu) = (\lambda - \mu)^{-1}$ for all $\mu \in \mathbb{R}$. By definition,

$$\psi(A)u = \int_{-\infty}^{\infty} \psi(\mu)\, dE_\mu u,$$

where $u \in D(\psi(A))$ iff $\int_{-\infty}^{\infty} |\psi(\mu)|^2 \, d\|E_\mu u\|^2 < \infty$. Since ψ is bounded on $\mathbb{R}$, we obtain $D(\psi(A)) = X$ and

$$\|\psi(A)u\|^2 = \int_{-\infty}^{\infty} |\psi(\mu)|^2 \, d\|E_\mu u\|^2$$

$$\leq C \int_{-\infty}^{\infty} d\|E_\mu u\|^2 = C\|u\|^2,$$

i.e., $\psi(A): X \to X$ is bounded. By $A_2(96v)$, it follows from $(\lambda - \mu)\psi(\mu) = 1$ and

$$|\psi(\mu)| \leq \text{constant} \qquad \text{for all} \quad \mu \in \mathbb{R}$$

that $(\lambda I - A)\psi(A) = I$ and $\psi(A)(\lambda I - A) \subseteq I$. Hence

$$\psi(A) = (\lambda I - A)^{-1},$$

i.e., λ belongs to the resolvent set of A.

19.7b. Suppose that there is a $c \in \mathbb{R}$ such that

$$(Au|u) \geq c(u|u) \qquad \text{for all} \quad u \in D(A).$$

Show that all real $\lambda < c$ belong to the resolvent set of A.

Solution:

(I) We first show that $E_\lambda = 0$ for all $\lambda \leq c$. If $E_\lambda u = u$, then

$$c\|u\|^2 \leq (Au|u) = \int_{-\infty}^{\infty} \mu \, d\|E_\mu u\|^2$$

$$= \int_{-\infty}^{\infty} \mu \, d\|E_\mu E_\lambda u\|^2 = \int_{]-\infty, \lambda]} \mu \, d\|E_\mu u\|^2 \leq \lambda \|u\|^2,$$

by $A_2(93)$. That means $E_\lambda = 0$ for $\lambda < c$. Since the spectral family $\{E_\lambda\}$ is continuous from the left, $E_\lambda = 0$ for $\lambda = c$.

(II) We fix $\lambda < c$ and set $\psi(\mu) = (\lambda - \mu)^{-1}$. Then

$$\psi(A)u = \int_{[c, \infty[} \psi(\mu)\, dE_\mu u.$$

We now use arguments similar to Problem 19.7a.

19.8. *The Friedrichs extension.*

19.8a. *Complex* H-*spaces*. Let $A: D(A) \subseteq X \to X$ be a linear symmetric and strongly positive operator on the complex H-space X, i.e.,

$$(Au|u) \geq c\|u\|^2 \quad \text{for all} \quad u \in D(A) \quad \text{and fixed} \quad c > 0. \tag{133}$$

Show that A can be extended to a self-adjoint operator $A_F: D(A_F) \subseteq X \to X$ such that

$$(A_F u|u) \geq c\|u\|^2 \quad \text{for all} \quad u \in D(A_F).$$

Solution: The proof runs analogously to Section 19.10 with the following modifications. Let X_E be the energetic space of A constructed as in Section 19.10. According to Section 18.11, let $\bar{X}_E^*$ denote the antidual space to X_E and let $\bar{A}_E: X_E \to \bar{X}_E^*$ be the corresponding antiduality map.

Replace $h: X \to X_E^*$ in Section 19.10 by the map $\bar{h}: X \to \bar{X}_E^*$, i.e., $\bar{h}(u) = \overline{h(u)}$, where the bar denotes the conjugate complex number. This yields the continuity of the embeddings

$$X_E \subseteq X \subseteq \bar{X}_E^*.$$

Moreover, replace the duality map $A_E: X_E \to X_E^*$ by $\bar{A}_E: X_E \to \bar{X}_E^*$. Then the Friedrichs extension is given by $D(A_F) = \bar{A}_E^{-1}(X)$ and $A_F u = \bar{A}_E u$ on $D(A_F)$.

19.8b. *Structure of the energetic space*. Let $A: D(A) \subseteq X \to X$ be given as in Problem 19.8a where X is an H-space over $\mathbb{K} = \mathbb{R}, \mathbb{C}$. Let A_F denote the Friedrichs extension of A and let X_E be the energetic space of A. Show that

$$X_E = D(A_F^{1/2}),$$

and $\|u\|_E = \|A_F^{1/2}u\|$ for all $u \in X_E$.

Solution: Set $J = [c, \infty[$. By (133) and Problem 19.7b,

$$A_F u = \int_J \mu \, dE_\mu u.$$

For $0 \leq \alpha \leq 1$, we obtain

$$A_F^\alpha u = \int_J \mu^\alpha \, dE_\mu u,$$

where $u \in D(A_F^\alpha)$ iff $\int_J \mu^{2\alpha} d\|E_\mu u\|^2 < \infty$. Let $B = A_F^{1/2}$. Then $D(A_F) \subseteq D(B)$. Since $A \subseteq A_F$, we obtain

$$\begin{aligned} \|u\|_E^2 = (Au|u) = (A_F u|u) &= \int_J \mu \, d\|E_\mu u\|^2 \\ &= \|Bu\|^2 \quad \text{for all} \quad u \in D(A). \end{aligned} \tag{134}$$

(I) We show that $X_E \subseteq D(B)$. Let $u \in X_E$. By construction of X_E in Section 19.10, there follows the existence of a sequence (u_n) in $D(A)$ with

$$u_n \to u \quad \text{in } X \qquad \text{as} \quad n \to \infty$$

and $(Au_n - Au_m|u_n - u_m) \to 0$ as $n, m \to \infty$. By (134), (Bu_n) converges in X. The operator B is self-adjoint and hence graph closed; therefore, $u \in D(B)$.

(II) We show that $D(B) = X_E$. By Section 19.10, $D(A_F) \subseteq X_E$. According to A_2(96vi), the set $D(A_F)$ is dense in $D(B)$ equipped with the graph norm $\|Bu\|$. Thus, it follows from (I) that $D(B) = X_E$.

19.9. *Nonexpansive semigroups.* Show that each nonlinear nonexpansive semigroup $\{S(t)\}$ is strongly continuous.

Solution: Let $t, h > 0$. Then

$$\|S(t+h)v - S(t)v\| = \|S(t)S(h)v - S(t)v\|$$
$$\leq \|S(h)v - v\| \to 0 \qquad \text{as} \quad h \to +0,$$

and

$$\|S(t-h)v - S(t)v\| = \|S(t-h)v - S(t-h)S(h)v\|$$
$$\leq \|v - S(h)v\| \to 0 \qquad \text{as} \quad h \to +0.$$

19.10. *Simple variants of the Sobolev embedding theorems and the Lipschitz continuity of nonlinear operators.* Let G be a bounded region in $\mathbb{R}^N$, $N \geq 1$. Let $\|\cdot\|_p$ and $\|\cdot\|$ denote the norm on $L_p(G)$ and $\mathring{W}_2^1(G)$, respectively, i.e.,

$$\|u\|_p = \left(\int_G |u|^p\,dx\right)^{1/p},$$

$$\|u\| = \left(\int_G \sum_{i=1}^N |D_i u|^2\,dx\right)^{1/2}.$$

In the following all the constants are denoted by c. Since $C_0^\infty(G)$ is dense in $\mathring{W}_2^1(G)$, in the following proofs we may restrict ourselves to functions from $C_0^\infty(G)$ (cf. Proposition 18.29).

19.10a. Let $N = 3$. Show that the embedding $\mathring{W}_2^1(G) \subseteq L_6(G)$ is continuous.

Solution: Let $x = (\xi, \eta, \zeta)$ and let $u \in C_0^\infty(G)$. Integration of $(u^4)_\xi = 4u^3 u_\xi$ yields

$$|u(x)|^4 = \int_{-\infty}^x 4u^3 u_\xi\,d\xi \leq 4\int_{-\infty}^\infty |u^3 u_\xi|\,d\xi.$$

Replacing ξ by η and ζ, we obtain

$$|u(x)|^6 \leq 2^3\left(\int_{-\infty}^\infty |u^3 u_\xi|\,d\xi\right)^{1/2}\left(\int_{-\infty}^\infty |u^3 u_\eta|\,d\eta\right)^{1/2}\left(\int_{-\infty}^\infty |u^3 u_\zeta|\,d\zeta\right)^{1/2}.$$

We integrate over G and apply the Hölder inequality based on $\frac{1}{2} + \frac{1}{2} = 1$. Then

$$\|u\|_6^6 \leq c\|u\|_6^{18/4}\|u\|^{6/4},$$

and hence

$$\|u\|_6 \leq c\|u\| \qquad \text{for all} \quad u \in C_0^\infty(G). \tag{135}$$

19.10b. Let $N = 3$. We set

$$f(u) = u^3.$$

Show that the operator $f\colon \mathring{W}_2^1(G) \to L_2(G)$ is locally Lipschitz continuous.

Solution: Let $u, v \in C_0^\infty(G)$ with $\|u\|, \|v\| \leq r$. By the Hölder inequality based on $\frac{1}{3} + \frac{2}{3} = 1$,

$$\Delta \stackrel{\text{def}}{=} \|f(u) - f(v)\|_2 = \|(u-v)(u^2 + uv + v^2)\|_2$$
$$\leq \|u - v\|_6 \|u^2 + uv + v^2\|_3^2$$
$$\leq \|u - v\|_6 (\|u\|_6^2 + \|u\|_6\|v\|_6 + \|v\|_6^2)^2.$$

By (135), $\Delta \leq c\|u - v\|$.

19.10c. Let $N = 1$. Show that the embedding $\mathring{W}_2^1(G) \subseteq C(\bar{G})$ is continuous.

Solution: Let $G =]a, b[$ and let $u \in C_0^\infty(G)$. Then

$$u(x) = \int_a^x u'(y)\,dy.$$

The Hölder inequality with 1-trick yields

$$\max_{a \le x \le b} |u(x)| \le c\left(\int_a^b |u'|^2\,dx\right)^{1/2} = c\,\|u\|.$$

19.10d. Let $N = 1$. We set

$$f(u) = uu'.$$

Show that the operator $f\colon \mathring{W}_2^1(G) \to L_2(G)$ is locally Lipschitz continuous.

Solution: Let $u, v \in C_0^\infty(G)$ with $\|u\|, \|v\| \le r$. By Problem 19.10c,

$$\begin{aligned}\|f(u) - f(v)\|_2 &= \|(u' - v')u + v'(u - v)\|_2\\ &\le \|(u' - v')u\|_2 + \|v'(u - v)\|_2\\ &\le \max|u(x)|\,\|u' - v'\|_2 + \max|u(x) - v(x)|\,\|v'\|_2\\ &\le c\,\|u\|\,\|u - v\| + c\,\|u - v\|\,\|v\| \le 2cr\,\|u - v\|.\end{aligned}$$

19.11. *Proof of Lemma 19.26.*

Solution: We consider the square

$$C = \{(\xi, \eta) \in \mathbb{R}^2\colon -R \le \xi, \eta \le R\}$$

with $R = \frac{1}{2}$. Let $u \in C^1(C)$, and let $x = (\xi, \eta)$ and $y = (\alpha, \beta)$. Then

$$u(x) - u(y) = \int_\alpha^\xi u_\xi(t, \beta)\,dt + \int_\beta^\eta u_\eta(\xi, t)\,dt.$$

By the inequality $(a + b)^2 \le 2(a^2 + b^2)$ and by the Hölder inequality with 1-trick,

$$\begin{aligned}(u(x) - u(y))^2 &= u(x)^2 + u(y)^2 - 2u(x)u(y)\\ &\le 2\int_{-1/2}^{1/2} [u_\xi(t, \beta)^2 + u_\eta(\xi, t)^2]\,dt.\end{aligned}$$

Integration over $C \times C$ with respect to (x, y) yields

$$2\int_C u^2\,dx \le 2\int_C (u_\xi^2 + u_\eta^2)\,dx + 2\left(\int_C u\,dx\right)^2.$$

This is the desired result.

For general R we use the transformation $x \mapsto 2Rx$. The proof proceeds analogously in $\mathbb{R}^N$.

19.12. *Functional calculus and one-parameter unitary groups.* Let $H\colon D(H) \subseteq X \to X$ be a self-adjoint operator on the complex H-space X. Show that $\{e^{-itH}\}$ is a one-parameter unitary group with the generator $-iH$.

Solution: We use the functional calculus summarized in $A_2(96)$. Set $\int = \int_{-\infty}^{\infty}$.

(I) We have

$$Hu = \int \mu\,dE_\mu u,$$

where $u \in D(H)$ iff $\int \mu^2 \, d\|E_\mu u\|^2 < \infty$. By definition,

$$e^{-itH}u = \int e^{-it\mu} \, dE_\mu u, \qquad t \in \mathbb{R}, \tag{136}$$

where $u \in D(e^{-itH})$ iff $\int |e^{-it\mu}|^2 \, d\|E_\mu u\|^2 < \infty$. Hence $D(e^{-itH}) = X$ and

$$\|e^{-itH}u\|^2 = \int |e^{-it\mu}|^2 \, d\|E_\mu u\|^2 = \int d\|E_\mu u\|^2 = \|u\|^2$$

for all $u \in X$. Since $e^{-it\mu}e^{it\mu} = 1$ and $|e^{-it\mu}| = 1$, we obtain

$$e^{-itH}e^{itH} = e^{itH}e^{-itH} = I$$

and $(e^{-itH})^* = e^{itH}$ by A_2(96v). Thus, the operator $e^{-itH}: X \to X$ is unitary. From $e^{-it\mu}e^{-is\mu} = e^{-i(t+s)\mu}$ it follows that

$$e^{-itH}e^{-isH} = e^{-i(t+s)H}$$

for all $t, s \in \mathbb{R}$, by A_2(96v).

(II) Let $u \in D(H)$. Then $\int \mu^2 \, d\|E_\mu u\|^2 < \infty$. Formal differentiation of (136) yields

$$\frac{d}{dt}e^{-itH}u = -i \int \mu e^{-it\mu} \, dE_\mu u. \tag{137}$$

For all t and $\mu \in \mathbb{R}$, $|\mu e^{-it\mu}|^2 \le \mu^2$. Hence, by the majorant criterion A_2(97), the derivative (137) exists at each point $t \in \mathbb{R}$.

(III) We set $\psi(\mu) = \mu e^{-it\mu}$ for all $\mu \in \mathbb{R}$ and fixed $t \in \mathbb{R}$. Since $|\psi(\mu)|^2 = \mu^2$ for all $\mu \in \mathbb{R}$, we have $D(\psi(H)) = D(H)$. From

$$|e^{-it\mu}| \le |\psi(\mu)| + 1, \qquad |\mu| \le |\psi(\mu)| \qquad \text{for all} \quad t, \mu \in \mathbb{R},$$

and from A_2(96v) we obtain

$$\psi(H) = He^{-itH} = e^{-itH}H.$$

Hence

$$\frac{d}{dt}e^{-itH}u = -iHe^{-itH}u \qquad \text{for all} \quad u \in D(H), \quad t \in \mathbb{R}. \tag{138}$$

(IV) Let C be the generator of $\{e^{-itH}\}$. Differentiation of

$$(e^{-itH}u | e^{-itH}v) = (u|v)$$

at $t = 0$ yields

$$(Cu|v) + (u|Cv) = 0 \qquad \text{for all} \quad u, v \in D(C),$$

i.e., C is skew-symmetric. By (138), $-iH \subseteq C$. The operator $-iH$ is skew-adjoint. Hence $C = -iH$, by Problem 19.5c.

19.13. *Functional calculus and semigroups.* Let $A: D(A) \subseteq X \to X$ be a linear self-adjoint operator on the H-space X over $\mathbb{K} = \mathbb{R}, \mathbb{C}$ with

$$(Au|u) \ge c\|u\|^2 \qquad \text{for all} \quad u \in D(A)$$

and fixed real c. Show that $\{e^{-tA}\}$ forms a linear strongly continuous semigroup which is nonexpansive in the case where $c \ge 0$.

Solution: Let $J = [c, \infty[$ and $t \geq 0$. By Problem 19.7b, we have

$$Au = \int_J \mu \, dE_\mu u,$$

where $u \in D(A)$ iff $\int_J \mu^2 \, d\|E_\mu u\|^2 < \infty$. By definition,

$$e^{-tA}u = \int_J e^{-t\mu} \, dE_\mu u, \tag{139}$$

where $u \in D(e^{-tA})$ iff $\int_J e^{-2t\mu} \, d\|E_\mu u\|^2 < \infty$. From

$$e^{-2t\mu} \leq e^{2t|c|} \qquad \text{for all} \quad \mu \in J$$

it follows that $D(e^{-tA}) = X$ and

$$\|e^{-tA}u\|^2 = \int_J e^{-2t\mu} d\|E_\mu u\|^2 \leq e^{2t|c|}\|u\|^2.$$

Let $h(\mu) = \mu e^{-t\mu}$ for all $\mu \in J$. By definition,

$$h(A)u = \int_J h(\mu) \, dE_\mu u.$$

According to A_2(96v), it follows from

$$|e^{-t\mu}| \leq |\mu e^{-t\mu}| + \text{constant} \qquad \text{for all} \quad \mu \in J$$

that

$$h(A) = Ae^{-tA} \qquad \text{and} \qquad h(A) \supseteq e^{-tA}A.$$

We set

$$u(t) = e^{-tA}w \qquad \text{for fixed} \quad w \in D(A) \quad \text{and all} \quad t \geq 0.$$

Formal differentiation of (139) yields

$$u'(t) = -\int_J h(\mu) \, dE_\mu w.$$

Since

$$h(\mu)^2 \leq e^{|c|T}\mu^2 \qquad \text{for all} \quad \mu \in J, \quad t \in [0, T],$$

the majorant criterion A_2(97) shows that $u'(t)$ exists on $\mathbb{R}_+$ for all $w \in D(A)$. Hence

$$u'(t) = -Ae^{-tA}w. \tag{140}$$

From $e^{-t\mu}e^{-s\mu} = e^{-(t+s)\mu}$ and A_2(96v), we obtain that

$$e^{-tA}e^{-sA} = e^{-(t+s)A} \qquad \text{for all} \quad t, s \geq 0.$$

Let B be the generator of $\{e^{-tA}\}$. The operator e^{-tA} is self-adjoint, i.e.,

$$(e^{-tA}u|v) = (u|e^{-tA}v) \qquad \text{for all} \quad u, v \in X.$$

Differentiation at $t = 0$ yields

$$(Bu|v) = (u|Bv) \qquad \text{for all} \quad u, v \in D(B).$$

By (140), $A \subseteq B$. Thus, the operator B is a symmetric extension of the self-adjoint operator A. Hence $A = B$, by Problem 19.5c.

19.14. *Limits of continuous functions and equicontinuity.* The following results will be used frequently. Let (f_n) be a sequence of functions

$$f_n\colon M \to Y,$$

where M is a metric space and Y is a B-space. We ask the following questions:

(i) When is the limit function of (f_n) continuous?

(ii) When is the convergence of (f_n) uniform?

(iii) Suppose we know that the limit

$$\lim_{n\to\infty} f_n(x) \tag{I}$$

exists for all x in a *dense* subset of M. When does the limit (I) exist for *all* $x \in M$?

The answers are based on the notion of *equicontinuity*.

19.14a. *Definition.* The sequence (f_n) is *equicontinuous* at the point x iff for each $\varepsilon > 0$, there exists a $\delta(\varepsilon) > 0$ such that

$$d(x, y) < \delta(\varepsilon) \qquad \text{implies} \qquad \|f_n(x) - f_n(y)\| < \varepsilon \qquad \text{for all } n.$$

The point is that δ does *not* depend on n. We call (f_n) equicontinuous on M iff (f_n) is equicontinuous at each point $x \in M$.

19.14b. *Equicontinuity and continuous limit functions.* Suppose that the limit

$$f(x) = \lim_{n\to\infty} f_n(x)$$

exists for all $x \in M$ and that (f_n) is *equicontinuous* at x_0. Then the limit function f is continuous at x_0.

19.14c. *Dense convergence.* Suppose that the limit

$$f(x) = \lim_{n\to\infty} f_n(x)$$

exists for all x in a *dense* subset of M and that (f_n) is *equicontinuous* on M. Then the limit exists for *all* $x \in M$ and the limit function $f\colon M \to Y$ is continuous.

19.14d. *Uniform convergence.* Suppose that all the functions $f_n\colon M \to Y$ are continuous and that we have the *uniform* convergence

$$f_n \rightrightarrows f \quad \text{on } M \qquad \text{as} \quad n \to \infty,$$

i.e., for each $\varepsilon > 0$, there exists an $n_0(\varepsilon)$ such that

$$\|f_n(x) - f(x)\| < \varepsilon \qquad \text{for all} \quad n \geq n_0 \quad \text{and } all \quad x \in M.$$

The point is that n_0 does *not* depend on x.

Then the limit function $f\colon M \to Y$ is continuous.

19.14e. *First criterion for uniform convergence.* Suppose that all the functions $f_n\colon M \to Y$ are continuous on the *compact* metric space M and that (f_n) is equicontinuous on M. Then the simple convergence

$$f_n(x) \to f(x) \qquad \text{as} \quad n \to \infty \qquad \text{for all} \quad x \in M$$

implies the uniform convergence

$$f_n \rightrightarrows f \quad \text{on } M \qquad \text{as} \quad n \to \infty,$$

and the limit function $f: M \to Y$ is continuous.

19.14f. *Second criterion for uniform convergence* (theorem of Dini). Suppose that all the functions $f_n: M \to \mathbb{R}$ are *monotone* increasing (resp. monotone decreasing) on the *compact* metric space M. Then the simple convergence

$$f_n(x) \to f(x) \qquad \text{as} \quad n \to \infty \qquad \text{for all} \quad x \in M$$

implies the uniform convergence

$$f_n \rightrightarrows f \quad \text{on } M \qquad \text{as} \quad n \to \infty$$

in the case where the limit function $f: M \to \mathbb{R}$ is continuous.

Hint: Use the corresponding definitions and the fact that Cauchy sequences in B-spaces are convergent. Cf. Dieudonné (1968, M), Vol. 1, Chapter 7.

19.15. *Commutativity of iterated limits on metric spaces.* We are interested in the relation

$$\lim_{x\to a}\lim_{y\to b} f(x, y) = \lim_{y\to b}\lim_{x\to a} f(x, y) = \lim_{(x,y)\to(a,b)} f(x, y) \tag{141}$$

for the mapping

$$f: X \times Y \to Z,$$

where X, Y, and Z are metric spaces (e.g., X and Y is a neighborhood of a and b in a B-space, respectively, and Z is a B-space). In this connection, the notion of *uniform* convergence is crucial.

19.15a. *Uniform convergence.* By definition, the limit

$$F(x) = \lim_{y\to b} f(x, y)$$

is uniform with respect to all $x \in M$ iff for each $\varepsilon > 0$ there is a $\delta(\varepsilon) > 0$ such that

$$d(y, b) < \delta(\varepsilon) \qquad \text{implies} \qquad d(f(x, y), F(x)) < \varepsilon \qquad \text{for all} \quad x \in M.$$

The point is that δ does not depend on x.

19.15b. *First criterion.* Suppose that:

(i) The inner limits $\lim_{y\to b} f(x, y)$ and $\lim_{x\to a} f(x, y)$ exist.

(ii) The inner limit $\lim_{y\to b} f(x, y)$ is *uniform* with respect to all x in a neighborhood $U(a)$ of a.

(iii) Z is a complete metric space (e.g., a B-space).

Then relation (141) holds where all the appearing limits exist.

19.15c. *Second criterion.* Suppose that:

(i) The inner limits $\lim_{y\to b} f(x, y)$ and $\lim_{x\to a} f(x, y)$ exist.

(ii) The limit $\lim_{(x,y)\to(a,b)} f(x, y)$ exists.

Then relation (141) holds where all the appearing limits exist.

19.15d. *Third criterion.* Suppose that:

(i) The inner limit $\lim_{y\to b} f(x, y)$ exists.

(ii) The limit $\lim_{(x,y)\to(a,b)} f(x, y)$ exists.

Then

$$\lim_{x\to a}\lim_{y\to b} f(x,y) = \lim_{(x,y)\to(a,b)} f(x,y). \tag{142}$$

Hint: Use the corresponding definitions and the fact that, in a complete metric space, each Cauchy sequence is convergent. Cf. Fichtenholz (1972, M), Sections 168 and 506.

19.16. *The Schmidt orthogonalization method.* Let X be a separable H-space, $X \neq \{0\}$. Show that there exists an at most countable complete orthonormal system in X.

Solution: By assumption, there exists an at most countable set $\{v_m\}$ which is dense in X. We may assume that $v_1 \neq 0$. Set

$$u_1 = v_1/\|v_1\|.$$

Suppose we have $u_1, \ldots, u_n$ constructed such that $\{u_1,\ldots,u_n\}$ forms an orthonormal system. Then let

$$\bar{u}_{n+1} = v_{n+1} - \sum_{k=1}^{n} (u_k|v_{n+1})u_k$$

and

$$u_{n+1} = \bar{u}_{n+1}/\|\bar{u}_{n+1}\|$$

in the case where $\bar{u}_{n+1} \neq 0$. Thus, $(u_k|u_{n+1}) = 0$ for all $k = 1,\ldots,n$. If $\bar{u}_{n+1} = 0$, then we use v_{n+2}, etc. This way we obtain an orthonormal system $\{u_m\}$.

By induction, it follows that all the v_m are finite linear combinations of the u_n. Hence the linear hull of $\{u_m\}$ is dense in X. By Theorem 19.A, $\{u_m\}$ is complete.

References to the Literature

Introduction: Riesz and Nagy (1956, M).

Semigroups and partial differential equations: Pazy (1983, M) (standard work), Friedman (1969, M), Reed and Simon (1971, M), Vols. 1, 2, Showalter (1977, M), Henry (1981, L), v. Wahl (1985, M), Benilan, Crandall, and Pazy (1989, M).

Functional calculus: Riesz and Nagy (1956, M), Reed and Simon (1971, M), Vol. 1, Triebel (1972, M).

Nonlinear hyperbolic differential equations: Christodoulou and Klainerman (1990, M).

Linear partial differential equations and their applications in mathematical physics: Dautray and Lions (1984, M), Vols. 1–3.

Nonlinear partial differential equations and their applications in mathematical physics: Chern (1984, S).

CHAPTER 20

Difference Methods and Stability

Consistency and stability imply convergence.

The golden rule of numerical analysis.

Difference methods are a proper tool in order to compute approximately the solutions of ordinary and partial differential equations and to prove the existence of generalized solutions.

Richard Courant, Kurt Friedrichs, and Hans Lewy (1928)

Our main concern here is with stability of difference methods rather than with accuracy.

John von Neumann and Robert Richtmyer (1947)

Well-posedness of the original differential equation problem and consistency imply the equivalence between stability and convergence of difference methods.

Peter Lax (1953)

Do not use unstable difference methods on computers! The rapidity of convergence depends on the order of consistency, i.e., on the accuracy of the approximation of the differential quotients by difference quotients.

Folclore

The advances that are now being made in various areas of computer science are forcing us to re-evaluate the way in which we do numerical computations. The realization that a different view is necessary has already created a new catch phrase, *scientific computation*, which is becoming recognized as a separate discipline transcending the old computational mathematics.

Mathematicians can play a significant role in this emerging discipline, if they are able to provide new paradigms for computation, but in order to do so much of what has been done up to now will have to be reconsidered. There are some signs that such a fundamental *rethink* is beginning. The work of Traub (1980) and Smale (1985) introduces ideas from information theory and computational complexity in an effort to understand more clearly what we can expect from numerical algorithms.... The full realization of the potential of computational mathematics requires the development of *new* and sophisticated analytical tools.

Peter Linz (1988)

A universally applicable method for the approximate treatment of differential equations are the difference methods. Here the basic idea consists of replacing differential quotients by difference quotients. The aim of this chapter is to explain the fundamental connection between *stability, consistency, and convergence.*

A principal difficulty in difference methods consists of the fact that one has *various* possibilities for approximating differential quotients by difference quotients, and this reduces to finding especially *favorable numerical approximations.* For the engineer and practitioner it is, in this connection, not so much the rapidity of convergence of the method that stands in the foreground but rather its stability, since unstable methods collapse or yield only very inexact results on computers. By *stability*, we understand roughly the situation that in the solution of the difference equation, errors can not grow indefinitely but rather remain uniformly bounded, independently of the width of the grid. We elucidate this using the simple first-order differential equation

$$y' = \lambda y \tag{1}$$

with the solution

$$y(x) = y(0)e^{\lambda x}.$$

First, we consider the stability of the differential equation. If y and z are two solutions of (1) then, for the difference

$$\Delta(x) = y(x) - z(x),$$

there holds

$$|\Delta(x)| = e^{\lambda x}|\Delta(0)|. \tag{2}$$

So that, for $x > 0$, the initial error $\Delta(0)$ does not become larger, the stability condition

$$\lambda \leq 0$$

must hold.

In order to formulate the difference method for (1) we set

$$y_p = y(ph) \qquad \text{for} \quad p = 0, 1, \ldots \quad \text{and fixed} \quad h > 0$$

and denote an approximate value for y_p by v_p. Then for (1) there first results the following two possibilities:

$$h^{-1}(v_{p+1} - v_p) = \lambda v_p, \qquad p = 0, 1, \ldots, \tag{3}$$

$$h^{-1}(v_{p+1} - v_p) = \lambda v_{p+1}, \qquad p = 0, 1, \ldots. \tag{4}$$

(i) *Explicit method.* From (3) we obtain

$$v_{p+1} = (1 + \lambda h)v_p, \qquad p = 0, 1, \ldots.$$

If w_p is another solution of this equation and we set

$$\Delta_p = v_p - w_p,$$

then $\Delta_{p+1} = (1 + \lambda h)\Delta_p$ holds and thus we get the *key* relation

$$|\Delta_{p+1}| = \alpha^{p+1}|\Delta_0|, \qquad p = 0, 1, \ldots, \tag{5}$$

with

$$\alpha = |1 + \lambda h|.$$

In order that the initial error Δ_0 does not increase in the course of the computation, we must have the *stability condition*

$$\alpha \leq 1.$$

For fixed $\lambda < 0$, this is a *condition* for the *grid step h*.

(ii) *Implicit method.* From (4) we obtain

$$v_{p+1} = (1 - \lambda h)^{-(p+1)} v_0.$$

This implies (5) with

$$\alpha = |1 - \lambda h|^{-1}.$$

In contrast to (3) there now arises, for $\lambda < 0$, *no* restriction on the grid step h, i.e., we obtain $\alpha < 1$ for all $\lambda < 0$ and $h > 0$. Therefore, we shall prefer (4) to (3). Method (3) is called an *explicit* method in contrast to the more favorable *implicit* method (4).

(iii) *Mean value method.* A still more favorable method results by using the mean value of (3) and (4), i.e.,

$$h^{-1}(v_{p+1} - v_p) = \lambda(v_{p+1} + v_p)/2. \tag{6}$$

Now $v_{p+1} = (1 + \lambda h/2)(1 - \lambda h/2)^{-1} v_p$ holds. Therefore, we obtain (5) with

$$\alpha = |1 + \lambda h/2||1 - \lambda h/2|^{-1}.$$

For $\lambda < 0$, $h > 0$, this value of is α smaller than in (3) and (4). Consequently, it follows from (5) that an initial error in (6) dies off more rapidly than in (3) or (4). For parabolic differential equations, (6) leads to the favorable Crank–Nicolson method in Section 20.4.

If the stability condition $\alpha \leq 1$ is violated in (5), i.e. $\alpha > 1$, then a round-off error for the initial condition will increase more and more at each step, and will completely falsify the result after a sufficiently large number of steps p. Therefore, *stability plays a decisive role* in the numerical realization of difference methods on computers. Whereas with an explicit method, as in (3), one obtains the solution of a difference equation directly; with an implicit method, as in (4), one has to solve an equation or a system of equations. However, in contrast to this greater demand on computer time with an implicit method, there is the decisive advantage that in general the stable grid step domain for h is *greater* in the case of an implicit method than with an explicit method. Therefore, as a rule, one will *prefer* implicit methods to explicit methods. In particular, one is interested in difference methods for which *all* grid steps are stable. In Example 20.5 we explain this situation on the basis of a numerical example.

20.1. Consistency, Stability, and Convergence

It is very remarkable that the proofs of convergence for difference methods can all be obtained from one unified principle which reads roughly as follows:

(C) *Consistency and stability imply convergence.*

In the following we explain this principle on the basis of an abstract model, and apply this model to concrete problems in Example 20.1 and the following sections. In Sections 20.6 and 20.7 we consider a certain converse of (C).

This principle reads as follows:

With consistent methods, for well-posed problems, stability and convergence are equivalent.

In this connection, *consistency* always means that the approximate equation converges in an appropriate sense to the original equation. A problem is said to be *well-posed* if for fixed data it possesses exactly one solution and the solutions depend continuously on the data in a certain way. We now study an important general abstract model.

We consider the equation

$$By = f, \qquad y \in Y, \tag{7}$$

and, for all $h \in]0, h_0]$, the approximate equation

$$B_h y_h = f_h, \qquad y_h \in Y_h, \tag{8}$$

together with the following approximation scheme:

$$\begin{array}{ccc} Y & \xrightarrow{B} & Z \\ {\scriptstyle P_h}\big\downarrow & & \big\downarrow{\scriptstyle Q_h} \\ Y_h & \xrightarrow{B_h} & Z_h \end{array} \tag{9}$$

Our aim is to investigate the convergence of this approximation method. We assume:

(H1) *Approximation scheme.* Let $h_0 > 0$ be fixed. For all $h \in]0, h_0]$, the spaces Y, Z, Y_h, and Z_h are given normed spaces over $\mathbb{K} = \mathbb{R}, \mathbb{C}$, and the operators B, B_h, P_h, and Q_h are given according to (9).

(H2) *Existence and uniqueness.* For given fixed $f \in Z$, the original equation (7) has exactly one solution y.

We are given a sequence (f_h), $0 < h < h_0$, such that the approximate equation (8) has a unique solution y_h for each $f_h \in Z_h$.

(H3) *Consistency.* There are positive constants α, β, and r, independent of h, such that

$$\|Q_h By - B_h P_h y\| \le \alpha h^r,$$

$$\|Q_h f - f_h\| \le \beta h^r \qquad \text{for all} \quad h \in]0, h_0].$$

(H4) *Stability*. There are positive constants γ and s, independent of h, such that

$$\|v - w\|_{Y_h}^s \le \gamma \|B_h v - B_h w\|_{Z_h} \tag{10}$$

for all $v, w \in Y_h$ and all $h \in]0, h_0]$.

Let $s = 1$. If B_h is a linear operator, then (10) is equivalent to

$$\|v\|_{Y_h} \le \gamma \|B_h v\|_{Z_h} \qquad \text{for all} \quad v \in Y_h \quad \text{and all} \quad h \in]0, h_0].$$

In turn, this is equivalent to the existence of the inverse operators B_h^{-1} on Z_h with the property

$$\|B_h^{-1}\| \le \gamma \qquad \text{for all} \quad h \in]0, h_0].$$

Theorem 20.A (Convergence). *From* (H1) *through* (H4) *we obtain the error estimate*

$$\|y_h - P_h y\|_{Y_h}^s \le (\alpha + \beta)\gamma h^r \qquad \textit{for all} \quad h \in]0, h_0]. \tag{11}$$

This implies the convergence of the approximation method, i.e.,

$$\|y_h - P_h y\|_{Y_h} \to 0 \qquad \textit{as} \quad h \to +0.$$

From this it follows, in particular, that, for fixed s in (H4), the rapidity of convergence with respect to h is determined by the consistency condition.

For all difference equations, the consistency condition results in an easy way from the Taylor theorem (see Section 20.2).

The verification of the stability condition requires a more precise analysis of the difference equations and can take on a knotty form. The stability condition (10) means that small changes on the right-hand side in $B_h y_h = f_h$ also lead to small changes in the solution y_h, i.e., small errors relative to f_h in (8) can not lead to a catastrophic falsification of the solution.

PROOF. We set $\tau_h = B_h P_h y - B_h y_h$. It follows from (7) and (8) that

$$\tau_h = B_h P_h y - Q_h B y + Q_h B y - B_h y_h = B_h P_h y - Q_h B y + Q_h f - f_h.$$

The triangle inequality and the consistency condition give

$$\|\tau_h\| \le \|B_h P_h y - Q_h B y\| + \|Q_h f - f_h\| \le (\alpha + \beta) h^r.$$

The stability condition yields

$$\|y_h - P_h y\|^s \le \gamma \|B_h y_h - B_h P_h y\| \le \gamma(\alpha + \beta) h^r. \qquad \square$$

STANDARD EXAMPLE 20.1. Let $\lambda < 0$. Parallel to (1) and (4), we consider the differential equation

$$\begin{aligned} Ly \overset{\text{def}}{=} y' - \lambda y &= g \qquad \text{on } [0, 1], \\ y(0) &= c \end{aligned} \tag{12}$$

along with the difference equation

$$L_p v \stackrel{\text{def}}{=} h^{-1}(v_{p+1} - v_p) - \lambda v_{p+1} = g_{p+1}, \qquad p = 0, 1, \ldots, N-1, \tag{13}$$
$$v_0 = c,$$

where $g \in C^1[0,1]$ and $c \in \mathbb{R}$ are given. Let y be the unique solution of (12). Then $y \in C^2[0,1]$.

We partition the interval $[0,1]$ by means of the partition points $0 = x_0 < x_1 < \cdots < x_N = 1$ with $x_p = ph$, and we set

$$y_p = y(ph), \qquad g_p = g(ph).$$

Let $\|y\|$ be the norm of y in $C^2[0,1]$, and let $\|g\|$ be the norm of g in $C^1[0,1]$.

(i) *Consistency.* Our point of departure is the formula

$$|L_p y - (Ly)(ph)| \leq (|\lambda| + \tfrac{1}{2})h\|y\|, \qquad p = 0, 1, \ldots, N-1. \tag{14}$$

This formula results, by the Taylor theorem, from

$$h^{-1}(y_{p+1} - y_p) - y'(ph) = hy''(\xi)/2,$$
$$\lambda(y_{p+1} - y(ph)) = \lambda h y'(\eta),$$

where ξ and η are suitable points. Moreover, we use the formula

$$|g_{p+1} - g_p| \leq h\|g\|, \qquad p = 0, 1, \ldots, N-1, \tag{15}$$

which follows from $g_{p+1} - g_p = hg'(\xi)$.

(ii) *The key equation (modified difference equation).* We set

$$e_p = y_p - v_p.$$

This quantity is crucial for our convergence proof. We want to obtain a difference equation for e_p. From the original differential equation (12) and the consistency condition (14) it follows that

$$L_p y = g_p + k_p, \qquad y_0 = c, \qquad p = 0, 1, \ldots, N-1 \tag{16}$$

with

$$|k_p| \leq (|\lambda| + \tfrac{1}{2})h\|y\|.$$

After subtracting (16) from the difference equation (13) we obtain the key equation

$$L_p e = m_p, \qquad e_0 = 0, \qquad p = 0, 1, \ldots, N-1 \tag{17}$$

with $m_p = g_p - g_{p+1} + k_p$.

(iii) *Stability.* We can easily solve (17) explicitly. Noting $\lambda < 0$, we obtain

$$e_{p+1} = (1 + |\lambda|h)^{-1}(e_p + hm_p), \qquad p = 0, \ldots, N-1.$$

Therefore, by induction, we get the crucial stability condition

$$\begin{aligned} |e_p| &\le ph(1+|\lambda|h)^{-1} \max_{0\le p\le N-1} |m_p| \\ &\le \max_{0\le p\le N-1} |m_p|, \qquad p = 0, \ldots, N. \end{aligned} \tag{18}$$

(iv) *Convergence*. Finally, (18) implies

$$\max_{0\le p\le N} |y(ph) - v_p| \le h(\|g\| + (|\lambda| + \tfrac{1}{2})\|y\|). \tag{19}$$

Letting $h \to 0$, we obtain the convergence of the difference method (13) for the original differential equation (12) in the case where $\lambda < 0$.

All convergence proofs for difference methods for linear ordinary and partial differential equations follow the method of Standard Example 20.1 above. The *basic idea* is the following:

(a) consider the difference

$$e_p = y_p - v_p$$

at the grid point p, where y denotes a solution of the original differential equation and v denotes a solution of the corresponding difference equation;

(b) construct a *modified* difference equation for e_p as in (17); and

(c) obtain *uniform* estimates for e_p as in (18) (stability).

In conclusion, we will yet show how the proof of convergence in Standard Example 20.1 corresponds to the situation of our abstract model in Theorem 20.A. To this end, we write the original differential equation (12) in the form

$$By = f, \qquad y \in Y, \qquad f \in Z,$$

with

$$By = (Ly, y(0)), \qquad f = (g, c)$$

and $Y = C^2[0,1]$, $Z = C^1[0,1] \times \mathbb{R}$.

Furthermore, let $y_h = (v_0, v_1, \ldots, v_N)$. We write the difference equation (13) in the form

$$B_h y_h = f_h, \qquad y_h \in Y_h, \quad f_h \in Z_h,$$

with $f_h = (g_1, \ldots, g_N, c)$ and $Y_h = Z_h = \mathbb{R}^{N+1}$. We choose the maximum norm $\|\cdot\|_\infty$ on Y_h and Z_h.

Standard Example 20.1 corresponds precisely to Theorem 20.A with the operators

$$P_h y = (y(0), y(h), \ldots, y(Nh)), \qquad Q_h f = f_h.$$

The consistency condition (H3) above corresponds to (14) and (15) and the stability condition (H4) corresponds to (18). The convergence statement (11) of Theorem 20.A coincides with (19).

20.2. Approximation of Differential Quotients

To verify the consistency condition one uses the *Taylor formula* in order to approximate differential quotients by means of difference quotients. In this connection, we also obtain in a simple way the order of approximation with respect to h, which is decisive for the rapidity of convergence of the difference method.

We define the *difference operator*

$$\nabla_h f(x) = \frac{f(x+h) - f(x)}{h}, \qquad h \gtrless 0,$$

and the *discrete* Laplace operator

$$\Delta_h = \nabla_{-h}\nabla_h, \qquad h > 0.$$

A short calculation shows that

$$\Delta_h f(x) = \frac{f(x+h) - 2f(x) + f(x-h)}{h^2}.$$

The following formulas are typical and important examples for approximating differential quotients by difference quotients via the Taylor formula:

$$\nabla_h f(x) = f'(x) + hf''(\xi)/2, \qquad \xi \in [x, x+h], \tag{20}$$

$$\Delta_h f(x) = f''(x) + hf^{(3)}(\xi)/3, \qquad \xi \in [x-h, x+h], \tag{21}$$

$$\Delta_h f(x) = f''(x) + h^2 f^{(4)}(\xi)/12, \qquad \xi \in [x-h, x+h]. \tag{22}$$

Observe that formula (22) corresponds to a higher accuracy of approximation with respect to h than (21).

In particular, it follows from the Taylor formula

$$f(x \pm h) = f(x) \pm hf'(x) + h^2 f''(x)/2 \pm h^3 f^{(3)}(\xi_\pm)/6, \qquad h > 0,$$

with $\xi_\pm \in [x-h, x+h]$, that the difference quotient

$$\tfrac{1}{2}(\nabla_h + \nabla_{-h}) = \frac{f(x+h) - f(x-h)}{2h}$$

is a better approximation of the derivative $f'(x)$ than ∇_h if f is C^3. In fact, we obtain that

$$\tfrac{1}{2}(\nabla_h + \nabla_{-h}) = f'(x) + h^2 f^{(3)}(\xi)/6, \qquad \xi \in [x-h, x+h],$$

if $f \in C^3[a,b]$ and $x, x+h \in [a,b]$. This approximation is of order h^2 whereas the approximation (20) is only of order h. In this connection, note that

$$\tfrac{1}{2}(f^{(3)}(\xi_+) + f^{(3)}(\xi_-)) = f^{(3)}(\xi)$$

for a suitable $\xi \in [x-h, x+h]$. This follows from the intermediate value theorem for continuous functions.

Table 20.1

Formula	(20)	(21)	(22)
Assumption on f	$f \in C^2[a,b]$	$f \in C^3[a,b]$	$f \in C^4[a,b]$
Assumption on x, h $(h \gtrless 0)$	$x, x+h \in [a,b]$	$x, x \pm h \in [a,b]$	

Generally, one has the following principle:

The smoother the functions are, the better is the approximation of differential quotients by suitable difference quotients.

By Theorem 20.A, the rapidity of the convergence of difference methods depends on the degree of consistency, i.e., on the accuracy of the approximation of differential quotients by difference quotients. Thus, we obtain:

The smoother the solutions of differential equations are, the higher is the rapidity of convergence of suitable difference methods.

This is an important general principle of numerical mathematics which is also valid for other approximation methods, for example, for the Ritz and Galerkin methods (see Chapters 21 and 22).

Proposition 20.2. *Let $[a,b]$ be a compact interval. Then the approximation formulas* (20) *through* (22) *are valid in the case where the assumptions of Table* 20.1 *are fulfilled.*

Proof. For instance, in order to obtain (22), we proceed from the Taylor formula with a remainder of fourth order

$$f(x \pm h) = f(x) \pm hf'(x) + h^2 f''(x)/2 \pm h^3 f^{(3)}(x)/6 + h^4 f^{(4)}(\xi_\pm)/24.$$

It follows from this, by addition, that

$$\Delta_h f(x) = f''(x) + h^4(f^{(4)}(\xi_+) + f^{(4)}(\xi_-))/24, \qquad \xi_\pm \in [x, x \pm h].$$

The intermediate value theorem for continuous functions yields

$$f^{(4)}(\xi_+) + f^{(4)}(\xi_-) = 2f^{(4)}(\xi)$$

with a suitable $\xi \in [x-h, x+h]$. □

20.3. Application to Boundary Value Problems for Ordinary Differential Equations

We consider the boundary value problem

$$-y'' + ay = f \quad \text{on }]0,1[, \qquad y(0) = y(1) = 0. \tag{23}$$

Let $a \geq 0$. In order to formulate the difference method, we partition the interval $[0, 1]$ by means of the partition points $0 < h < 2h < \cdots < Nh = 1$. We set

$$y_p = y(ph), \qquad f_p = f(ph)$$

and denote an approximate value of y_p by v_p. Motivated by (21), the difference equation for (23) reads as follows:

$$\begin{aligned} -h^{-2}(v_{p+1} - 2v_p + v_{p-1}) + av_p = f_p, \qquad p = 1, \ldots, N-1, \\ v_0 = v_N = 0, \end{aligned} \tag{24}$$

i.e., we replace y'' by $\Delta_h y$. In matrix notation, equation (24) means

$$\frac{1}{h^2}\begin{pmatrix} 2 + ah^2 & -1 & & 0 \\ -1 & 2 + ah^2 & -1 & \\ & \ddots & \ddots & \ddots \\ & -1 & 2 + ah^2 & -1 \\ 0 & & -1 & 2 + ah^2 \end{pmatrix}\begin{pmatrix} v_1 \\ v_2 \\ \vdots \\ \cdot \\ v_{N-1} \end{pmatrix} = \begin{pmatrix} f_1 \\ f_2 \\ \vdots \\ \cdot \\ f_{N-1} \end{pmatrix}. \tag{25}$$

For this we write

$$B_h(a)v_h = f_h. \tag{25*}$$

The coefficient matrix $B_h(a)$ of (25) is a symmetric tridiagonal matrix.

For arbitrary starting element $y_h^{(0)}$, the *total step method* for an iterative solution of (25), according to Section 1.8, reads as follows:

$$y_h^{(n+1)} = A_h y_h^{(n)} + b, \qquad n = 0, 1, \ldots \tag{26}$$

with

$$A_h = \begin{pmatrix} 0 & \alpha & & & 0 \\ \alpha & 0 & \alpha & & \\ & \ddots & \ddots & \ddots & \\ & & \alpha & 0 & \alpha \\ 0 & & & \alpha & 0 \end{pmatrix}, \qquad \alpha = \frac{1}{2 + ah^2}, \qquad b = \frac{h^2}{2 + ah^2} f_h.$$

Proposition 20.3. *If $f \in C^1[0, 1]$, then the boundary value problem (23) possesses exactly one solution $y \in C^3[0, 1]$. For each $N = 2, 3, \ldots$, the difference equation (24) has exactly one solution v and*

$$\max_{1 \leq p \leq N-1} |y(ph) - v_p| \leq \frac{h}{24} \max_{0 \leq x \leq 1} |y^{(3)}(x)|. \tag{27}$$

If $f \in C^2[0, 1]$, then $y \in C^4[0, 1]$ and

$$\max_{1 \leq p \leq N-1} |y(ph) - v_p| \leq \frac{h^2}{96} \max_{0 \leq x \leq 1} |y^{(4)}(x)|. \tag{28}$$

This shows that the difference method converges as $h \to 0$, and that the rapidity of the convergence depends on the *smoothness* of the solution y of the original boundary value problem (23).

Remark 20.4 (Iterative Solution of the Difference Equation). By Problem 20.1, the eigenvalues $\lambda_{j,h}$ and the eigenvectors $y_{j,h}$ of the matrix $B_h(a)$ are given by

$$\lambda_{j,h} = a + j^2\pi^2 \frac{\sin^2(j\pi h/2)}{(j\pi h/2)^2}, \qquad j = 1, \ldots, N-1, \tag{29}$$
$$y_{j,h} = (\sin j\pi h, \sin 2j\pi h, \ldots, \sin(N-1)j\pi h).$$

As $h \to 0$, they tend to the eigenvalues $\lambda_j = a + j^2\pi^2$ and the eigenfunctions $y_j = \sin jx$ of the corresponding boundary eigenvalue problem

$$-y'' + ay = \lambda y \quad \text{on }]0,1[, \qquad y(0) = y(1) = 0.$$

By Problem 20.1, we obtain, for the eigenvalues of the matrix A_h,

$$\mu_{j,h} = \frac{2\cos j\pi h}{2 + ah^2}, \qquad j = 1, \ldots, N-1,$$

i.e., the *spectral radius* of A_h is smaller than one. Consequently, the total step method (26) *converges* to the solution of the difference equation (24), by Proposition 1.15. However, note that by Table 1.3, the optimal relaxation method converges more rapidly.

PROOF OF PROPOSITION 20.3. We assume that $y \in C^4[0,1]$.

(I) Consistency. Let $e_p = y(ph) - v_p$. It follows immediately from (22), (23), and (24) that

$$-h^{-2}(e_{p+1} - 2e_p + e_{p-1}) + ae_p = \tau_p, \qquad p = 1, \ldots, N-1, \tag{30}$$

with

$$|\tau_p| \le \frac{h^2}{12} \max_{0 \le x \le 1} |y^{(4)}(x)|.$$

(II) Stability. Using the coefficient matrix $B_h(a)$ of (25), equation (30) means

$$B_h(a)e = \tau,$$

i.e., $e = B_h(a)^{-1}\tau$. Let $B_h(a)^{-1} = (c_{pq}(a))$. Then

$$e_p = \sum_{q=1}^{N-1} c_{pq}(a)\tau_q, \qquad p = 1, \ldots, N-1. \tag{31}$$

By Problem 20.1,

$$0 \le c_{pq}(a) \le c_{pq}(0)$$

for all $a \ge 0$ and all p, q. This implies the crucial *stability relation*

$$\max_{1 \le p \le N-1} |e_p| \le \beta \max_{1 \le p \le N-1} |\tau_p| \tag{32}$$

with

$$\beta = \max_{1 \le p \le N-1} \sum_{q=1}^{N-1} c_{pq}(0).$$

(III) Convergence. It follows from (32) that

$$\max_{1 \le p \le N-1} |e_p| \le \frac{\beta h^2}{12} \max_{0 \le x \le 1} |y^{(4)}(x)|.$$

This is assertion (28) in the case where we will yet show that $\beta \le \frac{1}{8}$.

To this end we choose $f(x) \equiv 1$. Then the function

$$y = \tfrac{1}{2}x(1 - x)$$

is a solution of the differential equation (23) with $a = 0$ and at the same time of the difference equation (24), since by (21) the discretization error depends only on the third derivative $y^{(3)}$ which vanishes here. Thus, equation (30) holds with $\tau_p = 1$ and

$$e_p = y(ph) = \tfrac{1}{2}ph(1 - ph),$$

therefore (31) also holds with $a = 0$. This implies

$$\beta = \max_{1 \le p \le N-1} \tfrac{1}{2}ph(1 - hp) \le \max_{0 \le x \le 1} \tfrac{1}{2}x(1 - x) = \tfrac{1}{8}.$$

Analogously, one obtains assertion (27) from (31). □

20.4. Application to Parabolic Differential Equations

In this section, by means of special difference methods for parabolic differential equations, we want to demonstrate the following two facts which are typical for general difference methods:

(i) For a given differential equation one can arrive at different difference methods by different approximations of the derivatives.

(ii) These different difference methods can exhibit completely *different* convergence behavior for the same given problem.

We consider the initial–boundary value problem

$$u_t - u_{xx} = g \quad \text{on } Q, \tag{33a}$$

$$\begin{aligned} u(x,0) &= \varphi(x) \qquad \text{for} \quad 0 \le x \le 1, \\ u(0,t) &= u(1,t) = 0 \qquad \text{for} \quad 0 \le t \le T, \end{aligned} \tag{33b}$$

where $Q = \{(x,t)\colon 0 < x < 1, 0 < t < T\}$. In order to formulate the difference method we choose the lattice points

$$x_p = p\Delta x, \qquad t_q = q\Delta t$$

with $p = 0, \ldots, N$ and $q = 0, \ldots, M$, so that $N\Delta x = 1$ and $M\Delta t = T$ (Fig. 20.1).

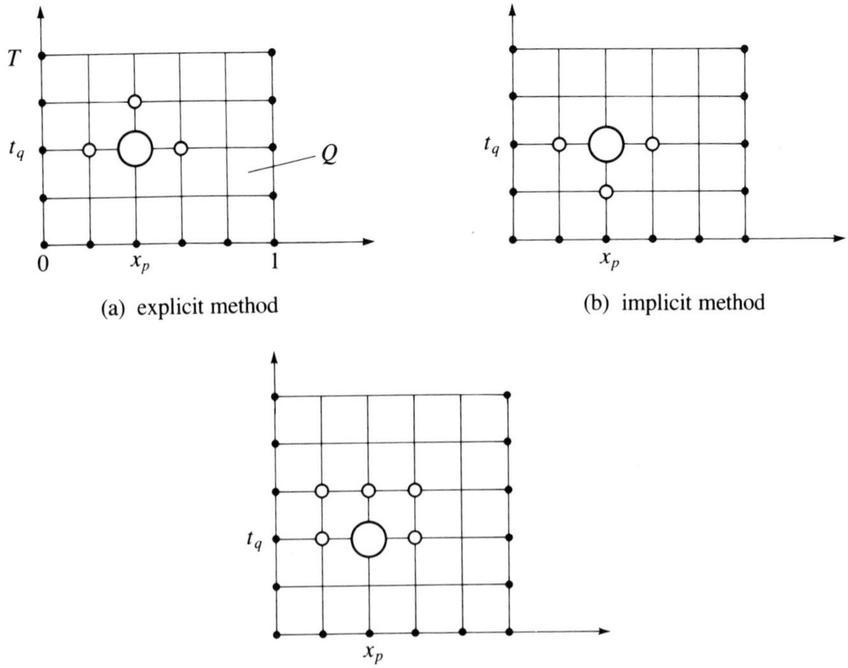

(a) explicit method (b) implicit method

(c) Crank–Nicolson method

Figure 20.1

We set

$$u_{p,q} = u(x_p, t_q)$$

and denote an approximate value for $u_{p,q}$ by $v_{p,q}$. Furthermore, let

$$g_{p,q} = g(x_p, t_q), \qquad \varphi_p = \varphi(x_p).$$

(A) *Explicit method*

$$\nabla_{\Delta t} v_{p,q} - \Delta_{\Delta x} v_{p,q} = g_{p,q}, \tag{34a}$$

$$\begin{aligned} v_{p,0} &= \varphi_p, \\ v_{0,q} = v_{N,q} &= 0. \end{aligned} \tag{34b}$$

In (34a) let $p = 1, \ldots, N-1$, $q = 0, \ldots, M-1$, and in (34b) let $p = 0, \ldots, N$, $q = 0, \ldots, M$. Equations (34a) (resp. (34b)) result from (33a) (resp. (33b)). By Section 20.2, the difference equation (34a) means

$$\frac{v_{p,q+1} - v_{p,q}}{\Delta t} - \frac{v_{p+1,q} - 2v_{p,q} + v_{p-1,q}}{(\Delta x)^2} = g_{p,q},$$

therefore, with $\rho = \Delta t/(\Delta x)^2$,

$$v_{p,q+1} = (1 - 2\rho)v_{p,q} + \rho(v_{p-1,q} + v_{p+1,q}) + g_{p,q}\Delta t. \tag{35}$$

In Figure 20.1(a) we have indicated symbolically by small circles which lattice points are involved in this equation, and by (34b) the values in the marked boundary lattice points are known. Let

$$1 - 2\rho \geq 0.$$

Using induction on $q = 0, 1, \ldots$, we obtain from (35) the decisive *stability estimate*

$$\begin{aligned} |v_{p,q}| &\leq q\Delta t \max_{p,q} |g_{p,q}| + \max_p |\varphi_p| \\ &\leq T \max_{p,q} |g_{p,q}| + \max_p |\varphi_p| \qquad \text{for all } p, q. \end{aligned} \tag{36}$$

(B) *Implicit method*

We replace the difference equation (34a) by

$$\nabla_{-\Delta t} v_{p,q} - \Delta_{\Delta x} v_{p,q} = g_{p,q}. \tag{37}$$

Here $p = 1, \ldots, N - 1$, $q = 1, \ldots, M$. The initial and boundary conditions coincide with (34b). In contrast to the explicit method (34a), the forward difference quotient $\nabla_{\Delta t}$ is replaced by the *backward difference* quotient $\nabla_{-\Delta t}$.

Analogous to (35), from (37) we obtain

$$(1 + 2\rho)v_{p,q} = v_{p,q-1} + \rho(v_{p-1,q} + v_{p+1,q}) + g_{p,q}\Delta t. \tag{38}$$

In Figure 20.1(b) we have indicated symbolically by small circles which lattice points are involved in this equation. Note that we have completely different situations in Figure 20.1(a) and (b).

We set $V_q = \max_p |v_{p,q}|$. It follows from (38) that

$$(1 + 2\rho)V_q \leq V_{q-1} + 2\rho V_q + \Delta t \max_{p,q} |g_{p,q}|,$$

therefore, $V_q \leq V_{q-1} + \Delta t \max_{p,q} |g_{p,q}|$. This yields the decisive *stability estimate*

$$\begin{aligned} |v_{p,q}| &\leq \max_p |\varphi_p| + q\Delta t \max_{p,q} |g_{p,q}| \\ &\leq \max_p |\varphi_p| + T \max_{p,q} |g_{p,q}| \qquad \text{for all } p, q. \end{aligned} \tag{39}$$

The following observation is important. Equation (38) represents a linear system of equations with as many equations as there are unknowns. By (39), if all φ_p and $g_{p,q}$ are equal to zero, then so are all $v_{p,q}$, i.e., the homogeneous system (38) has only the trivial solution. Consequently, the inhomogeneous system (38) has a unique solution.

This general argument underlines the importance of stability estimates for the solvability of difference equations.

(C) *Crank–Nicolson method*

We now replace the difference equation (34a) by

$$\nabla_{\Delta t} v_{p,q} - 2^{-1}(\Delta_{\Delta x} v_{p,q+1} + \Delta_{\Delta x} v_{p,q}) = 2^{-1}(g_{p,q+1} + g_{p,q}) \tag{40}$$

for $p = 1, \ldots, N-1$, $q = 0, \ldots, M-1$. The initial and boundary conditions coincide with (34b). In the introduction to this chapter we discussed the good behavior of method (6). In fact, (40) corresponds to (6). Thus we expect that (40) has good behavior too.

Equation (40) means

$$-\rho v_{p-1,q+1} + (2+2\rho)v_{p,q+1} - \rho v_{p+1,q+1}$$
$$= \rho v_{p-1,q} + (2-2\rho)v_{p,q} + \rho v_{p+1,q} + \Delta t(g_{p,q} + g_{p,q+1}). \quad (41)$$

In Figure 20.1(c) we have indicated by small circles which lattice points are involved in this equation.

We occupy ourselves, in Problem 20.2, with the unique solvability and the iterative solution of (41).

Example 20.5. In order to demonstrate the different behavior of the explicit method and the Crank–Nicolson method we choose

$$\varphi(x) = \begin{cases} 2x & \text{for} \quad 0 \le x \le \frac{1}{2}, \\ 2(1-x) & \text{for} \quad \frac{1}{2} \le x \le 1. \end{cases}$$

By means of Fourier series we obtain, in this special case, the explicit solution of the original problem (33) with $g \equiv 0$, namely,

$$u(x,t) = \frac{8}{\pi^2} \sum_{n=1}^{\infty} \frac{\sin(n\pi/2)}{n^2} \sin n\pi x \exp(-n^2\pi^2 t).$$

Table 20.2 shows that for $\rho = 1$ the explicit method fails. The reason lies in the fact that in (36) we have proved *stability* only for $\rho \le \frac{1}{2}$, whereas the implicit method and the Crank–Nicolson method are stable for all $\rho > 0$. The restrictive condition $\rho \le \frac{1}{2}$ in the explicit method has the decisive disadvantage that one can choose only very small time steps Δt. Thus, this simple

Table 20.2

Method	Explicit $\Delta x = 10^{-1}$, $\Delta t = 10^{-3}$		Explicit $\Delta x = 10^{-1}$, $\Delta t = 10^{-2}$	Crank–Nicolson $\Delta x = 10^{-1}$, $\Delta t = 10^{-2}$	
$\rho = \Delta t/(\Delta x)^2$	$\rho = 10^{-1}$		$\rho = 1$	$\rho = 1$	
Approximate solution	$v(0.5, t)$	Relative error	$v(0.5, t)$	$v(0.5, t)$	Relative error
$t = 0.000$	1.0000	0	1.0000	1.0000	0
$t = 0.005$	0.8597	$2.3 \cdot 10^{-2}$	—	—	—
$t = 0.01$	0.7867	$1.6 \cdot 10^{-2}$	0.6	0.7691	$-7 \cdot 10^{-3}$
$t = 0.02$	0.6891	$1.2 \cdot 10^{-2}$	1.0	0.6921	$1.6 \cdot 10^{-2}$
$t = 0.10$	0.3056	$1.2 \cdot 10^{-2}$	—	0.3069	$1.6 \cdot 10^{-2}$
Comment	100 times steps to reach $t = 0.10$		not usable	10 times steps to reach $t = 0.10$	

Table 20.3

Method	Explicit	Implicit	Crank–Nicolson
Condition on $\rho = \Delta t/(\Delta x)^2$	$0 < \rho \le \frac{1}{2}$	$\rho > 0$	$\rho > 0$
Stability	Stable	Stable	Stable
Error estimate $e_{p,q} = u(x_p, t_q) - v_{p,q}$	$\|e\| \le \|e\|_\infty \le K((\Delta x)^2 + \Delta t)$		$\sqrt{\Delta x}\, \|e\|_\infty \le \|e\| \le L((\Delta x)^2 + (\Delta t)^2)$
Convergence for $\Delta x \to 0$, $\Delta t \to 0$	Convergence for fixed ρ with $0 < \rho \le \frac{1}{2}$	Convergence for fixed ρ with $\rho > 0$	Convergence for fixed $\rho > 0$ or for fixed $(\Delta t)^2/(\Delta x)^2$
$\|e\| = \max_q (\sum_p e_{p,q}^2 \Delta x)^{1/2}$,	$\|e\|_\infty = \max_{p,q} \|e_{p,q}\|$		

example already shows that the stability investigation is not only of theoretical interest but also of great *practical* interest.

Proposition 20.6. *If the original initial–boundary value problem* (33) *has a solution* $u \in C^4(\bar{Q})$ *(resp.* $u \in C^5(\bar{Q})$*), then all the assertions in Table* 20.3 *are valid for the explicit and the implicit method (resp. for the Crank–Nicolson method).*

Here the constants K and L depend only on certain derivatives of u. They are independent of Δx and Δt. By the maximum principle for parabolic equations, u is the unique solution of (33) (cf. Problem 7.2d). Moreover, all the difference equations are uniquely solvable.

Proof for the Explicit Method. Let $e_{p,q} = u(x_p, t_q) - v_{p,q}$ and let $\|u\|$ be the norm of u in $C^4(\bar{Q})$.

(I) Consistency. It follows from (20) and (22) that

$$\nabla_{\Delta t} e_{p,q} - \Delta_{\Delta x} e_{p,q} = \tau_{p,q}$$

with $|\tau_{p,q}| \le (2^{-1}\Delta t + 12^{-1}(\Delta x)^2)\,\|u\|$.

(II) Stability. By (36) we obtain

$$|e_{p,q}| \le T \max_{p,q} |\tau_{p,q}|.$$

This yields the convergence. □

One argues in an analogous manner for the implicit method. We discuss the proof for the Crank–Nicolson method in Problem 20.3.

20.5. Application to Elliptic Differential Equations

Let $Q = \{(x, y) \in \mathbb{R}^2 : 0 < x < a, 0 < y < b\}$. We consider the boundary value problem

$$u_{xx} + u_{yy} = g \quad \text{on } Q, \tag{42a}$$

$$u = \varphi \quad \text{on } \partial Q. \tag{42b}$$

In order to formulate the difference method we choose the lattice points

$$x_p = p\Delta x, \qquad y_q = q\Delta y$$

with $\Delta x = \Delta y = h$ and $N\Delta x = a$, $M\Delta y = b$ (Fig. 20.2). We denote by $v_{p,q}$ an approximate value for $u(x_p, y_q)$. Furthermore, let

$$g_{p,q} = g(x_p, y_q), \qquad \varphi_{p,q} = \varphi(x_p, y_q).$$

As the difference equation we choose

$$\Delta_{\Delta x} v_{p,q} + \Delta_{\Delta y} v_{p,q} = g_{p,q} \quad \text{at all interior lattice points } (x_p, y_q) \text{ of } Q, \tag{43a}$$

$$v_{p,q} = \varphi_{p,q} \quad \text{at boundary lattice points.} \tag{43b}$$

By Section 20.2, equation (43a) means

$$4^{-1}(v_{p+1,q} + v_{p,q+1} + v_{p-1,q} + v_{p,q-1}) - v_{p,q} = 4^{-1}(\Delta x)^2 g_{p,q}. \tag{44}$$

In Figure 20.2 we indicate with little circles which lattice points are involved in this equation. The values at all the boundary lattice points are known.

If $g_{p,q} = 0$ for all p, q, then it follows from (44) that $v_{p,q}$ is the mean of four lattice point values (cf. Figure 20.2 where $v_{p,q}$ corresponds to the full circle). This immediately implies the following discrete maximum principle:

(M_d) *If the function* $(p, q) \mapsto v_{p,q}$ *is a solution of* (43) *with* $g_{p,q} = 0$ *for all* p, q, *then this function takes on its maximum and minimum on the boundary* ∂Q.

In order to solve (43), observe the following. The linear system of equations (43) has as many equations as unknowns. The corresponding homogeneous

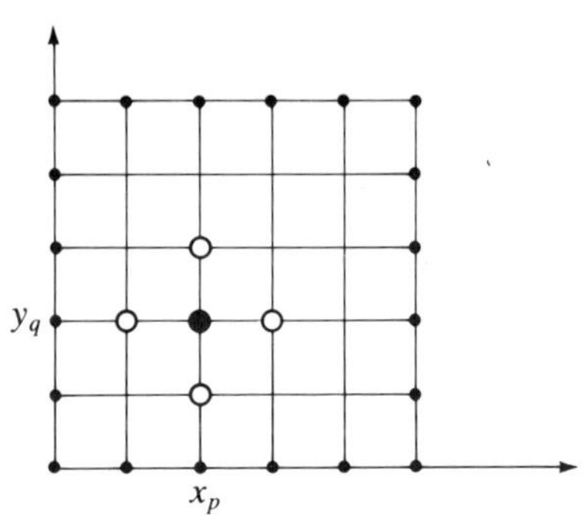

Figure 20.2

system (43) with $g_{p,q} = 0$ and $\varphi_{p,q} = 0$ for all p, q has only the trivial solution $v_{p,q} = 0$ for all p, q, by (M_d). Thus, the inhomogeneous system (43) has a *unique* solution.

We deal with the important iterative solution of (44) in Problem 20.4.

In connection with the discrete maximum principle (M_d), recall the classical maximum principle:

(M) *If the function* $u \in C^2(Q) \cap C(\bar{Q})$ *is a solution of* (42) *with* $g \equiv 0$, *then this function takes on its maximum and minimum on the boundary* ∂Q.

From (M) it follows immediately that the original problem (42) has at most one solution $u \in C^2(Q) \cap C(\bar{Q})$.

Proposition 20.7. *If the original boundary value problem* (42) *has a solution* $u \in C^4(\bar{Q})$, *then*

$$\max_{p,q} |u(x_p, y_q) - v_{p,q}| \leq Kh^2. \tag{45}$$

The constant K depends only on Q and the fourth partial derivatives of u.

Thus, the difference method (43) converges of order h^2 as $h \to 0$.

PROOF.

(I) Consistency. Let

$$e_{p,q} = u(x_p, y_q) - v_{p,q}.$$

It follows from (22), (43), and (44) that

$$\Delta_{\Delta x} e_{p,q} + \Delta_{\Delta y} e_{p,q} = \tau_{p,q} \tag{46}$$

with

$$|\tau_{p,q}| \leq \frac{h^2}{6} \max_{P \in \bar{Q}} \left| \frac{\partial^4 u(P)}{\partial x^4} \right| + \left| \frac{\partial^4 u(P)}{\partial y^4} \right|.$$

(II) Stability. We use a *discrete maximum principle* to show that

$$\max_{p,q} |e_{p,q}| \leq 2^{-1}\alpha(a^2 + b^2) \tag{46*}$$

with

$$\alpha = \max_{p,q} |\tau_{p,q}|.$$

In order to prove this let $w(x, y) = 4^{-1}\alpha(x^2 + y^2)$ and $w_{p,q} = w(x_p, y_q)$. We have

$$w_{xx} + w_{yy} = \alpha.$$

The function w has vanishing third derivatives. Therefore, by (21), we also have

$$\Delta_{\Delta x} w_{p,q} + \Delta_{\Delta y} w_{p,q} = \alpha.$$

Letting $z_{p,q} = e_{p,q} + w_{p,q}$ we obtain

$$\Delta_{\Delta x} z_{p,q} + \Delta_{\Delta y} z_{p,q} = \tau_{p,q} + \alpha \geq 0.$$

After multiplication by h^2, this is equivalent to the key formula

$$z_{p,q} \leq 4^{-1}(z_{p+1,q} + z_{p,q+1} + z_{p-1,q} + z_{p,q-1}).$$

By the same argument as in the proof of the discrete maximum principle (M_d) above, this implies that the function $(p, q) \mapsto z_{p,q}$ takes on its maximum at boundary lattice points. However, on the boundary ∂Q, we have

$$e_{p,q} = 0 \qquad \text{and} \qquad w_{p,q} \leq 4^{-1}\alpha(a^2 + b^2);$$

therefore, $z_{p,q} \leq 4^{-1}\alpha(a^2 + b^2)$ holds for all p, q. That is,

$$e_{p,q} \leq 2^{-1}\alpha(a^2 + b^2) \qquad \text{for all } p, q.$$

Similarly, if we define $z_{p,q} = e_{p,q} - w_{p,q}$, then we obtain

$$e_{p,q} \geq -2^{-1}\alpha(a^2 + b^2) \qquad \text{for all } p, q.$$

Consequently, (46*) holds.

(III) Convergence. From (46*) one immediately obtains (45). □

20.6. The Equivalence Between Stability and Convergence

We consider the operator equation

$$Au = b, \qquad u \in X, \tag{47}$$

along with the approximate equations

$$A_n u_n = b_n, \qquad u_n \in X, \quad n = 1, 2, \ldots, \tag{48}$$

and we formulate the following conditions:

(H1) *Well-posedness of the problem.* Let X and Y be B-spaces over $\mathbb{K} = \mathbb{R}, \mathbb{C}$. The operators

$$A, A_n\colon X \to Y, \qquad n = 1, 2, \ldots,$$

are linear, continuous, and bijective.

It follows from the open mapping theorem $\mathrm{A}_1(36)$ that the inverse operators

$$A^{-1}, A_n^{-1}\colon Y \to X, \qquad n = 1, 2, \ldots,$$

are continuous. This implies the *a priori* estimates

$$\begin{aligned} \|Au\| &\geq c\|u\| && \text{for all } u \in X, \\ \|A_n u\| &\geq c_n\|u\| && \text{for all } u \in X, \quad n = 1, 2, \ldots, \end{aligned} \tag{49}$$

where c and all the c_n are positive constants. Hence condition (H1) tells us that equation (47) (resp. (48)) has a unique solution for each $b \in Y$ (resp. $b_n \in Y$), and this solution depends continuously on b (resp. b_n).

(H2) *Consistency*. We have

$$A_n u \to Au \quad \text{for all} \quad u \in X \quad \text{as} \quad n \to \infty,$$
$$b_n \to b \quad \text{as} \quad n \to \infty.$$

(H3) *Stability*. There is a constant d such that

$$\|A_n^{-1}\| \le d \quad \text{for all } n.$$

(H4) *Convergence*. From

$$b_n \to b \quad \text{in } Y \quad \text{as} \quad n \to \infty$$

it follows that:

$$u_n \to u \quad \text{in } X \quad \text{as} \quad n \to \infty.$$

Here u and u_n denote the solution of (47) and (48), respectively.

Theorem 20.B (Kantorovič (1948)). *From well-posedness* (H1) *and consistency* (H2) *it follows that*:

$$\textit{stability}\text{ (H3)} \quad \Leftrightarrow \quad \textit{convergence}\text{ (H4)}.$$

PROOF. The key to the proof is the uniform boundedness theorem.

(I) (H3) $\Rightarrow$ (H4). By (H2) and (H3),

$$\|u - u_n\| \le d\|A_n(u - u_n)\| = d\|A_n u - b_n\|$$
$$\to \|Au - b\| \quad \text{as} \quad n \to \infty.$$

From $Au = b$ it follows that $u_n \to u$ as $n \to \infty$.

(II) (H4) $\Rightarrow$ (H3). We set $b = b_n$ for all n. Then (H4) implies that

$$A_n^{-1} b \to A^{-1} b \quad \text{as} \quad n \to \infty \quad \text{for all} \quad b \in Y.$$

The uniform boundedness principle A_1(35) tells us that

$$\sup_n \|A_n^{-1}\| < \infty. \qquad \square$$

20.7. The Equivalence Theorem of Lax for Evolution Equations

We want to translate the fundamental equivalence principle of Theorem 20.B to evolution equations. To this end, we consider the *initial value problem*

$$\begin{aligned} u'(t) &= Au(t), \qquad 0 \le t \le T < \infty, \\ u(0) &= u_0 \in D \end{aligned} \tag{50}$$

for fixed $T > 0$ together with the abstract *difference method*

$$v(h) = B(h)u_0,$$
$$v(2h) = B(h)v(h) = B(h)^2 u_0, \ldots,$$

that is,

$$v(nh) = B(h)^n u_0, \qquad n = 1, 2, \ldots, \quad 0 < h \le h_0. \tag{51}$$

We formulate the following conditions:

(H1) *Well-posedness of the problem.* Let X be a B-space and let

$$A: D(A) \subseteq X \to X$$

be a linear operator where $D \subseteq D(A)$ and D is a dense subset of X. For each initial value $u_0 \in D$, problem (50) has a unique solution $u = u(t)$, and this solution depends continuously on the initial values, i.e., there is a constant c such that

$$\sup_{0 \le t \le T} \|u(t) - \bar{u}(t)\| \le c\|u(0) - \bar{u}(0)\| \qquad \text{for all} \quad u(0), \bar{u}(0) \in D. \tag{52}$$

For each $h \in \,]0, h_0]$, the operator

$$B(h): X \to X$$

is linear and continuous with $\sup_{0 < h \le h_0} \|B(h)\| < \infty$.

In order to define the notion of a generalized solution of problem (50) we define the linear operator $S(t)$ by

$$u(t) = S(t)u_0,$$

where $u_0 \in D$ and $u = u(t)$ denote the unique solution of (50). It follows from (52) that

$$\sup_{0 \le t \le T} \|S(t)u_0 - S(t)\bar{u}_0\| \le c\|u_0 - \bar{u}_0\| \qquad \text{for all} \quad u_0, \bar{u}_0 \in D. \tag{53}$$

By the extension principle in Section 18.12, the operator $S(t): D \to X$ can be uniquely extended to a linear continuous operator

$$S(t): X \to X \qquad \text{for all} \quad t \in [0, T]$$

so that (53) remains valid for all $u_0, \bar{u}_0 \in X$. We call

$$u(t) = S(t)u_0, \qquad 0 \le t \le T,$$

a *generalized solution* of (50) if $u_0 \in X$. In the case where $u_0 \in D$, the generalized solution is also a classical solution. From (53) we get

$$\sup_{0 \le t \le T} \|S(t)\| \le c. \tag{54}$$

(H2) *Consistency.* There exists a dense subset D_1 of the B-space X so that

$$\lim_{h \to 0} \left\| \frac{B(h)u(t) - u(t+h)}{h} \right\| = 0 \qquad \text{uniformly on } [0, T]$$

for all generalized solutions $u = u(t)$ of (50) with $u(0) \in D_1$.

(H3) *Stability.* There exists a constant b such that

$$\|B(h)^n\| \le b$$

for all $h \in]0, h_0]$ and $n \in \mathbb{N}$ with $nh \le T$.

(H4) *Convergence of the difference method.* For all $t \in [0, T]$ and all initial values $u_0 \in X$ we have

$$\lim_{k \to \infty} \|B(h_k)^{n_k} u_0 - u(t)\| = 0.$$

Here u denotes the generalized solution of (50) with $u(0) = u_0$. Moreover, (h_k) is an arbitrary zero sequence, and (n_k) is an arbitrary monotone sequence of natural numbers with

$$\lim_{k \to \infty} n_k h_k = t.$$

Recall that $v(nh) = B(h)^n u_0$ denotes the corresponding solution of the difference method.

Theorem 20.C (Lax (1953)). *From well-posedness* (H1) *and consistency* (H2) *it follows that*:

$$\textit{stability (H3)} \quad \Leftrightarrow \quad \textit{convergence (H4)}.$$

Corollary 20.8 (Rapidity of Convergence). *If the order of consistency is h^p, $p > 0$, then the rapidity of convergence is also h^p.*

More precisely, let u be a generalized solution of (50) *with $u(0) \in D_1$. Suppose that* (H1), (H2), (H3) *hold with*

$$\sup_{0 \le t \le T} \left\| \frac{B(h)u(t) - u(t+h)}{h} \right\| \le Ch^p$$

for all $h \in]0, h_0]$ where C is a constant. Then we obtain

$$\sup_{0 \le t \le T} \|B(h)^n u(0) - u(t)\| \le Kh^p$$

for all $h \in]0, h_0]$ and $n \in \mathbb{N}$ with $nh = t$. Here K is a constant.

PROOF.

(I) We show that every generalized solution $u = u(t)$ of (50) is continuous on $[0, T]$. Indeed, this is true if $u(0) \in D$. In the case where $u(0) \in X$, use the density of D in X, condition (54), and the triangle inequality. More precisely, we begin with the decomposition

$$S(t)u_0 - S(s)u_0 = S(t)v - S(s)v + S(t)(u_0 - v) - S(s)(u_0 - v).$$

Then for each $\varepsilon > 0$, we can choose a point $v \in D$, sufficiently close to u_0, such that

$$\|S(t)u_0 - S(s)u_0\| \le \|S(t)v - S(s)v\| + 2 \sup_t \|S(t)\| \, \|u_0 - v\| < \varepsilon$$

if $|t - s| < \delta(\varepsilon)$.

(II) (H3) $\Rightarrow$ (H4). We set

$$d = B(h)^n u(0) - u(t),$$

where $nh \le T$ and $nh \to t$ as $n \to \infty$. The *key* to the proof is the decomposition

$$d = \sum_{k=0}^{n-1} B(h)^k [B(h)u((n-1-k)h) - u((n-k)h)] + u(nh) - u(t).$$

(II-1) Let $u(0) \in D_1$. By (H2) and (H3),

$$\|d\| \le \sup_{h,k} \|B(h)^k\| \sup_t nh \left\| \frac{B(h)u(t) - u(t+h)}{h} \right\|$$
$$+ \|u(nh) - u(t)\| \to 0 \qquad \text{as} \quad n \to \infty.$$

Note that $nh \le T$ and $nh \to t$ as $n \to \infty$. This is (H4).

The same argument yields Corollary 20.8.

(II-2) Let $u(0) \in X$ and let $u(t) = S(t)u(0)$. We will use a similar approximation argument as in (I). To this end, we begin with the decomposition

$$d = B(h)^n v - S(t)v + B(h)^n(u(0) - v) - S(t)(u(0) - v)).$$

For each $\varepsilon > 0$ we can choose a $v \in D_1$, sufficiently close to $u(0)$, such that

$$\|d\| \le \|B(h)^n v - S(t)v\| + \left(\sup_{h,n} \|B(h)^n\| + \|S(t)\| \right) \|u(0) - v\| < \varepsilon$$

if $n \ge n_0(\varepsilon)$. This is (H4).

(III) (H4) $\Rightarrow$ (H3). Suppose that (H3) is not true. Then there exist sequences (h_k) and (n_k) with

$$\lim_{k\to\infty} \|B(h_k)^{n_k}\| = \infty \tag{55}$$

and $n_k h_k \le T$ for all k. By compactness we may assume that (h_k) is convergent. If (n_k) is bounded, then we obtain

$$\sup_k \|B(h_k)^{n_k}\| \le \sup_k \|B(h_k)\|^{n_k} < \infty,$$

by (H1). But this contradicts (55). Thus (n_k) is unbounded, and hence $h_k \to 0$ as $k \to \infty$.

By the convergence condition (H4),

$$\sup_k \|B(h_k)^{n_k} u_0\| < \infty \qquad \text{for all} \quad u_0 \in X.$$

The uniform boundedness principle $A_1(35)$ tells us that

$$\sup_k \|B(h_k)^{n_k}\| < \infty.$$

This contradicts (55). □

PROBLEMS

20.1. *Special tridiagonal matrices.* Let $b \in \mathbb{R}$ and $M \geq 2$.

20.1a. Show that the $M \times M$-matrix

$$A_M(b) = \begin{pmatrix} b & -1 & & & 0 \\ -1 & b & -1 & & \\ & \ddots & \ddots & \ddots & \\ & & -1 & b & -1 \\ 0 & & & -1 & b \end{pmatrix} \tag{56}$$

has the eigenvalues

$$\lambda_j = b - 2\cos\alpha_j, \qquad j = 1, \ldots, M,$$

and the eigenvectors

$$y_j = (\sin\alpha_j, \sin 2\alpha_j, \ldots, \sin M\alpha_j)^T, \qquad j = 1, \ldots, M,$$

where $\alpha_j = j\pi/(M+1)$.

Solution: Explicit computation. Cf. Zurmühl (1964, M), p. 229.

20.1b. Let $A = (a_{ij})$ and $B = (b_{ij})$. We write

$$0 \leq A \leq B \qquad \text{iff} \qquad 0 \leq a_{ij} \leq b_{ij} \qquad \text{for all } i, j.$$

Show that

$$0 \leq A_M(b)^{-1} \leq A_M(2)^{-1} \qquad \text{for all} \quad b \geq 2.$$

Solution: Let $b \geq 2$. We consider the $M \times M$-matrix

$$B(b) = \begin{pmatrix} 0 & b^{-1} & 0 & & 0 \\ b^{-1} & 0 & b^{-1} & & \\ & \ddots & \ddots & \ddots & \\ & & b^{-1} & 0 & b^{-1} \\ 0 & & 0 & b^{-1} & 0 \end{pmatrix}.$$

Obviously,

$$A_M(b) = b(I - B(b)).$$

By Problem 20.1a, the eigenvalues of the matrix $A_M(2)$ satisfy $0 < \lambda_j < 4$ for all j. Hence the eigenvalues of $B(2) = I - 2^{-1}A_M(2)$ satisfy $-1 < \mu_j < 1$ for all j. Thus, the Neumann series

$$(I - B(2))^{-1} = I + B(2) + B(2)^2 + \cdots \geq 0$$

converges. Since $0 \leq B(b) \leq B(2)$, we also obtain the convergence of the series

$$(I - B(b))^{-1} = I + B(b) + B(b)^2 + \cdots \leq (I - B(2))^{-1}.$$

This implies

$$A_M(b)^{-1} = b^{-1}(I - B(b))^{-1} \leq 2^{-1}(I - B(2))^{-1} = A_M(2)^{-1}.$$

20.2. *Crank–Nicolson method and the iterative solution of the corresponding difference equations.* Parallel to Problem 1.6, investigate standard iterative methods for solving the difference equations (41).

Solution: For fixed $q + 1$, equation (41) reads as follows:

$$v_{p,q+1} = \rho(2 + 2\rho)^{-1}(v_{p-1,q+1} + v_{p+1,q+1}) + b_{p,q}, \qquad p = 1, \ldots, N - 1. \quad (57)$$

Here $b_{p,q}$ depends only on $v_{r,q}$ and $g_{r,s}$. Therefore we can calculate $v_{p,q}$ successively for the time lines $q = 0, 1, \ldots$. In fact, if $v_{p,q}$ is known up to the qth time line, then $b_{p,q}$ is known and the values of v on the $(q + 1)$th time line can be obtained from (57).

In (i)–(iii) below let $n = 0, 1, \ldots$ and $v^{(0)} = 0$.

(i) The *total step method* reads as follows:

$$v_{p,q+1}^{(n+1)} = \rho(2 + 2\rho)^{-1}(v_{p-1,q+1}^{(n)} + v_{p+1,q+1}^{(n)}) + b_{p,q}. \quad (58)$$

Here $q + 1$ is fixed and $p = 1, \ldots, N - 1$.

(ii) The *overrelaxation method* reads as follows:

$$v_{p,q+1}^{(n+1)} = \omega[\rho(2 + 2\rho)^{-1}(v_{p-1,q+1}^{(n+1)} + v_{p+1,q+1}^{(n)}) + b_{p,q}] - (\omega - 1)v_{p,q+1}^{(n)}, \quad (59)$$

where $0 < \omega < 2$.

Note that both the equations (58) and (59) are *equivalent* to the original equation (57) if we simply replace both $v^{(n)}$ and $v^{(n+1)}$ with v.

(iii) The *single step method* results from (59) for $\omega = 1$ which, in contrast to (58), always makes use of the newest known iteration values.

If we write (58) in the form

$$v^{(n+1)} = Av^{(n)} + b,$$

then $A = -\rho(2 + 2\rho)^{-1}A_{N-1}(0)$. By Problem 20.1, the eigenvalues of A are equal to $\lambda_j = \rho(1 + \rho)^{-1}\cos j\pi/N, j = 1, \ldots, N - 1$; therefore, we obtain for the *spectral radius* r of the matrix A,

$$r = \rho(1 + \rho)^{-1}\cos(\pi/N),$$

i.e., $r < 1$. Thus, the system (57) has *exactly one solution* and the iterative method (58) *converges* by Proposition 1.15.

According to Table 1.2 the corresponding spectral radius for the single step method is r^2 and the optimal relaxation parameter results for

$$\omega_{\text{opt}} = 1/(1 + \sqrt{1 - r^2})$$

with the spectral radius $\omega_{\text{opt}} - 1$ which is *less* than r^2.

Therefore, the methods converge faster and faster in the sequence: total step method, single step method, and optimal relaxation method. The corresponding asymptotic rates of convergence R_∞ are equal to $\log_{10} r^{-1}, 2\log_{10} r^{-1}, \log_{10}(\omega_{\text{opt}} - 1)^{-1}$ (cf. Definition 1.11).

20.2a. *An example for the rapidity of convergence.* We consider Example 20.5 once more with $\Delta x = 0.1$ and $\Delta t = 0.01$. Let n be the number of iteration steps needed to calculate $v_{5,1}$ on the first time line with four place precision. Show that $n = 11$ (total step method), $n = 7$ (single step method), and $n = 5$ (optimal overrelaxation).

20.3. *Proof of convergence for the Crank–Nicolson method.* Let $u \in C^5(\bar{Q})$.

Solution:

(I) Consistency. We denote the difference equation (40) by

$$L_\Delta v = g.$$

Let $e_{p,q} = u(x_p, t_q) - v_{p,q}$. As in Section 20.2, there results from the Taylor theorem that

$$L_\Delta e_{p,q} = \tau_{p,q} \tag{60}$$

with

$$\tau_{p,q} = 2^{-1}\Delta t[u_{tt}(x_p, t_q) - u_{xxt}(x_p, t_q) - g_t(x_p, t_q)] + O((\Delta x)^2 + (\Delta t)^2).$$

Since u satisfies the differential equation $u_t - u_{xx} = g$, the term with Δt drops out. That leads to an *increase* in the rapidity of convergence.

(II) Stability. Our goal is the inequality

$$\|e\| \le T\|\tau\| \tag{61}$$

with $\|e\| = \max_q(\sum_p e_{p,q}^2 \Delta x)^{1/2}$ and an analogous expression for $\|\tau\|$. Parallel to (41), there results from (60) the system of equations

$$\begin{aligned}(2 + 2\rho)e_{p,q+1} &- \rho(e_{p-1,q+1} + e_{p+1,q+1})\\ &= (2 - 2\rho)e_{p,q} + \rho(e_{p-1,q} + e_{p+1,q}) + \Delta t(\tau_{p,q} + \tau_{p,q+1})\end{aligned}$$

for $q = 1, \ldots, M - 1$ and $p = 1, \ldots, N - 1$ with $e_{0,q} = e_{N,q} = 0$. If we denote the column vector $(e_{1,q}, \ldots, e_{N-1,q})^T$ by e_q and the identity matrix by I, then we obtain

$$(2I + \rho T_{N-1})e_{q+1} = (2I - \rho T_{N-1})e_q + \Delta t(\tau_q + \tau_{q+1})$$

with $T_{N-1} = A_{N-1}(2)$ (cf. (56)); therefore,

$$\begin{aligned}\sqrt{\Delta x}\, e_{q+1} &= (2I + \rho T_{N-1})^{-1}(2I - \rho T_{N-1})\sqrt{\Delta x}\, e_q\\ &\quad + \Delta t(2I + \rho T_{N-1})^{-1}\sqrt{\Delta x}(\tau_q + \tau_{q+1}).\end{aligned} \tag{62}$$

By A_1(59a), the norm of a symmetric matrix with respect to the Euclidean norm $\|\cdot\|_2$ is equal to the spectral radius. According to Problem 20.1 and the spectral mapping theorem A_1(60c), the assertions in Table 20.4 are valid.

Table 20.4

Matrix	T_{N-1}	$A_\pm = (2I \pm \rho T_{N-1})$	A_+^{-1}	$A_+^{-1}A_-$
Eigenvalues ($s = 1, \ldots, N - 1$)	$\lambda_s = 4\sin^2(s\pi/2N)$	$\lambda_s^\pm = 2 \pm 4\rho\sin^2(s\pi/2N)$	$1/\lambda_s^+$	λ_s^-/λ_s^+
Spectral radius $r = \max_s\lvert\lambda_s\rvert$			$r \le \frac{1}{2}$	$r < 1$

It thus follows from (62) that

$$\|\sqrt{\Delta x}\, e_{q+1}\|_2 \le \|\sqrt{\Delta x}\, e_q\|_2 + 2^{-1}\Delta t(\|\sqrt{\Delta x}\,\tau_q\|_2 + \|\sqrt{\Delta x}\,\tau_{q+1}\|_2).$$

For $q = 0$, we have $e_q = 0$. Induction with respect to $q = 0, 1, \ldots$ yields

$$\|\sqrt{\Delta x}\, e_q\|_2 \le q\Delta t \max_q \|\sqrt{\Delta x}\,\tau_q\|_2.$$

This implies (61).

(III) Convergence. From (61) and $\|\tau\| = O((\Delta x)^2 + (\Delta t)^2)$ there follow the error estimates in Table 20.3.

20.4. *Iterative solution of the difference equations for the Poisson equation.* Investigate (44), i.e., study the system

$$v_{p-1,q} + v_{p+1,q} + v_{p,q-1} + v_{p,q+1} - 4v_{p,q} = (\Delta x)^2 g_{p,q} \tag{63}$$

for the rectangle in Figure 20.2, where $p = 1, \ldots, N-1$ and $q = 1, \ldots, M-1$.

Solution: We write (63) in the form

$$v = Av + b,$$

where

$$A = \begin{pmatrix} B & I & & 0 \\ I & B & \ddots & \\ & \ddots & \ddots & I \\ 0 & & I & B \end{pmatrix}, \qquad B = \begin{pmatrix} -4 & 1 & & 0 \\ 1 & -4 & \ddots & \\ & \ddots & \ddots & 1 \\ 0 & & 1 & -4 \end{pmatrix}. \tag{64}$$

Here, I denotes the identity matrix. According to the row sum criterion in Section 1.8, we obtain $r(A) \le \frac{1}{2}$ for the spectral radius of the matrix A. Consequently, the total step method

$$v^{(n+1)} = Av^{(n)} + b, \qquad n = 0, 1, \ldots,$$

converges to the solution v of (63) as $n \to \infty$ for each given $v^{(0)}$. The same holds true for the single step method, the optimal relaxation method, and the block iteration method with the rapidity of convergence given in Tables 1.2 and 1.3.

For nonrectangular regions, there result tridiagonal block matrices that have a more complicated structure in contrast to (64).

Today one uses the highly effective *multigrid methods* on computers (cf. Problem 20.9).

We recommend Varga (1962, M), Chapter 6, Meis and Marcowitz (1978, M), and Hackbusch (1985, M) (multigrid methods). A software system can be found in Rice and Boisvert (1984, M).

20.5. *Applications to reactor technology.* The calculation of neutron flows in nuclear reactors leads to partial differential equations with piecewise constant coefficients that correspond to the various reactor materials. The difference method leads to systems of equations with 10^5 and more unknowns whose matrices, however, contain many zeros (sparse matrices). In this connection, study Varga (1962, M), p. 302 (optimal overrelaxation block iterations), Hageman and Young (1981, M) (sparse matrices), and Hackbusch (1985, M) (multigrid methods).

The most challenging problem in technology is controlled nuclear fusion. Here, the basic equations of magnetohydrodynamics and the stability of plasma play a fundamental role. In this connection, numerical methods can be found in Bauer, Betancourt, and Garabedian (1978, M), (1984, M), (1986), Teller (1981, M), Vols. 1, 2, and Betancourt (1988).

20.6. *Difference methods in gas dynamics.* Shock waves represent a very interesting physical and mathematical phenomenon. They occur in supersonic flow of gases and correspond to jumps of velocity, pressure, entropy, temperature, etc.

(see Part V). The sonic booms that are caused by supersonic aircraft belong to these. Till now, exact proofs of existence and uniqueness for the full equations of gas dynamics are lacking. One has, however, set up sophisticated difference methods for numerical simulations.

In this connection, study Smoller (1983, M) (the stochastic Glimm difference method), Richtmyer and Morton (1967, M) (the Lax–Wendroff difference method), and Roždestvenskii and Janenko (1978, M) (comprehensive presentation). See also Chapter 86 from Part V.

Fundamental results on the design of shock-free airfoils based on numerical methods can be found in Bauer, Garabedian, and Korn (1972/77, M), Vols. 1–3.

20.7. *The stability test of John von Neumann by means of finite Fourier series.* In this chapter we used *a priori* estimates and the spectral properties of matrices in order to study the stability of difference methods. There is a more explicit method for the investigation of stability which represents the solutions of the difference equations by finite Fourier series.

In order to have something definite at hand, we consider the explicit method (34) for the heat equation in Section 20.4. Let $v^{(1)}$ and $v^{(2)}$ be two solutions of the difference equation (34) that differ only by different initial values. Let the boundary values and the inhomogeneous member $g_{p,q}$ be equal for both. We set $w = v^{(1)} - v^{(2)}$. Then

$$\nabla_{\Delta t} w_{p,q} - \Delta_{\Delta x} w_{p,q} = 0 \tag{65}$$

holds for $p = 1, \ldots, N-1$ and $q = 0, \ldots, M-1$. We interpret w as the error in the solution $v^{(1)}$ and ask ourselves when the initial error $w_{p,0}$ at time $t = 0$ remains bounded for all time steps. The *basic idea* consists of the trial

$$w_{p,q} = \sum_{n=0}^{N} C_n \xi_n^p \eta_n^q \tag{66}$$

with

$$\xi_n = e^{i\pi n\Delta x}, \qquad \eta_n = e^{\alpha_n \Delta t},$$

and with the constants C_n to be determined. If $\operatorname{Re}\alpha_n \leq 0$, i.e.,

$$|\eta_n| \leq 1 \qquad \text{for all } n, \tag{67}$$

then

$$|w_{p,q}| \leq \sum_n |C_n| \qquad \text{for all } p, q,$$

i.e., the errors remain *uniformly* bounded. In the case of (67), the difference method is said to be *stable* in the sense of John von Neumann.

20.7a. *Application to the explicit method for the heat equation.* Show that for equation (65) the coefficients C_n are uniquely determined from the initial errors $w_{p,0}$ and that for

$$\Delta t/(\Delta x)^2 \leq \tfrac{1}{2}$$

the method is stable in the sense of John von Neumann. This stability assertion coincides with Table 20.3, which refers to our general notion of stability.

Solution: Let $\rho = \Delta t/(\Delta x)^2$. The coefficient determinant for the determination of C_n is a Vandermonde determinant and thus not equal to zero. From (65) and (66) we obtain

$$w_{p,q+1} - w_{p,q} = \rho(w_{p-1,q} - 2w_{p,q} + w_{p+1,q}),$$

therefore $\eta_n - 1 = \rho(\xi_n^{-1} - 2 + \xi_n)$, i.e.,

$$\eta_n = 1 - 4\rho \sin^2(\pi n \Delta x/2).$$

Thus, if $\rho \leq \frac{1}{2}$, then

$$|\eta_n| \leq 1 \qquad \text{for all } n.$$

20.7b. *Application to the implicit method for the heat equation.* Show that the implicit method (37) is stable in the sense of John von Neumann for all $\Delta t/(\Delta x)^2 > 0$.

Solution: One obtains

$$\eta_n = (1 + 4\rho \sin^2(\pi n \Delta x/2))^{-1}.$$

20.7c. *Application to the wave equation.* To solve the wave equation

$$u_{tt} - u_{xx} = g, \tag{68}$$

we make use of the difference method

$$\Delta_{\Delta t} v_{p,q} - \Delta_{\Delta x} v_{p,q} = g_{p,q},$$

i.e.,

$$v_{p,q+1} - 2v_{p,q} + v_{p,q-1} = r^2(v_{p+1,q} - 2v_{p,q} - v_{p-1,q}) + g_{p,q}(\Delta t)^2, \tag{69}$$

with $r = \Delta t/\Delta x$. In Figure 20.3 we have marked the points involved in (69) by circles. As usual, $v_{p,q}$ denotes an approximate value for $u(x_p, t_q)$, where $x_p = p\Delta x$ and $t_q = q\Delta t$ with $p = 0, \pm 1, \pm 2, \ldots$ and $q = 0, 1, 2, \ldots$. In addition, we set $g_{p,q} = g(x_p, t_q)$.

More precisely, we want to solve the *initial value problem* for the wave equation (68), i.e., we have to add the following initial condition to (68):

$$u(x, 0) = a(x) \qquad \text{and} \qquad u_t(x, 0) = b(x) \qquad \text{for all} \quad x \in \mathbb{R},$$

where a and b are given functions.

This means that the values $v_{p,q}$ are prescribed for the time lines $q = 0$ and $q = 1$, namely, we set

$$v_{p,0} = a(x_p) \qquad \text{and} \qquad v_{p,1} = v_{p,0} + b(x_p)\Delta t \qquad \text{for all} \quad p.$$

From the difference equation (69) we obtain successively the values $v_{p,q}$ for the time lines $q = 2, 3, \ldots$.

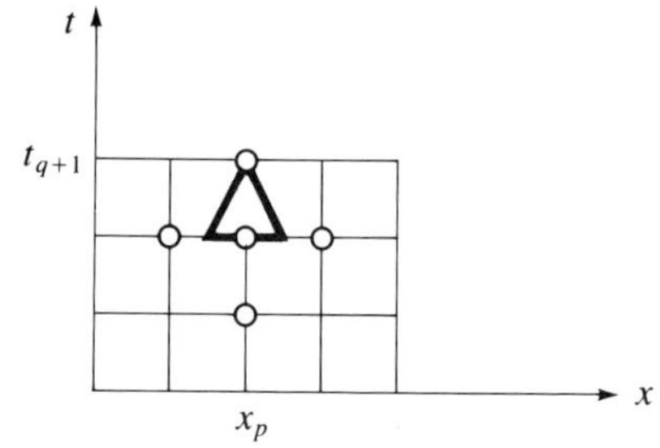

Figure 20.3

Show that for $r \leq 1$ the difference method (69) is stable in the sense of John von Neumann.

Hint: One obtains

$$\eta_n^2 - 2\eta_n(1 - 2r^2 \sin(\pi n \Delta x)) + 1 = 0.$$

Cf. Smith (1965, M), Example 3.2. This result permits an interpretation that is generally important for hyperbolic differential equations. The theory of characteristics for the wave equation shows that the value $u(x_p, t_{q+1})$ of a solution u of (68) depends only on the values of u and u_t at all points (x, t_q) with

$$x \in [x_p - \Delta t, x_p + \Delta t]$$

(Fig. 20.3). Because of the stability condition $r \leq 1$, therefore $\Delta t \leq \Delta x$, we obtain the following:

The dependence region of the hyperbolic differential equation must be smaller than that of the corresponding difference equation.

This is a natural general principle since the solution of the difference equation can, roughly speaking, not get away with less information than the solution of the differential equation.

A complete investigation of the wave equation can be found in Forsythe and Wasow (1960, M) and in Isaacson and Keller (1966, M). In this connection, the wave equation is handled with the Fourier method, and the corresponding difference equation completely parallel thereto with the discrete Fourier method (66).

A more exact study of the John von Neumann stability test can be found in Richtmyer and Morton (1967, M).

20.8. *Stability investigations for ordinary differential equations.* In this connection, study Albrecht (1979, M). Special caution is advisable in dealing with so-called *stiff systems* for which the eigenvalues of the matrix exhibit extreme differences. Such stiff systems appear frequently in microelectronics, chemical reaction kinetics, and so forth. There, special careful stability investigations are required. Cf. concerning these problems also Gear (1971, M), Willoughby (1974, P), Stoer and Bulirsch (1976, M), Shampine and Gear (1979, S), Miranker (1981, M), and Dekker and Verwer (1984, M).

20.9. *Multigrid methods.* The goal of these methods is to obtain very effective and very *fast* difference methods for solving linear and nonlinear elliptic differential equations. The basic idea is the following:

(i) Error smoothing by relaxation (see Problem 1.6).

(ii) Because of (i), corrections for the approximate solutions can be calculated on coarser grids. This idea can be used, recursively, employing coarser and coarser grids.

(iii) Combination with nested iteration, i.e., use of coarser grids to get good initial approximations on finer grids.

This procedure yields asymptotically optimal methods, i.e., the computational effort for getting the discretization accuracy is proportional to the number of the unknowns of the difference equation.

Study the survey articles Stüben and Trottenberg (1982) and Brandt (1982) as well as the monograph by Hackbusch (1985).

20.10. *The defect correction method.* The basic idea of this very effective modern numerical method is the following. Suppose we want to solve the equation $Au = 0$.

(i) Coarse discretization. Compute a rough approximate solution s by means of a *low-order* discretization method: $A_1 s = 0$.

(ii) Fine discretization. Use a *high-order* discretization method in order to compute the defect d of s: $d = A_2 s$.

(iii) Use d and a *low-order* discretization in order to compute the correction c by means of $A_1 c = d$.

(iv) Compute the corrected approximate solution $s + c$.

We recommend the survey by Böhmer and Stetter (1984). The multigrid method in Problem 20.9 is closely related to this method.

References to the Literature

Classical works: Courant, Friedrichs, and Lewy (1928) (fundamental paper), v. Neumann and Richtmyer (1947), Lax (1953), Lax and Richtmyer (1956).

Introduction: Smith (1965, M), Stoer and Bulirsch (1976, M), Meis and Marcowitz (1978, L).

Handbook of numerical analysis: Ciarlet and Lions (1988, M), Vols. 1ff (finite element method, difference method, etc.).

Classical monographs: Forsythe and Wasow (1960), Isaacson and Keller (1966), Richtmyer and Morton (1967), Birkhoff and Lynch (1984).

Monographs: Collatz (1960), Varga (1962), Ames (1977), Ansorge (1978), Albrecht (1979), Botha (1983), Marčuk and Shaidurov (1983), Shokin (1983), Samarskii (1984), Törnig (1985), Reinhardt (1985), Hackbusch (1985), (1986).

Sparse matrix technology: Hageman and Young (1981, M), Coleman (1984, M), Varga (1984, S), Duff (1986, M), Pissanetzky (1986, M).

Difference methods on irregular networks: Heinrich (1987, M).

Survey of the important modern defect correction method: Böhmer and Stetter (1984).

Multigrid methods and fast solvers: Stüben and Trottenberg (1982, S) (introduction), Brandt (1982, S), Hackbusch (1985, M) (comprehensive representation).

Multistep multiderivative methods: Gekeler (1984, L).

Functional analysis, discrete Sobolev spaces, and difference methods: Teman (1970, M), (1977, M), Aubin (1972, M), Ladyženskaja (1973, M), Wloka (1982, M), Rheinhardt (1985, M).

Shock waves in gas dynamics and difference methods: Smoller (1983, M) (recommended as an introduction), Lax and Wendroff (1964), Glimm (1965) (fundamental paper on stochastic difference schemes), Richtmyer and Morton (1967, M), Roždestvenskii and Janenko (1978, M), DiPerna (1983).

Design of shock-free transonic airfoils and numerical methods: Bauer, Garabedian, and Korn (1972/77, M), Vols. 1–3, Jameson (1988).

Controlled nuclear fusion, stability of plasma, and numerical methods: Bauer, Betancourt, and Garabedian (1978, M), (1984, M), (1986), Teller (1981, M), Vols. 1, 2, Betancourt (1988).

Ordinary differential equations and difference methods: Stoer and Bulirsch (1976, M) (introduction), Stetter (1973, M) (general theory), Albrecht (1979, M).

Stiff systems of ordinary differential equations: Stoer and Bulirsch (1976, M) (introduction), Shampine and Gear (1979, S) (a user's guide), Gear (1971, M), Willoughby

(1974, P) (numerous scientific applications), Albrecht (1979, M), Miranker (1981, M), Dekker and Verwer (1984, M).

Software: Rice (1983, M), Rice and Boisvert (1984, M) (solving elliptic problems using ELLPACK), Birkhoff and Schoenstadt (1984, P) (fast elliptic problem solvers), Hackbusch (1985, M).

Software system for general iteration methods: ITPACK (cf. Birkhoff (1989)).

Software system for solving partial differential equations and other general classes of mathematical problems IMSL (1987).

Pursue the journal *ACM Trans. Math. Software.*

Simulation: Bratley (1987, M).

Efficiency of algorithms of analysis: Smale (1985, S).

Computational complexity: Traub and Wozniakowski (1980, M), Calude (1984, M).

The art of computer programming: Knuth (1968, M), Vols. 1–3, Knuth (1982, M), (1989, M) (foundations of computer science).

Handbook of artificial intelligence: Cohen and Feigenbaum (1981, M), Vols. 1–3.

Supercomputer methods: Murman (1985, P), Lichnewsky and Saguez (1987, S), Martin (1988, S).

LINEAR MONOTONE PROBLEMS

That it is not enough to look for the most general solutions of partial differential equations, but rather that the question of the fulfillment of special boundary conditions and initial conditions had to be closely studied, was convincingly shown in 1822 by the appearance of Fourier's "Théorie de la chaleur" (heat theory), and this concept was, for example, explicitly expressed by Riemann around 1860.

The validity of theorems on eigenfunctions can be made plausible by the following observation made by Daniel Bernoulli (1700–1782). A mechanical system of n degrees of freedom possesses exact n eigenoscillations. A membrane is, however, a system with an infinite number of degrees of freedom. This system will, therefore, have an infinite number of eigenoscillations.

Arnold Sommerfeld (1900)

This book originated as a course of lectures that I have delivered since 1949 at the Department of Mathematics and Physics of Leningrad University in the areas of mathematical physics and partial differential equations.

The content of these lectures varied as my own understanding of the subject developed. However, the pivotal idea, which determined the style of the lectures, was clearly formulated from the very beginning. This idea consists of replacing classical formulations of boundary value problems by generalized formulations ....

The idea of introducing *generalized* solutions first started to penetrate mathematical physics in the 1920's. It came from two sources. The first source was that of two-dimensional variational problems. Investigation of these necessitated the extension of classes of functions among which a minimum is sought, and admission for consideration not only of continuously differentiable functions but also of continuous functions possessing so-called generalized derivatives (Tonelli classes) ....

The second source of generalized solutions was that of nonstationary problems, first the wave equation $u_{tt} = c^2 \Delta u$ and then the equations of hydrodynamics. Discontinuous solutions were introduced for both a long time ago: for the former, plane and spherical waves with a strong discontinuity on the front moving at velocity c; and for the latter, solutions describing shock waves. In the

1920's, researchers were trying to understand which discontinuous solutions should be considered as "admissible" ....

The 1930's led to a further development of the above-mentioned trends. In K. O. Friedrich's papers (1934), (1939), devoted to finding the minimum of the quadratic functional

$$J(u) = \int_G \left(\sum_{i,j=1}^{N} a_{ij} D_i u D_j u + au^2 - 2fu \right) dx$$

under the boundary condition $u = 0$ on ∂G and the ellipticity condition,

$$\sum_{i,j=1}^{N} a_{ij} \xi_i \xi_j \geq c \sum_{i=1}^{N} \xi_i^2, \qquad c > 0,$$

the considerations were carried out in the class of functions which subsequently was designated by $\mathring{W}_2^1(G)$....

In 1936, S. L. Sobolev obtained generalized solutions of the initial value problem for second-order hyperbolic equations with sufficiently smooth coefficients and homogeneous initial conditions but rather "poor" free terms.... In the late 1930's, Sobolev and his associate, V. I. Kondračov, obtained the most complete results, for that time, for the spaces $W_p^m(G)$. These were preceded by results obtained by F. Rellich (1930) on the compactness of embedding $W_2^1(G)$ in $L_2(G)$....

Also, in the 1930's, J. Leray went beyond the scope of classical solutions for boundary value problems. The paper by Leray (1934) concerns nonstationary solutions of the equations of hydrodynamics for viscous, incompressible fluids. The same applies to N. M. Gunter's work on the first boundary value problem for the equation $\Delta u = f$ and on the initial–boundary value problem for the equation of the oscillation of a nonhomogeneous spring....

In the late 1940's, I proposed that generalized solutions of boundary value and initial–boundary value problems for various types of equations (elliptic, parabolic, and hyperbolic) should be determined with the help of *integral identities* replacing the equation and, sometimes, a part of the initial and boundary conditions. Furthermore, the importance of the fact that for each one can introduce various classes of generalized solutions, defined by that functional space W, to which the generalized solution which is sought should belong, was also duly noted.... The choice of the space W is up to the researcher and could be limited only by smoothness of data. The only requirement which must be satisfied is that of the admissibility of the introduced extension, i.e., the requirement that the uniqueness theorem be preserved in the class W provided that this uniqueness theorem is in agreement with the "spirit" of the problem and occurs in the class of classical solutions....

Thus, the definition of a generalized solution of a problem was *separated* from any method of obtaining it (as it was earlier in the papers of K. O. Friedrichs and S. L. Sobolev) and, even more, from any analytic representation of it (as in the work of J. Leray and N. M. Gunter)....

At first, I established the existence of admissible generalized solutions using the method of finite differences. ... The entire problem under discussion was realized in the book by Ladyženskaja (1953), which is mainly devoted to initial boundary value problems for hyperbolic equations that were most difficult and were the least investigated at that time.

Olga Aleksandrovna Ladyženskaja (1973)

In recent years, progress in the theory of partial differential equations has been tremendous, often in unexpected directions, while also solving classical problems in more general settings. New fields have been added, like the study of variational inequalities, of solitons, of wave front sets, of pseudodifferential operators, of differential forms on manifolds, etc. Much of the progress has been made by the use of functional analysis. However, in the process, much of the original simplicity of the theory has been lost. This is perhaps connected with the emphasis on *solving problems*, which often requires piling up mountains of *a priori* inequalities and the skillful juggling of function spaces to make ends meet. It is good to remember that mathematics is *not* only concerned with solving problems, but also with studying the structure and behavior of objects that it creates. One of the best examples is the classical theory of functions of a complex variable. It, incidentally, does solve problems as in the Riemann mapping theorem. But much of its beauty lies in statements that can hardly be considered as "solving" anything, like the calculus of residues, or Picard's theorem, or Cauchy's formula

$$f(z) = \frac{1}{2\pi i} \int_C \frac{f(w)}{w - z} \, dw. \tag{C}$$

The only "problem" solved by (C) is the improper one of determing f from its values on the curve C, which generally has no solution. Formula (C) is not only strikingly beautiful but also extremely useful. It shows immediately that each analytic function f can be differentiated infinitely often and can be represented by convergent power series.

Fritz John (1984)

In the following Chapters 21 through 24 we consider operator equations and evolution equations with linear monotone operators. In Chapters 25 through 36 this linear theory will be generalized to nonlinear problems.

CHAPTER 21

Auxiliary Tools and the Convergence of the Galerkin Method for Linear Operator Equations

> I am not able to learn any mathematics unless I can see some problem I am going to solve with mathematics, and I don't understand how anyone can teach mathematics without having a battery of problems that the student is going to be inspired to want to solve and then see that he or she can use the tools for solving them.
>
> Steve Weinberg (1986)
> (Winner of the Nobel Prize in physics in 1979)

In this chapter we consider the following auxiliary tools, which will be used very frequently in the remaining chapters of Parts II–V:

(i) generalized derivatives and Sobolev spaces;
(ii) the Sobolev embedding theorems;
(iii) weak and weak* convergence;
(iv) relation between linear operators and bilinear forms;
(v) projection operators, bases, and Galerkin schemes with applications to finite elements and polynomial bases.

In order to display the basic ideas of the crucial Sobolev embedding theorems as clearly as possible, we first prove the Sobolev embedding theorems in $\mathbb{R}^1$ and after that in $\mathbb{R}^N$. This approach is well-situated for an *introductory lecture*. Indeed, we will consider two different important approaches to Sobolev spaces:

Elementary inequalities for integrals based on the Hölder inequality and the Young inequality (Section 21.4);
Fourier transform (Section 21.20).

Furthermore, we consider the following:

(a) Abstract *Fredholm alternatives* and the Riesz–Schauder theory for compact perturbations of invertible operators.

(b) The main theorem on the approximation-solvability of linear operator equations and the convergence of the *Galerkin method.*
(c) Interpolation inequalities, a general convergence trick, and the convergence of refined *iteration methods.*
(d) The fundamental Gagliardo–Nirenberg interpolation inequalities for Sobolev spaces, the Moser-type calculus inequalities, and the weak sequential continuity of nonlinear differential operators in Sobolev spaces.
(e) Banach algebras and Sobolev spaces.

The general theorem in (b) applies to

k-contractive perturbations of the identity, and
compact perturbations of linear strongly monotone operators.

These two classes of operators allow many applications because they are related to

the Banach fixed-point theorem,
the Riesz–Schauder theory, and
the theory of monotone operators.

Thus, the convergence of the Galerkin method can be established for broad classes of operators.

More precisely, the main theorem on the approximation-solvability of linear operator equations in Section 21.16 shows that

$$\|u - u_n\| \leq \text{const}\,\text{dist}(u, X_n),$$

where u is the solution of the original equation $Au = b$, and u_n is the solution of the corresponding Galerkin equation in the finite-dimensional space X_n. Hence we obtain the following fundamental result:

The rapidity of convergence of the Galerkin method depends on $\text{dist}(u, X_n)$.

Observe the following crucial fact. The methods of approximation theory allow us to estimate $\text{dist}(u, X_n)$. In particular, we will consider such estimates for polynomials (the theorem of Jackson) in Section 21.13, and for finite elements in Section 21.14.

The point is that the estimates for $\text{dist}(u, X_n)$ do not depend on the specific form of u, but only on the *smoothness class* of u.

Roughly speaking, we obtain the following: The convergence of the Galerkin method is faster, the smaller $\text{dist}(u, X_n)$ is. Moreover, $\text{dist}(u, X_n)$ is smaller, the smoother the solution u is. This implies that:

The convergence of the Galerkin method is faster, the smoother the solution u is.

Clearly, this result is fundamental for numerical analysis. In Chapter 20 we have discovered a similar principle, namely: The convergence of an appropriate difference method is faster, the smoother the solution of the corresponding differential equation is.

Concerning (c) above, we will prove a refined version of the Banach fixed-point theorem in Section 21.18 via tuples of B-spaces

$$X \subseteq Y \subseteq Z$$

and via interpolation inequalities. This allows important applications to mathematical physics (quasi-linear symmetric hyperbolic equations in Chapter 83). The use of *several* B-spaces in connection with interpolation inequalities represents an important *modern strategy* in the theory of nonlinear partial differential equations.

The reader who wants to learn, as quickly as possible, applications of the tools from this section to partial differential equations may *pass* to Chapter 22 after reading Sections 21.1 through 21.3 (Sobolev spaces) and Section 21.10 (bilinear forms).

21.1. Generalized Derivatives

The simple point of departure for the definition of generalized derivatives is the classical *integration by parts formula*

$$\int_G uD^\alpha v\,dx = (-1)^{|\alpha|}\int_G (D^\alpha u)v\,dx \tag{1}$$

for fixed $u \in C^{|\alpha|}(G)$ and all $v \in C_0^\infty(G)$. Here, G is a nonempty open set in $\mathbb{R}^N$ with $N \geq 1$ and D^α denotes a partial derivative of order $|\alpha|$. In the special case $N = 1$ with $G =]a, b[$, formula (1) corresponds to

$$\int_a^b uv^{(m)}\,dx = (-1)^m \int_a^b u^{(m)}v\,dx$$

for $u \in C^m(a, b)$ and all $v \in C_0^\infty(a, b)$.

Definition 21.1. Let $x = (\xi_1, \ldots, \xi_N)$ and $D_j = \partial/\partial\xi_j$. By a *multiindex* $\alpha = (\alpha_1, \ldots, \alpha_N)$, we understand a tupel of nonnegative integers $\alpha_1, \ldots, \alpha_N$. We set $|\alpha| = \alpha_1 + \cdots + \alpha_N$ and

$$D^\alpha u = D_1^{\alpha_1}\cdots D_N^{\alpha_N}u,$$

i.e.,

$$D^\alpha u = \frac{\partial^{|\alpha|}u}{\partial\xi_1^{\alpha_1}\cdots\partial\xi_N^{\alpha_N}}.$$

For $\alpha = 0$, we set $D^0 u = u$.

Formula (1) is obtained by a repeated application of the integration by parts formula (18.8). Note that $v \in C_0^\infty(G)$ vanishes on a boundary strip of G. Thus, in contrast to (18.8), no boundary integrals appear.

Now to the point. In formula (1) we set

$$w = D^\alpha u.$$

Hence we obtain the *key formula*

$$\int_G u D^\alpha v \, dx = (-1)^{|\alpha|} \int_G wv \, dx \qquad \text{for all} \quad v \in C_0^\infty(G). \tag{2}$$

The idea of the following definition is that relation (2) remains true for certain nonsmooth functions u and w.

Definition 21.2. Let G be a nonempty open set in $\mathbb{R}^N$ with $N \geq 1$ and let u, $w \in L_{1,\mathrm{loc}}(G)$. Then, w is called a *generalized derivative* of u of type D^α iff (2) holds. We write $w = D^\alpha u$.

This is one of the most important definitions of modern analysis.

Proposition 21.3. *If $u \in C^m(G)$, then all the classical partial derivatives of u up to order m are also generalized derivatives.*

Each generalized derivative $w = D^\alpha u$ is unique up to a change of the values of w on a set of measure zero.

Proof. The first statement follows from (1) and (2).

Suppose that $w = D^\alpha u$ and $\bar{w} = D^\alpha u$. By (2),

$$\int_G (w - \bar{w}) v \, dx = 0 \qquad \text{for all} \quad v \in C_0^\infty(G).$$

By Proposition 18.36, this implies $w(x) = \bar{w}(x)$ for almost all $x \in G$. □

The following examples describe typical generalized derivatives. Roughly speaking, we shall obtain the following:

Piecewise continuous derivatives are generalized derivatives.

This principle is important for the construction of finite elements (piecewise polynomial functions), which play a fundamental role in modern numerical analysis as basis functions for the Ritz and Galerkin methods.

Example 21.4. Let $u(x) = |x|$. Define

$$w(x) = \begin{cases} 1 & \text{if } x > 0, \\ -1 & \text{if } x < 0, \\ \alpha & \text{if } x = 0, \end{cases}$$

where α is a fixed, but otherwise arbitrary, real number. Then the function w is the generalized derivative of u on each open interval $]a, b[$.

Note that $w(x) = u'(x)$ if $x \neq 0$, i.e., $w(x)$ coincides with the classical derivative of u. At the point $x = 0$, where the classical derivative of u does not exist, the value of w can be chosen arbitrarily.

PROOF. Suppose first that $0 \in]a, b[$, where $-\infty < a < b < \infty$. For all $v \in C_0^\infty(a, b)$, integration by parts yields

$$\begin{aligned}\int_a^b uv'\,dx &= \int_a^0 uv'\,dx + \int_0^b uv'\,dx \\ &= uv\Big|_a^0 + uv\Big|_0^b - \int_a^0 u'v\,dx - \int_0^b u'v\,dx \\ &= -\int_a^b wv\,dx.\end{aligned}$$

Note that $v(a) = v(b) = 0$ since $v \in C_0^\infty(a, b)$.

In the case where $a = -\infty$ or $b = +\infty$, integrate over a sufficiently large bounded interval J with the property that v vanishes at the boundary points of J. □

STANDARD EXAMPLE 21.5. Let the function $u: [a, b] \to \mathbb{R}$ be continuous and piecewise continuously differentiable on the compact interval $[a, b]$.

Let C be the set of points x, where the classical derivative $u'(x)$ exists. Define

$$w(x) = \begin{cases} u'(x) & \text{if } x \in C, \\ \text{arbitrary} & \text{otherwise.} \end{cases}$$

Then the function w is the generalized derivative of u on $]a, b[$.

More precisely, we assume that the function u has the following properties:

(i) u is continuous on $[a, b]$.
(ii) There exist finitely many points a_i with $a = a_0 < a_1 < \cdots < a_n = b$ such that u is continuously differentiable on the open subintervals $]a_i, a_{i+1}[$ and the derivative u' can be extended continuously to the closed subintervals $[a_i, a_{i+1}]$ for all i (cf. Fig. 21.1).

PROOF. Divide the interval $[a, b]$ into the subintervals $[a_i, a_{i+1}]$ and use the same argument as in the proof of Example 21.4. □

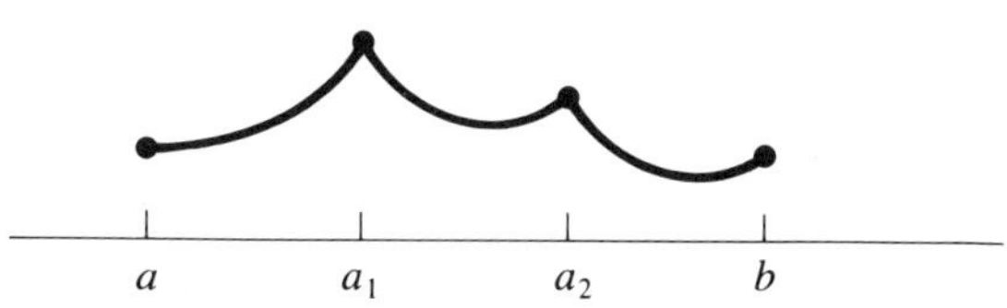

Figure 21.1

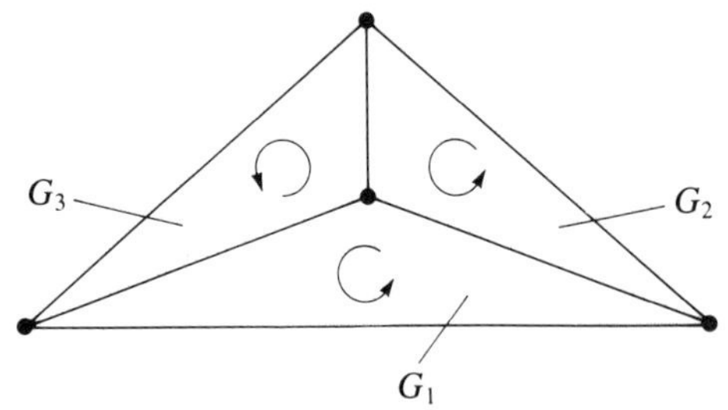

Figure 21.2

EXAMPLE 21.6. Let G be an open triangle in $\mathbb{R}^2$. Divide G into three open subtriangles G_1, G_2, G_3 as pictured in Figure 21.2. Suppose that the function $u: \bar{G} \to \mathbb{R}$ has the following properties:

(i) u is continuous on $\bar{G}$.
(ii) u has continuous partial derivatives $D_1 u$ and $D_2 u$ on G_1, G_2, G_3, which can be continuously extended to $\bar{G}_1, \bar{G}_2, \bar{G}_3$, respectively.

Define

$$w_j(x) = \begin{cases} D_j u(x) & \text{if } x \in G_i \text{ for some } i, \\ \text{arbitrary} & \text{otherwise.} \end{cases}$$

Then the function w_j is a generalized partial derivative of u on G.

PROOF. For all $v \in C_0^\infty(G)$, integration by parts yields

$$\begin{aligned} \int_G u D_j v \, dx &= \sum_i \int_{G_i} u D_j v \, dx \\ &= \sum_i \int_{\partial G_i} u v n_j \, ds - \sum_i \int_{G_i} v D_j u \, dx \\ &= -\int_G v D_j u \, dx. \end{aligned}$$

Note that the boundary integrals drop out since they have pairwise different signs because of the opposite directions of the outer unit normal $n = (n_1, n_2)$. In this connection, observe the continuity of u. □

EXAMPLE 21.7. Obviously, the preceding example can be generalized immediately to much more general situations. Roughly speaking, we need the following:

(i) G is a bounded region in $\mathbb{R}^N$.
(ii) $u \in C^k(G)$ for fixed $k = 0, 1, \ldots$.
(iii) All the partial derivatives $D^\alpha u$ with $|\alpha| = k + 1$ are piecewise continuous. Here, we assume that the surfaces of discontinuity are sufficiently regular, i.e., they belong to $C^{0,1}$.

Then the functions $D^\alpha u$ with $|\alpha| = k + 1$ are generalized $(k + 1)$-th partial derivatives of u on G. In this connection, the values of $D^\alpha u$ can be chosen arbitrarily on the surfaces of discontinuity.

Remark 21.8 (Distributions). Let $u(x) = |x|$ and consider the generalized derivative $w = u'$ as in Example 21.4. Then there does not exist any function v which is the generalized derivative of w on $\mathbb{R}$. However, we obtain that

$$w' = 2\delta,$$

where δ denotes the so-called Dirac δ-distribution at $x = 0$. This will be explained in the Appendix in A_2(66b). If G is a nonempty open set in $\mathbb{R}^N$, then each function $u \in L_{1,\mathrm{loc}}(G)$ has derivatives of arbitrary order in the sense of the theory of distributions (cf. A_2(65), (66)).

In what follows we do not need the theory of distributions since we only work with generalized derivatives which can be represented by functions.

21.2. Sobolev Spaces

Let $1 \le p < \infty$. We set

$$\|u\|_p = \left(\int_G |u(x)|^p \, dx\right)^{1/p} \tag{3}$$

and

$$\|u\|_{m,p} = \left(\sum_{0 \le |\alpha| \le m} \int_G |D^\alpha u(x)|^p \, dx\right)^{1/p}. \tag{4}$$

Here, the sum is taken over all the partial derivatives of u up to order m. Recall that $D^0 u = u$. Moreover, we set

$$(u|v)_2 = \int_G uv \, dx$$

and

$$(u|v)_{m,2} = \sum_{0 \le |\alpha| \le m} \int_G D^\alpha u D^\alpha v \, dx. \tag{5}$$

Note that

$$\|u\|_{m,p} = \left(\sum_{0 \le |\alpha| \le m} \|D^\alpha u\|_p^p\right)^{1/p},$$

$$(u|v)_{m,2} = \sum_{0 \le |\alpha| \le m} (D^\alpha u | D^\alpha v)_2.$$

For $m = 0$, we obtain that $\|u\|_{0,p} = \|u\|_p$ and $(u|v)_{0,2} = (u|v)_2$.

Definition 21.9. Let G be a nonempty open set in $\mathbb{R}^N$ with $N \ge 1$. Let $1 \le p < \infty$ and $m = 1, 2, \ldots$.

(i) The *Lebesgue space* $L_p(G)$ is the set of all measurable functions $u\colon G \to \mathbb{R}$ with $\|u\|_p < \infty$.
(ii) The *Sobolev space* $W_p^m(G)$ is the set of all functions $u \in L_p(G)$, which have generalized derivatives up to order m such that

$$D^\alpha u \in L_p(G) \qquad \text{for all} \quad \alpha\colon |\alpha| \le m.$$

For $m = 0$, we set $W_p^0(G) = L_p(G)$.

The spaces $W_\infty^m(G)$ will be defined at the end of this section.

Proposition 21.10. *Let* $1 \le p < \infty$ *and* $m = 0, 1, \ldots$.

(a) *The set* $W_p^m(G)$ *together with the norm* $\|\cdot\|_{m,p}$ *becomes a real* B-*space provided we identify any two functions which differ only on a set of* N-*dimensional Lebesgue measure zero.*
(b) *For* $p = 2$, $W_2^m(G)$ *is an* H-*space with the scalar product* $(\cdot|\cdot)_{m,2}$.

PROOF. Ad(a). We use the well-known fact that $L_p(G)$ becomes a B-space with the norm $\|\cdot\|_p$, provided two functions are identified whose values differ only on a set of N-dimensional Lebesgue measure zero.

It follows from the Minkowski inequality

$$\left(\sum_{i=1}^n |\xi_i + \eta_i|^p\right)^{1/p} \le \left(\sum_{i=1}^n |\xi_i|^p\right)^{1/p} + \left(\sum_{i=1}^n |\eta_i|^p\right)^{1/p}$$

for all real ξ_i, η_i that

$$\|u + v\|_{m,p} \le \|u\|_{m,p} + \|v\|_{m,p},$$

i.e., $\|\cdot\|_{m,p}$ is a norm on $W_p^m(G)$. In order to prove the completeness of $W_p^m(G)$, let (u_n) be a Cauchy sequence in $W_p^m(G)$. Then (u_n) and $(D^\alpha u_n)$ are Cauchy sequences in $L_p(G)$ for $|\alpha| \le m$, and hence

$$u_n \to u \quad \text{in } L_p(G) \quad \text{as} \quad n \to \infty,$$

$$D^\alpha u_n \to w_\alpha \quad \text{in } L_p(G) \quad \text{as} \quad n \to \infty.$$

In this connection, note that $\|D^\alpha v\|_p \le \|v\|_{m,p}$. By Example 18.15, we obtain that

$$u_n \to u \quad \text{in } L_1(H) \quad \text{as} \quad n \to \infty,$$

$$D^\alpha u_n \to w_\alpha \quad \text{in } L_1(H) \quad \text{as} \quad n \to \infty,$$

for each compact subset H of G.

Thus, it follows from

$$\int_G u_n D^\alpha v\,dx = (-1)^{|\alpha|} \int_G (D^\alpha u_n) v\,dx \qquad \text{for all} \quad v \in C_0^\infty(G)$$

that

$$\int_G u D^\alpha v\,dx = (-1)^{|\alpha|} \int_G w_\alpha v\,dx \qquad \text{for all} \quad v \in C_0^\infty(G).$$

This implies $w_\alpha = D^\alpha u$ and $w_\alpha \in L_p(G)$; therefore,

$$u_n \to u \quad \text{in } W_p^m(G) \qquad \text{as} \quad n \to \infty.$$

Consequently, each Cauchy sequence in $W_p^m(G)$ is convergent, i.e., $W_p^m(G)$ is a B-space.

Ad(b). This follows from the well-known fact that $L_2(G)$ is an H-space with the scalar product $(\cdot|\cdot)_2$. □

Definition 21.11. $\mathring{W}_p^m(G)$ denotes the closure of $C_0^\infty(G)$ in $W_p^m(G)$.

Explicitly, this means the following. Let $u \in W_p^m(G)$. Then, u belongs to $\mathring{W}_p^m(G)$ iff there exists a sequence (u_n) in $C_0^\infty(G)$ with $\|u_n - u\|_{m,p} \to 0$ as $n \to \infty$.

Proposition 21.12. *$\mathring{W}_p^m(G)$ is a B-space with the norm $\|\cdot\|_{m,p}$ and $\mathring{W}_2^m(G)$ is an H-space with the scalar product $(\cdot|\cdot)_{m,2}$.*

Proof. Note that $C_0^\infty(G)$ is a linear subspace of $W_p^m(G)$. Hence $\mathring{W}_p^m(G)$ is a closed linear subspace of $W_p^m(G)$. □

Let G be a nonempty open set in $\mathbb{R}^N$, $N \geq 1$. The space $L_\infty(G)$ has been introduced in Section 18.6. We set

$$\|u\|_{m,\infty} = \sum_{|\alpha| \leq m} \|D^\alpha u\|_\infty.$$

By definition, the *Sobolev space* $W_\infty^m(G)$ consists of all functions $u \in L_\infty(G)$ which have generalized derivatives up to order m such that

$$D^\alpha u \in L_\infty(G) \qquad \text{for all} \quad \alpha\colon |\alpha| \leq m.$$

Proposition 21.10 remains true for the space $W_p^m(G)$ with $p = \infty$.

21.3. The Sobolev Embedding Theorems

Definition 21.13. Let X and Y be B-spaces over $\mathbb{K} = \mathbb{R}, \mathbb{C}$ with $X \subseteq Y$. The embedding operator $j\colon X \to Y$ is defined by $j(u) = u$ for all $u \in X$.

(i) The embedding $X \subseteq Y$ is called *continuous* iff j is continuous, i.e.,

$$\|u\|_Y \leq \text{const}\,\|u\|_X \qquad \text{for all} \quad u \in X. \tag{6}$$

(ii) The embedding $X \subseteq Y$ is called *compact* iff j is compact, i.e., (6) holds and each bounded sequence (u_n) in X has a subsequence $(u_{n'})$ which is convergent in Y.

More generally, we speak of an embedding if there are two B-spaces X and Y over $\mathbb{K}$ and an injective linear operator $j\colon X \to Y$. Since j is injective, we can identify u with $j(u)$. In this sense, we write $X \subseteq Y$.

Recall that two norms $\|\cdot\|_1$ and $\|\cdot\|_2$ on the B-space X are called *equivalent* iff there exist positive constants c and d with

$$\|u\|_1 \le c\|u\|_2 \quad \text{and} \quad \|u\|_2 \le d\|u\|_1 \quad \text{for all} \quad u \in X.$$

The importance of this notion results from the fact that many properties are invariant under a passage to equivalent norms, e.g., this is true for "convergence, B-space, continuous operator, compact operator, bounded set, etc."

We set

$$\begin{aligned} \|u\|_{1,2} &= \left(\int_G \left(u^2 + \sum_{j=1}^N (D_j u)^2\right) dx\right)^{1/2}, \\ \|u\|_{1,2,0} &= \left(\int_G \sum_{j=1}^N (D_j u)^2\, dx\right)^{1/2}. \end{aligned} \tag{7}$$

Proposition 21.14. *Let G be a bounded region in $\mathbb{R}^N$ with $N \ge 1$. Then:*

(a) *The norms $\|\cdot\|_{1,2}$ and $\|\cdot\|_{1,2,0}$ are equivalent on $\mathring{W}_2^1(G)$.*
(b) *The embedding $\mathring{W}_2^1(G) \subseteq L_2(G)$ is compact.*
(c) *The embeddings*

$$L_2(G) \supseteq \mathring{W}_2^1(G) \supseteq \mathring{W}_2^2(G) \supseteq \mathring{W}_2^3(G) \supseteq \cdots$$

are compact.

It follows from (a) that $\mathring{W}_2^1(G)$ is an H-space with respect to the so-called energetic scalar product

$$(u|v)_{1,2,0} = \int_G \sum_{i=1}^N D_i u D_i v\, dx.$$

We now set

$$\begin{aligned} \|u\|_{1,2}^{*} &= \left(\int_G \sum_{i=1}^N (D_i u)^2\, dx + \left|\int_G u\, dx\right|^2\right)^{1/2}, \\ \|u\|_{1,2}^{**} &= \left(\int_G \sum_{i=1}^N (D_i u)^2\, dx + \int_{\partial G} u^2\, dO\right)^{1/2}. \end{aligned} \tag{8}$$

In $\mathbb{R}^1$ we set $\int_{\partial G} u^2\, dO = u(a)^2 + u(b)^2$, where $G =]a, b[$.

Theorem 21.A. *Let G be a bounded region in $\mathbb{R}^N$ with $N \ge 1$ and sufficiently smooth boundary,*[1] *i.e., $\partial G \in C^{0,1}$. Then:*

(a) Density. *$C^\infty(\overline{G})$ is dense in $W_2^1(G)$.*
(b) Compact embedding. *The embedding $W_2^1(G) \subseteq L_2(G)$ is compact.*
(c) Equivalent norms. *Each of the two norms in* (8) *is an equivalent norm on $W_2^1(G)$.*

[1] If $N = 1$, then G is a bounded open interval, as in Proposition 21.14.

(d) Regularity. *For $m - j > N/2$, the embedding*

$$W_2^m(G) \subseteq C^j(\bar{G})$$

is continuous, i.e., each function $u \in W_2^m(G)$ belongs to $C^j(\bar{G})$ after changing the values of u on a set of N-dimensional Lebesgue measure zero, if necessary.

(e) Generalized boundary function. *There exists exactly one linear continuous operator*

$$B: W_2^1(G) \to L_2(\partial G)$$

with the property that, for each $u \in C^1(\bar{G})$, the function

$$Bu: \partial G \to \mathbb{R}$$

is the classical boundary function to u, i.e., Bu is the restriction of $u: \bar{G} \to \mathbb{R}$ to the boundary ∂G.

In the case where $u \in W_2^1(G)$, the function $Bu \in L_2(\partial G)$ is called the generalized boundary function to u. The values of Bu are uniquely determined on ∂G up to changing the values on a set of surface measure zero.

If $u \in \mathring{W}_2^1(G)$, then $Bu = 0$ in $L_2(\partial G)$, i.e.,

$$Bu(x) = 0 \quad \text{for almost all} \quad x \in \partial G. \tag{9}$$

In the following, instead of (9) we will simply write:

$$u = 0 \quad \text{on } \partial G. \tag{10}$$

Corollary 21.15. *Let G be a bounded region in $\mathbb{R}^N$ with $N \geq 1$ and $\partial G \in C^{0,1}$. Then:*

(a) *$C^\infty(\bar{G})$ is dense in $W_2^m(G)$ for $m = 0, 1, 2, \ldots$.*

(b) *The embeddings*

$$L_2(G) \supseteq W_2^1(G) \supseteq W_2^2(G) \supseteq W_2^3(G) \supseteq \cdots$$

are compact.

(c) *If $u \in \mathring{W}_2^m(G)$ with $m \geq 1$, then $B(D^\alpha u) = 0$ in $L_2(\partial G)$ for $|\alpha| \leq m - 1$, i.e.,*

$$D^\alpha u = 0 \quad \text{on } \partial G \qquad \text{for all} \quad \alpha\colon |\alpha| \leq m - 1 \tag{11}$$

in the sense of (10).

The proofs of Proposition 21.14, Theorem 21.A, and Corollary 21.15 will be given in the next section. These embedding theorems are the prototypes of more general theorems, which will be summarized in $A_2(41)$ ff.

Remark 21.16 (The Importance of the Sobolev Embedding Theorems for Boundary Value Problems). In Chapter 22 we shall apply the Sobolev embedding theorems to boundary value problems for elliptic differential equations. In this connection, note the following.

(i) For solving the *first* boundary value problem for linear second-order elliptic equations, we need the equivalence of the norms $\|\cdot\|_{1,2}$ and $\|\cdot\|_{1,2,0}$ on $\mathring{W}_2^1(G)$. This corresponds to the classical inequality of Poincaré–Friedrichs.

(ii) For solving the *second* (resp. third) boundary value problem, we need the equivalence of the norms $\|\cdot\|_{1,2}$ and $\|\cdot\|_{1,2}^*$ (resp. $\|\cdot\|_{1,2}$ and $\|\cdot\|_{1,2}^{**}$) on the Sobolev space $W_2^1(G)$. This also corresponds to classical inequalities due to Poincaré and Friedrichs.

In order to prove the equivalence of norms on Sobolev spaces, one uses frequently the compactness of certain embeddings. In particular, the proof of Theorem 21.A(c) will be based on the compactness of the embedding $W_2^1(G) \subseteq L_2(G)$.

(iii) For solving *eigenvalue* problems, one needs the compactness of the embedding $\mathring{W}_2^1(G) \subseteq L_2(G)$.

(iv) The *regularity* statement (d) of Theorem 21.A will be used in order to prove the regularity of generalized solutions u. The basic idea is to show that u belongs to the Sobolev space $W_2^m(G)$ for sufficiently large m, i.e., u has *generalized* derivatives up to order m. Then, for $j < m - N/2$, we obtain $u \in C^j(\bar{G})$, i.e., u has *classical* derivatives up to order j. In particular, in $\mathbb{R}^1$ we have that

$$W_2^m(G) \subseteq C^{m-1}(\bar{G}), \qquad m = 1, 2, \ldots,$$

and in $\mathbb{R}^2$ and $\mathbb{R}^3$ we have that

$$W_2^m(G) \subseteq C^{m-2}(\bar{G}), \qquad m = 2, 3, \ldots.$$

(v) The boundary operator B is crucial for the formulation of *boundary conditions* in the generalized sense. For example, let $u \in W_2^1(G)$ and $g \in W_2^1(G)$. Then the boundary condition

$$u = g \quad \text{on } \partial G$$

is to be understood in the sense

$$Bu = Bg \quad \text{in } L_2(\partial G),$$

i.e., $Bu(x) = Bg(x)$ for almost all $x \in \partial G$.

The construction of B proceeds as follows. First, we prove the *key inequality*:

$$\|u\|_{L_2(\partial G)} \leq \text{const}\, \|u\|_{1,2} \qquad \text{for all} \quad u \in C^1(\bar{G}). \tag{12}$$

This will be done in the following section. That means

$$\int_{\partial G} u^2\, dO \leq \text{const} \int_G \left(u^2 + \sum_{i=1}^N (D_i u)^2\right) dx.$$

Now, let Bu denote the restriction of $u\colon \bar{G} \to \mathbb{R}$ to the boundary ∂G in the case where $u \in C^1(\bar{G})$. Then, (12) can be written as

$$\|Bu\|_{L_2(\partial G)} \leq \text{const}\, \|u\|_{1,2} \qquad \text{for all} \quad u \in C^1(\bar{G}). \tag{12*}$$

Since $C^1(\bar{G})$ is dense in $W_2^1(G)$, the linear operator

$$B: C^1(\bar{G}) \subseteq W_2^1(G) \to L_2(\partial G)$$

can be uniquely extended to a linear continuous operator

$$B: W_2^1(G) \to L_2(\partial G)$$

such that (12*) remains true for all $u \in W_2^1(G)$. This follows from the extension principle in Section 18.12.

Explicitly, we obtain Bu in the following way. Let $u \in W_2^1(G)$. We choose a sequence (u_n) in $C^1(\bar{G})$ such that

$$\|u_n - u\|_{1,2} \to 0 \qquad \text{as} \quad n \to \infty.$$

Then $\|Bu_n - Bu\|_{L_2(\partial G)} \to 0$, i.e.,

$$\int_{\partial G} (u_n - Bu)^2 \, dO \to 0 \qquad \text{as} \quad n \to \infty.$$

In particular, if $u \in \mathring{W}_2^1(G)$, then $C_0^\infty(G)$ is dense in $\mathring{W}_2^1(G)$, i.e., we can choose (u_n) from $C_0^\infty(G)$. Since $u_n = 0$ on ∂G, we obtain $Bu = 0$ in $L_2(\partial G)$, i.e., $Bu(x) = 0$ for almost all $x \in \partial G$.

If $u \in \mathring{W}_2^m(G)$, then $D^\alpha u \in \mathring{W}_2^1(G)$ for $|\alpha| \leq m - 1$. Hence $B(D^\alpha u) = 0$ in $L_2(\partial G)$, i.e., we obtain (11).

Relation (11) is important for the first boundary value problem for linear strongly elliptic equations of order $2m$.

Corollary 21.15(b) is an easy consequence of Theorem 21.A(b). To show this, let (u_n) be a bounded sequence in $W_2^m(G)$. Then $(D^\alpha u_n)$ is a bounded sequence in $W_2^1(G)$ provided $|\alpha| \leq m - 1$. Since the embedding $W_2^1(G) \subseteq L_2(G)$ is compact, there exists a subsequence, again denoted by (u_n), so that

$$u_n \to u \quad \text{in } L_2(G) \qquad \text{as} \quad n \to \infty,$$

$$D^\alpha u_n \to w_\alpha \quad \text{in } L_2(G) \qquad \text{as} \quad n \to \infty$$

for $|\alpha| \leq m - 1$. Hence $w_\alpha = D^\alpha u$. This implies

$$u_n \to u \quad \text{in } W_2^{m-1}(G) \qquad \text{as} \quad n \to \infty,$$

i.e., the embedding $W_2^m(G) \subseteq W_2^{m-1}(G)$ is compact.

In the same way, Proposition 21.14(c) follows from Proposition 21.14(b).

21.4. Proof of the Sobolev Embedding Theorems

In order to make the basic ideas of the Sobolev embedding theorems as transparent as possible, we begin with extremely simple proofs in $\mathbb{R}^1$. Furthermore, we shall use the following strategy of proof:

(i) Let $\mathscr{W}_p^m(G)$ denote the closure of $C^\infty(\bar{G})$ in $W_p^m(G)$.
(ii) We first prove the embedding theorems for $\mathscr{W}_2^m(G)$.
(iii) Finally, we show that $C^\infty(\bar{G})$ is dense in $W_2^m(G)$, i.e., $\mathscr{W}_2^m(G) = W_2^m(G)$, provided G is a bounded region in $\mathbb{R}^N$, $N \geq 1$, with $\partial G \in C^{0,1}$.

This approach has been chosen for didactic reasons. In an *introductory lecture*, it suffices to work with the spaces $\mathscr{W}_p^m(G)$ instead of $W_p^m(G)$. This simplifies the approach considerably, because $C^\infty(\bar{G})$ is automatically dense in $\mathscr{W}_p^m(G)$, $1 \leq p < \infty$.

Note that $\mathscr{W}_p^m(G) = W_p^m(G)$ holds since we assume that $\partial G \in C^{0,1}$. For regions G with *pathological* boundaries, we have $\mathscr{W}_p^m(G) \subset W_p^m(G)$, i.e., $C^\infty(\bar{G})$ is *not* dense in $W_p^m(G)$.

However, in the case of the spaces $\mathring{W}_p^m(G)$, the structure of the boundary ∂G does not play any role. This depends on the fact that, by construction, $C_0^\infty(G)$ is dense in $\mathring{W}_p^m(G)$ and each function $u \in C_0^\infty(G)$ vanishes on a boundary strip.

In the following, we shall see that the Sobolev embedding theorems result from simple *integral inequalities*, which are based on the Hölder inequality and the main formula of calculus:

$$u(x) - u(x_0) = \int_{x_0}^{x} u'(t)\,dt.$$

An alternative approach to the Sobolev spaces $W_2^m(G)$ via Fourier transform will be considered in Section 21.20.

21.4a. Proof of Proposition 21.14 in $\mathbb{R}^1$

Let $G =]a, b[$ with $-\infty < a < b < \infty$. In the following all positive constants are denoted by the same symbol c. Let $\int = \int_a^b$. We use the following norms

$$\|u\|_2 = \left(\int u^2\,dx\right)^{1/2}, \qquad \|u\|_C = \max_{a \leq x \leq b} |u(x)|,$$

$$\|u\|_{1,2} = \left(\int (u^2 + u'^2)\,dx\right)^{1/2}, \qquad \|u\|_{1,2,0} = \left(\int u'^2\,dx\right)^{1/2}.$$

Step 1: The embedding $\mathring{W}_2^1(G) \subseteq C(\bar{G})$ is continuous.

Let $u \in C_0^\infty(G)$. Since $u(a) = 0$, we obtain the *key formula*

$$u(x) = \int_a^x u'(t)\,dt. \tag{13}$$

By the Hölder inequality with 1-trick in Section 18.6,

$$\begin{aligned} \|u\|_C &\leq \int |u'|\,dt \leq (b-a)^{1/2}\left(\int u'^2\,dt\right)^{1/2} \\ &= (b-a)^{1/2}\|u\|_{1,2,0} \leq (b-a)^{1/2}\|u\|_{1,2}. \end{aligned} \tag{14}$$

This inequality shows that the embedding operator

$$j\colon C_0^\infty(G) \subseteq \mathring{W}_2^1(G) \to C(\bar{G})$$

is continuous. The set $C_0^\infty(G)$ is dense in $\mathring{W}_2^1(G)$. By the extension principle in Section 18.12, there exists a unique linear continuous extension

$$j\colon \mathring{W}_2^1(G) \to C(\bar{G}).$$

The typical argument (*A*). Let $u \in \mathring{W}_2^1(G)$. Then there exists a sequence (u_n) in $C_0^\infty(G)$ with

$$u_n \to u \quad \text{in } \mathring{W}_2^1(G) \qquad \text{as} \quad n \to \infty.$$

Hence

$$j(u_n) \to j(u) \quad \text{in } C(\bar{G}) \qquad \text{as} \quad n \to \infty.$$

Since $j(u_n) = u_n$ and $u_n \to u$ in $L_2(G)$ as $n \to \infty$, we get $u = j(u)$ in $L_2(G)$, and hence

$$u(x) = j(u)(x) \qquad \text{for almost all} \quad x \in G.$$

Thus, the operator $j\colon \mathring{W}_2^1(G) \to C(\bar{G})$ is injective. If we identify $u \in \mathring{W}_2^1(G)$ with $j(u) \in C(\bar{G})$, then we can write $\mathring{W}_2^1(G) \subseteq C(\bar{G})$ and

$$\|u\|_C \le c\|u\|_{1,2} \qquad \text{for all} \quad u \in \mathring{W}_2^1(G).$$

Step 2: The inequality of Poincaré–Friedrichs.
For all $u \in C_0^\infty(G)$,

$$\begin{aligned}\|u\|_{1,2}^2 &= \int u^2\,dt + \|u\|_{1,2,0}^2 \le (b-a)\|u\|_C^2 + \|u\|_{1,2,0}^2\\ &\le c\|u\|_{1,2,0}^2,\end{aligned}$$

by (14). A passage to the limit shows that $\|u\|_{1,2}^2 \le c\|u\|_{1,2,0}^2$ holds for all $u \in \mathring{W}_2^1(G)$. Hence $\|\cdot\|_{1,2}$ and $\|\cdot\|_{1,2,0}$ are equivalent norms on $\mathring{W}_2^1(G)$.

Step 3: The embedding $\mathring{W}_2^1(G) \subseteq C^{1/2}(\bar{G})$ is continuous.
For all $u \in C_0^\infty(G)$, it follows from (13) and the Hölder inequality that

$$\begin{aligned}|u(y) - u(x)| &= \left|\int_x^y u'\,dt\right|\\ &\le \left(\int_x^y dt\right)^{1/2}\left(\int_x^y u'^2\,dt\right)^{1/2} \le |y-x|^{1/2}\|u\|_{1,2}\end{aligned}$$

for all $x, y \in \bar{G}$. For the Hölder constant $H(u)$, we thus obtain $H(u) \le \|u\|_{1,2}$. Hence

$$\|u\|_{C^{1/2}} = \|u\|_C + H(u) \le c\|u\|_{1,2}.$$

As in Step 1, this implies the continuity of the embedding $\mathring{W}_2^1(G) \subseteq C^{1/2}(\bar{G})$.

Step 4: The embedding $C^{1/2}(\bar{G}) \subseteq C(\bar{G})$ is compact.
Let (u_n) be a bounded sequence in $C^{1/2}(\bar{G})$, i.e., $\|u_n\|_{C^{1/2}} \le c$ for all n. Then,

for all $x, y \in \bar{G}$ and all n,

$$|u_n(x) - u_n(y)| \le c|x - y|^{1/2}, \qquad |u_n(x)| \le c.$$

Hence (u_n) is equicontinuous and uniformly bounded. By the Arzelà–Ascoli theorem $A_1(24g)$, (u_n) has a subsequence which is convergent in $C(\bar{G})$.

Step 5: The embedding $C(\bar{G}) \subseteq L_2(G)$ is continuous.
This follows from

$$\|u\|_2 = \left(\int u^2\,dt\right)^{1/2} \le (b-a)^{1/2}\|u\|_C \qquad \text{for all} \quad u \in C(\bar{G}).$$

It follows from Steps 3–5 that the embeddings

$$\mathring{W}_2^1(G) \subseteq C(\bar{G}) \qquad \text{and} \qquad \mathring{W}_2^1(G) \subseteq L_2(G)$$

are compact.

21.4b. Proof of Theorem 21.A in $\mathbb{R}^1$

Let $u \in C^\infty(\bar{G})$. By the mean value theorem,

$$(b-a)^{-1}\int_a^b u\,dt = u(x_0) \qquad \text{for some} \quad x_0 \in [a,b].$$

Instead of (13) we now use the *key formula*:

$$u(x) = \int_{x_0}^x u'\,dt + (b-a)^{-1}\int_a^b u\,dt. \tag{15}$$

Step 1: The embedding $\mathscr{W}_2^1(G) \subseteq C(\bar{G})$ is continuous.
For all $u \in C^\infty(\bar{G})$, by the Hölder inequality, it follows from (15) that

$$\begin{aligned}\|u\|_C &\le c\left(\int u'^2\,dt\right)^{1/2} + c\left(\int u^2\,dt\right)^{1/2}\\ &\le 2c\,\|u\|_{1,2}.\end{aligned}$$

Since $C^\infty(\bar{G})$ is dense in $\mathscr{W}_2^1(G)$, the assertion follows as in Step 1 of Section 21.4a.

Step 2: The embedding $\mathscr{W}_2^1(G) \subseteq C^{1/2}(\bar{G})$ is continuous.
If $u \in C^\infty(\bar{G})$, then (15) yields

$$u(y) - u(x) = \int_x^y u'\,dt.$$

Now, the assertion follows as in Step 3 of Section 21.4a.

Step 3: Generalized boundary values.
For all $u \in C^\infty(\bar{G})$,

$$u(a)^2 + u(b)^2 \le 2\|u\|_C^2 \le 8c^2\|u\|_{1,2}^2. \tag{16}$$

By Step 1, if $u \in \mathscr{W}_2^1(G)$, then there exists a *unique continuous* function

$$j(u) \in C(\bar{G})$$

with $u(x) = j(u)(x)$ for almost all $x \in \bar{G}$. The generalized boundary values of u are given by

$$j(u)(a) \qquad \text{and} \qquad j(u)(b).$$

For brevity, we write $u(a) = j(u)(a)$ and $u(b) = j(u)(b)$.

Note the following. Since the function $u \in \mathscr{W}_2^1(G)$ can be changed on a set of measure zero, we obtain the uniquely determined generalized boundary values of u after changing u on a subset of $\bar{G}$ of measure zero, if necessary.

Since $C^\infty(\bar{G})$ is dense in $\mathscr{W}_2^1(G)$, a passage to the limit shows that (16) remains true for all $u \in \mathscr{W}_2^1(G)$, where $u(a)$ and $u(b)$ are to be understood as generalized boundary values.

We now set

$$\|u\|_{1,2}^* = \left(\int u'^2\,dx + \left|\int u\,dx\right|^2\right)^{1/2},$$

$$\|u\|_{1,2}^{**} = \left(\int u'^2\,dx + u(a)^2 + u(b)^2\right)^{1/2}.$$

Step 4: $\|\cdot\|_{1,2}^*$ is equivalent to $\|\cdot\|_{1,2}$ on $\mathscr{W}_2^1(G)$.

Let $u \in C^\infty(\bar{G})$. From $(A + B)^2 \le 2A^2 + 2B^2$ and (15) it follows that

$$u(x)^2 \le 2\left(\int u'\,dt\right)^2 + 2(b-a)^{-2}\left|\int u\,dt\right|^2,$$

and hence

$$\int u^2\,dt \le c\int u'^2\,dt + c\left|\int u\,dt\right|^2,$$

$$\left|\int u\,dt\right|^2 \le c\int u^2\,dt.$$

This implies

$$c\|u\|_{1,2} \le \|u\|_{1,2}^* \le d\|u\|_{1,2}. \tag{17}$$

A passage to the limit shows that this inequality remains true for all $u \in \mathscr{W}_2^1(G)$.

Step 5: $\|\cdot\|_{1,2}^{**}$ is equivalent to $\|\cdot\|_{1,2}$ on $\mathscr{W}_2^1(G)$.

Let $u \in C^\infty(\bar{G})$. From

$$u(x) = \int_a^x u'\,dt + u(a)$$

it follows that

$$\|u\|_C \le c\left(\int u'^2\,dt\right)^{1/2} + |u(a)|,$$

and hence

$$\int u^2\,dt \le c\left(\int u'^2\,dt + u(a)^2\right).$$

Together with (16) this implies

$$c\|u\|_{1,2} \le \|u\|_{1,2}^{**} \le d\|u\|_{1,2}.$$

A passage to the limit shows that this inequality holds for all $u \in \mathscr{W}_2^1(G)$.

21.4c. Proof of Theorem 21.A in $\mathbb{R}^N$

Note that Proposition 21.14 follows from the inequality of Poincaré–Friedrichs in Section 18.9 and the compactness theorem of Rellich in Section 19.13.

In order to prove Theorem 21.A, let G be a bounded region in $\mathbb{R}^N$ with $N \ge 2$ and $\partial G \in C^{0,1}$.

Step 1: The embedding $\mathscr{W}_2^1(G) \subseteq L_2(G)$ is compact.
This has been proved in Section 19.13.

Step 2: The generalized boundary operator.
For all $u \in C^1(\bar{G})$, we show that

$$\int_{\partial G} u^2\,dO \le c\int_G \left[u^2 + \sum_{j=1}^N (D_j u)^2\right]dx. \tag{18}$$

This implies Theorem 21.A(e) according to Remark 21.16.

To simplify notation, we consider the case $N = 2$. The general case proceeds analogously. In a neighborhood of a boundary point $x \in \partial G$, we choose a local (ξ, η)-coordinate system, where the boundary has the local representation

$$\eta = b(\xi), \qquad -\alpha \le \xi \le \alpha,$$

with the $C^{0,1}$-function b, i.e., b is Lipschitz continuous (Fig. 21.3).

Then there exists a $\beta > 0$ such that all the points (ξ, η) with

$$-\alpha \le \xi \le \alpha, \qquad b(\xi) - \beta \le \eta \le b(\xi)$$

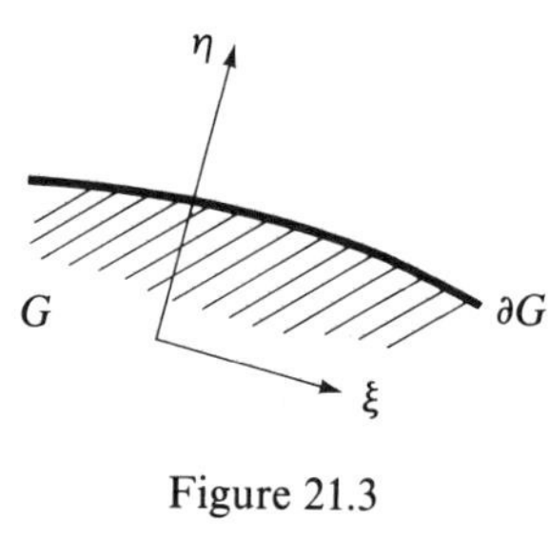

Figure 21.3

belong to $\bar{G}$. Let $u \in C^1(\bar{G})$. Then

$$u(\xi, b(\xi)) = \int_t^{b(\xi)} u_\eta(\xi, \eta)\, d\eta + u(\xi, t),$$

where $b(\xi) - \beta \le t \le b(\xi)$. It follows from the inequality $(A + B)^2 \le 2A^2 + 2B^2$ and from the Hölder inequality with 1-trick that

$$u(\xi, b(\xi))^2 \le 2\beta \int_{b(\xi)-\beta}^{b(\xi)} u_\eta(\xi, \eta)^2\, d\eta + 2u(\xi, t)^2.$$

Integration with respect to t yields

$$\beta u(\xi, b(\xi))^2 \le 2 \int_{b(\xi)-\beta}^{b(\xi)} [\beta^2 u_\eta(\xi, \eta)^2 + u(\xi, \eta)^2]\, d\eta.$$

Finally, integration over the interval $[-\alpha, \alpha]$ yields the *key estimate*:

$$\int_{-\alpha}^{\alpha} \beta u(\xi, b(\xi))^2\, d\xi \le 2 \int_S (\beta^2 u_\eta^2 + u^2)\, dx, \tag{19}$$

where S denotes a local boundary strip.

First suppose that $b(\cdot)$ is C^1. Then the differential of arc length is given by

$$ds = (1 + b'^2)^{1/2}\, d\xi.$$

Addition of the local inequalities (19) yields the assertion (18).

If $b(\cdot)$ is merely Lipschitz continuous, then the derivative $b'(\xi)$ exists for almost all all ξ and b' is bounded. Thus, we also obtain (18).

Step 3: Equivalent norms on $\mathscr{W}_2^1(G)$.

In order to get insight, we will prove a more general result than Theorem 21.A(c). To this end, we set

$$\|u\| = \left(\int_G \sum_{j=1}^N (D_j u)^2\, dx + p(u)^2\right)^{1/2},$$

where $p\colon \mathscr{W}_2^1(G) \to \mathbb{R}$ is a seminorm with the following two properties:

(i) *There is a positive constant d with*

$$p(u) \le d\|u\|_{1,2} \qquad \textit{for all} \quad u \in \mathscr{W}_2^1(G).$$

(ii) *If $u =$ constant, then $p(u) = 0$ implies $u = 0$.*

We shall show that:

$$\|\cdot\| \textit{ is an equivalent norm on } \mathscr{W}_2^1(G). \tag{20}$$

Recall that a seminorm p has the same properties as a norm with the only exception that $p(u) = 0$ does not necessarily imply $u = 0$. By (i), it is sufficient to prove that:

$$\|u\|_{1,2} \le c\|u\| \qquad \text{for all} \quad u \in \mathscr{W}_2^1(G). \tag{20*}$$

In the two special cases

$$p(u) = \left|\int_G u\,dx\right| \qquad \text{and} \qquad p(u) = \left(\int_{\partial G} u^2\,dO\right)^{1/2},$$

we obtain Theorem 21.A(c) from (20).

The proof of (20*) will be based on the *compactness of the embedding* $\mathscr{W}_2^1(G) \subseteq L_2(G)$. Suppose that (20*) is not true. Then there exists a sequence (u_n) in $\mathscr{W}_2^1(G)$ such that

$$\|u_n\|_{1,2} = 1 \qquad \text{and} \qquad 1 > n\|u_n\| \qquad \text{for all } n.$$

Since (u_n) is bounded in the H-space $\mathscr{W}_2^1(G)$, there exists a subsequence, again denoted by (u_n), such that

$$u_n \rightharpoonup u \quad \text{in } \mathscr{W}_2^1(G) \qquad \text{as} \quad n \to \infty,$$

by Theorem 21.D. This implies

$$u_n \to u \quad \text{in } L_2(G) \qquad \text{as} \quad n \to \infty,$$

because of the compactness of the embedding $\mathscr{W}_2^1(G) \subseteq L_2(G)$ (cf. Proposition 21.35(d)).

It follows from $\|u_n\| \to 0$ as $n \to \infty$ that

$$D_j u_n \to 0 \quad \text{in } L_2(G) \qquad \text{as} \quad n \to \infty \quad \text{for all } j,$$

$$p(u_n) \to 0 \qquad \text{as} \quad n \to \infty.$$

Hence $D_j u = 0$ on G for all j. By Problem 21.3a, $u = \text{const}$. Since $p: \mathscr{W}_2^1(G) \to \mathbb{R}$ is convex and continuous, by (i) and $|p(v) - p(w)| \le p(v - w)$, it follows from $u_n \rightharpoonup u$ in $\mathscr{W}_2^1(G)$ that

$$p(u) \le \underline{\lim}_{n\to\infty}\, p(u_n),$$

by Proposition 25.20, and hence $p(u) = 0$, because $p(\cdot)$ is nonnegative. By (ii), $u = 0$. From $D_j u_n \to 0$ and $u_n \to 0$ in $L_2(G)$ we obtain that

$$\|u_n\|_{1,2} \to 0 \qquad \text{as} \quad n \to \infty,$$

contradicting $\|u_n\|_{1,2} = 1$ for all n.

Step 4: The embedding $\mathscr{W}_2^k(G) \subseteq C(\bar{G})$ is continuous in the case where $k > N/2$.

This implies Theorem 21.A(d). In fact, if $u \in \mathscr{W}_2^{k+1}(G)$, then $D_j u \in \mathscr{W}_2^k(G)$ and hence $D_j u \in C(\bar{G})$ for all j; therefore $u \in C^1(\bar{G})$, etc.

As we shall show, the crucial condition "$k > N/2$" results from

$$\int_B \frac{dx}{|x|^{2(N-k)}} < \infty, \tag{21}$$

where B is a ball in $\mathbb{R}^N$.

Let C be a closed spherical cone with vertex at the boundary point $x \in \partial G$ and $C \subseteq \bar{G}$ (Fig. 21.4). Moreover, let $R > 0$ and $\Omega > 0$ be the radius and the

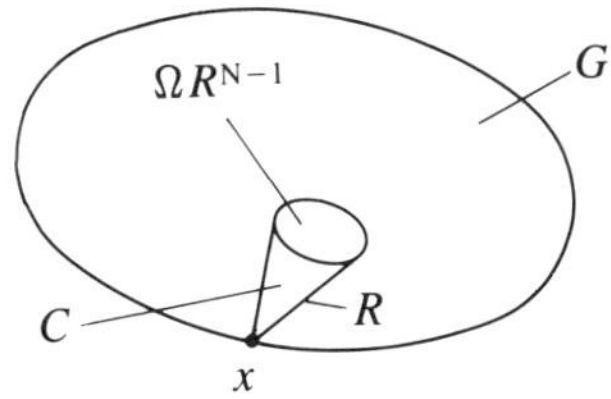

Figure 21.4

solid angle of C, respectively. We want to prove the *key inequality*:

$$|u(x)|^2 \leq c(R,\Omega)^2 \sum_{|\alpha|\leq k} \int_C |D^\alpha u|^2\,dx \qquad \text{for all} \quad u \in C^\infty(\bar{G}). \tag{22}$$

Since $\partial G \in C^{0,1}$, there exist numbers $R > 0$ and $\Omega > 0$ so that, at *each* boundary point $x \in \partial G$, there exists such a cone C with radius R and solid angle Ω. Thus, from (22) we obtain

$$\max_{x\in\bar{G}} |u(x)| \leq c(R,\Omega)\|u\|_{k,2} \qquad \text{for all} \quad u \in C^\infty(\bar{G}).$$

This implies the continuity of the embedding $W_2^k(G) \subseteq C(\bar{G})$, by using exactly the same argument (A) as in Step 1 of Section 21.4a.

We now prove (22). To simplify notation, let $N = 3$, $k = 2$, and $x = 0$. We choose a C^∞-function $h\colon \mathbb{R} \to \mathbb{R}$ with

$$h(t) = \begin{cases} 1 & \text{for } t \leq \frac{1}{2}, \\ 0 & \text{for } t \geq 1, \end{cases}$$

and we set

$$\varphi(r) = h(r/R).$$

Then $\varphi(0) = 1$ and $\varphi'(0) = \varphi(R) = \varphi'(R) = 0$. Integration by parts yields

$$u(0) = -\int_0^R (\varphi u)_r\,dr = \int_0^R r^{-1}(\varphi u)_{rr} r^2\,dr.$$

Noting $dx = r^2\,dr\,d\omega$, integration over the solid angle ω yields

$$\Omega u(0) = \int_C r^{-1}(\varphi u)_{rr}\,dx.$$

By the Hölder inequality,

$$|\Omega u(0)|^2 \leq \int_C r^{-2}\,dx \int_C (\varphi u)_{rr}^2\,dx.$$

This yields (22).

If $N \geq 2$, then we use (21) and the formula

$$u(0) = (-1)^k \int_0^R r^{k-N} \frac{\partial^k(\varphi u)}{\partial r^k} r^{N-1}\,dr.$$

21.4d. Density of Smooth Functions in Sobolev Spaces

We prove that

$$\mathscr{W}_2^1(G) = W_2^1(G) \tag{23}$$

provided $\partial G \in C$, i.e., we prove that $C^\infty(\bar{G})$ is dense in $W_2^1(G)$. The boundary property $\partial G \in C$ was introduced in Section 6.2. In particular, the condition $\partial G \in C$ is fulfilled in the case where $\partial G \in C^{0,1}$. We first consider the case $N = 1$ and $G =]a, b[$ with $-\infty < a < b < \infty$. Let $u \in W_2^1(G)$.

Step 1: Localization of u.

As in Figure 21.5(a), let $\{U_j\}_{j=1,2,3}$ be a covering of $\bar{G}$, where the sets U_j are nonempty and open, and the boundary points are covered by U_1 and U_3. Let $\{\psi_j\}$ be a partition of unity with respect to $\{U_j\}$, i.e.,

$$\psi_j \in C_0^\infty(U_j) \qquad \text{and} \qquad \sum_k \psi_k(x) = 1, \qquad 0 \le \psi_j(x) \le 1,$$

for all $x \in \bar{G}$ and all j. We set

$$u_j = u\psi_j.$$

By the product rule for generalized derivatives (cf. Problem 21.3d), we get

$$u_j' = u'\psi_j + u\psi_j'.$$

Hence $u_j \in W_2^1(G)$ for all j.

Step 2: Translation of the localizations u_j at the boundary points.

For sufficiently small $k > 0$, the function

$$v(x) = u_1(x + k)$$

has the following two properties:

(i) $v \in W_2^1(H)$ for a sufficiently small neighborhood H of $\bar{G}$.
(ii) $v \to u_1$ in $W_2^1(G)$ as $k \to 0$.

Property (i) follows from

$$v'(x) = u_1'(x + k) \qquad \text{for all} \quad x \in H.$$

In fact, for all $\varphi \in C_0^\infty(H)$, we have

$$\begin{aligned}\int_H u_1'(x + k)\varphi(x)\,dx &= \int_G u_1'(x)\varphi(x - k)\,dx \\ &= -\int_G u_1(x)\varphi'(x - k)\,dx = -\int_H u_1(x + k)\varphi'(x)\,dx.\end{aligned}$$

In this connection, note that $x \mapsto \varphi(x - k)$ belongs to $C_0^\infty(G)$ for sufficiently small neighborhoods H of $\bar{G}$.

Property (ii) follows from

$$\int_G (u_1(x + k) - u_1(x))^2 + (u_1'(x + k) - u_1'(x))^2\,dx \to 0 \qquad \text{as} \quad k \to 0,$$

according to the mean continuity of L_2-functions (cf. $A_2(32)$).

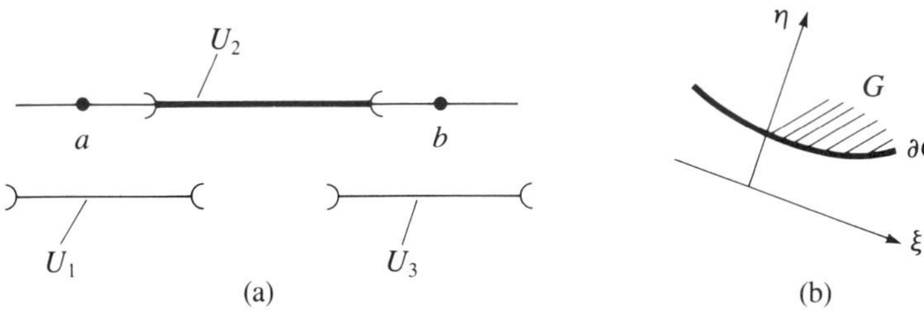

Figure 21.5

Step 3: Smoothing of v.
We set

$$v_\varepsilon = S_\varepsilon v.$$

By Proposition 18.35, it follows from $v \in W_2^1(H)$ and $\bar{G} \subset H$ that $v_\varepsilon \in C^\infty(\mathbb{R})$ for small $\varepsilon > 0$ and

$$v_\varepsilon \to v \quad \text{in } W_2^1(G) \qquad \text{as} \quad \varepsilon \to +0.$$

Using an analogous construction for the other localizations u_j, we find that $C^\infty(\bar{G})$ is dense in $W_2^1(G)$ if $G \subset \mathbb{R}^1$.

The general case $G \subset \mathbb{R}^N$ can be treated completely similarly by using local coordinate systems for the local representation of the boundary ∂G. Then the translation in Step 2 corresponds to a translation with respect to the local coordinate system. For example, if $N = 2$, then we use $v(\xi, \eta) = u_j(\xi, \eta + k)$ (cf. Fig. 21.5(b)).

In the same way we prove the density of $C^\infty(\bar{G})$ in $W_p^k(G)$ with $1 \le p < \infty$, $k = 1, 2, \ldots$, in the case where G is a bounded region in $\mathbb{R}^N$ with $N \ge 1$ and $\partial G \in C$. Hence

$$\mathscr{W}_p^k(G) = W_p^k(G). \tag{24}$$

This completes the proof of all the assertions stated in Section 21.3. □

21.5. Duality in B-Spaces

We summarize some well-known facts about B-spaces. Let X be a B-space over $\mathbb{K} = \mathbb{R}, \mathbb{C}$. By definition, a *linear continuous functional* on X is a linear continuous map $f: X \to \mathbb{K}$. We set

$$\langle f, u \rangle = f(u) \qquad \text{for all} \quad u \in X$$

and

$$\|f\| = \sup_{\|u\| \le 1} |\langle f, u \rangle|.$$

Hence

$$|\langle f, u \rangle| \le \|f\| \, \|u\| \qquad \text{for all} \quad u \in X.$$

Let X^* denote the set of all linear continuous functionals on X. Then X^*, together with the norm $\|f\|$, becomes a B-space over $\mathbb{K}$, which is called the *dual space* to X. We set $X^{**} = (X^*)^*$.

Let $u \in X$. If we set

$$U(f) = \langle f, u \rangle \qquad \text{for all} \quad f \in X,$$

then $U: X^* \to \mathbb{K}$ is a linear continuous functional, i.e., $U \in X^{**}$ and $\|U\| = \|u\|$. Letting $j(u) = U$, we obtain a linear isometric map

$$j: X \to X^{**}.$$

Since j is injective, we can identify X with the subset $j(X)$ of X^{**}. In this sense, we write $X \subseteq X^{**}$. The B-space X is called *reflexive* iff j is surjective. Then j is a normisomorphism from X onto X^{**}, and we can write $X = X^{**}$. Moreover, if we identify u with $U = j(u)$, then we can write

$$\langle f, u \rangle = \langle u, f \rangle \qquad \text{for all} \quad f \in X^*, \quad u \in X.$$

Let X and Y be two B-spaces over $\mathbb{K}$ and let

$$A: D(A) \subseteq X \to Y$$

be a linear operator. Recall that this means that $D(A)$ is a linear subspace of X and

$$A(\alpha u + \beta v) = \alpha A u + \beta A v \qquad \text{for all} \quad u, v \in D(A), \quad \alpha, \beta \in \mathbb{K}.$$

Suppose that $D(A)$ is dense in X. Then the *dual operator*

$$A^*: D(A^*) \subseteq Y^* \to X^*$$

is defined by the relation

$$\langle A^* v, u \rangle = \langle v, Au \rangle \qquad \text{for all} \quad v \in D(A^*), \quad u \in D(A). \tag{25}$$

To be precise, by definition, v belongs to $D(A^*)$ iff there exists a $w \in X^*$ with

$$\langle w, u \rangle = \langle v, Au \rangle \qquad \text{for all} \quad u \in D(A).$$

Then we set $A^* v = w$.

Since $D(A)$ is dense in X, the element w is uniquely determined by v. For all $u, v \in D(A)$ and all $\alpha, \beta \in \mathbb{K}$,

$$A^*(\alpha u + \beta v) = \alpha A^* u + \beta A^* v. \tag{26}$$

The operator A^* is graph closed.

If $A: X \to Y$ is linear and continuous, then the dual operator $A^*: Y^* \to X^*$ is also linear and continuous, and we have

$$\|A^*\| = \|A\|. \tag{27}$$

If, in addition, X and Y are reflexive, then

$$A^{**} = A$$

in the sense of $X^{**} = X$ and $Y^{**} = Y$.

The linear continuous operator $A: X \to Y$ is compact iff $A^*: Y^* \to X^*$ is compact (theorem of Schauder (1930)).

The most important theorem on dual operators in B-spaces is the following closed range theorem of Banach.

Let

$$^{\perp}N(A) = \{v \in X^*: \langle v, u\rangle = 0 \text{ for all } u \in N(A)\},$$

$$N(A^*)^{\perp} = \{v \in Y: \langle u, v\rangle = 0 \text{ for all } u \in N(A^*)\}.$$

Theorem 21.B (Banach 1932). *Let $A: D(A) \subseteq X \to Y$ be a linear, densely defined, graph closed operator, where X and Y are* B-*spaces over $\mathbb{K}$. Then the following four conditions are mutually equivalent*:

(i) *$R(A)$ is closed in Y.*
(ii) *$R(A^*)$ is closed in X^*.*
(iii) *$R(A) = N(A^*)^{\perp}$.*
(iv) *$R(A^*) = {}^{\perp}N(A)$.*

This theorem is a generalization of the existence principle in Section 19.30. In particular, if the range $R(A)$ of A is closed, then we have $R(A) = N(A^*)^{\perp}$, i.e., the equation

$$Au = b, \qquad u \in D(A), \tag{28}$$

has a solution iff

$$\langle w, b\rangle = 0$$

for all w with $A^*w = 0$. This is an abstract *Fredholm alternative*. As in Section 19.30, the closedness of $R(A)$ follows from the *a priori* estimate:

$$\|Au\| \geq c\|u\| \qquad \text{for all} \quad u \in D(A) \quad \text{and fixed} \quad c > 0.$$

Note that A is graph closed.

The proofs of all the statements above can be found in Yosida (1965, M) and Kato (1966, M).

21.6. Duality in H-Spaces

Let $A: X \to X$ be a linear continuous operator on the real H-space X. By Section 21.5, the *dual* operator $A^*: X^* \to X^*$ satisfies

$$\langle A^*v, u\rangle = \langle v, Au\rangle \qquad \text{for all} \quad v \in D(A^*), \quad u \in D(A).$$

By Section 19.8, the adjoint operator $A^{*'}: X \to X$ satisfies

$$(A^{*'}v|u) = (v|Au) \qquad \text{for all} \quad v \in D(A^{*'}), \quad u \in D(A).$$

We want to show that there is a close connection between X and X^* as well as A^* and $A^{*'}$. This is given by the following commutative diagram:

$$\begin{array}{ccc} X & \xrightarrow{A^*} & X \\ {\scriptstyle J}\downarrow & & \uparrow{\scriptstyle J^{-1}} \\ X^* & \xrightarrow{A^*} & X^*. \end{array} \tag{29}$$

Furthermore, we need the following two formulas:

$$\langle Ju, v\rangle = (u|v) \qquad \text{for all} \quad u, v \in X, \tag{30}$$

$$\langle A^*Ju, v\rangle = (A^{*'}u|v) \qquad \text{for all} \quad u, v \in X. \tag{31}$$

Recall that $L(X, Y)$ denotes the set of all linear continuous operators A: $X \to Y$.

Proposition 21.17. *Let X be a real* H-*space. Then*:

(a) *For each $u \in X$, there exists a unique linear continuous functional Ju on X with* (30). *The operator J: $X \to X^*$ is linear, bijective, and $\|Ju\| = \|u\|$ for all $u \in X$.*
(b) *If $A \in L(X, X)$, then* (31) *holds and $A^{*'} = J^{-1}A^*J$.*

The operator J is called the *duality map* of the H-space X. Duality maps of B-spaces will be considered in Chapter 32. Such duality maps can be multivalued.

Proof. Ad(a). This is another formulation of the Riesz theorem in Section 18.11.

Ad(b). For all $u, v \in X$,

$$\langle A^*Ju, v\rangle = \langle Ju, Av\rangle = (u|Av) = (A^{*'}u|v) = \langle JA^{*'}u, v\rangle.$$

Hence $A^*J = JA^{*'}$. □

Identification Principle 21.18 ($X^* = X$). *Let X be a real* H-*space.*

(i) *Since the duality map J: $X \to X^*$ is a normisomorphism, we can identify Ju with u. In this sense, we write $X^* = X$. Then, we have*

$$\langle u, v\rangle = (u|v) \qquad \textit{for all} \quad u, v \in X, \tag{32}$$

and J becomes the identity operator on X.
(ii) *For $A \in L(X, X)$, we obtain $A^* = A^{*'}$, and each operator $A \in L(X, X^*)$ can be identified with an operator $A \in L(X, X)$.*

In particular, we obtain that each H-space is reflexive.

21.7. The Idea of Weak Convergence

For the development of Functional Analysis, the most important concepts introduced by Hilbert (1906a) were what he calls "continuity" and "complete continuity," which correspond to what will later be called the "strong" and "weak" topologies on Hilbert space.

Jean Dieudonné (1981)

Theorem 21.C (Riesz (1918)). *A* B-*space is finite-dimensional iff the closed unit ball is compact.*

The proof will be given in Problem 21.2. This famous result tells us that in infinite-dimensional B-spaces there exist bounded sequences (u_n), which have *no* convergent subsequence. This lack of compactness in infinite-dimensional B-spaces is responsible for many difficulties in the calculus of variations and the theory of partial differential equations. In order to overcome this difficulty, Hilbert (1906a) introduced the concept of weak convergence in H-space.

Definition 21.19. A sequence (u_n) in the B-space X is called *weakly convergent*, i.e.,

$$u_n \rightharpoonup u \quad \text{as} \quad n \to \infty,$$

iff

$$\langle f, u_n \rangle \to \langle f, u \rangle \quad \text{as} \quad n \to \infty \quad \text{for all} \quad f \in X^*.$$

The following result is crucial for proving existence theorems in the calculus of variations and the theory of monotone operators.

Theorem 21.D (Eberlein (1947), Šmuljan (1940)). *Each bounded sequence in a reflexive* B-*space has a weakly convergent subsequence.*

PROOF.

(I) Let X be a separable H-space and let (u_n) be a bounded sequence in X. We choose a countable set $\{v_k\}$, which is dense in X, and we use the *diagonal procedure*:

$$(u_{11}|v_1), (u_{12}|v_1), (u_{13}|v_1), \ldots \to a_1,$$

$$(u_{21}|v_2), (u_{22}|v_2), (u_{23}|v_2), \ldots \to a_2, \ldots.$$

To be precise, since $|(u_n|v_1)| \le \|u_n\| \, \|v_1\|$, there exists a subsequence of (u_n), denoted by (u_{1n}), such that $(u_{1n}|v_1) \to a_1$ as $n \to \infty$. Furthermore, there exists a subsequence (u_{2n}) of (u_{1n}) such that $(u_{2n}|v_2) \to a_2$ as $n \to \infty$, etc.

The "diagonal sequence" (w_n) with $w_n = u_{nn}$ has the property that

$$(w_n|v_k) \to a_k \quad \text{as} \quad n \to \infty \quad \text{for all } k.$$

Moreover, there exist numbers $a(v)$ such that

$$(w_n|v) \to a(v) \quad \text{as} \quad n \to \infty \quad \text{for all} \quad v \in X. \tag{33}$$

This follows from

$$\begin{aligned}|(w_n - w_m|v)| &= |(w_n - w_m|v - v_k) + (w_n - w_m|v_k)| \\ &\leq \|w_n - w_m\| \, \|v - v_k\| + |(w_n - w_m|v_k)| < \varepsilon\end{aligned}$$

for suitable v_k and all $n, m \geq n_0(\varepsilon)$. Note that (w_n) is bounded and $\{v_k\}$ is dense in X.

Obviously, the map $v \mapsto a(v)$ is linear and, by (33),

$$|a(v)| \leq \|v\| \sup_n \|w_n\| \quad \text{for all} \quad v \in X.$$

By the *Riesz theorem*, there exists a $w \in X$ with $a(v) = (w|v)$ for all $v \in X$. From (33) it follows that

$$w_n \rightharpoonup w \quad \text{as} \quad n \to \infty.$$

(II) The proof for the general case can be found in Yosida (1965, M), Chapter 5. □

Definition 21.20. A B-space X is called *strictly convex* iff

$$\|tu + (1 - t)v\| < 1$$

provided $\|u\| = \|v\| = 1$, $u \neq v$, and $0 < t < 1$.

A B-space X is called *locally uniformly convex* iff for each ε, $0 < \varepsilon \leq 2$, and for each $u \in X$, $\|u\| = 1$, there exists a $\delta(\varepsilon, u) > 0$ such that the following holds for all $u, v \in X$:

$$\|u - v\| \geq \varepsilon, \quad \|u\| = \|v\| = 1 \quad \text{implies} \quad \|2^{-1}(u + v)\| \leq 1 - \delta(\varepsilon, u).$$

A B-space X is called *uniformly convex* iff it is locally uniformly convex and δ can be chosen to be independent of u.

We have the following implications:

$$\begin{aligned}\textit{X is uniformly convex} \quad &\Rightarrow \quad \textit{X is locally uniformly convex} \\ &\Rightarrow \quad \textit{X is strictly convex.}\end{aligned}$$

Each uniformly convex B-space is reflexive (cf. Section 10.1).

In every reflexive B-space X, an *equivalent norm* can be introduced so that both X and X^* are locally uniformly convex (cf. Troyanski (1971)). This result simplifies some proofs in the theory of monotone operators.

The importance of these notions with respect to weak convergence will follow from Proposition 21.23(d) below. Intuitively, in a uniformly convex B-space, the unit sphere is round (cf. Fig. 10.1).

Example 21.21. Each H-space is uniformly convex and hence locally uniformly convex.

Proof. This follows from the parallelogram identity

$$\|\tfrac{1}{2}(u-v)\|^2 + \|\tfrac{1}{2}(u+v)\|^2 = \tfrac{1}{2}(\|u\|^2 + \|v\|^2). \qquad \square$$

Example 21.22. Let $1 < p < \infty$, and let G be a nonempty open set in $\mathbb{R}^N$, $N \geq 1$. Then, for all f, $g \in L_p(G)$, we have the generalized parallelogram inequality:

$$\|\tfrac{1}{2}(f-g)\|^p + \|\tfrac{1}{2}(f+g)\|^p \leq \tfrac{1}{2}(\|f\|^p + \|g\|^p), \tag{34a}$$

provided $2 \leq p < \infty$, and

$$\|\tfrac{1}{2}(f-g)\|^q + \|\tfrac{1}{2}(f+g)\|^q \leq [\tfrac{1}{2}(\|f\|^p + \|g\|^p)]^{q-1}, \tag{34b}$$

provided $1 < p < 2$. Here, $\|\cdot\|$ denotes the norm in $L_p(G)$, and $p^{-1} + q^{-1} = 1$.

This inequality is due to Clarkson (1936). Hence the Lebesgue space $L_p(G)$ is *uniformly convex* for all $1 < p < \infty$.

More generally, we have the following result. Let Y be a B-space over $\mathbb{K} = \mathbb{R}, \mathbb{C}$. For fixed p with $1 < p < \infty$, we set

$$\|u\| = \left(\sum_{k=1}^{M} \|u_k\|^p\right)^{1/p},$$

where $M = 1, 2, \ldots$ or $M = \infty$. Moreover, define

$$X = \left\{u \in \prod_{k=1}^{M} Y \colon \|u\| < \infty\right\}.$$

If we now equip X with the norm $\|\cdot\|$, then X is a uniformly convex B-space over $\mathbb{K}$ iff Y is uniformly convex.

In particular, this implies that the B-spaces $\mathbb{R}^N$ and $\mathbb{C}^N$, $N \geq 1$, are uniformly convex with respect to the norm

$$\|x\|_p = \left(\sum_{k=1}^{N} |\xi_k|^p\right)^{1/p},$$

where $x = (\xi_1, \ldots, \xi_N)$.

Moreover, it follows from the uniform convexity of the space $L_p(G)$, $1 < p < \infty$, that the Sobolev spaces $W_p^m(G)$ are uniformly convex for $1 < p < \infty$ and $m = 0, 1, \ldots$. In this connection, note that the Sobolev space $W_p^m(G)$ is norm-isomorphic to a closed linear subspace of the product space $L_p(G)^M$ by means of the mapping $u \mapsto (D^\alpha u)_{|\alpha| \leq m}$.

The proofs can be found in Köthe (1969, M), Vol. 1.

Further important material on strictly convex and locally uniformly convex B-spaces can be found in the Appendix to Part III.

Recall that $u_n \to u$ as $n \to \infty$ in a B-space X means $\|u_n - u\| \to 0$ as $n \to \infty$. This usual convergence in X is sometimes called the strong convergence in contrast to the weak convergence. The following results will be used frequently.

Proposition 21.23 (Properties of Weak Convergence). *Let (u_n) be a sequence in the* B-*space X over $\mathbb{K} = \mathbb{R}, \mathbb{C}$. Then:*

(a) *The strong convergence $u_n \to u$ as $n \to \infty$ implies the weak convergence $u_n \rightharpoonup u$ as $n \to \infty$.*

(b) *If* $\dim X < \infty$, *then the weak convergence $u_n \rightharpoonup u$ as $n \to \infty$ implies the strong convergence $u_n \to u$ as $n \to \infty$.*

(c) *If $u_n \rightharpoonup u$ as $n \to \infty$, then (u_n) is bounded and*

$$\|u\| \leq \varliminf_{n \to \infty} \|u_n\| \tag{35}$$

(theorem of Banach and Steinhaus (1927)).

(d) *If X is locally uniformly convex, then $u_n \rightharpoonup u$ and $\|u_n\| \to \|u\|$ as $n \to \infty$ implies $u_n \to u$ as $n \to \infty$.*

(e) *If $u_n \rightharpoonup u$ as $n \to \infty$, then there exists a sequence (v_n) in the closed convex hull of $\{u_n\}$ such that $v_n \to u$ as $n \to \infty$ (theorem of Mazur* (1933)).

(f) *If $u_n \rightharpoonup u$ as $n \to \infty$, then u lies in the closed convex hull of $\{u_n\}$.*

(g) *If (u_n) is bounded in X and if there exist a $u \in X$ and a dense set D in X^* such that*

$$\langle f, u_n \rangle \to \langle f, u \rangle \quad \text{as} \quad n \to \infty \quad \text{for all} \quad f \in D,$$

then $u_n \rightharpoonup u$ as $n \to \infty$.

(h) *If X is reflexive and the real sequence $(\langle v, u_n \rangle)$ converges for all $v \in X^*$, then there exists a $u \in X$ such that $u_n \rightharpoonup u$ as $n \to \infty$.*

(i) *If (u_n) is a bounded sequence in the reflexive* B-*space X, then (u_n) has a weakly convergent subsequence.*

If, in addition, each weakly convergent subsequence of (u_n) has the same limit u, then $u_n \rightharpoonup u$ as $n \to \infty$.

(j) *It follows from*

$$u_n \rightharpoonup u \quad \text{in } X \quad \text{as} \quad n \to \infty,$$

$$f_n \to f \quad \text{in } X^* \quad \text{as} \quad n \to \infty,$$

that $\langle f_n, u_n \rangle \to \langle f, u \rangle$ as $n \to \infty$.

(k) *If X is reflexive, then it follows from*

$$u_n \to u \quad \text{in } X \quad \text{as} \quad n \to \infty,$$

$$f_n \rightharpoonup f \quad \text{in } X^* \quad \text{as} \quad n \to \infty,$$

that $\langle f_n, u_n \rangle \to \langle f, u \rangle$ as $n \to \infty$.

Proof. Ad(a). If $u_n \to u$, then $\langle f, u_n \rangle \to \langle f, u \rangle$ for all $f \in X^*$, since f is continuous.

Ad(b). Choosing a basis in X, we can identify X with $\mathbb{K}^n$. Hence the weak convergence means convergence of the components. This implies the strong convergence.

Ad(c). This is a special case of the general Banach–Steinhaus theorem $A_1(35)$. The proof can be found in Yosida (1965, M), Chapter 5.

Ad(d). For $u = 0$ the assertion is obvious. Let $u \neq 0$. We set

$$v = u/\|u\| \qquad \text{and} \qquad v_n = u_n/\|u_n\|$$

for sufficiently large n. Hence $\|v_n\| = \|v\| = 1$ and

$$v_n \rightharpoonup v \qquad \text{as} \quad n \to \infty.$$

By (c), it follows from $v_n + v \rightharpoonup 2v$ as $n \to \infty$ that

$$\begin{aligned} 2 = \|2v\| &\leq \varliminf_{n\to\infty} \|v_n + v\| \leq \varlimsup_{n\to\infty} \|v_n + v\| \\ &\leq \lim_{n\to\infty} \|v_n\| + \|v\| = 2. \end{aligned}$$

Hence $\|v_n + v\| \to 2$ as $n \to \infty$. By Definition 21.20,

$$\|v_n - v\| \to 0 \qquad \text{as} \quad n \to \infty.$$

This implies $u_n \to u$ as $n \to \infty$, since $\|u_n\| \to \|u\|$.

Ad(e). The proof, which is based on the Hahn–Banach theorem, can be found in Yosida (1965, M), Chapter 5.

Ad(f). This follows from (e).

Ad(g). Let $g \in X^*$. Use

$$\begin{aligned} |\langle g, u_n\rangle - \langle g, u\rangle| &= |\langle g - f, u_n\rangle + \langle f, u_n - u\rangle + \langle f - g, u\rangle| \\ &\leq \|g - f\|(\|u_n\| + \|u\|) + |\langle f, u_n - u\rangle| < \varepsilon \end{aligned}$$

for suitable $f \in D$ and all $n \geq n_0(\varepsilon)$.

Ad(h). By the Banach–Steinhaus theorem $A_1(35)$, the sequence (u_n) is bounded. Let

$$a(v) = \lim_{n\to\infty} \langle v, u_n\rangle \qquad \text{for all} \quad v \in X^*.$$

From $|a(v)| \leq \|v\| \sup_n \|u_n\|$ it follows that $a \in X^{**}$. Since X is reflexive, there exists a $u \in X$ with $a(v) = \langle v, u\rangle$ for all $v \in X^*$.

Ad(i). This follows from Theorem 21.D and the convergence principle (Proposition 10.13(4)).

Ad(j). Since (u_n) is bounded by (c), we get

$$\begin{aligned} |\langle f_n, u_n\rangle - \langle f, u\rangle| &= |\langle f_n - f, u_n\rangle + \langle f, u_n - u\rangle| \\ &\leq \|f_n - f\| \sup_n \|u_n\| + |\langle f, u_n - u\rangle| \to 0 \qquad \text{as} \quad n \to \infty. \end{aligned}$$

Ad(k). Use (j) and $X^{**} = X$. $\square$

21.8. The Idea of Weak* Convergence

The crucial convergence theorem of Eberlein and Šmuljan (Theorem 21.D) is only valid in reflexive B-spaces. More precisely, a B-space is reflexive iff every bounded sequence has a weakly convergent subsequence. In nonreflexive B-spaces one can frequently replace weak convergence by weak* convergence.

Definition 21.24. Let X be a B-space. A sequence (f_n) in X^* is called *weakly* convergent*, i.e.,

$$f_n \overset{*}{\rightharpoonup} f \qquad \text{as} \quad n \to \infty,$$

iff

$$\langle f_n, u \rangle \to \langle f, u \rangle \qquad \text{as} \quad n \to \infty \quad \text{for all} \quad u \in X.$$

Theorem 21.E. *Let X be a separable* B-*space. Then, each bounded sequence (f_n) in X^* has a weakly* convergent subsequence.*

PROOF. Let $\{v_k\}$ be a countable dense subset of X. As in (33), we obtain a subsequence (w_n) of (f_n) with

$$\langle w_n, v \rangle \to a(v) \qquad \text{as} \quad n \to \infty \quad \text{for all} \quad v \in X.$$

From $|\langle a, v \rangle| \leq \|v\| \sup_n \|w_n\|$ it follows that $a \in X^*$. □

EXAMPLE 21.25. Let G be a nonempty bounded open set in $\mathbb{R}^N$, $N \geq 1$. Let

$$X = L_1(G).$$

Then, $X^* = L_\infty(G)$. We have

$$f_n \overset{*}{\rightharpoonup} f \quad \text{in } X^* \qquad \text{as} \quad n \to \infty$$

iff

$$\int_G f_n u \, dx \to \int_G f u \, dx \qquad \text{as} \quad n \to \infty \quad \text{for all} \quad u \in X.$$

Since $C_0^\infty(G)$ is dense in $L_1(G)$, the space X is separable. However, X^* is not separable and both X and X^* are not reflexive.

Let (f_n) be a bounded sequence in X^*, i.e., the functions $f_n: G \to \mathbb{R}$ are measurable and there exists a constant C such that

$$|f_n(x)| \leq C \qquad \text{for all } n \text{ and almost all } x \in G.$$

By Theorem 21.E, there exists a subsequence $(f_{n'})$ of (f_n) with $f_{n'} \overset{*}{\rightharpoonup} f$ as $n \to \infty$.

The proofs of the other statements can be found in Kufner, John, and Fučik (1977, M), Chapter 2.

Proposition 21.26 (Properties of Weak* Convergence). *Let X be a* B-*space over* $\mathbb{K}$, *and let (f_n) be a sequence in the dual space X^*. Then:*

(a) *If* $f_n \to f$ *in* X^* *as* $n \to \infty$, *then* $f_n \overset{*}{\rightharpoonup} f_n$ *as* $n \to \infty$.
(b) *If* $f_n \overset{*}{\rightharpoonup} f$ *as* $n \to \infty$, *then* (f_n) *is bounded in* X^* *and*

$$\|f\| \leq \varliminf_{n\to\infty} \|f_n\|$$

(theorem of Banach–Steinhaus (1927)).
(c) *If* (f_n) *is bounded in* X^* *and if there exists an* $f \in X^*$ *and a dense subset* D *of* X *with*

$$\langle f_n, u\rangle \to \langle f, u\rangle \quad \text{as} \quad n \to \infty \quad \text{for all} \quad u \in D,$$

then $f_n \overset{*}{\rightharpoonup} f$ *as* $n \to \infty$.
(d) *If the real sequences* $(\langle f_n, u\rangle)$ *are convergent for all* $u \in X$, *then there exists an* $f \in X^*$ *such that* $f_n \overset{*}{\rightharpoonup} f$ *as* $n \to \infty$.
(e) *It follows from*

$$u_n \to u \quad \text{in } X \quad \text{as} \quad n \to \infty,$$

$$f_n \overset{*}{\rightharpoonup} f \quad \text{in } X^* \quad \text{as} \quad n \to \infty,$$

that $\langle f_n, u_n\rangle \to \langle f, u\rangle$ *as* $n \to \infty$.
(f) *If* X *is reflexive then* $f_n \overset{*}{\rightharpoonup} f$ *as* $n \to \infty$ *is equivalent to* $f_n \rightharpoonup f$ *as* $n \to \infty$.

Proof. This follows analogously to the proof of Proposition 21.23 above. □

21.9. Linear Operators

Let $A: X \to Y$ be a linear operator, where X and Y are B-spaces over $\mathbb{K}$. Recall that the following two conditions are equivalent:

(i) A is bounded, i.e., there exists a constant $d > 0$ *such that*

$$\|Au\| \leq d\|u\| \qquad \text{for all} \quad u \in X.$$

(ii) A is continuous, i.e., $u_n \to u$ as $n \to \infty$ implies $Au_n \to Au$ as $n \to \infty$.

Proposition 21.27. *Let* X *and* Y *be* B*-spaces over* $\mathbb{K}$. *If* $A: X \to Y$ *is linear and continuous, then* A *is weakly sequentially continuous, i.e.,*

$$u_n \rightharpoonup u \quad \text{as} \quad n \to \infty \qquad \text{implies} \qquad Au_n \rightharpoonup Au \quad \text{as} \quad n \to \infty.$$

Proof. Let $u_n \rightharpoonup u$. From

$$\langle v, Au_n\rangle = \langle A^*v, u_n\rangle, \qquad \langle v, Au\rangle = \langle A^*v, u\rangle$$

it follows that

$$\langle v, Au_n\rangle \to \langle v, Au\rangle \quad \text{as} \quad n \to \infty \quad \text{for all} \quad v \in Y^*,$$

i.e., $Au_n \rightharpoonup Au$ as $n \to \infty$. □

Proposition 21.28. *Let $A: X \to X^*$ be a linear demicontinuous operator on the* B-*space X, i.e.,*

$$u_n \to u \quad \text{as} \quad n \to \infty \qquad \text{implies} \qquad Au_n \rightharpoonup Au \quad \text{as} \quad n \to \infty.$$

Then A is continuous.

Proof. If A is not continuous, then there exist an $\varepsilon > 0$ and a sequence (u_n) with $u_n \to 0$ as $n \to \infty$ and $\|Au_n\| \geq \varepsilon$ for all n. Let $a_n = \|u_n\|^{-1/2}$ and $v_n = a_n u_n$. Then, $v_n \to 0$ and

$$\|Av_n\| \geq a_n \varepsilon \to +\infty \quad \text{as} \quad n \to \infty.$$

Otherwise, it follows from $v_n \to 0$ that $Av_n \rightharpoonup 0$, i.e., (Av_n) is bounded. This is a contradiction. □

Proposition 21.29. *Let $A: X \to Y$ be a linear operator, where X and Y are* B-*spaces over $\mathbb{K}$. Then:*

(a) *If A is compact, then A is strongly continuous, i.e.,*

$$u_n \rightharpoonup u \quad \text{as} \quad n \to \infty \qquad \text{implies} \qquad Au_n \to Au \quad \text{as} \quad n \to \infty.$$

(b) *Conversely, if A is strongly continuous and X is reflexive, then A is compact.*

Proof. Ad(a). Let $u_n \rightharpoonup u$. Then (u_n) is bounded. Since A is compact, there exists a subsequence $(u_{n'})$ such that $Au_{n'} \to w$ as $n \to \infty$. By Proposition 21.27, $w = Au$. The convergence principle (Proposition 10.13(1)) yields the convergence of the entire sequence, i.e., $Au_n \to w$ as $n \to \infty$.

Ad(b). Let (u_n) be bounded. Since X is reflexive, there exists a subsequence with $u_{n'} \rightharpoonup u$ as $n \to \infty$. Hence $Au_{n'} \to Au$ as $n \to \infty$, i.e., A is compact. □

21.10. Bilinear Forms

We want to investigate the relation between linear continuous operators $A: X \to X^*$ on real B-spaces X and bilinear forms $a: X \times X \to \mathbb{R}$. The *key* formula is

$$\langle Au, v \rangle = a(u, v) \qquad \text{for all} \quad u, v \in X. \tag{36}$$

Definition 21.30. Let X be a real B-space. By a bilinear form on X, we understand a map $a: X \times X \to \mathbb{R}$ with

$$a(u, \alpha v + \beta w) = \alpha a(u, v) + \beta a(u, w),$$

$$a(\alpha v + \beta w, u) = \alpha a(v, u) + \beta a(w, u),$$

for all u, v, $w \in X$ and all α, $\beta \in \mathbb{R}$. In the following, c and d are positive constants and u and v are arbitrary elements in X.

(i) a is called *bounded* iff $|a(u,v)| \leq d\|u\| \|v\|$.
(ii) a is called *positive* iff $a(u,u) \geq 0$.
(iii) a is called *strictly positive* iff $a(u,u) > 0$ for all $u \neq 0$.
(iv) a is called *strongly positive* iff $a(u,u) \geq c\|u\|^2$.
(v) a is called *symmetric* iff $a(u,v) = a(v,u)$.
(vi) a is called *compact* iff, as $n \to \infty$,

$$u_n \rightharpoonup u \quad \text{and} \quad v_n \rightharpoonup v \qquad \text{implies} \qquad a(u_n, v_n) \to a(u,v).$$

Let X be a *complex* B-space. Then the bilinear form $a\colon X \times X \to \mathbb{C}$ is called positive iff

$$\operatorname{Re} a(u,u) \geq 0 \qquad \text{for all} \quad u \in X.$$

Similarly, a is called strongly positive (resp. strictly positive) iff

$$\operatorname{Re} a(u,u) \geq c\|u\|^2 \qquad \text{for all} \quad u \in X \quad \text{and} \quad \text{fixed } c > 0$$

(resp. $\operatorname{Re} a(u,u) > 0$ for all $u \in X$ with $u \neq 0$).

Proposition 21.31. *Let X be a real* B-*space. Then:*

(a) *There exists a one-to-one correspondence between linear continuous operators $A\colon X \to X^*$ and bilinear forms $a\colon X \times X \to \mathbb{R}$, which is given by formula* (36) *above.*
(b) *If X is reflexive, then A is compact* iff *a is compact.*

Proof. Ad(a). Let $A\colon X \to X^*$ be linear and continuous. We set $a(u,v) = \langle Au, v\rangle$. Then

$$|a(u,v)| \leq \|Au\| \|v\| \leq \|A\| \|u\| \|v\|,$$

i.e., $a\colon X \times X \to \mathbb{R}$ is bilinear and bounded.

Conversely, if $a\colon X \times X \to \mathbb{R}$ is bilinear and bounded, then, for fixed $u \in X$, the map $v \mapsto a(u,v)$ is linear and continuous on X. Thus, there exists an element in X^*, denoted by Au, such that

$$\langle Au, v\rangle = a(u,v) \qquad \text{for all} \quad v \in X.$$

From $|\langle Au, v\rangle| \leq d\|u\| \|v\|$ it follows that $\|Au\| \leq d\|u\|$. Hence the operator $A\colon X \to X^*$ is linear and continuous.

Ad(b). Let A be compact. By Proposition 21.29(a), $u_n \rightharpoonup u$ implies $Au_n \to Au$. Hence $u_n \rightharpoonup u$ and $v_n \rightharpoonup v$ implies $\langle Au_n, v_n\rangle \to \langle Au, v\rangle$, i.e., $a(u_n, v_n) \to a(u,v)$.

Conversely, let a be compact. Then the map $v \mapsto a(u,v)$ is linear and continuous on X. Thus, for each $u \in X$, there exists an element in X^*, denoted by Au, such that

$$\langle Au, v\rangle = a(u,v) \qquad \text{for all} \quad v \in X.$$

The operator $A\colon X \to X^*$ is linear. Let $u_n \rightharpoonup u$ as $n \to \infty$. We want to show that

$Au_n \to Au$, i.e.,

$$\sup_{\|v\|=1} |\langle Au_n - Au, v\rangle| \to 0 \quad \text{as} \quad n \to \infty$$

if $X \neq \{0\}$. Otherwise, there is an $\varepsilon > 0$ and there are subsequences, again denoted by (u_n) and (v_n), such that

$$|\langle Au_n - Au, v_n\rangle| \geq \varepsilon \quad \text{and} \quad \|v_n\| = 1 \quad \text{for all } n.$$

Since X is reflexive, there exists a subsequence, again denoted by (v_n), such that $v_n \rightharpoonup v$ as $n \to \infty$. This yields the contradiction $\langle Au_n - Au, v_n\rangle = a(u_n - u, v_n) \to 0$ as $n \to \infty$.

By Proposition 21.29, A is compact. □

Definition 21.32. Let $A\colon X \to X^*$ be a linear continuous operator on the real B-space X.

(i) A is called *positive* iff the corresponding bilinear form a is positive, i.e., $\langle Au, u\rangle \geq 0$ for all $u \in X$.
(ii) A is called strictly positive, strongly positive, or symmetric iff a has the corresponding property.

Corollary 21.33. *Let X be a real* H-*space. Then:*

(a) *There exists a one-to-one correspondence between linear continuous operators $A\colon X \to X$ and bilinear forms $a\colon X \times X \to \mathbb{R}$, which is given by the fomula*

$$(Au|v) = a(u, v) \qquad \textit{for all} \quad u, v \in X. \tag{37}$$

(b) *A is compact* iff *a is compact.*

Proof. This follows from Proposition 21.31 and from $X = X^*$ by the Identification Principle 21.18. □

Definition 21.34. Let $A\colon X \to X$ be a linear continuous operator on the real H-space X.

(i) A is called *positive* iff a in (37) is positive, i.e., $(Au|u) \geq 0$ for all $u \in X$.
(ii) A is called strictly positive, strongly positive, or symmetric iff a in (37) has the corresponding property.

Let $A\colon X \to X^*$ be a linear continuous operator on the real B-space X, or let $A\colon X \to X$ be a linear continuous operator on the real H-space X. By Section 18.7i, we have:

(a) *A is positive* iff *A is monotone*;
(b) *A is strictly positive* iff *A is strictly monotone*;
(c) *A is strongly positive* iff *A is strongly monotone.*

21.11. Application to Embeddings

Recall Definition 21.13 on embeddings.

Proposition 21.35. *Let X, Y, Z be* B*-spaces over* $\mathbb{K}$. *Then:*

(a) *If the embeddings $X \subseteq Y$ and $Y \subseteq Z$ are continuous, then so is $X \subseteq Z$.*
If, in addition, one of the embeddings $X \subseteq Y$ or $Y \subseteq Z$ is compact, then so is $X \subseteq Z$.

(b) *If the embedding $X \subseteq Y$ is continuous with the embedding operator $j: X \to Y$, and if X is dense in Y, then the embedding $Y^* \subseteq X^*$ is also continuous with the embedding operator $i: Y^* \to X^*$, where $i = j^*$.*
If, in addition, the embedding $X \subseteq Y$ is compact, then so is $Y^* \subseteq X^*$.

(c) *If the embedding $X \subseteq Y$ is continuous, then, as $n \to \infty$,*

$$u_n \to u \quad \text{in } X \qquad \text{implies} \qquad u_n \to u \quad \text{in } Y$$

and

$$u_n \rightharpoonup u \quad \text{in } X \qquad \text{implies} \qquad u_n \rightharpoonup u \quad \text{in } Y.$$

(d) *If the embedding $X \subseteq Y$ is compact, then, as $n \to \infty$,*

$$u_n \rightharpoonup u \quad \text{in } X \qquad \text{implies} \qquad u_n \to u \quad \text{in } Y.$$

(e) *If the embedding $X \subseteq Y$ is continuous and X is dense in Y, then Y^* is dense in X^* provided the* B*-space X is reflexive.*

Proof. Suppose that the embedding $X \subseteq Y$ is continuous with the embedding operator $j: X \to Y$. Then

$$\langle v, j(u) \rangle = \langle v, u \rangle \qquad \text{for all} \quad v \in Y^*, \quad u \in X.$$

The embedding $Y^* \subseteq X^*$ is to be understood in the natural sense that each linear continuous functional $v: Y \to \mathbb{K}$ is also a linear continuous functional on X (cf. Problem 18.6). Let

$$i: Y^* \to X^*$$

be the corresponding embedding operator. Then

$$\langle i(v), u \rangle = \langle v, u \rangle \qquad \text{for all} \quad v \in Y^*, \quad u \in X.$$

Hence $\langle i(v), u \rangle = \langle v, j(u) \rangle$ for all $v \in Y^*$ and $u \in X$, i.e., $i = j^*$.

The remaining assertions follow from our results on linear operators in Sections 21.5 and 21.9, and from Problem 18.6(ii). □

21.12. Projection Operators

Projection operators are useful in order to describe the Galerkin method in terms of operators. Furthermore, the functional analytic formulation of

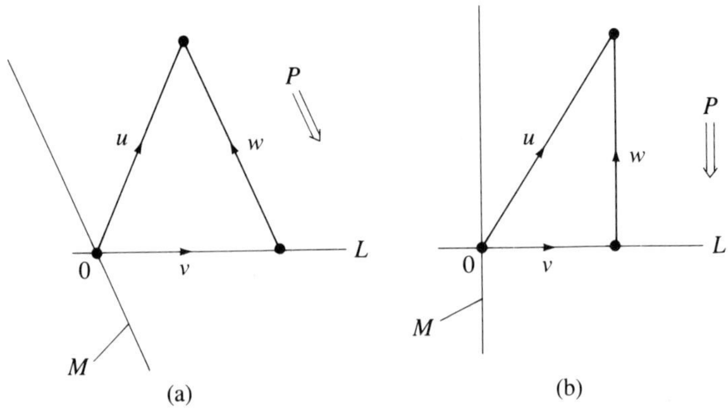

Figure 21.6

interpolation methods leads to projection operators (e.g., the Lagrange interpolation). As an important application we will consider the method of finite elements.

Definition 21.36. Let X be a linear space (e.g. a B-space), and let L and M be linear subspaces of X. We write

$$X = L \oplus M$$

iff each element $u \in X$ allows a unique decomposition of the form:

$$u = v + w, \qquad v \in L, \quad w \in M. \tag{38}$$

We call $X = L \oplus M$ a *direct sum* (Fig. 21.6).

If X is an H-space and if (38) holds with

$$(v|w) = 0 \qquad \text{for all} \quad v \in L, \quad w \in M,$$

then $X = L \oplus M$ is called an *orthogonal direct sum* (Fig. 21.6(b)).

The following proposition shows that there is a one-to-one correspondence between direct sums and linear operators P with

$$P^2 = P.$$

The simple geometrical interpretation of P will be given in Example 21.39 below.

Proposition 21.37.

(a) *Let* $X = L \oplus M$ *with* (38). *If we set*

$$Pu = v \qquad \textit{for all} \quad u \in X, \tag{39}$$

then we obtain $(I - P)u = w$ *for all* $u \in X$, *and the operator* $P: X \to X$ *is linear and* $P^2 = P$.

Moreover, we have

$$v \in L \quad \text{iff} \quad Pv = v \qquad \textit{and} \qquad w \in M \quad \text{iff} \quad (I - P)w = w. \tag{40}$$

(b) *Conversely, let* $P: X \to X$ *be a linear operator with* $P^2 = P$. *Then we obtain*

$$X = P(X) \oplus (I - P)(X),$$

where (38) *holds with* $v = Pu$ *and* $w = (I - P)u$ *for all* $u \in X$.

PROOF. Ad(a). From $v = v + 0$ we obtain $Pv = v$ for all $v \in L$. Conversely, if $Pv = v$, then $v \in L$, since $P(X) \subseteq L$. This implies

$$P^2 u = Pv = v = Pu, \qquad \text{for all} \quad u \in X,$$

i.e., $P^2 = P$. From $w = 0 + w$ we obtain $Pw = 0$ for all $w \in M$, and hence $(I - P)w = w$. Conversely, if $(I - P)w = w$, then $Pw = 0$, and hence $w \in M$, by (38) with $u = w$.

Ad(b). Let $P^2 = P$. Obviously,

$$u = Pu + (I - P)u \qquad \text{for all} \quad u \in X.$$

To show that this decomposition is unique, suppose that

$$u = v + w, \qquad v \in P(X), \quad w \in (I - P)(X).$$

This implies $v = Px$ for some $x \in X$, and hence

$$Pv = P^2 x = Px = v.$$

Moreover, we obtain that $w = (I - P)y$ for some $y \in X$, and hence

$$(I - P)w = (I - P)^2 y = (I - 2P + P^2)y = (I - P)y = w.$$

From $Pv = v$ and $Pw = 0$ we get

$$Pu = Pv + Pw = v,$$

and hence $w = u - v = (I - P)u$. □

Definition 21.38. Let X be a B-space. By a *projection operator* on X we understand a linear *continuous* operator $P: X \to X$ with $P^2 = P$. Then $P(X)$ is called the corresponding projection space.

The direct sum $X = M \oplus L$ is called a *topological* direct sum iff the corresponding operator P in (39) is a projection operator on X.

Let X be an H-space. By an *orthogonal projection operator* on X we understand a projection operator $P: X \to X$ with

$$(Pu|(I - P)z) = 0 \qquad \text{for all} \quad u, z \in X. \tag{41}$$

If $P: X \to X$ is a projection operator (resp. orthogonal projection operator), then $I - P: X \to X$ is also a projection operator (resp. orthogonal projection operator). This follows from $(I - P)^2 = I - 2P + P^2 = I - P$ and (41).

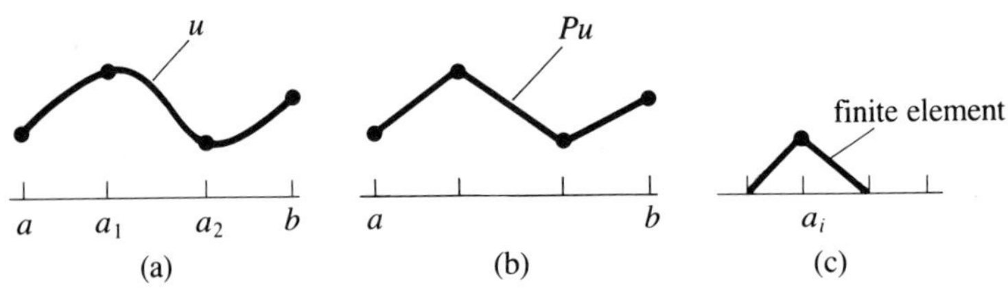

Figure 21.7

If $X = L \oplus M$ is a topological direct sum, then the linear subspaces L and M are closed. This follows immediately from (40) and the continuity of P.

Example 21.39 (Geometrical Interpretation). In $\mathbb{R}^2$ and $\mathbb{R}^3$, projection operators and orthogonal projection operators correspond to parallel projection and orthogonal projection, respectively (Fig. 21.6).

Example 21.40 (Linear Interpolation and Finite Elements). Let $X = C[a,b]$ with $-\infty < a < b < \infty$ and let $a = a_0 < a_1 < a_2 < \cdots < a_n = b$. For $u \in X$, we set

$$w = Pu,$$

where w is the uniquely determined *piecewise linear* function which satisfies

$$w(a_i) = u(a_i) \qquad \text{for all } i \text{ (Fig. 21.7(b)).}$$

Then $P: X \to X$ is a projection operator.

All the piecewise linear functions $w: [a,b] \to \mathbb{R}$ with

$$w(a_i) = 1 \qquad \text{for fixed } i$$

and $w(a_j) = 0$ for all $j \neq i$ are called *finite elements* (Fig. 21.7(c)). The set of these finite elements forms a basis in the projection space $P(X)$.

Example 21.41 (Lagrange Interpolation). We consider the same situation as in Example 21.40. But now we set

$$w = Pu,$$

where w is the uniquely determined polynomial of degree n which satisfies $w(a_i) = u(a_i)$ for all i. Here, $n \geq 1$.

Explicitly, we define

$$g_0(x) = \frac{(x - a_1)(x - a_2)\dots(x - a_n)}{(a_0 - a_1)(a_0 - a_2)\dots(a_0 - a_n)}$$

and analogously $g_1, \dots, g_n$. Then

$$w = u(a_0)g_0 + \cdots + u(a_n)g_n,$$

and $P: X \to X$ is a projection operator. In fact, the *uniqueness* of the interpolation polynomial implies $P(Pu) = Pu$, i.e., $P^2 = P$.

EXAMPLE 21.42 (Orthogonal Projection onto Finite-Dimensional Subspaces of H-Spaces). Let L be a finite-dimensional subspace of the H-space X. We choose an orthonormal basis $\{u_1, \ldots, u_n\}$ in L and we set

$$Pu = \sum_{i=1}^{n} (u_i|u)u_i.$$

Then $P: X \to X$ is an orthogonal projection operator onto L.

EXAMPLE 21.43 (Projection onto Finite-Dimensional Subspaces of B-Spaces). Let L be a finite-dimensional subspace of the B-space X. We choose a basis $\{u_1, \ldots, u_n\}$ in L, and we choose linear continuous functionals $v_1, \ldots, v_n$ on X with

$$\langle v_i, u_j \rangle = \delta_{ij} \qquad \text{for} \quad i, j = 1, \ldots, n. \tag{42}$$

Such functionals always do exist. In fact, since $u_1, \ldots, u_n$ are linearly independent, we find linear functionals $v_1, \ldots, v_n$ on L with (42) by solving a linear system of equations. Because $\dim L < \infty$, those functionals are continuous on L and by the Hahn–Banach theorem $A_1(34)$, they can be extended to linear continuous functionals on X.

If we set

$$Pu = \sum_{i=1}^{n} \langle v_i, u \rangle u_i,$$

then $P: X \to X$ is a projection operator onto L.

Proposition 21.44 (Orthogonal Projection). *Let L be a closed linear subspace of the* H-*space X, and let $L^\perp$ denote the orthogonal complement to L. Then*:

(a) *The operator $P: X \to X$ is an orthogonal projection operator* iff *it is a self-adjoint projection operator.*
(b) *We have the orthogonal direct sum $X = L \oplus L^\perp$.*
(c) *There exists exactly one orthogonal projection operator $P: X \to X$ onto L. In this case, $I - P: X \to X$ is the orthogonal projection operator onto $L^\perp$. Furthermore,*

$$\|u - Pu\| = \inf_{v \in L} \|u - v\| \qquad \textit{for all} \quad u \in X.$$

(d) *If $P: X \to X$ is an orthogonal projection operator, then $P(X)$ is a closed linear subspace of X, and we obtain the orthogonal direct sum*

$$X = P(X) \oplus (I - P)(X).$$

(e) *Each orthogonal direct sum is also a topological direct sum.*

PROOF. Ad(a). The condition $(Pu|(I-P)z) = 0$, i.e.,

$$(Pu|(I-P)z) = 0 \qquad \text{and} \qquad ((I-P)z|Pu) = 0 \qquad \text{for all} \quad u, z \in X$$

is equivalent to $(Pu|z) = (u|Pz)$ for all $u, z \in X$ provided $P^2 = P$.

Ad(b), (c), (d). By Corollary 18.27, for each $u \in X$, we have the unique decomposition $u = v + w$, $v \in L$, $w \in L^\perp$. Now use Theorem 18.C and Proposition 21.37.

Ad(e). Let $X = A \oplus B$ be an orthogonal direct sum, where A and B are linear subspaces of X. Let $P: X \to \bar{A}$ be the orthogonal projection operator onto the closure $\bar{A}$ of A. From

$$u = v + w, \qquad v \in A, \quad w \in B,$$

and $A \subseteq \bar{A}$, as well as $B \subseteq (\bar{A})^\perp$, it follows that $v = Pu$. Hence $A = \bar{A}$ and $B = A^\perp$. □

EXAMPLE 21.45 (Compact Operators). Let $A: X \to X$ be a linear continuous operator on the H-space X over $\mathbb{K}$ with the adjoint operator A^*. Then:

(a) A is compact iff there exists a sequence (A_n) of linear continuous operators $A_n: X \to X$ with $\dim R(A_n) < \infty$ for all n and

$$\lim_{n\to\infty} \|A - A_n\| = 0. \tag{43}$$

(b) A is compact iff A^* is compact.

PROOF. Ad(a). Suppose that (43) holds. Since $\dim R(A_n) < \infty$, the operator A_n is compact. Let M be a bounded set in X. Since

$$\|Au - A_n u\| \le \varepsilon \qquad \text{for all} \quad u \in M, \quad n \ge n_0(\varepsilon),$$

and since the set $A_n(M)$ is relatively compact, it follows, by the ε-net criterion $A_1(21b)$, that $A(M)$ is relatively compact, i.e., A is compact.

Conversely, suppose that A is compact and let B denote the closed unit ball in X. Since the set $A(B)$ is relatively compact, there exists a finite-dimensional subspace L_n with

$$\sup_{u\in B} \operatorname{dist}(Au, L_n) \le n^{-1}, \tag{44}$$

by the ε-net criterion $A_1(21b)$. Let $P_n: X \to L_n$ be the orthogonal projection operator onto L_n. We set $A_n = P_n A$. By Example 21.42,

$$A_n u = \sum_i (A^* u_i | u) u_i \qquad \text{for all} \quad u \in X$$

and

$$A_n^* v = \sum_i (u_i | v) A^* u_i \qquad \text{for all} \quad v \in X,$$

since $(A_n^* v | u) = (v | A_n u)$ for all $u, v \in X$. By (44),

$$\sup_{u\in B} \|Au - P_n Au\| \le n^{-1}.$$

This yields (43).

Ad(b). If A is compact, then

$$\|A^* - A_n^*\| = \|A - A_n\| \to 0 \qquad \text{as} \quad n \to \infty.$$

Since $\dim R(A_n^*) < \infty$, the operator A_n^* is compact. Hence A^* is compact by (a).

If A^* is compact, then so is A, since $A = (A^*)^*$. □

21.13. Bases and Galerkin Schemes

The following will be important for the proof of the convergence of Galerkin methods. Recall that $\operatorname{dist}(u, Y) = \inf_{v \in Y} \|u - v\|$.

Definition 21.46. Let X be a B-space. By a *Galerkin scheme* in X, we understand a sequence (Y_n) of finite-dimensional nonzero subspaces Y_n of X with

$$\lim_{n\to\infty} \operatorname{dist}(u, Y_n) = 0 \qquad \text{for all} \quad u \in X.$$

By a *basis* of X, we understand an at most countable sequence (w_j) of elements $w_j \in X$, where finitely many $w_1, \ldots, w_n$ are always linearly independent and

$$X = \overline{\bigcup_n X_n}$$

with $X_n = \operatorname{span}\{w_1, \ldots, w_n\}$.

In Section 21.16 we shall show that the *rapidity of convergence* of the Galerkin method depends on $\operatorname{dist}(u, Y_n)$. Therefore, in what follows we are interested in estimating $\operatorname{dist}(u, Y_n)$. Such estimates will be given in (45) for polynomials and in (46), (47) for finite elements.

EXAMPLE 21.47. Let $\{u_1, u_2, \ldots\}$ be a complete orthonormal system in the separable H-space X. Then (u_j) is a basis in X. If we set

$$P_n u = \sum_{j=1}^{n} (u_j|u)u_j,$$

then $P_n\colon X \to X$ is an orthogonal projection operator onto $X_n = \operatorname{span} \times \{u_1, \ldots, u_n\}$, and (X_n) is a Galerkin scheme in X, since

$$\operatorname{dist}(u, X_n) = \|u - P_n u\| \to 0 \qquad \text{as} \quad n \to \infty \quad \text{for all} \quad u \in X.$$

By the Schmidt orthogonalization method, in each separable H-space $X \neq \{0\}$ one can construct explicitly a complete orthonormal system (cf. Problem 19.16).

EXAMPLE 21.48 (Polynomial Bases). We set

$$w_k(x) = x^{k-1}, \qquad k = 1, 2, \ldots$$

and $X_n = \operatorname{span}\{w_1, \ldots, w_n\}$, i.e., X_n consists of all real polynomials of degree

$\leq n-1$. Then (w_n) is a *basis* and (X_n) is a Galerkin scheme in the spaces

$$X = C[a,b], \qquad C^m[a,b], \qquad L_p(a,b), \qquad W_p^m(a,b).$$

Here, $m = 1, 2, \ldots, 1 \leq p < \infty$, and $-\infty < a < b < \infty$. This follows from the fact that the set of polynomials is dense in all those spaces.

Let $X = C[a,b]$. The famous classical *approximation theorem of Jackson* (1912) says that

$$\operatorname{dist}_X(u, X_{n+1}) \leq \frac{C}{n^k} \|u^{(k)}\|_X \tag{45}$$

for all $u \in C^k[a,b]$ with $k = 1, \ldots, n$. The constant C depends only on a, b, and k (cf. Cheney (1967, M), p. 147). Here,

$$\operatorname{dist}_X(u, X_{n+1}) = \inf_{v \in X_{n+1}} \|u - v\|_X.$$

Formula (45) tells us that

$$\lim_{n\to\infty} \operatorname{dist}(u, X_{n+1}) = 0,$$

and the convergence is faster, the smoother u is.

Proposition 21.49 (Existence of Bases and Galerkin Schemes). *Let X be a separable* B-*space. Then*:

(a) *X has a basis.*
(b) *If (w_n) is a basis in X, then (X_n) with $X_n = \operatorname{span}\{w_1, \ldots, w_n\}$ is a Galerkin scheme in X.*
(c) *If (Y_n) is a Galerkin scheme in X, then we can construct a basis in X by means of (Y_n).*

Proof. Ad(a). Let $X \neq \{0\}$. Then there exists a countable dense set $\{u_j\}$ in X with $u_1 \neq 0$. We set $w_1 = u_1$. If $u_2 \notin \operatorname{span}\{w_1\}$, then let $w_2 = u_2$, etc.

Ad(b). This follows from Definition 21.46.

Ad(c). Note that $\bigcup_n Y_n$ is dense in X and use (a). □

The following approximation formula (46) below plays a fundamental role in the theory of *finite elements*, as we will show in the next section.

Example 21.50. Let G be a bounded region in $\mathbb{R}^N$ with $N \geq 1$ and $\partial G \in C^{0,1}$. Let $m = 1, 2, \ldots$. We set

$$|u|_m = \left(\int_G \sum_{|\alpha|=m} |D^\alpha u|^2 \, dx\right)^{1/2}.$$

Let P_{m-1} denote the set of all real polynomials on G of degree $\leq m-1$. Then there exists a constant $C > 0$ such that

$$\operatorname{dist}(u, P_{m-1}) \leq C|u|_m \qquad \text{for all} \quad u \in W_2^m(G), \tag{46}$$

i.e.,

$$\inf_{p \in P_{m-1}} \|u - p\|_{m,2} \le C|u|_m \qquad \text{for all} \quad u \in W_2^m(G).$$

PROOF. Let $X = W_2^m(G)$ and $\dim P_{m-1} = J$. By (42), there exist linear continuous functionals $f_1, \ldots, f_J$ on X such that, for $p \in P_{m-1}$,

$$f_j(p) = 0 \quad \text{for all } j \qquad \text{implies} \qquad p = 0. \tag{46*}$$

As in Step 3 of Section 21.4c, the *compactness* of the embedding $W_2^m(G) \subseteq W_2^{m-1}(G)$ implies that

$$\|u\| = \left(|u|_m^2 + \sum_j |f_j(u)|^2\right)^{1/2}$$

is an equivalent norm on X. By (46*), for given $u \in X$, there exists a unique $p \in P_{m-1}$ with

$$f_j(p) = f_j(u), \qquad j = 1, \ldots, J.$$

In this connection, note that in terms of a basis this is a system of J real equations with J unknowns. Hence

$$\|u - p\|_{m,2} \le C\|u - p\| = C|u|_m. \qquad \square$$

21.14. Application to Finite Elements

Finite elements play a fundamental role in modern numerical analysis. Our goal is to explain the simple basic ideas. In particular, we want to show that finite elements form Galerkin schemes. Moreover, we want to obtain information about the *accuracy* of approximation in terms of the mesh size h. Estimates of the type (47) below are very important in order to obtain information about the *rapidity* of convergence of the Galerkin method, as will be shown in Section 21.16. We consider the situation of Example 21.40 and we make the following assumption:

(H) Let $-\infty < a = a_0 < a_1 < a_2 < \cdots < a_n = b < \infty$ and let $h = \max_i (a_{i+1} - a_i)$.

A *finite element* is a piecewise linear function $f: [a, b] \to \mathbb{R}$ with $f(a_i) = 1$ for fixed i and $f(a_j) = 0$ for all $j \neq 1$ (Fig. 21.7(c)).

The corresponding *finite element space F* consists of the real linear hull of all the finite elements, i.e., F consists of all real linear combinations of finite elements (Fig. 21.7(b)). By Example 21.5, F is a finite-dimensional *subspace* of the Sobolev space $X = W_2^1(a, b)$.

Proposition 21.51. *Suppose that* (H) *holds. Then:*

(a) *There exists a constant* $D > 0$ *such that*

$$\mathrm{dist}_X(u, F) \le h \cdot D\|u\|_{2,2} \qquad \textit{for all} \quad u \in W_2^2(a, b). \tag{47}$$

Here, $\mathrm{dist}_X(u, F) = \inf_{v \in F} \|u - v\|_X$.

(b) *If we consider a sequence of partitions of the interval* $[a, b]$, *then the corresponding sequence* (F_n) *of finite element spaces forms a Galerkin scheme in* $X = W_2^1(a, b)$ *in the case where* $h_n \to 0$ *as* $n \to \infty$, *i.e., the mesh size goes to zero.*

Note that (b) is an immediate consequence of (a) since $W_2^2(a, b)$ is dense in X. The proof of (a) will be based on the following abstract result of Proposition 21.52 below. We set

$$|u|_m = \left(\int_c^d |u^{(m)}(x)|^2\,dx\right)^{1/2}, \qquad |u|_m^* = \left(\int_0^1 |u^{(m)}|^2\,dx\right)^{1/2}.$$

Furthermore, we consider the affine transformation

$$x \mapsto \alpha x + \beta$$

and define the operator

$$(Au)(x) = u(\alpha x + \beta).$$

Our assumptions are the following.

(H1) Let $k = 1, 2, \ldots$ and $m = 0, 1, \ldots$ with $k > m$. The operator $\Pi\colon W_2^k(c, d) \to W_2^m(c, d)$ is linear and continuous for all c, d: $-\infty < c < d < \infty$.
(H2) $\Pi u = u$ for all polynomials of degree $\le k - 1$.
(H3) Π is affinely invariant, i.e., $A\Pi = \Pi A$ for all A.

Proposition 21.52. *Suppose that* (H1) *through* (H3) *hold. Then there exists a constant* C *depending on* k *and* m *such that*

$$|u - \Pi u|_m \le C(d - c)^{k-m}|u|_k \qquad \text{for all} \quad u \in W_2^k(c, d).$$

PROOF.

(I) First let $c = 0$, $d = 1$, and denote the norm on $W_2^m(0, 1)$ by $\|u\|_{m,2}^*$. Let P_{k-1} be the set of all real polynomials of degree $\le k - 1$. Let $v \in W_2^k(0, 1)$. By (H2),

$$v - \Pi v = (v - p) - \Pi(v - p) \qquad \text{for all} \quad p \in P_{k-1}.$$

Hence

$$|v - \Pi v|_m^* \le \|v - \Pi v\|_{m,2}^* \le \|I - \Pi\|\,\|v - p\|_{k,2}^*.$$

By Example 21.50,

$$|v - \Pi v|_m^* \le \|I - \Pi\| \inf_{p \in P_{k-1}} \|v - p\|_{k,2}^* \le C|v|_k^*. \tag{48}$$

(II) Affine transformation from $[0, 1]$ onto $[c, d]$. Let

$$(Av)(x) = v\left(\frac{x}{d - c} - \frac{c}{d - c}\right), \qquad x \in [c, d].$$

The substitution rule for integrals shows that

$$|Av|_m = (d - c)^{-m+1/2}|v|_m^*.$$

Letting $u = Av$ and noting (H3), we obtain from (48) that

$$|u - \Pi u|_m (d-c)^{m-1/2} \leq C|u|_k (d-c)^{k-1/2}. \qquad \square$$

PROOF OF PROPOSITION 21.51. Let $u \in W_2^2(a,b)$. We will use Proposition 21.52 with $c = a_i$, $d = a_{i+1}$, $k = 2$, $m = 0, 1$.

The embedding $W_2^2(a,b) \subseteq C[a,b]$ is continuous. Thus it makes sense to define the operator

$$\Pi: W_2^2(a_i, a_{i+1}) \to W_2^m(a_i, a_{i+1})$$

in the following way: Πu is the polynomial of degree one which satisfies

$$(\Pi u)(x) = u(x) \qquad \text{for} \quad x = a_i, a_{i+1}.$$

Obviously, if u is a polynomial of degree one, then $\Pi u = u$. By Proposition 21.52, for all $u \in W_2^2(a_i, a_{i+1})$,

$$|u - \Pi u|_m \leq C(a_{i+1} - a_i)^{2-m}|u|_2.$$

Hence

$$|u - \Pi u|_m^2 \leq C^2 h^{2(2-m)}|u|_2^2.$$

Summation over i and $m = 0, 1$ yields the desired estimate

$$\|u - \Pi u\|_{W_2^1(a,b)}^2 \leq C^2(h^2 + h^4)\|u\|_{W_2^2(a,b)}^2$$

for all $u \in W_2^2(a,b)$. This implies (47). Note that $\Pi u \in F$. $\square$

Finite elements in $\mathbb{R}^N$ can be treated analogously. This will be considered in the Appendix, $A_2(58)$.

21.15. Riesz–Schauder Theory and Abstract Fredholm Alternatives

We consider the operator equation

$$Bu + Cu = b, \qquad u \in X, \tag{49}$$

and the corresponding adjoint equation

$$B^*v + C^*v = 0, \qquad v \in X, \tag{49*}$$

where B^* and C^* denote adjoint operators. We make the following assumptions:

(H1) The operator $B: X \to X$ is linear, continuous, and bijective on the H-space X over $\mathbb{K} = \mathbb{R}, \mathbb{C}$ (e.g., $B = I$).

(H2) The operator $C: X \to X$ is linear and compact.

Theorem 21.F (Riesz (1918), Schauder (1930)). *Suppose that* (H1) *and* (H2) *hold. Then*:

(a) *Equation* (49) *has a solution iff* $(v|b) = 0$ *for all solutions* v *of* (49*).
(b) *Equation* (49) *with* $b = 0$ *and equation* (49*) *have the same finite number of linearly independent solutions.*
(c) *If* $Bu + Cu = 0$ *implies* $u = 0$, *then, for each* $b \in X$, *equation* (49) *has a unique solution and the inverse operator* $(B + C)^{-1}: X \to X$ *is continuous.*

Statement (c) shows that

uniqueness implies existence.

Moreover, statement (b) includes the fact that (49) with $b = 0$ has the unique solution $u = 0$ iff (49*) has the unique solution $v = 0$.

Theorem 21.F is a special case of the general theory of linear Fredholm operators in Section 8.4. In fact, since B is bijective, we have $\dim N(B) = \operatorname{codim} R(B) = 0$. Hence B is a Fredholm operator of index zero. Then the compact perturbation $B + C$ of B is a Fredholm operator of index zero, too. This yields Theorem 21.F.

In the following we want to give a simple direct proof, which can be used in an introductory lecture. In this connection, we shall use the open mapping theorem $A_1(36)$, which says that each linear continuous bijective operator $L: X \to Y$ has a continuous inverse operator provided X and Y are B-spaces over $\mathbb{K}$.

PROOF. We set

$$S = B + C, \qquad N = N(S), \qquad R = R(S).$$

Obviously, the null space N of the operator S is closed. Below we shall show that the range R of S is also closed. Therefore, we have the orthogonal direct sums:

$$X = N \oplus N^\perp = R \oplus R^\perp,$$

i.e., $\operatorname{codim} R = \dim R^\perp$ (cf. $A_1(22k)$). Recall that

$$\operatorname{ind} S = \dim N - \operatorname{codim} R.$$

Step 1: $\dim N < \infty$. Let (u_n) be a bounded sequence in N, i.e., $Bu_n + Cu_n = 0$ for all n. Since C is compact, there is a subsequence $(u_{n'})$ such that $Cu_{n'} \to v$ as $n \to \infty$. Hence $u_{n'} \to -B^{-1}v$. By Theorem 21.C, $\dim N < \infty$.

Step 2: $\dim N(S^*) < \infty$. Note that $S^* = B^* + C^*$. Since C is compact, so is C^*, by Example 21.45. Furthermore, $B^{*-1} = (B^{-1})^*$. The same argument as in Step 1 yields $\dim N(S^*) < \infty$.

Step 3: We have the direct orthogonal sum $X = \bar{R} \oplus N(S^*)$. In fact, it follows from

$$(Su|v) = (u|S^*v) \qquad \text{for all} \quad u, v \in X$$

that $N(S^*) \subseteq \bar{R}^\perp$ and $\bar{R}^\perp \subseteq N(S^*)$.

Using $(S^*)^* = S$, we obtain the orthogonal direct sum

$$X = \overline{R(S^*)} \oplus N(S). \tag{50}$$

Step 4: $R(S)$ is closed. By the proof of Corollary 19.59 and (50), it is sufficient to prove the *a priori* estimate

$$\|Su\| \geq c\|u\| \qquad \text{for all} \quad u \in \overline{R(S^*)} \quad \text{and fixed} \quad c > 0. \tag{51}$$

If (51) is not true, then there exists a sequence (u_n) in $\overline{R(S^*)}$ with $\|u_n\| = 1$ for all n and $Su_n \to 0$ as $n \to \infty$. The same argument as in Step 1 shows the existence of a subsequence with $u_{n'} \to w$ as $n \to \infty$. Hence

$$w \in N(S), \qquad w \in \overline{R(S^*)}, \qquad \|w\| = 1.$$

This contradicts (50).

Step 5: The fundamental *invariance* of the index against small perturbations. We show that if the operator $D: X \to X$ is linear and compact and if $\|D - C\|$ is sufficiently small, then

$$\operatorname{ind} S = \operatorname{ind} T, \qquad \text{where} \quad S = B + C \qquad \text{and} \qquad T = B + D.$$

The decisive *trick* consists in the construction of the operator

$$\hat{T}(u, v) = Tu + v \qquad \text{for all} \quad u \in N^\perp, \quad v \in R^\perp.$$

Then $\hat{T}: N^\perp \times R^\perp \to X$ is linear and continuous. If $T = S$, then $\hat{S}$ is bijective. This is the key observation. In fact, it follows from $\hat{S}(u, v) = 0$ that

$$Su + v = 0, \qquad u \in N^\perp, \quad v \in R^\perp,$$

and hence $v = 0$, $Su = 0$. This implies $u = 0$.

If $\|\hat{T} - \hat{S}\|$ is sufficiently small, then it follows from Problem 1.7 that $\hat{T}$ is also bijective. Thus, if $\|D - C\|$ is sufficiently small, $\hat{T}$ is *bijective*. Hence we obtain the two *key relations*:

$$\dim T(N^\perp) = \operatorname{codim} R^\perp, \qquad N(T) \subseteq N,$$

and hence

$$\operatorname{codim} T(N^\perp) = \operatorname{codim} R.$$

In this connection, note that $Tu = 0$, $u \in N^\perp$, implies $\hat{T}(u, 0) = 0$, and hence $u = 0$. Therefore, $N(T) \subseteq N$.

We now choose a linear subspace M of N such that

$$N = N(T) \oplus M.$$

Then $X = (N(T) \oplus M) \oplus N^\perp$. Hence T is injective on $M \oplus N^\perp$. Thus, we get

$$R(T) = T(M) \oplus T(N^\perp) \qquad \text{and} \qquad \dim T(M) = \dim M,$$

i.e.,

$$X = (T(M) \oplus T(N^\perp)) \oplus R(T)^\perp.$$

This immediately implies

$$\operatorname{codim} R = \operatorname{codim} T(N^\perp) = \operatorname{codim} R(T) + \dim M,$$

$$\dim N = \dim N(T) + \dim M.$$

Hence $\operatorname{ind} S = \operatorname{ind} T$.

Step 6: $\operatorname{ind}(B + C) = \operatorname{ind} B = 0$. In fact, the function $t \mapsto B + tC$ is continuous from $[0, 1]$ to $L(X, X)$. By Step 5, $\operatorname{ind}(B + tC) = \text{constant}$ for all t.

Step 7: By Steps 3 and 4,

$$X = R(S) \oplus N(S^*).$$

This yields Theorem 21.F(a).

From $\operatorname{ind} S = 0$ it follows that $\dim N(S) = \operatorname{codim} R(S) = \dim N(S^*)$. This is Theorem 21.F(b).

Finally, if $\dim N(S) = 0$, then $\operatorname{codim} R(S) = 0$, i.e., $S\colon X \to X$ is bijective and $S^{-1}\colon X \to X$ is continuous, by the open mapping theorem $A_1(36)$. This yields Theorem 21.F(c). □

Corollary 21.53. *Suppose that* (H1) *and* (H2) *hold. Let* $\mu \in \mathbb{K}$. *Then*:

(a) *If* μ *is not a characteristic number of the pair* (B, C), *i.e.*, $Bu + \mu Cu = 0$ *implies* $u = 0$, *then there exists the continuous inverse operator* $(B + \mu C)^{-1}$: $X \to X$.
(b) *The set of characteristic numbers of* (B, C) *has no finite accumulation point.*

PROOF. Ad(a). Replace C with μC and use Theorem 21.F(c).

Ad(b).

(I) Let $B = I$. Suppose that there is a sequence (μ_n) of characteristic numbers with $\mu_n \to \mu$ as $n \to \infty$, $\mu \in \mathbb{K}$, and $\mu_n \neq \mu_m$ if $n \neq m$. Let $(I + \mu_n C)u_n = 0$ with $u_n \neq 0$ for all n. Note that $\mu_n \neq 0$ and set $\lambda_n = \mu_n^{-1}$. Then $u_1, \ldots, u_m$ are linearly independent for all m. To show this suppose that $u_1, \ldots, u_n$ are linearly independent and

$$u_{n+1} = a_1 u_1 + \cdots + a_n u_n.$$

Application of the operator C to this equation yields

$$(\lambda_{n+1} - \lambda_1) a_1 u_1 + \cdots + (\lambda_{n+1} - \lambda_n) a_n u_n = 0$$

and hence $a_i = 0$ for all i. This contradicts $u_{n+1} \neq 0$.

By the Schmidt orthogonalization method in Problem 19.16, we construct elements v_i such that $(v_i | v_j) = \delta_{ij}$ for all i, j and

$$v_i = \sum_{k=1}^{i} b_{ik} u_k$$

for suitable b_{ik}. If $i > j$, then

$$Cv_i - Cv_j = -\lambda_i v_i + z, \qquad z \in \operatorname{span}\{u_1, \ldots, u_{i-1}\}.$$

Hence $z \in \operatorname{span}\{v_1, \ldots, v_{i-1}\}$. This implies

$$\|Cv_i - Cv_j\|^2 = |\lambda_i|^2 + \|z\|^2 \geq |\lambda_i|^2.$$

Thus, (Cv_n) does not contain a convergent subsequence, since $\inf_i |\lambda_i| > 0$. This contradicts the compactness of the operator C.

(II) In the general case, apply (I) to the operator

$$B + \mu C = B(I + \mu B^{-1}C). \qquad \square$$

Applications of the results in this section to differential and integral equations will be considered in Chapter 22.

21.16. The Main Theorem on the Approximation-Solvability of Linear Operator Equations, and the Convergence of the Galerkin Method

We consider the operator equation

$$Au = b, \qquad u \in X, \tag{52}$$

together with the Galerkin equations

$$P_n A u_n = P_n b, \qquad u_n \in X_n, \quad n = 1, 2, \ldots. \tag{53}$$

We make the following assumptions:

(H1) X is a real separable infinite-dimensional H-space, and (w_n) is a basis in X.

(H2) Let $X_n = \operatorname{span}\{w_1, \ldots, w_n\}$, and let $P_n\colon X \to X_n$ be the orthogonal projection operator from X onto X_n.

Since P_n is self-adjoint, the Galerkin equation (53) is equivalent to

$$(Au_n|w_j) = (b|w_j), \qquad u_n \in X_n, \quad j = 1, \ldots, n. \tag{53*}$$

The following notion is crucial for numerical analysis.

Definition 21.54. For given $b \in X$, equation (52) is called *uniquely approximation-solvable* iff the following hold:

(i) The original equation (52) has a unique solution u.

(ii) There exists a number n_0 such that, for all $n \geq n_0$, the Galerkin equation (53) has a unique solution u_n.

(iii) The Galerkin method converges, i.e., $u_n \to u$ as $n \to \infty$.

Theorem 21.G. *Assume* (H1) *and* (H2). *Then, for each $b \in X$, equation* (52) *is uniquely approximation-solvable in the case where the linear operator $A\colon X \to X$ has one of the following four properties*:

(a) *$A = I + K$, where $K\colon X \to X$ is linear and k-contractive, i.e., $\|K\| < 1$.*

(b) *$A = I + C$, where $C\colon X \to X$ is linear and compact, and $Au = 0$ implies $u = 0$.*

(c) *$A: X \to X$ is linear, continuous, and strongly monotone, i.e.,*

$$(Au|u) \geq c\|u\|^2 \qquad \textit{for all} \quad u \in X \textit{ and fixed} \quad c > 0.$$

(d) *$A = B + C$, where $B: X \to X$ is linear, continuous, and strongly monotone, $C: X \to X$ is linear and compact, and $Au = 0$ implies $u = 0$.*

In cases (a) *and* (c), *$n_0 = 1$. In case* (b), *n_0 is independent of b.*

Corollary 21.55. *In cases* (a), (b), (c) *above, we have the error estimates* (54a, b, c), *respectively, i.e., for all $n \geq n_0$,*

$$\|u - u_n\| \leq (1 - \|K\|)^{-1} \operatorname{dist}(u, X_n), \tag{54a}$$

$$\|u - u_n\| \leq \operatorname{const} \operatorname{dist}(u, X_n), \tag{54b}$$

$$c\|u - u_n\| \leq \|Au_n - b\|. \tag{54c}$$

This shows that the rapidity of convergence of the Galerkin method depends on $\operatorname{dist}(u, X_n)$ in cases (a) and (b). Estimates for $\operatorname{dist}(u, X_n)$ were obtained in Example 21.48 (polynomials) and Proposition 21.51 (finite elements). Roughly speaking, one has the following important principle:

The convergence of the Galerkin method is faster, the smoother the solution u is.

Theorem 21.G contains important classes of operators and it allows many applications. Note that (a)–(d) in Theorem 21.G correspond to the Banach fixed-point principle, the Riesz–Schauder theory, and the theory of monotone operators.

In Chapter 34 we shall consider generalizations of Theorem 21.G to nonlinear operators. There we shall recognize the fundamental connection between existence, stability, consistency, A-proper maps, and the unique approximation-solvability of operator equations.

Lemma 21.56. *Assume* (H1) *and* (H2). *Then, for all n and all $u \in X$:*

(a) $\|P_n\| = 1$.
(b) *$P_u u \to u$ as $n \to \infty$.*
(c) $\sup_{u \in M} \|u - P_n u\| \to 0$ as $n \to \infty$ *provided the nonempty set M is compact.*

PROOF. Ad(a). Note that $u = P_n u + (I - P_n)u$ and

$$\|u\|^2 = \|P_n u\|^2 + \|(I - P_n)u\|^2.$$

Ad(b). This follows from

$$\|u - P_n u\| = \inf_{v \in X_n} \|u - v\| = \operatorname{dist}(u, X_n),$$

and $\operatorname{dist}(u, X_n) \to 0$ as $n \to \infty$, by (H1).

Ad(c). By (a), $\|P_n u - P_n v\| \le \|u - v\|$ for all $u, v \in X$, i.e., (P_n) is equicontinuous on X. From Problem 19.14e it follows that the convergence in (b) is uniform on M. □

Proof of Theorem 21.G. Ad(a). Note that $P_n u_n = u_n$ for $u_n \in X_n$. Since $\|P_n K\| \le \|K\| < 1$, equations (52) and (53), i.e.,

$$u + Ku = b, \quad u \in X \qquad \text{and} \qquad u_n + P_n K u_n = P_n b, \quad u_n \in X_n \tag{55}$$

have unique solutions by the Banach fixed-point principle (Theorem 1.A). Furthermore,

$$\begin{aligned}\|(I + P_n K)^{-1}\| &= \left\| \sum_{m=0}^{\infty} (P_n K)^m \right\| \le \sum_{m=0}^{\infty} \|K\|^m \\ &= (1 - \|K\|)^{-1}.\end{aligned}$$

By (55),

$$(I + P_n K)(u - u_n) = u - P_n u.$$

Hence

$$\|u - u_n\| \le (1 - \|K\|)^{-1} \|u - P_n u\|.$$

This yields $u_n \to u$ as $n \to \infty$ and the estimate (54a).

Ad(b). By Theorem 21.F, the original equation

$$(I + C)u = b, \qquad u \in X,$$

has a unique solution and $(I + C)^{-1}\colon X \to X$ is continuous. Since $P_n u_n = u_n$ for $u_n \in X_n$, the Galerkin equation is equivalent to

$$(I + P_n C)u_n = P_n b, \qquad u_n \in X. \tag{56}$$

Let $B = \{u \in X\colon \|u\| \le 1\}$ and $M = \overline{C(B)}$.

Since the operator C is compact, the set M is compact. By Lemma 21.56(c), $\|(I - P_n)C\| \to 0$ as $n \to \infty$. Thus, according to Problem 1.7, there exists an n_0 such that the operator

$$(I + P_n C) = (I + C) + (P_n C - C)$$

has a continuous inverse operator on X for all $n \ge n_0$ and $\sup_n \|(I + P_n C)^{-1}\| < \infty$. Consequently, (56) has a unique solution. The error estimates and $u_n \to u$ as $n \to \infty$ are obtained as in (a).

Ad(c). For all $u \in X_n$, we have $P_n u = u$ and hence

$$(P_n A u | u) = (Au | P_n u) \ge c\|u\|^2. \tag{57}$$

Thus, the operator $P_n A\colon X_n \to X_n$ is strongly monotone. By Theorem 18.E, the two operator equations

$$Au = b, \qquad u \in X, \tag{58}$$

$$P_n A u_n = P_n b, \qquad u_n \in X_n, \tag{59}$$

have unique solutions.

If $n \geq j$, then it follows from (59) that

$$(Au_n|w_j) = (b|w_j), \tag{60}$$

$$(Au_n|u_n) = (b|u_n). \tag{61}$$

By (57),

$$c\|u_n\|^2 \leq (b|u_n) \leq \|b\| \|u_n\|.$$

This yields the *a priori estimate*:

$$c\|u_n\| \leq \|b\|,$$

i.e., (u_n) is bounded.

Let $(u_{n'})$ be a weakly convergent subsequence with

$$u_{n'} \rightharpoonup v \quad \text{as} \quad n \to \infty.$$

By (60),

$$(Au_{n'}|w) \to (b|w) \quad \text{as} \quad n \to \infty \quad \text{for all} \quad w \in \bigcup_n X_n.$$

Since $\bigcup_n X_n$ is dense in X and (Au_n) is bounded, we obtain

$$Au_{n'} \rightharpoonup b \quad \text{as} \quad n \to \infty,$$

by Proposition 21.23(g). Since $A\colon X \to X$ is linear and continuous,

$$Au_{n'} \rightharpoonup Av \quad \text{as} \quad n \to \infty,$$

by Proposition 21.27. Hence $Av = b$, i.e., $v = u$. Since the weak limit v is the same for all weakly convergent subsequences of (u_n), we get

$$u_n \rightharpoonup u \quad \text{as} \quad n \to \infty,$$

according to Proposition 21.23(i). It follows from

$$\begin{aligned} c\|u_n - u\|^2 &\leq (Au_n - Au|u_n - u) \\ &= (b|u_n) - (Au_n|u) - (Au|u_n - u) \to 0 \quad \text{as} \quad n \to \infty \end{aligned}$$

that

$$u_n \to u \quad \text{as} \quad n \to \infty.$$

Finally, the error estimates follow from

$$c\|u_n - u\|^2 \leq \|Au_n - Au\| \|u_n - u\|$$

and $Au = b$.

Ad(d). By Theorem 18.E, the operator $B\colon X \to X$ is bijective. Thus, it follows from Theorem 21.F that the equation

$$Bu + Cu = b, \qquad u \in X,$$

has a unique solution and $(B + C)^{-1}\colon X \to X$ is continuous. Let $A_t u = Bu + Cu - tb$. It follows from $A_t u = 0$ that $\|u\| \leq \|(B + C)^{-1}\| |t| \|b\|$. Now, the assertion is a special case of Theorem 29.B, by using $X = X^*$ in the sense of the Identification Principle 21.18. Note that A_t satisfies condition $(S)_0$, by Examples 27.2 and 27.3. □

21.17. Interpolation Inequalities and a Convergence Trick

We make the following assumption:

(H) The embeddings

$$X \subseteq Y \subseteq Z$$

are continuous, where X, Y, and Z are B-spaces over $\mathbb{K} = \mathbb{R}, \mathbb{C}$. There exist constants $C > 0$ and $0 < \theta < 1$ such that

$$\|u\|_Y \leq C\|u\|_X^{1-\theta}\|u\|_Z^\theta \qquad \text{for all} \quad u \in X. \tag{62}$$

Inequality (62) is called an *interpolation inequality*. Such inequalities play a fundamental role in the modern theory of nonlinear partial differential equations. Motivated by Standard Example 21.58 below, we call X a "very nice" space and Z a "poor" space. In applications, the functions in X are smoother than those in Y and, in turn, the functions in Y are smoother than those in Z. The following simple result is crucial.

Proposition 21.57 (Convergence Trick). *Suppose that* (H) *holds. Then*:

(a) *If (u_n) is a bounded sequence in the "very nice" space X, and (u_n) is convergent in the "poor" space Z, then (u_n) is also convergent in the "nice" space Y, i.e., more precisely,*

$$u_n \to u \quad \text{in } Z \qquad \text{as} \quad n \to \infty$$

implies $u_n \to u$ in Y as $n \to \infty$.

(b) *If the embedding $X \subseteq Z$ is compact, then so is $X \subseteq Y$.*

Proof. Ad(a). It follows from

$$\begin{aligned}\|u_n - u_m\|_Y &\leq C\|u_n - u_m\|_X^{1-\theta}\|u_n - u_m\|_Z^\theta \\ &\leq D\|u_n - u_m\|_Z^\theta\end{aligned}$$

that (u_n) is convergent in Y, i.e., $u_n \to v$ in Y as $n \to \infty$. Hence $u_n \to v$ in Z as $n \to \infty$, since the embedding $Y \subseteq Z$ is continuous. Therefore, $v = u$.

Ad(b). If (u_n) is bounded in X, then there exists a subsequence with $u_{n'} \to u$ in Z as $n \to \infty$. By (a), $u_{n'} \to u$ in Y as $n \to \infty$. □

Standard Example 21.58 (Hölder Spaces). Let G be a bounded region in $\mathbb{R}^N$, $N \geq 1$, and let $0 < \beta < \alpha < 1$. We set

$$X = C^\alpha(\bar{G}), \qquad Y = C^\beta(\bar{G}), \qquad Z = C(\bar{G}).$$

Then the interpolation inequality (62) holds with $1 - \theta = \beta/\alpha$.

A bounded sequence (u_n) in X is equicontinuous by the Hölder condition. Thus, (u_n) is relatively compact in Z by the Arzelà–Ascoli theorem $A_1(24g)$. Hence the embedding $X \subseteq Z$ is compact. From Proposition 21.57 we obtain that the embedding $X \subseteq Y$ is also compact.

All the results above remain true for $\alpha = 1$, i.e., $X = C^{0,1}(\bar{G})$.

Proof. We show that (62) holds. By definition, the Hölder constant $H_\alpha(u)$ is the smallest constant such that

$$|u(x) - u(y)| \leq H_\alpha(u)|x - y|^\alpha \qquad \text{for all} \quad x, y \in \bar{G}.$$

The norms on X, Y, and Z are given by

$$\|u\|_Z = \max_{x \in \bar{G}} |u(x)|, \qquad \|u\|_X = \|u\|_Z + H_\alpha(u),$$

$$\|u\|_Y = \|u\|_Z + H_\beta(u).$$

Let $u \in X$. From

$$|u(x) - u(y)| = |u(x) - u(y)|^{1-\theta}|u(x) - u(y)|^\theta$$
$$\leq H_\alpha(u)^{1-\theta}|x - y|^\beta 2^\theta \|u\|_Z^\theta$$

we obtain $H_\beta(u) \leq 2^\theta H_\alpha(u)^{1-\theta}\|u\|_Z^\theta$. Hence

$$\|u\|_Y \leq (1 + 2^\theta)\|u\|_X^{1-\theta}\|u\|_Z^\theta. \qquad \square$$

Standard Example 21.59 (Lebesgue Spaces). Let G be a bounded measurable set in $\mathbb{R}^N$, $N \geq 1$, and let $1 \leq r < p < q < \infty$. Then there exists a θ: $0 < \theta < 1$ with

$$\frac{1}{p} = \frac{1-\theta}{q} + \frac{\theta}{r}.$$

We set

$$X = L_q(G), \qquad Y = L_p(G), \qquad Z = L_r(G).$$

Then the interpolation inequality (62) holds.

Proof. By the Hölder inequality,

$$\| |u|^{1-\theta}|u|^\theta \|_p \leq \|u\|_q^{1-\theta}\|u\|_r^\theta. \qquad \square$$

Standard Example 21.60 (Sobolev Spaces). Let $G = \mathbb{R}^N$, $N \geq 1$, or let G be a bounded region in $\mathbb{R}^N$ with $\partial G \in C^{0,1}$. Let $1 \leq p \leq \infty$, and let k, m be integers with $0 \leq k < m$. For $G = \mathbb{R}^N$, let $p < \infty$. If we set

$$X = W_p^m(G), \qquad Y = W_p^k(G), \qquad Z = L_p(G),$$

then the interpolation inequality (62) holds with $\theta = 1 - k/m$, i.e.,

$$\|u\|_{k,p} \leq C\|u\|_{m,p}^{k/m}\|u\|_p^{1-k/m} \qquad \text{for all} \quad u \in W_p^m(G). \tag{63}$$

In particular, we have

$$\|u\|_{1,p} \le C\|u\|_{2,p}^{1/2}\|u\|_p^{1/2} \qquad \text{for all} \quad u \in W_p^2(G).$$

The constant C depends on G, k, m, p.

PROOF. This is a special case of the Gagliardo–Nirenberg inequalities which will be proved in Section 21.19. In fact, it follows from (74) that

$$\sum_{|\alpha|=k-r} \|D^\alpha u\|_p \le \text{const}\,\|u\|_{m,p}^{(k-r)/m}\|u\|_p^{1-(k-r)/m}.$$

Noting $\|u\|_p \le \|u\|_{m,p}$ and summing over $r = 0, \ldots, k$, we obtain (63). □

STANDARD EXAMPLE 21.61 (C^m-Spaces). Let G be a bounded region in $\mathbb{R}^N$ with $\partial G \in C^{0,1}$ and $N \ge 1$, and let k, m be integers with $0 \le k < m$. If we set

$$X = C^m(\bar{G}), \qquad Y = C^k(\bar{G}), \qquad Z = C(\bar{G}),$$

then the interpolation inequality (62) holds with $\theta = 1 - k/m$, i.e.,

$$\|u\|_{C^k(\bar{G})} \le C\|u\|_{C^m(\bar{G})}^{k/m}\|u\|_{C(\bar{G})}^{1-k/m} \qquad \text{for all} \quad u \in C^m(\bar{G}). \tag{64}$$

PROOF. Use (63) with $p = \infty$. □

STANDARD EXAMPLE 21.62 (C-Spaces). Let $G = \mathbb{R}^N$, $N \ge 1$, or let G be a bounded region in $\mathbb{R}^N$ with $\partial G \in C^{0,1}$. If we set

$$X = W_2^N(G), \qquad Y = C(\bar{G}), \qquad Z = L_2(G),$$

then the embedding $X \subseteq Y$ is continuous and (62) holds for $\theta = \frac{1}{2}$, i.e.,

$$\|u\|_{C(\bar{G})} \le C\|u\|_{N,2}^{1/2}\|u\|_2^{1/2} \qquad \text{for all} \quad u \in W_2^N(G). \tag{65}$$

Here, $C(\bar{G})$ denotes the set of all continuous functions $u: \bar{G} \to \mathbb{R}$ such that $\|u\|_{C(\bar{G})} = \sup_{x \in \bar{G}} |u(x)| < \infty$.

The proof will be given in Section 21.20 via Fourier transform.

21.18. Application to the Refined Banach Fixed-Point Theorem and the Convergence of Iteration Methods

We consider the operator equation

$$u = Au, \qquad u \in Y, \tag{66}$$

together with the iteration method

$$u_{n+1} = Au_n, \qquad n = 0, 1, \ldots. \tag{67}$$

We make the following assumptions:

(H1) *Interpolation inequality.* The embeddings

$$X \subseteq Y \subseteq Z$$

are continuous, where X, Y, and Z are B-spaces over $\mathbb{K} = \mathbb{R}, \mathbb{C}$. There are constants $C > 0$ and $0 < \theta < 1$ such that

$$\|u\|_Y \leq C\|u\|_X^{1-\theta}\|u\|_Z^{\theta} \qquad \text{for all} \quad u \in X.$$

(H2) *Boundedness* with respect to the "very nice" space X. We have $\sup_n \|u_n\|_X < \infty$.

(H3) *k-Contractivity* with respect to the "poor" space Z. For all n,

$$\|u_{n+1} - u_n\|_Z \leq k\|u_n - u_{n-1}\|_Z + a_n,$$

where $0 < k < 1$, $a_n \geq 0$, and $\sum_n a_n < \infty$.

(H4) *Continuity* with respect to the "nice" space Y. The operator $A: Y \to Z$ is continuous.

Theorem 21.H. *Suppose that* (H1) *through* (H4) *hold. Then* (u_n) *converges in* Y *to a solution of the original equation* (66).

Proof. From (H3) it follows that

$$\|u_2 - u_1\|_Z \leq k\|u_1 - u_0\|_Z + a_1,$$

$$\|u_3 - u_2\|_Z \leq k^2\|u_1 - u_0\|_Z + ka_1 + a_2,$$

$$\|u_4 - u_3\|_Z \leq k^3\|u_1 - u_0\|_Z + k^2a_1 + ka_2 + a_3, \quad \text{etc.}$$

Hence

$$\sum_n \|u_{n+1} - u_n\|_Z < \infty.$$

Thus, (u_n) is a Cauchy sequence in Z, i.e.,

$$u_n \to u \quad \text{in } Z \qquad \text{as} \quad n \to \infty.$$

By (H1) and (H2),

$$u_n \to u \quad \text{in } Y \qquad \text{as} \quad n \to \infty.$$

Letting $n \to \infty$, it follows from (67) and (H4) that $Au = u$. □

Theorem 21.H can be used in order to give existence proofs for systems of conservation laws (see Section 33.6).

21.19. The Gagliardo–Nirenberg Inequalities

The following interpolation inequalities represent an important tool for the investigation of nonlinear partial differential equations. Our first goal is the following inequality:

$$\|D_j u\|_{2r} \le C\|u\|_{2,p}^{1/2}\|u\|_q^{1/2}, \qquad j = 1, \ldots, N, \tag{68}$$

where

$$p^{-1} + q^{-1} = r^{-1}, \qquad 1 \le p, q, r \le \infty, \tag{69}$$

with the convention $\infty^{-1} = 0$. In particular, this means $p = r$ if $q = \infty$. Note that $x = (\xi_1, \ldots, \xi_N)$ and $D_j = \partial/\partial\xi_j$. The following simple result is basic for the general approach.

Lemma 21.63. *Let $N \ge 1$. Under the assumption* (69), *we have*

$$\|D_j u\|_{2r} \le C\|D_j^2 u\|_p^{1/2}\|u\|_q^{1/2} \qquad \text{for all} \quad u \in C_0^\infty(\mathbb{R}^N), \tag{70}$$

where $j = 1, \ldots, N$. The norm refers to $\mathbb{R}^N$, and C is a constant.

PROOF.

(I) First let $p, r < \infty$ and $1 \le q \le \infty$. We set $D = D_j$. Integrating the identity

$$(Du)^{2r} = D(uDu(Du)^{2r-2}) + (1 - 2r)uD^2u(Du)^{2r-2}$$

over $\mathbb{R}^N$, we get

$$\int (Du)^{2r}\,dx = (1 - 2r)\int uD^2u(Du)^{2r-2}\,dx.$$

By the Hölder inequality for three factors,

$$\int (Du)^{2r}\,dx \le (1 + 2r)\|u\|_q\|D^2u\|_p\left(\int (Du)^{2r}\,dx\right)^{1-1/r}.$$

This yields (70).

(II) Let $p = r = \infty$. A more careful elementary argument shows that the constant C in (70) is independent of $r, p, q \in \mathbb{R}$ (cf. Friedman (1969, M), (9.14)). Letting $r, p \to \infty$ and $q \to \infty$ in (70), we obtain the assertion. In this connection, note that

$$\lim_{s\to\infty}\|f\|_s = \|f\|_\infty$$

if $f \in L_\infty(K)$ and meas $K < \infty$. Moreover, note that the function $u \in C_0^\infty(\mathbb{R}^N)$ vanishes outside a compact set K. □

Proposition 21.64. *Let $G = \mathbb{R}^N$, $N \ge 1$, or let G be a bounded region in $\mathbb{R}^N$ with $\partial G \in C^{0,1}$. Under the assumption* (69), *inequality* (68) *holds for each*

$$u \in W_p^2(G) \cap L_q(G).$$

The constant C in (68) *depends on G, p, q.*

PROOF.

(I) Let $G = \mathbb{R}^N$. We set $D = D_j$. By Lemma 21.63,

$$\|Du\|_{2r} \le C\|u\|_{2,p}^{1/2}\|u\|_q^{1/2} \tag{71}$$

for all $u \in C_0^\infty(\mathbb{R}^N)$. By using a density argument, we want to show that (71) holds for more general functions u.

(I-1) Let $p, r < \infty$ and $q \leq \infty$. Let $u \in W_p^2(\mathbb{R}^N) \cap L_q(\mathbb{R}^N)$. The set $C_0^\infty(\mathbb{R}^N)$ is dense in $W_p^2(\mathbb{R}^N)$. To be precise, Problem 21.6 tells us that there is a sequence (u_n) in $C_0^\infty(\mathbb{R}^N)$ such that

$$\|u_n - u\|_{2,p} \to 0 \quad \text{as} \quad n \to \infty,$$

and since $u \in L_q(\mathbb{R}^N)$, we also obtain that

$$\|u_n\|_q \leq \text{const}\, \|u\|_q \quad \text{for all } n.$$

By (71),

$$\|Du_n - Du_m\|_{2r} \leq C \|u_n - u_m\|_{2,p}^{1/2} \|u_n - u_m\|_q^{1/2}$$

for all n, m, and hence

$$Du_n \to v \quad \text{in } L_{2r}(\mathbb{R}^N) \quad \text{as} \quad n \to \infty.$$

This implies $v = Du$ by Problem 21.4. Again by (71),

$$\|Du_n\|_{2r} \leq \text{const}\, \|u_n\|_{2,p}^{1/2} \|u\|_q^{1/2} \quad \text{for all } n.$$

Letting $n \to \infty$, we obtain (73).

(I-2) Let $q < \infty$, and hence $r < \infty$. In this case we choose a sequence (u_n) in $C_0^\infty(\mathbb{R}^N)$ such that

$$\|u_n - u\|_q \to 0 \quad \text{as} \quad n \to \infty \quad \text{and} \quad \|u_n\|_{2,p} \leq \text{const}\, \|u\|_{2,p} \quad \text{for all } n.$$

(I-3) Let $p = q = \infty$, and hence $r = \infty$. Let $u \in W_\infty^2(\mathbb{R}^N)$. Since $C_0^\infty(\mathbb{R}^N)$ is *not* dense in $L_\infty(\mathbb{R}^N)$, we need another argument as above. By Problem 21.6, there exists a sequence (u_n) in $C_0^\infty(\mathbb{R}^N)$ such that

$$\|u_n\|_{2,\infty} \leq \text{const}\, \|u\|_{2,\infty} \quad \text{for all } n \tag{72}$$

and

$$Du_n(x) \to Du(x) \quad \text{as } n \to \infty \quad \text{for almost all } x \in \mathbb{R}^N.$$

From this convergence and $\sup_n \|Du_n\|_\infty < \infty$, it follows that

$$\|Du\|_\infty \leq \varliminf_{n \to \infty} \|Du_n\|_\infty,$$

according to Problem 21.10. By (71),

$$\|Du_n\|_\infty \leq C \|u_n\|_{2,\infty}^{1/2} \|u_n\|_\infty^{1/2} \leq \text{const}\, \|u\|_{2,\infty}^{1/2} \|u\|_\infty^{1/2},$$

and hence $\|Du\|_\infty \leq \text{const}\, \|u\|_{2,\infty}^{1/2} \|u\|_\infty^{1/2}$.

(I-4) Summarizing we obtain that the inequality

$$\|Du\|_{2r} \leq \text{const}\, \|u\|_{2,p}^{1/2} \|u\|_q^{1/2} \tag{73}$$

holds for each

$$u \in W_p^2(\mathbb{R}^N) \cap L_q(\mathbb{R}^N). \tag{73*}$$

(II) Extension trick for $G \neq \mathbb{R}^N$. Let

$$v \in W_p^2(G) \cap L_q(G).$$

By the universal extension theorem from Problem 21.8b, there exists a function $u: \mathbb{R}^N \to \mathbb{R}$ with (73*) such that

$$u(x) = v(x) \quad \text{on } G$$

and

$$\|u\|_{2,p} \le \text{const}\, \|v\|_{2,p}, \qquad \|u\|_q \le \text{const}\, \|v\|_q,$$

where the norm for u and v refers to $\mathbb{R}^N$ and G, respectively. The constants are independent of u and v. From (73) we get

$$\|Dv\|_{2r} \le \text{const}\, \|v\|_{2,p}^{1/2} \|v\|_q^{1/2},$$

since $\|Dv\|_{2r} \le \|Du\|_{2r}$. This is the original inequality (68). $\square$

Remark 21.65 (A General Strategy). The proof of Proposition 21.64 is based on a general strategy via density and extension arguments. Roughly speaking, by using this strategy, it is enough to prove inequalities for functions $u \in C_0^\infty(\mathbb{R}^N)$ in order to obtain inequalities for Sobolev spaces with respect to both $\mathbb{R}^N$ and bounded subregions G of $\mathbb{R}^N$.

The following results should explain the *usefulness* of this strategy.

EXAMPLE 21.66. Let G be given as in Proposition 21.64. Let $1 \le p \le \infty$, and let k, m be integers with $0 \le k < m$. Then

$$\sum_{|\alpha|=k} \|D^\alpha u\|_p \le \text{const}\, \|u\|_{m,p}^{k/m} \|u\|_p^{1-k/m} \tag{74}$$

for all $u \in W_p^m(G)$. The constant depends on G, p, k, m.

PROOF. Let $u \in C_0^\infty(\mathbb{R}^N)$. We set

$$f(s) = \sum_{|\alpha|=s} \|D^\alpha u\|_p.$$

By Lemma 21.63 with $p = q$,

$$f(s) \le \text{const}\, f(s+1)^{1/2} f(s-1)^{1/2}, \qquad s = 1, \ldots, m-1.$$

By induction, this implies

$$f(k) \le \text{const}\, f(m)^{k/m} f(0)^{1-k/m}.$$

This is (74), since $f(m) \le \|u\|_{m,p}$.

Now proceed as in the proof of Proposition 21.64. $\square$

EXAMPLE 21.67. Let G be given as in Proposition 21.64, and let k, m be integers with $0 \le k < m$. Then

$$\sum_{|\alpha|=k} \|D^\alpha u\|_{2m/k} \le \text{const}\, \|u\|_{m,2}^{k/m} \|u\|_\infty^{1-k/m} \tag{75}$$

for all $u \in W_2^m(G) \cap L_\infty(G)$. The constant depends on G, k, m.

PROOF. Set

$$f(s) = \sum_{|\alpha|=s} \|D^\alpha u\|_{2m/s}$$

and conclude as in the proof of Example 21.66. □

More general Gagliardo–Nirenberg interpolation inequalities are summarized in the Appendix $A_2(54)$.

21.20. The Strategy of the Fourier Transform for Sobolev Spaces

Let $\hat{u}$ denote the Fourier transform of u, i.e.,

$$u(x) = (2\pi)^{-N/2} \int_{\mathbb{R}^N} e^{i\langle x|y\rangle} \hat{u}(y)\, dy.$$

Important classical properties of the Fourier transform are summarized in the Appendix $A_2(74)$. We set

$$\|u\|_{m,2}^* = \left(\int_{\mathbb{R}^N} (1 + |y|^{2m}) |\hat{u}(y)|^2 \, dy \right)^{1/2}.$$

Our first goal is the following crucial inequality:

$$\|u\|_{m,2}^* \le \|u\|_{m,2} \le c\|u\|_{m,2}^* \qquad \text{for all} \quad u \in C_0^\infty(\mathbb{R}^N). \tag{76}$$

Lemma 21.68. *Let $N \ge 1$, $m = 0, 1, \ldots$. Then there is a constant c depending on N and m such that* (76) *holds.*

PROOF. Let $N = 1$. The general case $N \ge 1$ proceeds analogously. Obviously, there is a $c > 0$ such that

$$1 \le \frac{1 + |y|^2 + \cdots + |y|^{2m}}{1 + |y|^{2m}} \le c \qquad \text{for all} \quad y \in \mathbb{R}.$$

Let $u \in C_0^\infty(\mathbb{R})$. The Fourier transform

$$u(x) = (2\pi)^{-1/2} \int_{\mathbb{R}} e^{ixy} \hat{u}(y)\, dy \tag{77}$$

yields

$$D^k u(x) = (2\pi)^{-1/2} \int_{\mathbb{R}} e^{ixy} (iy)^k \hat{u}(y)\, dy.$$

By the Parseval identity $A_2(74b)$,

$$\int_{\mathbb{R}} |D^k u(x)|^2 \, dx = \int_{\mathbb{R}} |(iy)^k \hat{u}(y)|^2 \, dy.$$

Summing over $k = 0, \ldots, m$, we obtain $\|u\|_{m,2}^2$, and hence (76). □

Remark 21.69 (A General Strategy). If we combine Lemma 21.68 with the extension strategy of Remark 21.65, then it is possible to prove quickly a number of important properties of Sobolev spaces. The following proposition should illustrate this.

Proposition 21.70. *Let $G = \mathbb{R}^N$, $N \geq 1$, or let G be a bounded region in $\mathbb{R}^N$ with $\partial G \in C^{0,1}$. Suppose that $m > N/2$. Then:*

(a) *The embedding $W_2^m(G) \subseteq C(\bar{G})$ is continuous, i.e.,*

$$\|u\|_{C(\bar{G})} \leq \text{const}\,\|u\|_{m,2} \qquad \text{for all} \quad u \in W_2^m(G). \tag{78}$$

(b) *For all $u \in W_2^N(G)$,*

$$\|u\|_{C(\bar{G})} \leq \text{const}\,\|u\|_{N,2}^{1/2}\,\|u\|_2^{1/2}. \tag{79}$$

More precisely, the functions u in (a) *and* (b) *belong to $C(\bar{G})$ after changing their values on a set of N-dimensional Lebesgue measure zero, if necessary.*

PROOF. Ad(a).

(I) Let $G = \mathbb{R}^N$. We will critically use that

$$\int_{\mathbb{R}^N} \frac{dy}{1 + |y|^{2m}} < \infty \qquad \text{if} \quad m > N/2.$$

Let $u \in C_0^\infty(\mathbb{R}^N)$. In the following all the integrals are taken over $\mathbb{R}^N$. The Fourier transform

$$u(x) = (2\pi)^{-N/2} \int e^{i\langle x|y\rangle} \hat{u}(y)\, dy$$

yields

$$(2\pi)^{N/2}|u(x)| \leq \int |\hat{u}(y)|\, dy = \int \frac{|\hat{u}(y)|(1 + |y|^{2m})^{1/2}\, dy}{(1 + |y|^{2m})^{1/2}}.$$

By the Hölder inequality and Lemma 21.68,

$$\begin{aligned}(2\pi)^{N/2}|u(x)| &\leq \left(\int |\hat{u}(y)|^2(1 + |y|^{2m})\, dy\right)^{1/2} \left(\int \frac{dy}{1 + |y|^{2m}}\right)^{1/2} \\ &\leq \text{const}\,\|u\|_{m,2}^* \leq \text{const}\,\|u\|_{m,2}.\end{aligned}$$

This implies (78) for all $u \in C_0^\infty(\mathbb{R}^N)$.

Since $C_0^\infty(\mathbb{R}^N)$ is dense in $W_2^m(\mathbb{R}^N)$, the assertion follows as in the proof of Theorem 21.A(d).

(II) Let $G \neq \mathbb{R}^N$. By the extension trick in the proof of Proposition 21.64, inequality (78) holds for all $u \in W_2^m(G)$. Now use the fact that $C^m(\bar{G})$ is dense in $W_2^m(G)$, by Theorem 21.A.

Ad(b). By Remark 21.65, it is sufficient to prove (79) for all $u \in C_0^\infty(\mathbb{R}^N)$. To this end, we use critically that:

$$\int_{\mathbb{R}^N} \frac{dy}{1 + s^2|y|^{2N}} = \frac{1}{s} \int_{\mathbb{R}^N} \frac{dx}{1 + |x|^{2N}} = \frac{C}{s}$$

for all $s > 0$. Let $u \in C_0^\infty(\mathbb{R}^N)$. The Fourier transform yields

$$\begin{aligned}(2\pi)^N|u(x)|^2 &\le \left(\int |\hat{u}(y)|\,dy\right)^2\\ &= \left(\int \frac{|\hat{u}(y)|(1+s^2|y|^{2N})^{1/2}}{(1+s^2|y|^{2N})^{1/2}}\right)^2\\ &\le \int |\hat{u}|^2(1+s^2|y|^{2N})\,dy \int \frac{dy}{1+s^2|y|^{2N}}.\end{aligned}$$

By Lemma 21.68,

$$\|u\|^2_{C(\mathbb{R}^N)} \le \text{const}(s^{-1}\|u\|_2^2 + s\|u\|_{N,2}^2)$$

noting that $\int |\hat{u}|^2\,dy = \int |u|^2\,dx = \|u\|_2^2$. Letting $s = \|u\|_2/\|u\|_{N,2}$, we obtain (79) for all $u \in C_0^\infty(\mathbb{R}^N)$. □

21.21. Banach Algebras and Sobolev Spaces

Definition 21.71. Let X be a function space. Then, X is called a generalized *Banach algebra* iff X is a B-space such that $u, v \in X$ implies $uv \in X$ and

$$\|uv\| \le C\|u\|\,\|v\|, \tag{80}$$

where C denotes a constant. If $C = 1$, then we speak of a Banach algebra.

If we set $\|u\|_* = C\|u\|$, then

$$\|uv\|_* \le \|u\|_*\|v\|_* \qquad \text{for all} \quad u, v \in X.$$

If X is a generalized Banach algebra, then $u \in X$ implies $u^k \in X$ and

$$\|u^k\| \le C^k\|u\|^k \qquad \text{for all} \quad k = 1, 2, \ldots. \tag{81}$$

The following result demonstrates the *usefulness* of Banach algebras for nonlinear problems.

Proposition 21.72. *Let X be a generalized Banach algebra. Let*

$$f(u) = a_0 + a_1u + a_2u^2 + \cdots$$

be an analytic function from $]-r, r[$ to $\mathbb{R}$, where $0 < r \le \infty$. Set $B = \{u \in X\colon \|u\| < r/C\}$, where $C > 0$ corresponds to (80). *Then the operator $F\colon B \to X$ given by*

$$u \mapsto f(u) \tag{82}$$

is analytic. For each $v \in B$ and $k = 1, 2, \ldots$, the k*th* F*-derivative $F^{(k)}(v)$ at v is given by*

$$(h_1, \ldots, h_k) \mapsto f^{(k)}(v)h_1 \ldots h_k$$

for all $h_1, \ldots, h_k \in X$.

PROOF. For all $u \in B$,

$$\|f(u)\| \leq \sum_{k=0}^{\infty} |a_k| C^k \|u\|^k < \infty.$$

Now use (16) in Chapter 8. □

STANDARD EXAMPLE 21.73 (C-Spaces and Hölder Spaces). Let G be a bounded region in $\mathbb{R}^N$, $N \geq 1$. Then all the spaces

$$C^k(\bar{G}), \qquad C^{k,\alpha}(\bar{G}), \qquad k = 0, 1, 2, \ldots, \quad 0 \leq \alpha \leq 1,$$

form generalized Banach algebras.

PROOF.

(I) Let $G =]0, 1[$, and let $u, v \in C[0, 1]$. Then

$$\|uv\|_C = \max_{0 \leq x \leq 1} |u(x)v(x)| \leq \|u\|_C \|v\|_C.$$

(II) If $u, v \in C^1[0, 1]$, then $(uv)' = u'v + uv'$, and hence

$$\|(uv)'\|_C \leq \|u'\|_C \|v\|_C + \|u\|_C \|v'\|_C.$$

This implies

$$\|uv\|_{C^1} = \|uv\|_C + \|(uv)'\|_C \leq \|u\|_{C^1} \|v\|_{C^1}.$$

(III) Let $u, v \in C^\alpha[0, 1]$, $0 < \alpha < 1$. Then

$$|u(x) - u(y)| \leq H_\alpha(u)|x - y|^\alpha \qquad \text{for all} \quad x, y \in [0, 1],$$

and hcnce

$$\begin{aligned}|u(x)v(x) - u(y)v(y)| &= |u(x)(v(x) - v(y)) + v(y)(u(x) - u(y))| \\ &\leq (\|u\|_C H_\alpha(v) + \|v\|_C H_\alpha(u))|x - y|^\alpha.\end{aligned}$$

This implies $H_\alpha(uv) \leq \|u\|_C H_\alpha(v) + \|v\|_C H_\alpha(u)$, and hence

$$\|uv\|_\alpha = \|uv\|_C + H_\alpha(uv) \leq \|u\|_\alpha \|v\|_\alpha.$$

The general case proceeds analogously. □

STANDARD EXAMPLE 21.74 (Sobolev Spaces). Let $G = \mathbb{R}^N$ or let G be a bounded region in $\mathbb{R}^N$ with $\partial G \in C^{0,1}$ and $N \geq 1$. Then the Sobolev space

$$X = W_p^m(G), \qquad 1 \leq p < \infty, \quad m = 1, 2, \ldots,$$

forms a generalized Banach algebra in the case where

$$mp > N. \tag{83}$$

$W_\infty^m(G)$ is a generalized Banach algebra for $m = 0, 1, \ldots$.

For example, the spaces

$$W_2^1(G), \quad G \subseteq \mathbb{R}^1, \qquad \text{and} \qquad W_2^2(G), \quad G \subseteq \mathbb{R}^2, \mathbb{R}^3,$$

form generalized Banach algebras.

PROOF. If $p = \infty$, then the assertion follows from $\|uv\|_\infty \leq \|u\|_\infty \|v\|_\infty$ and

$$\|u\|_{m,\infty} = \sum_{|\alpha| \leq m} \|D^\alpha u\|_\infty.$$

The proof for $p < \infty$ will be given in Problem 21.5, by using the continuity of the embedding $X \subseteq C(\bar{G})$. □

Remark 21.75. If the crucial condition (83) is *violated*, then the investigation of nonlinear operators has to be based on the Sobolev embedding theorems and the Gagliardo–Nirenberg estimates. A typical example will be considered in Section 21.23.

COUNTEREXAMPLE 21.76 (Lebesgue Spaces). The function $u(x) = 1/\sqrt[4]{x}$ has the property

$$u \in L_2(0, 1) \qquad \text{and} \qquad u^2 \notin L_2(0, 1).$$

Thus, $L_2(0, 1)$ is *not* a generalized Banach algebra. Consequently, the operator

$$u \mapsto u^2$$

is *not* explained on the entire space $L_2(0, 1)$.

21.22. Moser-Type Calculus Inequalities

Our goal is the following fundamental inequality:

$$\|D^\alpha u D^\beta v\|_2 \leq \text{const}(\|u\|_{m,2}\|v\|_\infty + \|v\|_{m,2}\|u\|_\infty), \tag{84}$$

for all $u, v \in W_2^m(G) \cap L_\infty(G)$, where

$$m = |\alpha| + |\beta| \geq 0,$$

and for all $u, v \in W_2^m(G)$ if $m > N/2$. More generally, we consider the inequality

$$\|D^{\alpha_1}u_1 \cdots D^{\alpha_r}u_r\|_2 \leq \text{const} \sum_{i=1}^r \|u_i\|_{m,2} \prod_{j=1, j\neq i}^r \|u_j\|_\infty, \tag{85}$$

where $m = |\alpha_1| + \cdots + |\alpha_r|$.

Proposition 21.77 (Moser (1966)). *Let $G = \mathbb{R}^N$ or let G be a bounded region in $\mathbb{R}^N$ with $\partial G \in C^{0,1}$ and $N \geq 1$. Then, inequality* (85) *holds for all*

$$u_1, \ldots, u_r \in W_2^m(G) \cap L_\infty(G), \qquad m = 0, 1, \ldots.$$

Corollary 21.78. *If $m > N/2$, then, for all*

$$u_1, \ldots, u_r \in W_2^m(G),$$

inequality (85) *holds and*

$$\|D^{\alpha_1}u_1 \cdots D^{\alpha_r}u_r\|_2 \le \text{const}\,\|u_1\|_{m,2} \cdots \|u_r\|_{m,2}. \tag{85*}$$

This corollary follows from the continuity of the embedding $W_2^m(G) \subseteq C(\bar{G})$.

PROOF.

(I) Let $r = 2$, i.e., we first prove (84). If $|\alpha| = 0$, then

$$\|D^\alpha u D^\beta v\|_2 = \|u D^\beta v\|_2 \le \|u\|_\infty \|D^\beta v\|_2.$$

Thus, we may assume that $|\alpha|, |\beta| \ge 1$. Set

$$p = m/|\alpha|, \qquad q = m/|\beta|, \qquad p^{-1} + q^{-1} = 1.$$

By the Hölder inequality,

$$\|D^\alpha u D^\beta v\|_2 \le \|D^\alpha u\|_{2p} \|D^\beta v\|_{2q}.$$

By the Gagliardo–Nirenberg inequality (75),

$$\|D^\alpha u\|_{2p} \le \text{const}\,\|u\|_{m,2}^{1/p} \|u\|_\infty^{1/q},$$

$$\|D^\beta v\|_{2q} \le \text{const}\,\|v\|_{m,2}^{1/q} \|v\|_\infty^{1/p}.$$

By the Young inequality

$$ab \le \frac{|a|^p}{p} + \frac{|b|^q}{q} \qquad \text{for all} \quad a, b \in \mathbb{R},$$

we obtain the assertion (84).

(II) In the general case, use the Hölder inequality and the Young inequality for r factors (cf. A_2(29), (30)). □

Let $u = (u_1, \ldots, u_k)$ and $v = (v_1, \ldots, v_k)$. The following simple inequalities are frequently used:

$$\|a(u)w\|_2 \le C(R)\|w\|_2, \tag{86}$$

$$\|(a(u) - a(v))w\|_2 \le C(R)\|u - v\|_\infty \|w\|_2, \tag{87}$$

where $\|u\|_\infty = \sum_{i=1}^k \|u_i\|_\infty$ and $\|u\|_\infty, \|v\|_\infty \le R$. Here, R is an arbitrary real number.

If G is bounded, then, in particular, we may set $w = 1$. In this case, $\|w\|_2 < \infty$.

Proposition 21.79 (Killing of Nonlinearities). *Let $G = \mathbb{R}^N$ or let G be a bounded region in $\mathbb{R}^N$, $N \ge 1$. Let $a\colon \mathbb{R}^k \to \mathbb{R}$ be locally Lipschitz continuous (e.g., a is*

C^1). *Then, the inequalities* (86) *and* (87) *hold for all*

$$w \in L_2(G) \qquad \textit{and} \qquad u_i, v_i \in L_\infty(G), \qquad i = 1, \ldots, k.$$

For (86) *it is sufficient that* $a(\cdot)$ *is continuous.*

PROOF. If $\|u\|_\infty, \|v\|_\infty \leq R$, then the function $x \mapsto a(u(x))$ is bounded on G, and hence

$$\left| \int_G (aw)^2 \, dx \right| \leq \text{const} \int_G w^2 \, dx.$$

Moreover, it follows from

$$|a(u(x)) - a(v(x))| \leq L(R)|u(x) - v(x)| \qquad \text{for all} \quad x \in G$$

that

$$\left| \int_G (a(u) - a(v))^2 w^2 \, dx \right| \leq L^2 \|u - v\|_\infty^2 \int_G w^2 \, dx. \qquad \square$$

Important inequalities for the nonlinear terms of partial differential equations can be obtained by combining (85), (85*) with (86), (87).

EXAMPLE 21.80. Let $G = \mathbb{R}^N$ or let G be a bounded region in $\mathbb{R}^N$ with $\partial G \in C^{0,1}$ and $N \geq 1$. Suppose that the function $a: \mathbb{R}^k \to \mathbb{R}$ is C^1. Let

$$m > N/2$$

and let $|\alpha_1| + \cdots + |\alpha_r| \leq m$. Set $u = (u_1, \ldots, u_k)$, $v = (v_1, \ldots, v_k)$, and $\|u\|_{m,2} = \sum_i \|u_i\|_{m,2}$. Then:

$$\|a(u) \prod_{j=1}^r D^{\alpha_j} w_j\|_2 \leq C(R) \prod_{j=1}^r \|w_j\|_{m,2}, \tag{88}$$

$$\|(a(u) - a(v)) \prod_{j=1}^r D^{\alpha_j} w_j\|_2 \leq C(R) \|u - v\|_{m,2} \prod_{j=1}^r \|w_j\|_{m,2}, \tag{89}$$

for all $u_i, v_i, w_j \in W_2^m(G)$ with $\|u\|_{m,2}, \|v\|_{m,2} \leq R$.

These inequalities are very convenient in applications.

PROOF. Since the embedding $W_2^m(G) \subseteq C(\bar{G})$ is continuous, we have

$$\|f\|_\infty \leq \text{const} \|f\|_{m,2} \qquad \text{for all} \quad f \in W_2^m(G).$$

Now use (85*) and (86), (87). $\square$

More general results can be found in Chapter 83.

21.23. Weakly Sequentially Continuous Nonlinear Operators on Sobolev Spaces

The operator $A: X \to Y$ between the B-spaces X and Y is called weakly sequentially continuous iff, as $n \to \infty$,

$$u_n \rightharpoonup u \qquad \text{implies} \qquad Au_n \rightharpoonup Au.$$

Such operators play an important role in the modern theory of nonlinear partial differential equations.

Proposition 21.81. *Let X, Y, Z be B-spaces over $\mathbb{K} = \mathbb{R}, \mathbb{C}$. Then:*

(a) *If $A: X \to Y$ is linear and continuous, then A is also weakly sequentially continuous.*
(b) *If $A: Y \to Z$ is continuous and the embedding $X \subseteq Y$ is compact, then $A: X \to Z$ is strongly continuous, and hence also weakly sequentially continuous.*

PROOF. Ad(a). Cf. Proposition 21.27.

Ad(b). If $u_n \rightharpoonup u$ in X, then $u_n \to u$ in Y, and hence $Au_n \to Au$ in Z. □

Lemma 21.82. *Let $G = \mathbb{R}^N$ or let G be a nonempty bounded open set in $\mathbb{R}^N$, $N \geq 1$. Let $(v_n w_n)$ be a bounded sequence in $W_p^m(G)$, where $m = 0, 1, 2, \ldots$, $1 < p < \infty$.*

In the case where $G = \mathbb{R}^N$, suppose that, for each open ball B in $\mathbb{R}^N$,

$$\begin{aligned} v_n &\to v \quad \text{in } C(\bar{B}) \quad \text{as} \quad n \to \infty, \\ w_n &\rightharpoonup w \quad \text{in } L_p(B) \quad \text{as} \quad n \to \infty, \end{aligned} \tag{90}$$

where $v, w: G \to \mathbb{R}$ are fixed functions. In the case where $G \neq \mathbb{R}^N$, suppose that (90) holds for $B = G$. Then

$$v_n w_n \rightharpoonup vw \quad \text{in } W_p^m(G) \quad \text{as} \quad n \to \infty.$$

PROOF. The B-space $W_p^m(G)$ is reflexive. If we assume that

$$v_n w_n \rightharpoonup z \quad \text{in } W_p^m(G) \quad \text{as} \quad n \to \infty, \tag{91}$$

then

$$\int_B \varphi v_n w_n \, dx \to \int_B \varphi z \, dx \quad \text{as} \quad n \to \infty \quad \text{for all} \quad \varphi \in C_0^\infty(B),$$

since $u \mapsto \int_B \varphi u \, dx$ is a linear continuous functional on $W_p^m(G)$, by the Hölder inequality. From

$$\varphi v_n \to \varphi v \quad \text{in } L_q(B) \quad \text{and} \quad w_n \rightharpoonup w \quad \text{in } L_p(B), \quad p^{-1} + q^{-1} = 1,$$

it follows, by Proposition 21.23(j), that

$$\int_B \varphi v_n w_n \, dx \to \int_B \varphi v w \, dx \quad \text{as} \quad n \to \infty,$$

and hence $\int_B \varphi(vw - z) \, dx = 0$ for all $\varphi \in C_0^\infty(B)$ and all B. Since $C_0^\infty(B)$ is dense in $L_q(B)$, we obtain that

$$z = vw \quad \text{almost everywhere on } G.$$

If (91) holds for a subsequence $(v_{n'} w_{n'})$, then we obtain again that $z = vw$. This yields the assertion, by the convergence principle (Prop. 21.23(i)). □

Lemma 21.83. *Let K be a nonempty compact set in $\mathbb{R}^N$, $N \geq 1$, and let $a: \mathbb{R} \to \mathbb{R}$ be continuous. If we set*

$$Au = a(u),$$

then the operator $A: C(K) \to C(K)$ is continuous.

PROOF. Let $u_n \to u$ in $C(K)$ as $n \to \infty$. Since (u_n) is bounded in $C(K)$ and $a(\cdot)$ is uniformly continuous on bounded sets, we get

$$\|Au_n - Au\| = \max_{x \in K} |a(u_n(x)) - a(u(x))| \to 0 \quad \text{as} \quad n \to \infty. \qquad \square$$

We now study the *typical* nonlinear operator

$$Au = a(u)D^\alpha u, \qquad |\alpha| \leq m. \tag{92}$$

Proposition 21.84. *Let $G = \mathbb{R}^N$ or let G be a bounded region in $\mathbb{R}^N$ with $\partial G \in C^{0,1}$ and $N \geq 1$. Set*

$$X = W_2^{m+k}(G), \qquad Y = W_2^k(G), \qquad m > N/2, \quad k = 0, 1, \ldots.$$

Let $a: \mathbb{R} \to \mathbb{R}$ be C^{k+1}. Then the operator

$$A: X \to Y,$$

defined by (92), is bounded, continuous, Lipschitz continuous on balls, and weakly sequentially continuous.

PROOF.

(I) We show that A is bounded and Lipschitz continuous on balls.

(I-1) Let $k = 0$, and let $\|u\|_{m,2}, \|v\|_{m,2} \leq R$. By Example 21.80,

$$\|Au\|_2 \leq C(R)\|u\|_{m,2}$$

and

$$\begin{aligned} \|Au - Av\|_2 &= \|(a(u) - a(v))D^\alpha u + a(v)(D^\alpha u - D^\alpha v)\|_2 \\ &\leq K(R)\|u - v\|_{m,2}. \end{aligned}$$

(I-2) Let $k = 1$. We will use the equivalent norm $\|v\|_{1,2}^* = \|v\|_2 + \sum_j \|D_j v\|_2$ on $Y = W_2^1(G)$. Note that

$$\|Au\|_{1,2}^* = \|Au\|_2 + \sum_{j=1}^N \|D_j Au\|_2$$

and

$$D_j Au = a'(u)D_j u D^\alpha u + a(u)D_j D^\alpha u.$$

Thus, the case $k = 1$ can be reduced to the known case $k = 0$.

(I-3) For $k \geq 2$, we use a similar argument.

(II) We show that $A: X \to Y$ is weakly sequentially continuous. To this end, let $B = G$ if G is bounded, and let B be an open ball if $G = \mathbb{R}^N$. Since the embedding $W_2^m(B) \subseteq C(\bar{B})$ is compact, it follows from

$$u_n \rightharpoonup u \quad \text{in } X = W_2^{m+k}(G) \qquad \text{as} \quad n \to \infty$$

that $u_n \to u$ in $C(\bar{B})$, and hence

$$a(u_n) \to a(u) \quad \text{in } C(\bar{B}) \qquad \text{as} \quad n \to \infty,$$

$$D^\alpha u_n \rightharpoonup D^\alpha u \quad \text{in } L_2(B) \qquad \text{as} \quad n \to \infty.$$

By Lemma 21.82,

$$Au_n \rightharpoonup Au \quad \text{in } Y = W_2^k(G) \qquad \text{as} \quad n \to \infty.$$

Note that (Au_n) is bounded in Y, since $A\colon X \to Y$ is bounded and (u_n) is bounded in X. □

Problems

21.1. *Almost orthogonal elements.* Let M be a closed linear subspace of the B-space X with $M \neq X$. Show that, for each $\varepsilon \in]0, 1[$, there exists an element u_ε with

$$\|u_\varepsilon\| = 1 \qquad \text{and} \qquad \operatorname{dist}(u_\varepsilon, M) \geq 1 - \varepsilon.$$

Solution: Let $v \in X - M$. Then $d = \operatorname{dist}(v, M) > 0$, since M is closed. We choose an element $m_\varepsilon \in M$ with $\|v - m_\varepsilon\| \leq d(1 - \varepsilon)^{-1}$. Then, $u_\varepsilon = (v - m_\varepsilon)/\|v - m_\varepsilon\|$ is the desired element. In fact, for all $m \in M$,

$$\|u_\varepsilon - m\| = \|v - m_\varepsilon\|^{-1} \|v - m_\varepsilon - \|v - m_\varepsilon\| m\|$$
$$\geq \|v - m_\varepsilon\|^{-1} d \geq 1 - \varepsilon.$$

21.2. *Proof of Theorem* 21.C.

Solution: Let $\dim X = \infty$. We choose w_1 with $\|w_1\| = 1$. Let $M = \operatorname{span}\{w_1\}$. By Problem 21.1 there exists a w_2 with $\|w_2\| = 1$ and $\|w_2 - w_1\| \geq \frac{1}{2}$. This way we obtain a sequence (w_n) with $\|w_n\| = 1$ for all n and $\|w_n - w_m\| \geq \frac{1}{2}$ for all $m > n$. Obviously, (w_n) has no convergent subsequence, i.e., the unit ball in X is not compact.

If $\dim X < \infty$, then it follows from the classical Bolzano–Weierstrass theorem that the closed unit ball in X is compact.

21.3. *Properties of generalized derivatives.* We want to show that generalized derivatives have properties similar to those of classical derivatives. In Problems 21.3b–21.3h below, let G be a nonempty open set in $\mathbb{R}^N$, $N \geq 1$. All derivatives are to be understood in the generalized sense.

21.3a. *Functions with vanishing generalized derivatives.* Let G be a region in $\mathbb{R}^N$ with $N \geq 1$ and let $u \in W_1^1(G)$. Show that

$$D_j u = 0 \quad \text{on } G \qquad \text{for} \quad j = 1, \ldots, N$$

implies $u(x) = \text{constant}$ for almost all $x \in G$.

Solution: Let $H \subset\subset G$, and let H be bounded. By Proposition 18.35,

$$D_j(S_\varepsilon u) = S_\varepsilon(D_j u) = 0 \quad \text{on } H$$

for all $\varepsilon \in]0, \varepsilon_0[$. Hence $S_\varepsilon u = \text{constant}$. By Proposition 18.33,

$$S_\varepsilon u \to u \quad \text{in } L_1(H) \qquad \text{as} \quad \varepsilon \to +0.$$

Since the constant functions form a finite-dimensional subspace in $L_1(H)$, we obtain $u = \text{constant}$ on H. Now, note that H can be chosen arbitrarily.

Analogously, we obtain that it follows from $u \in W_1^k(G)$ and

$$D^\alpha u = 0 \quad \text{on } G \qquad \text{for all} \quad \alpha\colon |\alpha| \le k$$

that $u(x) = (\text{polynomial of degree} \le k - 1)$ for almost all $x \in G$.

21.3b. *Weak limits.* Let α be a fixed multi-index. Show that it follows from

$$u_n \rightharpoonup u \quad \text{in } L_1(G) \qquad \text{as} \quad n \to \infty,$$

$$D^\alpha u_n \rightharpoonup v \quad \text{in } L_1(G) \qquad \text{as} \quad n \to \infty,$$

that $D^\alpha u = v$ on G.

Solution: Letting $n \to \infty$, it follows from

$$\int_G u_n D^\alpha \varphi \, dx = (-1)^{|\alpha|} \int_G (D^\alpha u_n)\varphi \, dx \qquad \text{for all} \quad \varphi \in C_0^\infty(G)$$

that

$$\int_G u D^\alpha \varphi \, dx = (-1)^{|\alpha|} \int_G v\varphi \, dx \qquad \text{for all} \quad \varphi \in C_0^\infty(G).$$

Consequently, the generalized derivative $D^\alpha u$ exists on G and $D^\alpha u = v$ on G.

21.3c. *Characterization of generalized derivatives.* Let $u, v \in L_{1,\text{loc}}(G)$, and let α be a fixed multi-index. Show that the following two statements are equivalent:

(i) The generalized derivative $D^\alpha u$ exists on G and $D^\alpha u = v$ on G.

(ii) There exists a sequence (u_n) in $C^\infty(G)$ such that

$$u_n \to u \quad \text{in } L_1(H) \qquad \text{as} \quad n \to \infty,$$

$$D^\alpha u_n \to v \quad \text{in } L_1(H) \qquad \text{as} \quad n \to \infty,$$

for all compact subsets H of G.

Solution: (i) $\Rightarrow$ (ii). Letting

$$u_n = S_{1/n} u \qquad \text{for} \quad n = 1, 2, \ldots,$$

statement (ii) follows immediately from Proposition 18.35(b).

(ii) $\Rightarrow$ (i). Use the same argument as in Problem 21.3b.

21.3d. *The product rule for generalized derivatives.* For fixed $i = 1, \ldots, N$, show that $D_i(uv)$ exists on G such that

$$D_i(uv) = (D_i u)v + u D_i v \quad \text{on } G,$$

provided the following assumptions are satisfied:

(i) $u \in L_{p,\text{loc}}(G)$ and $v \in L_{q,\text{loc}}(G)$, where $1 < p < \infty$ and $p^{-1} + q^{-1} = 1$.

(ii) The generalized derivatives $D_i u$ and $D_i v$ exist on G such that $D_i u \in L_{p,\text{loc}}(G)$ and $D_i v \in L_{q,\text{loc}}(G)$.

Solution: We set $u_n = S_{1/n}u$ and $v_n = S_{1/n}v$ for all n. By Propositions 18.33 and 18.35, we get $u_n, v_n \in C^\infty(G)$ for all n, as well as

$$u_n \to u \quad \text{and} \quad D_i u_n \to D_i u \quad \text{in } L_p(H) \quad \text{as} \quad n \to \infty,$$

$$v_n \to v \quad \text{and} \quad D_i v_n \to D_i v \quad \text{in } L_q(H) \quad \text{as} \quad n \to \infty,$$

for all compact subsets H of G. Letting $n \to \infty$, we obtain from

$$\int_G u_n v_n D_i \varphi\, dx = -\int_G (v_n D_i u_n + u_n D_i v_n)\varphi\, dx \qquad \text{for all} \quad \varphi \in C_0^\infty(G),$$

that

$$\int_G uv D_i \varphi\, dx = -\int_G (v D_i u + u D_i v)\varphi\, dx \qquad \text{for all} \quad \varphi \in C_0^\infty(G).$$

21.3e. *The chain rule for generalized derivatives.* For fixed $i = 1, \ldots, N$, show that the generalized derivative $D_i(f \circ u)$ exists on G such that

$$D_i(f \circ u) = (f' \circ u) D_i u \quad \text{on } G,$$

provided the following assumptions are satisfied:

(i) $u \in L_{1,\text{loc}}(G)$.

(ii) The generalized derivative $D_i u$ exists on G and $D_i u \in L_{1,\text{loc}}(G)$.

(iii) $f \in C^1(\mathbb{R})$ and $\sup_{x \in \mathbb{R}} |f'(x)| < \infty$.

Solution: Let H be a compact subset of G. We choose a sequence (u_n) in $C^\infty(G)$ such that

$$u_n \to u \quad \text{and} \quad D_i u_n \to D_i u \quad \text{in } L_1(H) \quad \text{as} \quad n \to \infty,$$

by Problem 21.3c. Then

$$\int_H |f(u_n) - f(u)|\, dx \le \sup_{x \in \mathbb{R}} |f'(x)| \int_H |u_n - u|\, dx \to 0 \qquad \text{as} \quad n \to \infty,$$

and

$$A_n \stackrel{\text{def}}{=} \int_H |f'(u_n) D_i u_n - f'(u) D_i u|\, dx$$

$$\le \sup_{x \in \mathbb{R}} |f'(x)| \int_H |D_i u_n - D_i u|\, dx + \int_H |f'(u_n) - f'(u)||D_i u|\, dx.$$

Since $u_n \to u$ in $L_1(H)$ as $n \to \infty$, there exists a subsequence, again denoted by (u_n), such that $u_n(x) \to u(x)$ as $n \to \infty$ for almost all $x \in H$, and hence

$$f'(u_n(x)) \to f'(u(x)) \qquad \text{as} \quad n \to \infty \qquad \text{for almost all} \quad x \in H.$$

This implies $A_n \to 0$ as $n \to \infty$, by the principle of majorized convergence $A_2(19)$. Hence

$$f \circ u_n \to f \circ u \quad \text{in } L_1(H) \qquad \text{as} \quad n \to \infty,$$

$$D_i(f \circ u_n) \to (f' \circ u) D_i u \quad \text{in } L_1(H) \qquad \text{as} \quad n \to \infty.$$

By Problem 21.3c, $D_i(f \circ u) = (f' \circ u) D_i u$.

21.3f. *Special generalized derivatives.* Let $u \in L_{1,\mathrm{loc}}(G)$, and suppose that, for fixed $i = 1, \ldots, N$, the generalized derivative $D_i u$ exists on G, where $D_i u \in L_{1,\mathrm{loc}}(G)$. We set

$$u^+(x) = \max(u(x), 0), \qquad u^-(x) = \min(u(x), 0),$$

and $|u|(x) = |u(x)|$. Show that the generalized derivatives $D_i u^\pm$ and $D_i|u|$ exist on G, and we have the following relations:

$$(D_i u^+)(x) = \begin{cases} D_i u(x) & \text{if } u(x) > 0, \\ 0 & \text{if } u(x) \le 0; \end{cases}$$

$$(D_i u^-)(x) = \begin{cases} 0 & \text{if } u(x) \ge 0, \\ D_i u(x) & \text{if } u(x) < 0; \end{cases}$$

$$(D_i|u|)(x) = \begin{cases} \pm D_i u(x) & \text{if } u(x) \gtrless 0, \\ 0 & \text{if } u(x) = 0. \end{cases}$$

Solution: For $\varepsilon > 0$, define

$$f_\varepsilon(u) = \begin{cases} (u^2 + \varepsilon^2)^{1/2} - \varepsilon & \text{if } u > 0, \\ 0 & \text{if } u \le 0. \end{cases}$$

By Problem 21.3e, for all $\varphi \in C_0^\infty(G)$,

$$\int_G f_\varepsilon(u) D_i\varphi \, dx = -\int_{G_+} u D_i u (u^2 + \varepsilon^2)^{-1/2} \varphi \, dx,$$

where $G_+ = \{x \in G : u(x) > 0\}$. Letting $\varepsilon \to 0$, we get

$$\int_G u^+ D_i \varphi \, dx = -\int_{G_+} (D_i u)\varphi \, dx.$$

This implies the claimed relation for $D_i u^+$. The other results follow from $u^- = -(-u)^+$ and $|u| = u^+ - u^-$.

21.3g. *A special property of the Dirichlet integral.* Let $u \in W_2^1(G)$, and let $v = |u|$. Show that $v \in W_2^1(G)$ and

$$\int_G |Dv|^2 \, dx \le \int_G |Du|^2 \, dx,$$

where $|Du|^2 = \sum_{i=1}^N (D_i u)^2$.

Solution: This follows immediately from Problem 21.3f.

21.3h. *Iterated generalized derivatives.* Let α, β be multi-indices. Show that the generalized derivative $D^\beta u$ exists on G such that

$$D^{\alpha+\beta} u = D^\alpha(D^\beta u) \quad \text{on } G,$$

provided the following hold:

(i) The generalized derivative $D^\beta u$ of u exists, where $u, D^\beta u \in L_{1,\mathrm{loc}}(G)$.

(ii) The generalized derivative $D^\alpha(D^\beta u)$ of $D^\beta u$ exists on G, where $D^\alpha(D^\beta u) \in L_{1,\mathrm{loc}}(G)$.

Solution: Let $v = D^\beta u$. For all $\varphi \in C_0^\infty(G)$,

$$\int_G uD^{\alpha+\beta}\varphi\,dx = \int_G uD^\beta(D^\alpha\varphi)\,dx$$

$$= (-1)^{|\beta|}\int_G vD^\alpha\varphi\,dx$$

$$= (-1)^{|\alpha|+|\beta|}\int_G (D^\alpha v)\varphi\,dx.$$

21.4. *A convergence theorem.* Let G be a nonempty open subset of $\mathbb{R}^N$, $N \geq 1$. Let $m = 1, 2, \ldots, 1 < p < \infty$, and $p^{-1} + q^{-1} = 1$. Show that

$$u_n \rightharpoonup u \quad \text{in } W_p^m(G) \qquad \text{as} \quad n \to \infty$$

implies

$$D^\alpha u_n \rightharpoonup D^\alpha u \quad \text{in } L_p(G) \qquad \text{as} \quad n \to \infty \quad \text{for all} \quad \alpha\colon |\alpha| \leq m.$$

Solution: Let $X = W_p^m(G)$. For fixed $v \in L_q(G)$, define

$$b(u) = \int_G (D^\alpha u)v\,dx \qquad \text{for all} \quad u \in X.$$

By the Hölder inequality,

$$|b(u)| \leq \|D^\alpha u\|_p\|v\|_q \leq \|u\|_X\|v\|_q.$$

Hence $b \in X^*$. Consequently, as $n \to \infty$, $u_n \rightharpoonup u$ in X implies $b(u_n) \to b(u)$ and hence $D^\alpha u_n \rightharpoonup D^\alpha u$ in $L_p(G)$ (cf. A_2(34b)).

21.5. *Sobolev spaces and Banach algebras.* Let $G = \mathbb{R}^N$ or let G be a bounded region in $\mathbb{R}^N$ such that $\partial G \in C^{0,1}$ and $N \geq 1$. Show that the Sobolev space

$$X = W_p^k(G), \qquad k = 1, 2, \ldots, \quad 1 \leq p < \infty,$$

forms a generalized Banach algebra in the case where $pk > N$.

This means that $u, v \in X$ implies $uv \in X$ and

$$\|uv\| \leq \text{const}\,\|u\|\,\|v\| \qquad \text{for all} \quad u, v \in X.$$

Use the embedding theorems from A_2(45). In particular, use the continuity of the embedding $W_p^k(G) \subseteq C(\bar{G})$ for $pk > N$.

Solution: Let $u, v \in X$. The product rule shows that the derivative

$$D^\alpha(uv), \qquad |\alpha| \leq k,$$

consists of finitely many summands of the form

$$D^\beta u D^\gamma v, \qquad |\beta| = j, \quad |\gamma| \leq k - j, \quad j = 0, \ldots, k.$$

We have to show that

$$\|D^\beta u D^\gamma v\|_p \leq \text{const}\,\|u\|_{k,p}\|v\|_{k,p}.$$

(I) Let $j = 0$. Then the continuous embedding $W_p^k(G) \subseteq C(\bar{G})$ implies $D^\beta u \in C(\bar{G})$ and

$$\|D^\beta u\|_\infty \equiv \sup_{x\in\bar{G}} |D^\beta u(x)| \leq \text{const}\,\|u\|_{k,p}.$$

Hence

$$\|D^\beta u D^\gamma v\|_p \le \text{const}\,\|D^\beta u\|_\infty \|D^\gamma v\|_p$$
$$\le \text{const}\,\|u\|_{k,p}\|v\|_{k,p}.$$

The case $j = k$ proceeds analogously.

(II) Let $1 \le j \le k - 1$. The continuous embeddings

$$W_p^k(G) \subseteq W_{rp}^j(G), \qquad W_p^k(G) \subseteq W_{sp}^{k-j}(G),$$

with $r = k/j$, $r^{-1} + s^{-1} = 1$, and the Hölder inequality imply

$$\int_G |D^\beta u D^\gamma v|^p\,dx \le \left(\int_G |D^\beta u|^{rp}\,dx\right)^{1/r}\left(\int_G |D^\gamma v|^{sp}\,dx\right)^{1/s}.$$

Hence

$$\|D^\beta u D^\gamma v\|_p \le \|D^\beta u\|_{rp}\|D^\gamma v\|_{sp}$$
$$\le \text{const}\,\|u\|_{k,p}\|v\|_{k,p}.$$

21.6. *A density theorem.* Let $m = 0, 1, \ldots, 1 \le p < \infty, N \ge 1$. Show that the set $C_0^\infty(\mathbb{R}^N)$ is dense in $W_p^m(\mathbb{R}^N)$. More precisely, the following is true.

(i) If $u \in W_p^m(\mathbb{R}^N)$, then there exists a sequence (u_n) in $C_0^\infty(\mathbb{R}^N)$ such that

$$\|u_n - u\|_{m,p} \to 0 \qquad \text{as} \quad n \to \infty.$$

(ii) If, in addition, $u \in W_q^k(\mathbb{R}^N)$ for fixed $k = 0, 1, \ldots, 1 \le q \le \infty$, then

$$\|u_n\|_{k,q} \le \text{const}\,\|u\|_{k,q} \qquad \text{for all } n,$$

and

$$D^\alpha u_n(x) \to D^\alpha u(x) \qquad \text{as} \quad n \to \infty \quad \text{for almost all} \quad x \in \mathbb{R}^N$$

in the case where $|\alpha| \le k$.

Solution:

(I) If we set

$$v_n = S_{1/n}u, \qquad n = 1, 2, \ldots,$$

then $v_n \in C^\infty(\mathbb{R}^N)$ for all n, and (i), (ii) are valid for (v_n), according to the properties of the smoothing operator S_ε in Section 18.14.

(II) Let $v \in C^\infty(\mathbb{R}^N) \cap W_p^m(\mathbb{R}^N) \cap W_q^k(\mathbb{R}^N)$. We choose a C^∞-function

$$\varphi\colon \mathbb{R}^N \to \mathbb{R} \qquad \text{such that} \qquad 0 \le \varphi(x) \le 1 \quad \text{on } \mathbb{R}^N$$

and

$$\varphi(x) = \begin{cases} 1 & \text{if } |x| \le 1, \\ 0 & \text{if } |x| \ge 2. \end{cases}$$

If we set

$$w_n(x) = v(x)\varphi(x/n),$$

then w_n belongs to $C_0^\infty(\mathbb{R}^N)$ for all n and

$$\|w_n - v\|_{m,p} \to 0 \qquad \text{as} \quad n \to \infty,$$
$$\|w_n\|_{k,q} \le \text{const}\,\|v\|_{k,q} \qquad \text{for all } n.$$

This follows from

$$|D^\alpha w_n(x)| \le \text{const} \sum_{|\gamma| \le |\alpha|} |D^\gamma v(x)| \qquad \text{for all} \quad x \in \mathbb{R}^N,$$

and hence

$$\int_{\mathbb{R}^N} |D^\alpha w_n - D^\alpha v|^p \, dx = \int_{|x| \geq n} |D^\alpha w_n - D^\alpha v|^p \, dx$$

$$\leq \text{const} \int_{|x| \geq n} \sum_{|\gamma| \leq |\alpha|} |D^\gamma v(x)|^p \, dx \to 0 \quad \text{as } n \to \infty.$$

(III) Putting (I) and (II) together, we obtain the assertion. Note that $\varphi(x/n) \to 1$ as $n \to \infty$.

21.7. *An extension principle for smooth functions.*

21.7a. *Main idea.* Let $f\colon [0, a[\to \mathbb{R}$ be C^m with $m \geq 0$ and $0 < a \leq \infty$. We set

$$g(x) = \begin{cases} f(x) & \text{if } x \geq 0, \\ \sum_{j=1}^{m+1} c_j f(-x/j) & \text{if } x < 0, \end{cases}$$

where the numbers c_j satisfy the linear system

$$\sum_{j=1}^{m+1} c_j(-1/j)^k = 1, \qquad k = 0, \ldots, m. \tag{93}$$

Show that $g\colon]-a, a[\to \mathbb{R}$ is a C^m-extension of f.

Solution: The determinant of (93) is a Vandermonde determinant, and hence it is different from zero. By (93), $g^{(k)}(0) = f^{(k)}(0)$ for $k = 0, 1, \ldots, m$.

21.7b. Let $1 \leq p \leq \infty$, $m = 0, 1, \ldots$, $N \geq 1$. Let G be a bounded region in $\mathbb{R}^N$ with $\partial G \in C^m$. Choose an open set G_0 such that $\bar{G} \subset G_0$ (e.g., $G_0 = \mathbb{R}^N$).

Show that given any function $u \in C^m(\bar{G})$, there exists a function $v \in C^m(G_0)$ such that

$$v = u \quad \text{on } G,$$

$\operatorname{supp} v \subset G_0$, and

$$\|v\|_{m,p,G_0} \leq \text{const} \, \|u\|_{m,p,G}.$$

The constant depends on G, G_0, m, p.

Hint: Consider Figure 21.8(a). Use a partition of unity. Map neighborhoods of boundary points locally to a half-space as indicated in Figure 21.8. Extend the corresponding functions from the half-space to the entire space $\mathbb{R}^N$ via Problem 21.7a. Transform the extended function back to G_0. Cf. Friedman (1969, M), p. 10.

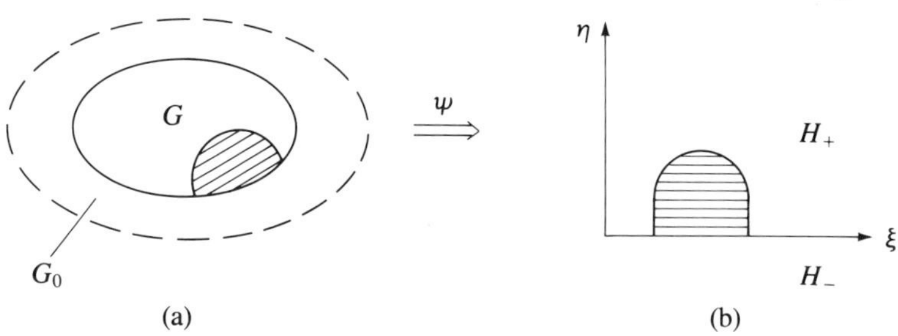

Figure 21.8

21.8. *Extension operator for Sobolev spaces.*

21.8a. *Smooth boundaries.* Let $m = 0, 1, \ldots, 1 \le p < \infty$, and $N \ge 1$. Let G be an open half-space or let G be a bounded region in $\mathbb{R}^N$ with $\partial G \in C^m$. Show that there exists a linear continuous operator

$$E\colon W_p^m(G) \to W_p^m(\mathbb{R}^N) \tag{94}$$

such that the function $Eu\colon \mathbb{R}^N \to \mathbb{R}$ is an extension of $u\colon G \to \mathbb{R}$.

Hint: Construct the function Eu via Problem 21.7b, and use the density of $C^m(\bar{G})$ in $W_p^m(G)$.

21.8b.* *Universal extension operator.* Let $m = 0, 1, \ldots, 1 \le p \le \infty$, and $N \ge 1$. Let G be an open half-space or let G be a bounded region in $\mathbb{R}^N$ with $\partial G \in C^{0,1}$. Show that the assertion of Problem 21.8a remains true, where the construction of the operator E is *independent* of m and p. Moreover, the function Eu is C^∞ on $\mathbb{R}^N - \bar{G}$.

Hint: Cf. Stein (1970, M), Chapter 6, and Burenkov (1976), (1985). The proof is based on a modification of the method in Problem 21.7a.

21.9. *An important equivalent norm on $L_2(G)$.* Let G be a bounded region in $\mathbb{R}^N$ with $N \ge 1$ and $\partial G \in C^{0,1}$. We set

$$B = \{u \in \mathring{W}_2^1(G)\colon \|u\|_{1,2} \le 1\}$$

and

$$\begin{aligned} \|u\|_{-1,2} &= \sup_{v \in B} \left| \int_G uv\,dx \right|, \\ \|D_i u\|_{-1,2} &= \sup_{v \in B} \left| \int_G u D_i v\,dx \right| \end{aligned} \tag{95}$$

as well as

$$\|u\|_2^* = \|u\|_{-1,2} + \sum_{i=1}^N \|D_i u\|_{-1,2}.$$

Use the Fourier transformation from $A_2(74)$ in order to show that $\|u\|_2^*$ is an equivalent norm on $L_2(G)$, i.e., there exist positive constants c and d with

$$c\|u\|_2 \le \|u\|_2^* \le d\|u\|_2 \qquad \text{for all} \quad u \in L_2(G).$$

In Chapter 65 we shall use this inequality in order to give a simple proof for the fundamental Korn inequality in elasticity theory.

Solution: The estimate $\|u\|_2^* \le d\|u\|_2$ is obvious. Thus, we only need to prove that

$$c\|u\|_2 \le \|u\|_2^* \qquad \text{for all} \quad u \in L_2(G). \tag{96}$$

The key to the proof is the simple identity

$$\frac{1}{1 + |y|^2} + \sum_{i=1}^N \frac{|\eta_i|^2}{1 + |y|^2} = 1. \tag{97}$$

(I) Proof of (96) for $G = \mathbb{R}^N$. Let $\int = \int_{\mathbb{R}^N}$. Define

$$\|u\|_{1,2}^* = \left(\int (1 + |y|^2)|\hat{u}|^2\,dy \right)^{1/2},$$

where $\hat{u}$ is the *Fourier transform* of u and $y = (\eta_1, \ldots, \eta_N)$. Note that $\mathring{W}_2^1(\mathbb{R}^N) = W_2^1(\mathbb{R}^N)$. By $A_2(74)$, we have:

$$u \in W_2^1(\mathbb{R}^N) \qquad \text{iff} \qquad u \in L_2(\mathbb{R}^N) \qquad \text{and} \qquad \|u\|_{1,2}^* < \infty.$$

Moreover, $\|\cdot\|_{1,2}^*$ is an equivalent norm on $W_2^1(\mathbb{R}^N)$. Thus, it is sufficient to prove (96) with B in (95) replaced by the ball

$$B = \{u \in W_2^1(\mathbb{R}^N) \colon \|u\|_{1,2}^* \le 1\}.$$

Using the Fourier transform, we obtain that

$$\begin{aligned}\|D_i u\|_{-1,2} &= \sup_{v \in B} \left| \int u D_i v \, dx \right| = \sup_{v \in B} \left| \int \hat{u} \eta_i \bar{\hat{v}} \, dy \right| \\ &= \sup_{v \in B} \left| \int \frac{\hat{u}\eta_i}{(1 + |y|^2)^{1/2}} (1 + |y|^2)^{1/2} \bar{\hat{v}} \, dy \right| \\ &= \sup_{\|w\|_2 \le 1} \left| \int \frac{\hat{u}\eta_i}{(1 + |y|^2)^{1/2}} \bar{w} \, dy \right| \\ &= \left(\int \frac{|\hat{u}|^2 |\eta_i|^2 \, dy}{1 + |y|^2} \right)^{1/2}.\end{aligned}$$

Similarly, we get

$$\|u\|_{-1,2} = \sup_{v \in B} \left| \int uv \, dx \right| = \left(\int \frac{|\hat{u}|^2 \, dy}{1 + |y|^2} \right)^{1/2}.$$

By (97),

$$\|u\|_{-1,2}^2 + \sum_i \|D_i u\|_{-1,2}^2 = \int |\hat{u}|^2 \, dy = \|u\|_2^2.$$

This yields (96).

(II) Proof of (96) for G = half-space H_+. Let

$$H_\pm = \{(\xi, \eta) \in \mathbb{R}^N \colon \eta \gtrless 0\},$$

where $\xi = (\eta_1, \ldots, \eta_{N-1})$ and $\eta = \eta_N$ (cf. Fig. 21.8 above). We set

$$w(\xi, \eta) = v(\xi, \eta) - 3v(\xi, -\eta) + 2v\left(\xi, -\frac{\eta}{2}\right)$$

and $B_0 = \{v \in C_0^\infty(\mathbb{R}^N) \colon \|v\|_{1,2,\mathbb{R}^N} \le 1\}$. Then, for all $v \in B_0$, we get $w(\xi, 0) \equiv 0$, and hence

$$w \in \mathring{W}_2^1(\mathbb{R}_+) \qquad \text{and} \qquad \sup_{v \in B_0} \|w\|_{1,2,H_+} < \infty.$$

Now let $u \in L_2(H_+)$. We extend the function u to $\mathbb{R}^N$ by letting

$$u(\xi, \eta) = -3u(\xi, -\eta) + 4u(\xi, -2\eta), \qquad \eta < 0.$$

We want to prove that $\|u\|_{2,\mathbb{R}^N}^* \le C\|u\|_{2,H_+}^*$, i.e.,

$$\|u\|_{-1,2,\mathbb{R}^N} + \sum_{i=1}^N \|D_i u\|_{-1,2,\mathbb{R}^N} \le C\left(\|u\|_{-1,2,H_+} + \sum_{i=1}^N \|D_i u\|_{-1,2,H_+} \right). \tag{98}$$

Then, by (I), we obtain (96) for $G = H_+$.

First, let $i = 1, \ldots, N-1$. Then, a simple integral substitution yields

$$\begin{aligned}\|D_i u\|_{-1,2,\mathbb{R}^N} &= \sup_{v \in B_0} \left| \int_{H_+} u D_i v \, dx + \int_{H_-} u D_i v \, dx \right| \\ &= \sup_{v \in B_0} \left| \int_{H_+} u D_i w \, dx \right| \leq \operatorname{const} \|u\|_{-1,2,H_+}.\end{aligned}$$

The other estimates in (98) are obtained analogously.

(III) Proof of (96) for bounded regions G with $\partial G \in C^{0,1}$. We use a covering of G and a corresponding partition of unity $\{\varphi_i\}$. Let $u \in L_2(G)$. We set $u_i = u\varphi_i$. If $\operatorname{supp} \varphi_i \subset G$, then we use (I). Furthermore, if $\operatorname{supp} \varphi_i \cap \partial G \neq \emptyset$, then we use a local rectification of the corresponding boundary part as indicated in Figure 21.8 above, and we apply (II). The simple technical details can be found in Nečas (1967, M).

21.10. *Two properties of the space $L_\infty(G)$.*

21.10a. Let G be a nonempty open set in $\mathbb{R}^N$, $N \geq 1$. Let $u_n \in L_\infty(G)$ for all n such that

$$u_n(x) \to u(x) \quad \text{as} \quad n \to \infty \quad \text{for almost all} \quad x \in G$$

and $\sup_n \|u_n\|_\infty < \infty$. Show that $u \in L_\infty(G)$ and

$$\|u\|_\infty \leq \varliminf_{n\to\infty} \|u_n\|_\infty.$$

Solution: Set $X = L_1(G)$ and

$$f_n(v) = \int_G v u_n \, dx, \qquad f(v) = \int_G v u \, dx \qquad \text{for all} \quad v \in X.$$

Then $f_n, f \in X^*$ and $\|f_n\| = \|u_n\|_\infty$, $\|f\| = \|u\|_\infty$. Moreover,

$$f_n(v) \to f(v) \qquad \text{as} \quad n \to \infty \quad \text{for all} \quad v \in C_0^\infty(G).$$

Since $C_0^\infty(G)$ is dense in X and $\sup_n \|f_n\| < \infty$, we obtain that $f_n \overset{*}{\rightharpoonup} f$ as $n \to \infty$ and hence $\|f\| \leq \varliminf_{n\to\infty} \|f_n\|$, by Proposition 21.26.

21.10b. Let G be a nonempty measurable set in $\mathbb{R}^N$, $N \geq 1$ with meas $G < \infty$. Show that

$$\lim_{p\to\infty} \|u\|_p = \|u\|_\infty \qquad \text{for all} \quad u \in L_\infty(G).$$

Hint: Cf. Kufner (1977, M), p. 84.

21.11. *Properties of integral operators.* We consider the integral operator

$$(Ku)(x) = \int_M k(x, y) u(y) \, dy. \tag{99}$$

Let M be a nonempty measurable set in $\mathbb{R}^N$, $N \geq 1$, and let $k \in L_2(M \times M)$, i.e., the function $k \colon M \times M \to \mathbb{R}$ is measurable and

$$\int_{M\times M} k(x, y)^2 \, dx \, dy < \infty. \tag{100}$$

21.11a. Show that the operator $K \colon L_2(M) \to L_2(M)$ is linear and continuous with

$$\|K\|^2 \leq \int_{M\times M} k(x, y)^2 \, dx \, dy.$$

Solution: By the Fubini theorem $A_2(23)$, it follows from (100) that

$$\int_M k(x, y)^2 \, dy < \infty$$

for almost all $x \in M$. Set $w = Ku$, where $u \in L_2(M)$. By the Hölder inequality,

$$w(x)^2 \le \int_M k(x, y)^2 \, dy \int_M u(y)^2 \, dy$$

for almost all $x \in M$. Hence $Ku \in L_2(M)$, and

$$\|Ku\|_2^2 = \int_M w(x)^2 \, dx \le \left(\int_{M \times M} k(x, y)^2 \, dx \, dy \right) \|u\|_2^2.$$

21.11b. Show that the adjoint operator $K^*: L_2(M) \to L_2(M)$ corresponds to the adjoint kernel $k^*(x, y) = k(y, x)$.

Solution: Set

$$(Lv)(x) = \int_M k^*(x, y)v(y) \, dy.$$

Then, for all $u, v \in L_2(M)$,

$$(u|Lv) = \int_{M \times M} v(y)k(y, x)u(x) \, dx \, dy = (Ku|v).$$

Hence $L = K^*$.

21.11c. Show that $K: L_2(M) \to L_2(M)$ is compact.

Solution: The linear hull of the set of all the functions

$$c(x, y) = a(x)b(y), \qquad a, b \in L_2(M),$$

is dense in $L_2(M \times M)$. This follows from the density of $C_0^\infty(\mathbb{R}^{2N})$ in $L_2(\mathbb{R}^{2N})$, and from the Weierstrass approximation theorem $A_1(23e)$. Thus, for each $\varepsilon > 0$, there are functions $a_i, b_i \in L_2(M)$ such that

$$\int_{M \times M} \left[k(x, y) - \sum_{i=1}^n a_i(x)b_i(y) \right]^2 dx \, dy < \varepsilon. \tag{101}$$

Set

$$(C_\varepsilon u)(x) = \int_M \sum_{i=1}^n a_i(x)b_i(y)u(y) \, dy$$

for all $u \in L_2(M)$. Then, $C_\varepsilon: L_2(M) \to L_2(M)$ is linear and continuous. From

$$C_\varepsilon u = \sum_{i=1}^n a_i \left(\int_M b_i(y)u(y) \, dy \right)$$

it follows that $\dim R(C_\varepsilon) < \infty$. By (101),

$$\|K - C_\varepsilon\|^2 < \varepsilon.$$

Now, the compactness of K follows from Example 21.45.

21.12. *An existence principle.* Let X and Y be B-spaces over $\mathbb{K}$, and let $A: X \to Y$ be a linear and continuous operator. Suppose that the equation

$$Au = b, \qquad u \in X, \tag{102}$$

has approximate solutions, i.e., there are constants $C > 0$ and $q \in [0, 1[$ such that, for each $b \in Y$, there exists a $u(b) \in X$ with

$$\|Au(b) - b\| \leq q\|b\|, \qquad \|u(b)\| \leq C\|b\|.$$

Show that, for each $b \in Y$, equation (102) has a solution u with $\|u\| \leq C(1-q)^{-1}\|b\|$.

Solution: Let $b_1 = b$. By assumption, there exists a u_1 with

$$\|Au_1 - b_1\| \leq q\|b_1\|, \qquad \|u_1\| \leq C\|b_1\|.$$

Set $b_2 = b_1 - Au_1$. Then we find a u_2 with

$$\|Au_2 - b_2\| \leq q\|b_2\| \leq q^2\|b_1\|,$$
$$\|u_2\| \leq C\|b_2\| \leq Cq\|b_1\|.$$

This way we get two sequences (u_n) and (b_n) with

$$b_{n+1} = b_n - Au_n, \qquad \|b_n\| \leq q^{n-1}\|b\|, \qquad \|u_n\| \leq Cq^{n-1}\|b\|.$$

Hence

$$b_{n+1} = b - A(u_1 + \cdots + u_n). \tag{103}$$

Since $\sum_n \|u_n\| \leq C(1-q)^{-1}\|b\|$, the series

$$u = \sum_{n=1}^{\infty} u_n$$

is convergent with $\|u\| \leq C(1-q)^{-1}\|b\|$. Letting $n \to \infty$ in (103), we obtain $0 = b - Au$.

21.13. *The approximation theorem of Kantorovič.* We consider the operator equation

$$u + Lu = b, \qquad u \in X, \tag{104}$$

together with the corresponding approximate equation

$$u_n + L_n u_n = P_n b, \qquad u_n \in X_n, \quad n = 1, 2, \ldots. \tag{105}$$

For all n, we make the following assumptions:

(i) X_n is a linear subspace of the B-space X with $\dim X_n = n$, and $P_n: X \to X_n$ is a projection operator onto X_n.

(ii) The operators $L: X \to X$ and $L_n: X_n \to X_n$ are linear and continuous, and $I + L: X \to X$ is bijective.

(iii) There exists a constant d_n with

$$\operatorname{dist}(Lu, X_n) \leq d_n\|u\| \qquad \text{for all} \quad u \in X.$$

(iv) Set $c_n = \|P_n L - L_n\|_{X_n}$ and suppose that, as $n \to \infty$,

$$c_n \to 0, \qquad \|P_n\| d_n \to 0, \qquad \|P_n\| \operatorname{dist}(b, X_n) \to 0,$$

for all $b \in X$.

Show that equation (104) is uniquely approximation-solvable. Moreover, one has the following error estimates:

$$\begin{aligned} \|u_n - u\| &\leq \text{const}(c_n + \|P_n\|(d_n + \operatorname{dist}(b, X_n))), \\ \|u_n - u\| &\leq \text{const}(c_n + \|P_n\| \operatorname{dist}(u, X_n)). \end{aligned} \tag{106}$$

The condition $c_n \to 0$ as $n \to \infty$ describes the consistency of this approximation method.

Hint: Use Problem 21.12. Cf. Kantorovič and Akilov (1964, M), Chapter 14.

In applications, L_n can be a discretization of the operator L (e.g., L is an integral operator and L_n is a sum operator obtained by a quadrature formula). The terms on the right-hand side of (106) can be estimated by using results of approximation theory.

21.14. *Collocation method as a special projection method.* We consider the boundary value problem

$$\begin{aligned} u''(x) + q(x)u(x) &= f(x), \qquad -1 < x < 1, \\ u(-1) = u(1) &= 0, \end{aligned} \tag{107}$$

where f and q are continuous on $[-1, 1]$. The idea of the so-called collocation method is to make the trial

$$u_n(x) = (1 - x^2)(c_0 + c_1 x + \cdots + c_{n-1} x^{n-1}) \tag{108}$$

and to use the partition $-1 < a_1 < a_2 < \cdots < a_n < 1$. The unknown coefficients $c_0, \ldots, c_{n-1}$ are determined by the n equations

$$u_n''(a_i) + q(a_i)u_n(a_i) = f(a_i), \qquad i = 1, \ldots, n. \tag{109}$$

21.14a. Show that this method is a special projection method.

Solution: We set

$$X = \{u \in C^2[-1, 1]: u(\pm 1) = 0\}, \qquad Y = C[-1, 1].$$

Let X_n be the set of all the polynomials of the form (108). Finally, let Q_n: $Y \to Y$ be the projection operator which assigns to the function f the unique interpolation polynomial $Q_n f$ of degree $\leq n - 1$ with the property

$$(Q_n f)(a_i) = f(a_i), \qquad i = 1, \ldots, n.$$

Then, (109) is equivalent to the operator equation

$$Au_n + Q_n(qu_n) = Q_n f, \qquad u_n \in X_n, \tag{110}$$

where the operator $A: X \to Y$ is given by $Au = u''$. The compact inverse operator A^{-1}: $Y \to X$ has the form

$$(A^{-1}f)(x) = \int_{-1}^{1} G(x, y) f(y)\, dy,$$

where G denotes the continuous Green function to the following boundary value problem:

$$u'' = f \quad \text{on }]-1, 1[, \qquad u(-1) = u(1) = 0.$$

From (110) we obtain

$$u_n + A^{-1}Q_n(qu_n) = A^{-1}Q_n f, \qquad u_n \in X_n. \tag{111}$$

The original equation (107) can be written as

$$u + A^{-1}(qu) = A^{-1}f, \qquad u \in X. \tag{111*}$$

21.14b.* Let $a_1, \ldots, a_n$ be the so-called Čebyšev interpolation points, i.e.,

$$a_{i+1} = \cos\frac{2i+1}{2n}\pi, \qquad i = 0, 1, \ldots, n-1.$$

Then

$$\|Q_n\| \leq A \ln n + B \qquad \text{for all } n,$$

where A and B are positive constants (cf. Natanson (1955, M), p. 390). Let f, $q \in C^1[-1, 1]$. Show that

$$\|u - u_n\|_X \leq \text{const}\frac{\ln n}{n} \qquad \text{for all } n,$$

i.e., the collocation method converges to the solution u of the original problem (107) as $n \to \infty$. The convergence of the collocation method is faster, the smoother the functions f and q are.

Hint: Use a modification of Problem 21.13, and use the approximation theorem of Jackson for polynomials from Example 21.48. Cf. Kantorovič and Akilov (1964, M), Chapter 14.

21.15. *General projection methods in nonseparable* B*-spaces.* We consider the operator equation

$$Au = b, \qquad u \in X, \quad b \in Y, \tag{112}$$

together with the general projection method

$$P_n A Q_n u = P_n b,$$

where $P_n\colon D(P_n) \subseteq Y \to Y$ and $Q_n\colon D(Q_n) \subseteq X \to X$ are (not necessarily continuous) projections, i.e., $P_n^2 = P_n$ and $Q_n^2 = Q_n$ with finite-dimensional projection spaces. If the spaces X and Y are *not* separable, then the projection method does not converge for each $b \in Y$. Then it is important to describe the set of all b for which convergence occurs. In this connection, a general theory may be found in Prößdorf and Silbermann (1977, L).

For example, if (112) represents a singular integral equation, then X and Y must be Hölder spaces in the smooth case. But note that Hölder spaces are not separable.

21.16. *Generalized Schwarz inequality.* Let $a\colon X \times X \to \mathbb{R}$ be a bilinear positive symmetric functional on the real linear space X. Show that a satisfies the inequality

$$a(u, v)^2 \leq a(u, u)a(v, v) \qquad \text{for all} \quad u, v \in X.$$

Solution: Use the same argument as in the proof of the Schwarz inequality in the Introduction on page 8. Replace the scalar product $(u|v)$ with $a(u, v)$.

References to the Literature

Classical works on Sobolev spaces: Rellich (1930), Friedrichs (1934), Sobolev (1936), (1950, M).

Standard work on Sobolev spaces: Adams (1975, M).

Sobolev spaces: Nečas (1967, M) and Kufner, John, and Fučik (1977, M) (introduction), Friedman (1969, M), Wloka (1982, M) (Sobolev spaces and Fourier transformation), Triebel (1983, M), Mazja (1985, M).

Inequalities of Gagliardo–Nirenberg: Gagliardo (1959), Nirenberg (1966), Friedman (1969, M), Adams (1975, M).

Moser-type calculus: Moser (1966), Klainerman and Majda (1981), Majda (1984).

Extension operators for Sobolev spaces: Stein (1970, M), Burenkov (1976), (1985), Wloka (1982, M).

Approximative solvability of linear operator equations: Kantorovič (1948) (classical work), Kantorovič and Akilov (1964, M), Anselone (1971, M), Krasnoselskii (1973, M), Vainikko (1976, L), Prößdorf and Silbermann (1977, L), Fenyö and Stolle (1982, M), Vol. 4.

Finite elements: Ciarlet (1977, M).

Introduction to approximation theory: Natanson (1955, M), Cheney (1967, M), Laurent (1972, M), Braess (1986, M), Ditzian and Totik (1987, M).

Weak and weak* convergence: Dunford and Schwartz (1958, M), Vol. 1, Yosida (1965, M).

(Cf. also the References to the Literature in the Appendix.)

CHAPTER 22

Hilbert Space Methods and Linear Elliptic Differential Equations

> The role of functional analysis has been decisive exactly in connection with classical problems. Almost all problems are on the applications, where functional analysis enables one to focus on a specific set of concrete analytical tasks and organize material in a clear and transparent form, so that you know what the difficulties are.
>
> Concrete and functional analysis exist today in an inextricable symbiosis. When someone writes down a system of axioms, no one is going to take them seriously unless they arise from some intuitive body of concrete subject matter that you would really want to study, and about which you really want to find out something.
>
> Felix E. Browder (1975)

> The perfection of mathematical beauty is such that whatsoever is most beautiful and regular is also found to be most useful and excellent.
>
> D'Arcy W. Thompson (1917)

Generalized boundary value problems for linear and quasi-linear elliptic differential equations lead to problems of the following form:

$$(\mathrm{P}_a) \qquad a(u, v) = b(v) \qquad \text{for all} \quad v \in X.$$

For given $b \in X^*$, we seek $u \in X$. Here, X is a real B-space. In this connection we want to explain the following two principles (1a) and (1b) below which will be used frequently. Our goal is the following equivalent operator equation:

$$(\mathrm{P}_A) \qquad Au = b, \qquad u \in X,$$

with the linear or nonlinear operator $A: X \to X^*$.

The Equivalent Operator Equation

Suppose that the map $a: X \times X \to \mathbb{R}$ is linear with respect to the *second* argument, and suppose that, for each $u \in X$, there exists a constant $C(u)$ such that

$$|a(u, v)| \leq C(u)\|v\| \qquad \text{for all} \quad u, v \in X.$$

Then, for each u, there exists a linear continuous functional on X denoted by Au such that

$$\langle Au, v\rangle = a(u, v) \qquad \textit{for all} \quad v \in X. \tag{1a}$$

This way we obtain the operator $A: X \to X^*$. The original problem (P_a) is equivalent to

$$\langle Au, v\rangle = \langle b, v\rangle \qquad \text{for all} \quad v \in X,$$

i.e., problem (P_a) is equivalent to (P_A).

Let $a: X \times X \to \mathbb{R}$ be *bilinear* and bounded. Then the operator $A: X \to X^*$ is linear and continuous. Moreover, a is positive (resp. strictly positive, strongly positive, symmetric) iff A has the same property, by Definition 21.32. Note that for the linear operator $A: X \to X^*$ the following hold:

(α) A is *monotone* iff A is positive, i.e., $\langle Au, u\rangle \geq 0$ for all $u \in X$.
(β) A is *strictly* monotone iff A is strictly positive, i.e., $\langle Au, u\rangle > 0$ for all nonzero $u \in X$.
(γ) A is *strongly* monotone iff A is strongly positive, i.e., there exists a $c > 0$ such that $\langle Au, u\rangle \geq c\|u\|^2$ for all $u \in X$.

If, in addition, X is reflexive, then:

a is compact iff A is compact
iff A is strongly continuous.

This follows from Sections 21.9 and 21.10.

Functions as Functionals on Sobolev Spaces

The right-hand side of equation (P_a), i.e., the functional $b \in X^*$ corresponds frequently to the right-hand side of elliptic differential equations. In this connection, we will use the following result.

Let $f \in L_q(G)$ with $1 < q < \infty$, where G is a bounded region in $\mathbb{R}^N$, $N \geq 1$. We set

$$b(v) = \int_G f(x)v(x)\,dx \qquad \text{for all} \quad v \in W_p^m(G),$$

where $m = 0, 1, \dots$ and $p^{-1} + q^{-1} = 1$. Recall that $W_p^m(G) = L_p(G)$ for $m = 0$.

Then, b is a linear continuous functional on the Sobolev space $W_p^m(G)$.

This follows immediately from the Hölder inequality:

$$\begin{aligned} |b(v)| &\leq \left(\int_G |f|^q\,dx\right)^{1/q}\left(\int_G |v|^p\,dx\right)^{1/p} \\ &\leq \|f\|_q\|v\|_{m,p} \qquad \text{for all} \quad v \in W_p^m(G). \end{aligned} \tag{1b}$$

In addition, we get

$$\|b\| \leq \|f\|_q,$$

in particular, $\|b\| = \|f\|_q$ in the case where $m = 0$.

The Main Results

Suppose that $a: X \times X \to \mathbb{R}$ is bilinear and bounded on the real H-space X, and suppose that $b: X \to \mathbb{R}$ is linear and continuous. We want to solve problem (P_a).

The two *key estimates* are given by

$$(\mathrm{G}_s) \qquad a(u, u) \geq c\|u\|_X^2 \quad \text{for all} \quad u \in X,$$

and

$$(\mathrm{G}) \qquad a(u, u) \geq c\|u\|_X^2 - C\|u\|_Z^2 \quad \text{for all} \quad u \in X.$$

Here, c and C are positive constants, and Z is a real H-space with the compact embedding $X \subseteq Z$. Condition (G_s) means that a is strongly positive. Furthermore, (G) is called an abstract *Gårding inequality*, and (G_s) is called a *strict* Gårding inequality.

(i) *Symmetry and quadratic variational problems.* If a is symmetric and strongly positive, then the original problem (P_a) is equivalent to the quadratic variational problem

$$(\mathrm{V}) \qquad \tfrac{1}{2}a(u, u) - b(u) = \min!, \qquad u \in X,$$

which has a unique solution (Section 22.1). This generalizes the classical Dirichlet principle.

(ii) *Nonsymmetric strongly positive problems.* If a is strongly positive, then (P_a) has a unique solution (Section 22.5). This corresponds to the main theorem on linear monotone operators.

(iii) *Gårding's inequality and Fredholm alternatives.* If a satisfies (G), but not (G_s), then (P_a) need not have a unique solution. In this case, we obtain a Fredholm alternative (Section 22.7). More precisely, the original problem (P_a) has a solution iff

$$b(w) = 0$$

for all solutions $w \in X$ of the dual problem

$$(\mathrm{P}_a^*) \qquad a(v, w) = 0 \quad \text{for all} \quad v \in X.$$

The corresponding linear operator $A: X \to X^*$, involved in the equivalent operator equation (P_A), has the following properties.

Case (i): A is strongly monotone and symmetric.

Case (ii): A is strongly monotone.

Case (iii): A is a compact perturbation of a strongly monotone operator, and hence A is a Fredholm operator of index zero.

Application to Boundary Value Problems

In order to explain Cases (i)–(iii) above with a simple, but typical example, let us consider the following boundary value problem

$$\text{(B)} \qquad \begin{aligned} -\Delta u + \alpha u_\xi + \beta u_\eta + \gamma u &= f \quad \text{on } G, \\ u &= 0 \quad \text{on } \partial G \end{aligned}$$

together with the dual problem

$$\text{(B*)} \qquad \begin{aligned} -\Delta w - \alpha w_\xi - \beta w_\eta + \gamma w &= 0 \quad \text{on } G, \\ w &= 0 \quad \text{on } \partial G. \end{aligned}$$

Here, G is a bounded region in $\mathbb{R}^2$, and α, β, γ are real numbers. Then, for example, we have the following situation:

Case (i): $\alpha = \beta = 0, \gamma \geq 0$.
Case (ii): $\alpha^2 + \beta^2 \neq 0$; $|\alpha|, |\beta|, |\gamma|$ are sufficiently small.
Case (iii): $\alpha, \beta, \gamma =$ arbitrary.

In the latter case, the principal elliptic part $-\Delta u$ corresponds to a strongly monotone operator, whereas the *lower order terms* correspond to a *compact* perturbation of that operator. Let $f \in L_2(G)$.

In Cases (i) and (ii), the boundary value problem (B) has a unique generalized solution $u \in \mathring{W}_2^1(G)$.

In Case (iii), problem (B) has a generalized solution $u \in \mathring{W}_2^1(G)$ iff

$$\int_G fw\,dx = 0$$

for all generalized solutions $w \in \mathring{W}_2^1(G)$ of the dual problem (B*).

Here, we set

$$a(u, v) = \int_G (u_\xi v_\xi + u_\eta v_\eta + \alpha u_\xi v + \beta u_\eta v + \gamma uv)\,dx,$$

$$b(v) = \int_G fv\,dx,$$

and the generalized problem corresponds to (P_a) with $X = \mathring{W}_2^1(G)$.

Eigenvalue Problems

Along with the problem (P_a), we will also consider eigenvalue problems of the form

$$(\mathrm{P}_b) \qquad \lambda a(u, v) = b(u, v) \qquad \text{for all} \quad v \in X,$$

where X is a real H-space. We seek a nonzero $u \in X$ and a real number λ which satisfy (P_b). We assume that $a, b\colon X \to \mathbb{R}$ are bilinear, *symmetric*, and bounded. Further assume that a is strongly positive and that b is *compact* and strictly positive.

Then we will show that problem (P_b) has an *infinite* number of eigenvalues in the case where $\dim X = \infty$.

The idea of proof is to reduce (P_b) to the equivalent operator equation

$$(\mathrm{P}_B) \qquad \lambda u = Bu,$$

where $B\colon X \to X$ is symmetric, *compact*, and strictly positive. Thus, it is possible to apply the *Hilbert–Schmidt theory* to (P_B) and hence to (P_b). The operator B is defined through the relation

$$b(u, v) = (Bu|v)_E, \qquad \text{for all} \quad u, v \in X,$$

where $(u|v)_E = a(u, v)$ is the so-called energetic scalar product on X, which is equivalent to the original scalar product $(u|v)$ on X.

In connection with (P_b) and (P_B), we will also study the Courant maximum–minimum principle. In Part III we will consider the Ljusternik–Schnirelman theory for nonlinear eigenvalue problems which is a generalization of that principle.

As a typical example let us consider the following boundary–eigenvalue problem:

$$-\lambda \Delta u = u \quad \text{on } G,$$

$$u = 0 \quad \text{on } \partial G,$$

where G is a bounded region in $\mathbb{R}^2$. Then the corresponding generalized problem is given by (P_b) with $X = \mathring{W}_2^1(G)$ and

$$a(u, v) = \int_G (u_\xi v_\xi + u_\eta v_\eta)\, dx,$$

$$b(u, v) = \int_G uv\, dx.$$

In this chapter, in addition to the problems mentioned above, we also investigate the following topics:

(a) Convergence of the Ritz method and the Galerkin method.
(b) Acceleration of the convergence of the Ritz method by means of the duality trick.
(c) Duality and *two-sided a posteriori* error estimates for the Ritz method.
(d) The method of finite elements.
(e) Fredholm alternatives for integral equations.
(f) Regularity of generalized solutions.

In connection with (f), we will obtain the following general principle for linear strongly elliptic differential equations:

The solutions are smoother, the smoother the data are, and, similarly, the convergence of the Ritz and Galerkin methods is faster, the smoother the solutions are.

The logical structure of this chapter is pictured in Figure 22.1.

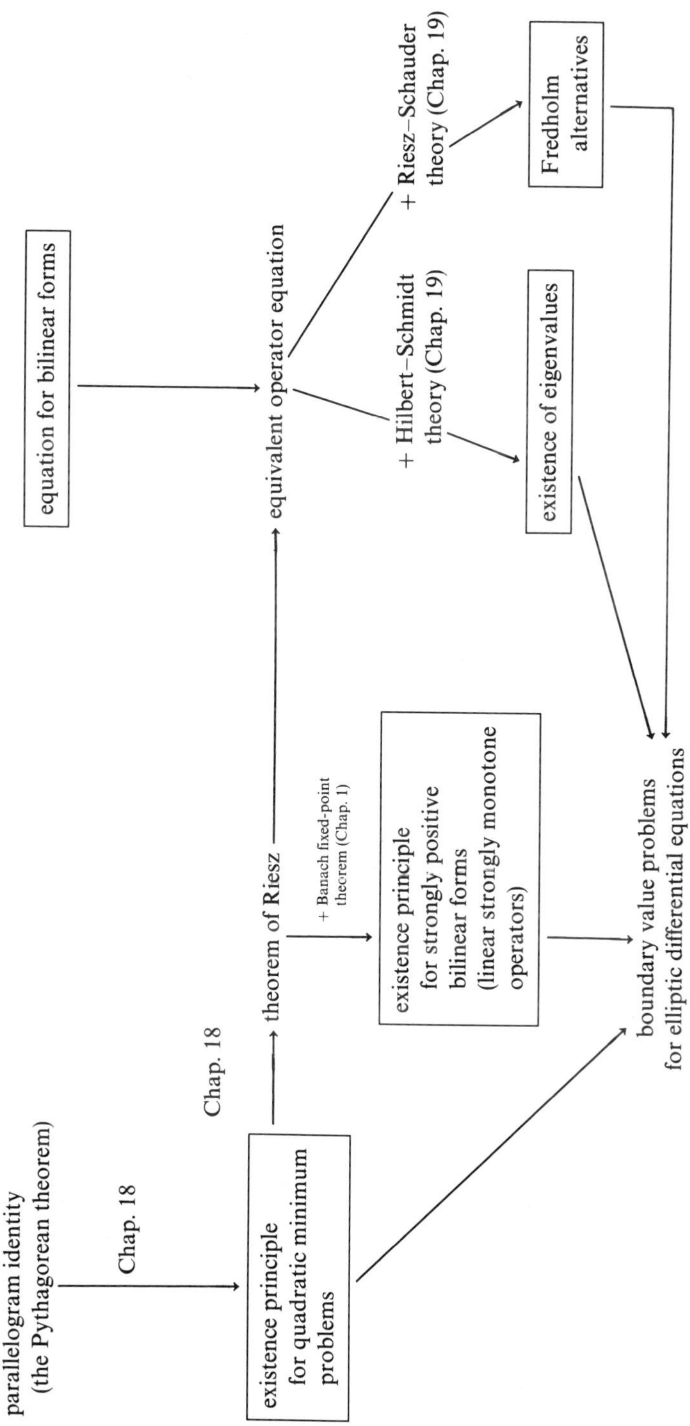

parallelogram identity (the Pythagorean theorem)
Chap. 18
existence principle for quadratic minimum problems
Chap. 18
theorem of Riesz
+ Banach fixed-point theorem (Chap. 1)
existence principle for strongly positive bilinear forms (linear strongly monotone operators)
boundary value problems for elliptic differential equations
equation for bilinear forms
equivalent operator equation
+ Hilbert–Schmidt theory (Chap. 19)
existence of eigenvalues
+ Riesz–Schauder theory (Chap. 19)
Fredholm alternatives

Figure 22.1

22.1. Main Theorem on Quadratic Minimum Problems and the Ritz Method

We consider the variational problem

$$\tfrac{1}{2}a(u,u) - b(u) = \min!, \qquad u \in X, \tag{2}$$

together with both the variational equation

$$a(u,v) = b(v) \qquad \text{for fixed} \quad u \in X \quad \text{and all} \quad v \in X, \tag{3}$$

and the operator equation

$$Au = b, \qquad u \in X. \tag{4}$$

As we shall see below, problems (2), (3), and (4) are mutually equivalent, where

$$\langle Au, v \rangle = a(u,v) \qquad \text{for all} \quad u, v \in X. \tag{5}$$

We call (3) and (4) the *Euler equation* to (2). Set

$$F(w) = \tfrac{1}{2}a(w,w) - b(w).$$

By definition, u is a *strict minimum* of (2) iff $F(w) > F(u)$ for all $w \in X$ with $w \neq u$.

Along with the variational problem (2) we consider the approximate *Ritz problem*

$$\tfrac{1}{2}a(u_n,u_n) - b(u_n) = \min!, \qquad u_n \in X_n, \tag{6}$$

together with the *Ritz equation*

$$a(u_n,v) = b(v) \qquad \text{for fixed} \quad u_n \in X_n \quad \text{and all} \quad v \in X_n,$$

which can be written in the following form:

$$a(u_n, w_{kn}) = b(w_{kn}), \qquad k = 1, \ldots, n'. \tag{7}$$

Here, we set

$$u_n = \sum_{k=1}^{n'} c_{kn} w_{kn}.$$

The elements w_{kn} are given. Then problem (7) represents a linear system of equations for the unknown real coefficients $c_{1n}, \ldots, c_{n'n}$. As we shall see below, problem (6) is equivalent to (7). We make the following assumptions:

(H1) Let X be a real H-space. The map $a: X \times X \to \mathbb{R}$ is bilinear, symmetric, bounded, and strongly positive, i.e., there exist positive constants c and d such that

$$|a(u,v)| \leq d\|u\|\,\|v\|,$$

$$a(u,u) \geq c\|u\|^2 \qquad \text{for all} \quad u, v \in X.$$

(H2) The functional $b: X \to \mathbb{R}$ is linear and continuous, i.e., $b \in X^*$.

(H3) Suppose that X is separable and infinite-dimensional, and let (X_n) be a Galerkin scheme in X, where $X_n = \operatorname{span}\{w_{1n}, \ldots, w_{n'n}\}$.

Theorem 22.A (Quadratic Minimum Problems). *Assume* (H1) *and* (H2). *Then*:

(a) Existence and uniqueness. *The original quadratic minimum problem* (2) *has a unique solution* u.
This solution corresponds to a strict minimum.
(b) Equivalence. *Problem* (2) *is equivalent to the Euler equation* (3).
(c) A priori estimate. *The solution* u *of the original problem* (2) *satisfies the inequality*

$$\|u\| \leq c^{-1}\|b\|. \tag{8}$$

(d) Convergence of the Ritz method. *In addition, assume* (H3). *Then, for each* $n \in \mathbb{N}$, *the Ritz equation* (7) *has a unique solution* u_n, *and, as* $n \to \infty$, *the sequence* (u_n) *converges in* X *to the unique solution* u *of the original minimum problem* (2).
The Ritz equation (7) *is equivalent to the Ritz problem* (6).

This theorem, together with the corollaries below, allows many important applications to elliptic boundary value problems (cf. Section 22.2) and to the theory of elasticity (cf. Part IV). In terms of elasticity, problem (2) allows the following physical interpretation:

$$u = \text{displacement of the elastic body},$$

$$\tfrac{1}{2}a(u,u) = \text{elastic potential energy of the body},$$

$$b(u) = \text{work of the outer forces}.$$

Then the original minimum problem (2) corresponds to the so-called principle of minimal potential energy.

By definition, the *energetic space* X_E is equal to the set X equipped with the so-called *energetic scalar product*

$$(u|v)_E = a(u,v) \qquad \text{for all} \quad u, v \in X.$$

This notation is motivated by the interpretation in terms of elasticity.

Corollary 22.1 (Equivalent Operator Equation).

(i) *Assume* (H1) *and* (H2). *Then the original quadratic minimum problem* (2) *is equivalent to the operator equation* (4). *The operator* $A: X \to X^*$ *given by* (5) *has the following properties*:

(P) *A is linear, continuous, symmetric, and strongly monotone.*

(ii) *Conversely, let* $A: X \to X^*$ *be an operator on the real* H-*space* X *with* (P). *Then, for each* $b \in X^*$, *the operator equation* $Au = b$ *has a unique solution* $u \in X$, *and this equation is equivalent to the quadratic minimum problem* (2).

Proof of Theorem 22.A. Ad(a), (b). This is a special case of Theorem 18.A in Section 18.8.

Ad(c). It follows from $a(u, v) = b(v)$ for all $v \in X$ that

$$c\|u\|^2 \le a(u, u) = b(u) \le \|b\| \, \|u\|,$$

and hence $c\|u\| \le \|b\|$.

Ad(d).

(I) Existence and uniqueness of the solution of the Ritz equation. Replacing X by X_n, it follows from Theorem 22.A(a), (b) that the Ritz problem (6) has a unique solution, and that (6) is equivalent to the Ritz equation (7).

(II) Convergence of the Ritz method. The *key* to the proof is the orthogonality relation (10) below.

Subtracting the Ritz equation (7) from the Euler equation (3), we obtain

$$a(u - u_n, v) = 0 \qquad \text{for all} \quad v \in X_n. \tag{9}$$

Letting $v = u_n$, we get $a(u - u_n, u_n) = 0$ and hence

$$a(u - u_n, u - u_n) = a(u - u_n, u - v) \qquad \text{for all} \quad v \in X_n. \tag{10}$$

This yields

$$\begin{aligned} c\|u - u_n\|^2 &\le a(u - u_n, u - u_n) \\ &= a(u - u_n, u - v) \le d\|u - u_n\| \, \|u - v\| \end{aligned}$$

for all $v \in X_n$. Hence

$$d^{-1}c\|u - u_n\| \le \inf_{v \in X_n} \|u - v\| = \text{dist}_X(u, X_n). \tag{11}$$

Since (X_n) is a Galerkin scheme in X, $\text{dist}_X(u, X_n) \to 0$ as $n \to \infty$. Hence $u_n \to u$ as $n \to \infty$. □

Corollary 22.1 follows immediately from (1a).

22.1a. Error Estimates

Recall that $F(w) = 2^{-1}a(w, w) - b(w)$.

Corollary 22.2. *Assume* (H1) *trough* (H3). *Let u and u_n be the solution of the original minimum problem* (2) *and the Ritz equation* (7), *respectively. Let $w \in X$ be given* (*e.g.*, $w = u_n$). *Then*:

(a) Upper bounds for the minimal value. *There holds $F(u) \le F(w)$.*

(b) Error estimates for u via lower bounds for $F(u)$. *If α is a lower bound for the minimal value $F(u)$, i.e., $\alpha \le F(u)$, then*

$$\frac{c}{2}\|u - w\|^2 \le F(w) - \alpha. \tag{12}$$

(c) *A posteriori* error estimates for u. *Suppose that there is a constant $C > 0$*

such that

$$|a(w,v)-b(v)|\le C\|v\| \qquad \textit{for all} \quad v\in X.$$

Then

$$c\|u-w\|\le C. \tag{13}$$

(d) *A priori* error estimates for the Ritz method. *For all* $n\in\mathbb{N}$,

$$\|u-u_n\|\le c^{-1}d\cdot \operatorname{dist}_X(u,X_n). \tag{14}$$

Remark 22.3 (Lower Bounds and Duality). According to (b) it is important to know lower bounds for the minimal value. In this connection, one uses the methods of general duality theory which will be considered in Part III. In Section 22.3 we will investigate the method of orthogonal projection which represents a special duality method.

It follows from (14) that the rapidity of convergence of the Ritz method depends on the quantity $\operatorname{dist}_X(u,X_n)$ of approximation theory. Thus, as in Section 21.16, we obtain the following rough principle:

The convergence of the Ritz method is faster, the smoother the solution u is.

Proof. Ad(a). Note that $F(u)\le F(w)$ for all $w\in X$.

Ad(b). A simple calculation shows that

$$F(u+v)=2^{-1}a(v,v)+(a(u,v)-b(v))+2^{-1}a(u,u)-b(u).$$

By (3), $a(u,v)-b(v)=0$. This implies

$$F(u+v)=2^{-1}a(v,v)+F(u),$$

and hence

$$F(u+v)-F(u)\ge 2^{-1}c\|v\|^2 \qquad \text{for all} \quad v\in X.$$

Ad(c). By (3), $a(u,u-w)=b(u-w)$. Hence

$$c\|u-w\|^2\le a(u-w,u-w)=b(u-w)-a(w,u-w)\le C\|u-w\|.$$

Ad(d). This follows from (11). □

22.1b. The Abstract Green Operator

As a special case of the Euler equation (3), we consider the equation

$$a(u,v)=b(v) \qquad \text{for fixed} \quad u\in X \quad \text{and all} \quad v\in X, \tag{15}$$

where

$$b(v)=(f|v)_Y \qquad \text{for all} \quad v\in Y,$$

and we assume:

(H) Assumption (H1) of Theorem 22.A is satisfied, where Y is a real H-space with the continuous embedding $X\subseteq Y$.

Let $f \in Y$ be given. Then

$$|b(v)| \leq \|f\|_Y \|v\|_Y \leq \text{const}\, \|f\|_Y \|v\|_X \tag{16}$$

for all $v \in X$. Hence $b \in X^*$. By Theorem 22.A, equation (15) has a unique solution $u \in X$, and we set

$$u = Kf.$$

The operator K is called the abstract *Green operator* or the solution operator to (15). If (15) corresponds to an elliptic boundary value problem, then K corresponds to an integral operator whose kernel is the classical Green function.

Corollary 22.4. *Assume* (H). *Then*:

(a) *The operator* $K: Y \to Y$ *is linear continuous, monotone, and symmetric.*
(b) *If the embedding* $X \subseteq Y$ *is compact, then* $K: Y \to Y$ *is compact.*

PROOF. By Theorem 22.A and (16),

$$\|Kf\|_X \leq c^{-1} \|b\| \leq \text{const}\, \|f\|_Y \qquad \text{for all} \quad f \in Y.$$

Hence $K: Y \to X$ is continuous. The continuity (resp. compactness) of the embedding $X \subseteq Y$ implies the continuity (resp. compactness) of $K: Y \to Y$.

The symmetry of $K: Y \to Y$ follows from

$$(f|Kg)_Y = a(Kf, Kg) = a(Kg, Kf) = (g|Kf)_Y \qquad \text{for all} \quad f, g \in Y.$$

Hence $(Kf|f)_Y = a(Kf, Kf) \geq 0$, i.e., K is monotone. □

22.1c. The Duality Trick and Refined Error Estimates for the Ritz Method

Let $u \in X$ be the unique solution of problem (15), i.e.,

$$a(u, v) = b(v) \qquad \text{for all} \quad v \in X, \tag{17}$$

with $b(v) = (f|v)_Y$, and let $u_n \in X_n$ be the solution of the corresponding Ritz equation

$$a(u_n, v) = b(v) \qquad \text{for all} \quad v \in X_n. \tag{18}$$

We make the following assumptions.

(A1) Assume (H1) and (H3) of Theorem 22.A concerning X and $a(\cdot, \cdot)$.
(A2) Let X, Y, and Z be real H-spaces with the continuous embedding $X \subseteq Y$.
(A3) *A priori estimate*. There is a constant $C > 0$ such that $u \in Z$ and

$$\|u\|_Z \leq C \|f\|_Y$$

for all $f \in Y$ and for all the corresponding solutions $u \in X$ of the original problem (17).

(A4) *Approximation property*. For all $z \in X \cap Z$ and $n = 1, 2, \ldots,$

$$\operatorname{dist}_X(z, X_n) \leq \alpha_n \|z\|_Z,$$

where α_n is a constant.

Corollary 22.5. *Assume* (A1)–(A4). *Then we obtain the following error estimates for the Ritz method*:

$$\|u - u_n\|_X \leq \alpha_n (c^{-1} dC) \|f\|_Y \qquad \text{for all} \quad f \in Y.$$

In the space Y, we get the stronger error estimates

$$\|u - u_n\|_Y \leq \alpha_n^2 (c^{-1} d^2 C^2) \|f\|_Y \qquad \text{for all} \quad f \in Y,$$

where, $n = 1, 2, \ldots$.

This result is a special case of Corollary 22.21 whose proof is based on a duality trick.

As we shall see in the next section, this result is responsible for the *high speed* of the convergence of the method of finite elements in the space Y. For example, in the case of the first boundary value problem

$$-\Delta u = f \quad \text{on } G, \qquad u = 0 \quad \text{on } \partial G,$$

where G is a bounded region in $\mathbb{R}^2$ with $\partial G \in C^\infty$, we may choose

$$Z = W_2^2(G), \qquad X = \mathring{W}_2^1(G), \qquad Y = L_2(G).$$

Then, for each $f \in Y$, we obtain that the solution u belongs to $X \cap Z$. Moreover, we obtain that α_n is proportional to the maximal *diameter* of the finite elements (triangles) in X_n. Note that the functions in Z are *smoother* than those in X.

Now it follows from Corollary 22.5 that the method of finite elements converges in Y like $\alpha_n^2 \to 0$ as $n \to \infty$, i.e., we have *quadratic* convergence in Y with respect to the maximal diameters of the triangles corresponding to the triangulation of the region G.

22.2. Application to Boundary Value Problems

In this section we use the following convention.

G is a bounded region in $\mathbb{R}^N$ with $N \geq 1$, $x = (\xi_1, \ldots, \xi_N)$, and $D_i = \partial/\partial \xi_i$. (19)

22.2a. The First Boundary Value Problem

We consider the homogeneous boundary value problem

$$\begin{aligned} -\Delta u &= f \quad \text{on } G, \\ u &= 0 \quad \text{on } \partial G. \end{aligned} \tag{20}$$

We set

$$a(u, v) = \int_G \sum_{i=1}^N D_i u D_i v \, dx,$$

$$b(v) = \int_G fv \, dx.$$

Definition 22.6. Assume (19). Let $X = \mathring{W}_2^1(G)$. Then the *generalized problem* to (20) reads as follows. For given $f \in L_2(G)$, we seek $u \in X$ with

$$a(u, v) = b(v) \qquad \text{for all} \quad v \in X. \tag{21}$$

This definition is motivated by Section 18.3. Formally, we obtain (21) from (20) by multiplying (20) with $v \in C_0^\infty(G)$ and by using subsequent integration by parts. Note that the functions u in $\mathring{W}_2^1(G)$ satisfy the boundary condition "$u = 0$ on ∂G" in the generalized sense. According to Section 22.1, problem (21) corresponds to the variational problem

$$\tfrac{1}{2} a(u, u) - b(u) = \min!, \qquad u \in X. \tag{22}$$

This is exactly the Dirichlet problem

$$\frac{1}{2} \int_G \left(\sum_i (D_i u)^2 - 2fu \right) dx = \min!, \qquad u \in \mathring{W}_2^1(G).$$

Proposition 22.7. *All the assertions of Section* 22.1 *hold for the two problems* (21) *and* (22).

In particular, for each $f \in L_2(G)$, the generalized first boundary value problem (21) *has a unique solution $u \in X$, which is also the unique solution of the variational problem* (22).

Corollary 22.8. *Let $Y = L_2(G)$. Denote by Kf the solution u of* (21). *Then the operator $K\colon Y \to Y$ is linear, monotone, compact, and symmetric.*

Classically, the solutions of (20) are given by

$$u(x) = \int_G k(x, y) f(y) \, dy,$$

where $k(\cdot, \cdot)$ is the symmetric Green function. The operator K generalizes this integral operator, i.e., $u = Kf$.

Proof of Proposition 22.7. By (1b), $b \in X^*$.

On X we have the norm $\|u\| = \|u\|_{1,2}$. Moreover, $a(u, u) = \|u\|_{1,2,0}^2$. Obviously, the map $a\colon X \times X \to \mathbb{R}$ is bilinear and symmetric. By the Hölder inequality,

$$\begin{aligned}|a(u,v)| &\le \sum_i \int_G |D_iuD_iv|\,dx\\ &\le \sum_i \left(\int_G |D_iu|^2\,dx\right)^{1/2}\left(\int_G |D_iv|^2\,dx\right)^{1/2}\\ &\le N\|u\|\,\|v\| \qquad \text{for all} \quad u, v \in X.\end{aligned}$$

Hence, a is bounded. By Section 21.3, the norm $\|\cdot\|_{1,2}$ is *equivalent* to the norm $\|\cdot\|_{1,2,0}$ on X. This implies

$$a(u,u) \ge c\,\|u\|^2 \qquad \text{for all} \quad u \in X,$$

i.e., a is *strongly positive.*

Now, Theorem 22.A shows that (21) and (22) have unique solutions. □

By Section 21.3, the embedding $X \subseteq Y$ is compact. Thus, Corollary 22.8 follows from Corollary 22.4.

In what follows we use the results of Section 22.1 in order to obtain additional information on the first boundary value problem (20). Recall that $N = \dim G$.

22.2b. Regularity of the Generalized Solutions

Let $u \in \mathring{W}_2^1(G)$ be the generalized solution of (20) for $f \in L_2(G)$. In Section 22.19 we will prove the following crucial result.

(i) *Sobolev spaces.* Let $k = 0, 1, \ldots$. Suppose that the boundary ∂G is smooth, i.e., $\partial G \in C^\infty$. Then,

$$f \in W_2^k(G) \qquad \text{implies} \qquad u \in W_2^{k+2}(G).$$

Furthermore, there is a constant D_k such that

$$\|u\|_{k+2,2} \le D_k\|f\|_{k,2} \qquad \text{for all} \quad f \in W_2^k(G).$$

In particular, $f \in L_2(G)$ implies $u \in W_2^2(G)$.

It follows from the Sobolev embedding theorems in Section 21.3 that

$$u \in W_2^{k+2}(G) \qquad \text{implies} \qquad u \in C^m(\bar{G})$$

in the case where $k + 2 > m + N/2$. If $k > N/2$, then $m \ge 2$, and u is a *classical* solution of the first boundary value problem (20).

In Chapter 6 we used the following sharp Schauder estimates.

(ii) *Hölder spaces.* Suppose that $\partial G \in C^{m+2,\beta}$, where $m = 0, 1, \ldots$ and $0 < \beta < 1$. Then,

$$f \in C^{m,\beta}(\bar{G}) \qquad \text{implies} \qquad u \in C^{m+2,\beta}(\bar{G}),$$

and

$$\|u\|_{C^{m+2,\beta}(\bar{G})} \le \text{const}\,\|f\|_{C^{m,\beta}(\bar{G})}.$$

Below we will show that these results are important for estimating the *rapidity* of convergence of the Ritz method.

22.2c. The Inequality of Poincaré–Friedrichs and *a posteriori* Error Estimates

The following error estimates depend on the constant c_0 of the Poincaré–Friedrichs inequality

$$c_0 \int_G u^2\,dx \le \int_G \sum_{i=1}^N (D_i u)^2\,dx \qquad \text{for all} \quad u \in \mathring{W}_2^1(G). \tag{23}$$

This implies

$$c \int_G \left(u^2 + \sum_{i=1}^N (D_i u)^2\right) dx \le \int_G \sum_{i=1}^N (D_i u)^2\,dx \tag{24}$$

for all $u \in \mathring{W}_2^1(G)$, where $c = c_0/(1 + c_0)$, i.e., $a(u, u) \ge c\|u\|_{1,2}^2$. Moreover, the defect of a function w with respect to the equation $\Delta u + f = 0$ on G is defined to be

$$\alpha = \left(\int_G (\Delta w + f)^2\,dx\right)^{1/2}.$$

Corollary 22.9. *Suppose that G satisfies* (19) *with $\partial G \in C^{0,1}$, and let $u \in \mathring{W}_2^1(G)$ be a generalized solution of the first boundary value problem* (20). *Choose an arbitrary function $w \in C^2(\bar{G})$ with $w = 0$ on ∂G. Then we obtain the following error estimates*:

$$\|u - w\|_2 \le \alpha/c_0, \qquad \|u - w\|_{1,2} \le \alpha/c. \tag{25}$$

Recall that $\|v\|_2^2 = \int_G v^2\,dx$ and $\|v\|_{1,2}^2 = \int_G (v^2 + \sum_i (D_i v)^2)\,dx$. By Section 22.13, the largest possible value of c_0 is equal to the smallest eigenvalue μ of the problem:

$$-\Delta v = \mu v \quad \text{on } G, \qquad v = 0 \quad \text{on } \partial G.$$

Sharp estimates for c_0 can be found in Michlin (1981, M).

PROOF.

(I) Integration by parts yields

$$|a(w, v) - b(v)| = \left|\int_G (\Delta w + f)v\,dx\right|$$
$$\le \alpha\|v\|_2 \le \alpha\|v\|_{1,2} \qquad \text{for all} \quad v \in \mathring{W}_2^1(G). \tag{26}$$

By (13), $c\|u - w\|_{1,2} \le \alpha$.

(II) It follows from (23) that

$$c_0\|u - w\|_2^2 \le a(u - w, u - w).$$

Since u is a generalized solution of (20), $a(u, u - w) = b(u - w)$. By (26),

$$c_0 \|u - w\|_2^2 \leq |b(u - w) - a(w, u - w)| \leq \alpha \|u - w\|_2. \qquad \square$$

22.2d. The Convergence of the Ritz Method

We make the following assumption.

(H) The set G satisfies (19), and (X_n) is a Galerkin scheme in the Sobolev space $X = \mathring{W}_2^1(G)$. Let $X_n = \operatorname{span}\{w_{1n}, \ldots, w_{n'n}\}$.

According to Section 22.1, the Ritz equation, corresponding to the first boundary value problem (20), is given by

$$a(u_n, v) = b(v) \qquad \text{for all} \quad v \in X_n. \tag{27}$$

We seek $u_n \in X_n$, i.e.,

$$u_n = c_{1n} w_{1n} + \cdots + c_{n'n} w_{n'n}.$$

Equation (27) is equivalent to the system

$$a(u_n, w_{kn}) = b(w_{kn}), \qquad k = 1, \ldots, n'.$$

Explicitly, that means

$$\int_G \sum_{i=1}^N D_i u_n D_i w_{kn} \, dx = \int_G f w_{kn} \, dx, \qquad k = 1, \ldots, n'. \tag{28}$$

This is a linear system for determing the unknown real coefficients $c_{1n}, \ldots, c_{n'n}$.

The following statement follows from Section 22.1a.

Corollary 22.10. *Assume* (H). *Let $u \in X$ be the generalized solution of the first boundary value problem* (20) *for given $f \in L_2(G)$. Then, for each $n \in \mathbb{N}$, the Ritz equation* (28) *has a unique solution u_n, and the Ritz method converges in X, i.e.,*

$$\lim_{n \to \infty} \|u - u_n\|_{1,2} = 0.$$

The rapidity of convergence is given by

$$\|u - u_n\|_{1,2} \leq c^{-1} N \operatorname{dist}_X(u, X_n), \tag{29}$$

where $N = \dim G$ and the constant c has been introduced in Section 22.2c.

Example 22.11. Let $N = 1$. We consider the first boundary value problem

$$-\Delta u = f \quad \text{on } G, \qquad u = 0 \quad \text{on } \partial G$$

in the special case of the finite interval $G =]\gamma, \delta[$. Let $f \in C^{m,\beta}(\bar{G})$, where $m = 0, 1, \ldots$ and $0 < \beta < 1$. Let X_n be the set of all the following polynomials

$$v(x) = (x - \gamma)(x - \delta)(a_0 + a_1 x + \cdots + a_n x^n).$$

Note that v satisfies the boundary condition.

Then the rapidity of convergence of the corresponding Ritz method (28) is given by

$$\|u - u_n\|_{1,2} \leq \frac{\text{const}}{n^{m+1+\beta}} \|f\|_{C^{m,\beta}(\bar{G})}, \qquad n = 1, 2, \ldots.$$

Hence, the convergence is faster, the smoother f is.

In this connection, we have

$$w_{kn}(x) = (x - \gamma)(x - \delta)x^{k-1}, \qquad k = 1, \ldots, n + 1.$$

PROOF. By Section 22.2b,

$$\|u\|_{C^{m+2,\beta}(\bar{G})} \leq \text{const} \|f\|_{C^{m,\beta}(\bar{G})}.$$

Moreover, approximation theory (cf. $A_2(57)$) tells us that

$$\text{dist}_X(u, X_n) \leq \frac{\text{const}}{n^{m+1+\beta}} \|u\|_{C^{m+2,\beta}(\bar{G})}. \qquad \square$$

22.2e. The Method of Finite Elements

Generally, the Ritz method with piecewise polynomial basis functions is called the finite-element method. This method has the following nice properties.

(a) The finite elements can be easily implemented on computers.
(b) The coefficient matrix of the Ritz equation is a sparse matrix, i.e., most entries are equal to zero.
(c) One may take the specifics of the problem into account. For example, one can use an extraordinarily fine triangulation in subregions, where a subtle behavior of the solution is expected (e.g., special parts of airplanes).

Today the method of finite elements is one of the most important numerical methods for solving problems in natural sciences and engineering.

To explain the basic idea of this method, we consider the first boundary value problem

$$-\Delta u = f \quad \text{on } G, \qquad u = 0 \quad \text{on } \partial G, \tag{30}$$

where G is a bounded region in $\mathbb{R}^2$. Let (T_n) be a sequence of triangulations of G which has the following two properties:

(a) $h_n \to 0$ as $n \to \infty$, where h_n denotes the maximal diameter of the triangles of T_n.
(b) $\inf_n \delta_n > 0$, where δ_n denotes the minimal angle of the triangles of T_n (cf. Fig. 22.2).

We consider the following special finite elements with respect to T_n:

$$w_{kn}(x) = \begin{cases} 0 & \text{for all the node points } x \in \partial G, \\ 1 & \text{for a } \textit{fixed} \text{ node point } x \in G, \\ 0 & \text{for the remaining node points.} \end{cases}$$

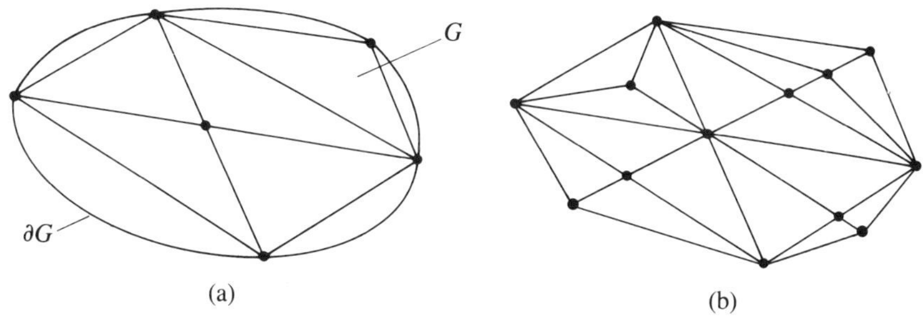

Figure 22.2

Further the function w_{kn} is extended to G via *linear interpolation.* We set $X_n = \text{span}\{w_{1n}, \ldots, w_{n'n}\}$. Then the space X_n consists exactly of all piecewise linear functions with respect to the triangulation T_n, which vanish at the boundary.

Let $f \in L_2(G)$. Moreover, let u be the generalized solution of (30), and let u_n be the solution of the Ritz equation (28) with respect to the finite element space X_n.

Suppose that $\partial G \in C^\infty$. We now use the following two important estimates.

(a) *Regularity.* By Section 22.2b, u belongs to the Sobolev space $W_2^2(G)$ and

$$\|u\|_{2,2} \leq \text{const}\,\|f\|_2.$$

(b) *Approximation theory.* Let $X = \mathring{W}_2^1(G)$. We prove in Appendix $A_2(59)$ that

$$\text{dist}_X(u, X_n) \leq \text{const} \cdot h_n \|u\|_{2,2}.$$

Corollary 22.12. *We have the following error estimates*:

$$\|u - u_n\|_{1,2} \leq \text{const} \cdot h_n \|f\|_2, \tag{31}$$

$$\|u - u_n\|_2 \leq \text{const} \cdot h_n^2 \|f\|_2, \qquad n = 1, 2, \ldots. \tag{32}$$

Consequently, the method of finite elements for (30) converges linearly in the space $\mathring{W}_2^1(G)$, and it converges *quadratically* in the space $L_2(G)$.

PROOF. Ad(31). This follows from Corollary 22.10 and from (a) and (b) above.

Ad(32). Set $Z = W_2^2(G)$, $X = \mathring{W}_2^1(G)$, and $Y = L_2(G)$. Then the assertion follows from the refined error estimates obtained in Corollary 22.5 and from (a) and (b) above. □

22.2f. The Second Boundary Value Problem

We consider the second boundary value problem

$$\begin{aligned} -\Delta u &= f \quad \text{on } G, \\ \frac{\partial u}{\partial n} &= g \quad \text{on } \partial G. \end{aligned} \tag{33}$$

We set

$$a(u, v) = \int_G \sum_{i=1}^{N} D_i u D_i v \, dx,$$

$$b(v) = \int_G fv \, dx + \int_{\partial G} gv \, dO.$$

Problem (33) is *not* uniquely solvable since $u = 1$ is a solution of the corresponding homogeneous problem. In order to get a unique solution, we add the normalization condition

$$\int_G u \, dx = 0. \tag{33*}$$

Definition 22.13. Let $X = \{u \in W_2^1(G): \int_G u \, dx = 0\}$, where G is a bounded region in $\mathbb{R}^N$, $N \geq 1$, with $\partial G \in C^{0,1}$. Then the *generalized problem* to (33) reads as follows. For given $f \in L_2(G)$ and $g \in L_2(\partial G)$ with

$$b(1) = \int_G f \, dx + \int_{\partial G} g \, dO = 0, \tag{34}$$

we seek $u \in X$ such that

$$a(u, v) = b(v) \qquad \text{for all} \quad v \in X. \tag{35}$$

This problem is motivated by Section 18.4. Formally, we obtain (35) from (33) by multiplying the first equation of (33) with $v \in C^\infty(\bar{G})$ and by using subsequent integration by parts. Letting $v = 1$, equation (35) yields $b(1) = 0$. This is (34).

The corresponding variational problem is given by

$$\tfrac{1}{2}a(u, u) - b(u) = \min!, \qquad u \in X. \tag{36}$$

Proposition 22.14. *All the assertions of Section* 22.1 *are valid for the two problems* (35) *and* (36).

In particular, the generalized second boundary value problem (35) *has a unique solution $u \in X$, which is also the unique solution of the variational problem* (36).

Proof. Since $W_2^1(G) \subseteq L_1(G)$, X is a closed linear subspace of $W_2^1(G)$. Thus, X is a B-space with the norm $\|u\|_{1,2}$.

(I) We show that $b \in X^*$. By Section 21.3, each function $v \in W_2^1(G)$ has generalized boundary values on ∂G belonging to the space $L_2(\partial G)$. Furthermore,

$$\int_{\partial G} v^2 \, dO \leq \text{const} \, \|v\|_{1,2}^2 \qquad \text{for all} \quad v \in W_2^1(G).$$

By the Hölder inequality, for all $v \in W_2^1(G)$,

$$\begin{aligned} |b(v)| &\leq \left|\int_G fv\,dx\right| + \left|\int_{\partial G} gv\,dO\right| \\ &\leq \|f\|_2\|v\|_2 + \text{const}\,\|g\|_{L_2(\partial G)}\|v\|_{1,2} \\ &\leq \text{const}\,\|v\|_{1,2}. \end{aligned}$$

(II) We investigate $a(\cdot,\cdot)$. As in the proof for Proposition 22.7, we obtain that $a: X \times X \to \mathbb{R}$ is bilinear, bounded, and symmetric.

By Section 21.3, the norm

$$\|u\|_{1,2}^* = \left(\int_G \sum_i (D_i u)^2\,dx + \left|\int_G u\,dx\right|^2\right)^{1/2}$$

is *equivalent* to the norm $\|\cdot\|_{1,2}$ on $W_2^1(G)$. For all $u \in X$, $\int_G u\,dx = 0$. Hence

$$\|u\|_{1,2}^{*2} = a(u,u) \geq c\,\|u\|_{1,2}^2 \qquad \text{for all} \quad u \in X,$$

i.e., a is *strongly positive*.

Theorem 22.A yields the existence of a unique solution $u \in X$ for (35) and (36). □

22.2g. The Third Boundary Value Problem

We consider the third boundary value problem

$$\begin{aligned} -\Delta u &= f \quad \text{on } G, \\ \frac{\partial u}{\partial n} + hu &= g \quad \text{on } \partial G. \end{aligned} \tag{37}$$

Here, it is important that

$$h(x) \geq h_0 > 0 \qquad \text{for all} \quad x \in \partial G \quad \text{and fixed } h_0. \tag{38}$$

We set

$$\begin{aligned} a(u,v) &= \int_G \sum_{i=1}^N D_i u D_i v\,dx + \int_{\partial G} huv\,dO, \\ b(v) &= \int_G fv\,dx + \int_{\partial G} gv\,dO. \end{aligned}$$

Definition 22.15. Let $X = W_2^1(G)$, where G is a bounded region in $\mathbb{R}^N$, $N \geq 1$, with $\partial G \in C^{0,1}$. Let $h \in C(\partial G)$ with (38). Then the *generalized problem* to (37) reads as follows. For given $f \in L_2(G)$ and $g \in L_2(\partial G)$, we seek $u \in X$ such that

$$a(u,v) = b(v) \qquad \text{for all} \quad v \in X. \tag{39}$$

This problem is motivated by Section 18.4. Formally, we obtain (39) from (37) by multiplying the first equation of (37) with $v \in C^\infty(\bar{G})$ and by using subsequent integration by parts.

The corresponding variational problem is given by

$$\tfrac{1}{2}a(u,u) - b(u) = \min!, \qquad u \in X. \tag{40}$$

Proposition 22.16. *All the assertions of Section 22.1 are valid for the two problems* (39) *and* (40).

In particular, the generalized third boundary problem (39) *has a unique solution $u \in X$, which is also the unique solution of the variational problem* (40).

PROOF.

(I) As in the proof of Proposition 22.14, it follows that $b \in X^*$.

(II) The map $a: X \times X \to \mathbb{R}$ is bounded because

$$\left|\int_{\partial G} huv\,dO\right| \le \max|h| \left|\int_{\partial G} uv\,dO\right|$$
$$\le \max|h|\,\|u\|_{2,\partial G}\|v\|_{2,\partial G} \le \text{const}\,\|u\|_{1,2}\|v\|_{1,2}.$$

Obviously, a is bilinear and symmetric.

By Section 21.3, the norm

$$\|u\|_{1,2}^{**} = \left(\int_G \sum_i (D_i u)^2\,dx + \int_{\partial G} u^2\,dO\right)^{1/2}$$

is *equivalent* to the norm $\|\cdot\|_{1,2}$ on X. Because $h \ge h_0 > 0$ on ∂G, we obtain

$$a(u,u) \ge c\,\|u\|_{1,2}^2 \qquad \text{for all} \quad u \in X,$$

i.e., a is *strongly positive.*

Theorem 22.A yields the existence of a unique solution for (39) and (40). □

22.2h. The Mixed Boundary Value Problem

The following problem

$$\begin{aligned} -\Delta u &= f \quad \text{on } G,\\ u &= g \quad \text{on } \partial_1 G,\\ \frac{\partial u}{\partial n} &= h \quad \text{on } \partial_2 G, \end{aligned} \tag{41}$$

is called the mixed boundary value problem. Here, we assume that $\partial G = \partial_1 G \cup \partial_2 G$ and $\partial_1 G \cap \partial_2 G = \varnothing$. In the special case where $\partial_2 G = \varnothing$ (resp.

$\partial_1 G = \emptyset$) we get the first (resp. second) boundary value problem. Generalized solutions of this problem will be investigated in Theorem 25.I in connection with linear and nonlinear conservation laws. Mixed boundary value problems also play an important role in elasticity. This will be considered in Chapter 62.

The first (resp. second and mixed) boundary value problem is also called the Dirichlet (resp. Neumann and Zaremba) problem.

22.3. The Method of Orthogonal Projection, Duality, and *a posteriori* Error Estimates for the Ritz Method

Our goal is to explain the basic idea of *duality theory* in simple geometrical terms. Our approach is based on the Pythagorean theorem in H-spaces (Fig. 22.3).

We set

$$F(u) = \|g - u\|^2,$$

$$F^*(v) = -\|g - v\|^2 + \|g\|^2.$$

Along with the original minimum problem

$$F(u) = \min!, \qquad u \in X, \tag{42}$$

we consider the *dual* maximum problem

$$F^*(v) = \max!, \qquad v \in Y. \tag{42*}$$

As we shall show below, problems (42) and (42*) have the same extremal value. This is important for obtaining error estimates for the original problem (42). The Ritz problems to (42) (resp. (42*)) are given by

$$F(u_n) = \min!, \qquad u_n \in X_n \tag{43}$$

(resp.

$$F^*(v_n) = \max!, \qquad v_n \in Y_n). \tag{43*}$$

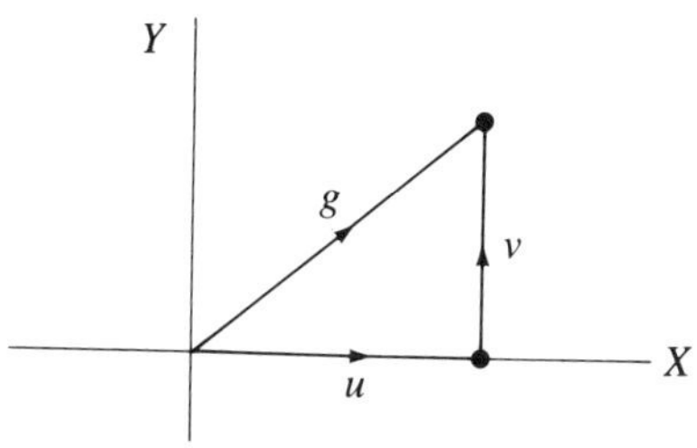

Figure 22.3

According to (7) the corresponding Ritz equations are obtained through

$$(u_n - g|u_{kn}) = 0, \qquad k = 1, \ldots, n', \tag{44}$$

with $u_n = c_{1n}u_{1n} + \cdots + c_{n'n}u_{n'n}$ (resp.

$$(v_n - g|v_{kn}) = 0, \qquad k = 1, \ldots, m', \tag{44*}$$

with $v_n = d_{1n}v_{1n} + \cdots + d_{m'n}v_{m'n}$). Equation (44) (resp. (44*)) is a linear system for the unknown real coefficients c_{kn} (resp. d_{kn}). We make the following assumptions.

(H1) The space X is a closed linear subspace of the real separable H-space Z.
(H2) The space Y is the orthogonal complement to X, i.e.,

$$Y = \{v \in Z \colon (u|v) = 0 \text{ for all } u \in X\}.$$

(H3) (X_n) (resp. (Y_n)) is a Galerkin scheme in X (resp. Y).
(H4) Let $g \in Z$ be given.

Theorem 22.B. *Assume* (H1) *through* (H4). *Then*:

(a) Existence and uniqueness. *The original problem* (42) *and the dual problem* (42*) *has the unique solution u and v, respectively. We have the extremal relation*

$$g = u + v, \tag{45}$$

i.e., u (resp. v) is the orthogonal projection of g to X (resp. Y). Furthermore, $F(u) = F^*(v)$.

(b) Convergence. *For each* $n \in \mathbb{N}$, *the Ritz equations* (44) *and* (44*) *have the unique solutions* u_n *and* v_n, *respectively, and*

$$\lim_{n\to\infty} \|u - u_n\|_X = 0, \qquad \lim_{n\to\infty} \|v - v_n\|_Y = 0.$$

(c) *A posteriori* error estimates. *The minimal value* $F(u)$ *of the original problem* (42) *satisfies*

$$F^*(v_n) \le F(u) \le F(u_n), \tag{46}$$

and the solution u of (42) *satisfies*

$$\|u - u_n\|_X^2 \le F(u_n) - F^*(v_n). \tag{47}$$

A more general result will be proved in Problem 22.1.

In contrast to Section 22.1, the solution of the two Ritz problems (44) and (44*) allows us to obtain *two-sided* error estimates for the minimal value $F(u)$. If we have computed the solutions u_n and v_n, say, by the finite-element method on a computer, then we also obtain estimates for u, by (47). In elasticity, $F(u)$ is the elastic potential energy of the elastic body, and u is the displacement. The Ritz equation (44*) is also called the *Trefftz equation*.

PROOF. The existence and uniqueness results as well as the convergence of the Ritz method follow from Theorem 22.A. In this connection note that

$$F(u) = (u|u) - 2(u|g) + (g|g), \qquad F^*(v) = -(v|v) + 2(v|g).$$

By Theorem 18.C, we have the unique orthogonal decomposition

$$g = u + v, \qquad u \in X, \quad v \in Y,$$

where u and v are solutions of (42) and (42*), respectively. Since

$$\|g\|^2 = \|u\|^2 + \|v\|^2,$$

we obtain that $F(u) = F^*(v)$. It follows from $Y_n \subseteq Y$ and $X_n \subseteq X$ that

$$F^*(v_n) \leq F^*(v) \qquad \text{and} \qquad F(u) \leq F(u_n).$$

This yields (46). By (12),

$$\|u - u_n\|^2 \leq F(u_n) - F(u).$$

Since $F(u) = F^*(v) \geq F^*(v_n)$, we get (47). □

22.4. Application to Boundary Value Problems

We consider the first boundary value problem

$$\begin{aligned} -\Delta v &= 0 \quad \text{on } G, \\ v &= g \quad \text{on } \partial G. \end{aligned} \tag{48}$$

The *key* to our approach is the fact that

$$(u|v) = \int_G \sum_{i=1}^N D_i u D_i v \, dx + \int_{\partial G} uv \, dO$$

is an equivalent scalar product on the Sobolev space $W_2^1(G)$, by Theorem 21.A. We set

$$Z = \left\{ v \in W_2^1(G) : \int_{\partial G} v \, dO = 0 \right\}.$$

Then Z is a real separable H-space equipped with the scalar product $(\cdot|\cdot)$. Moreover, we set

$$X = \mathring{W}_2^1(G) \qquad \text{and} \qquad Z = X \oplus Y,$$

i.e., Y is the orthogonal complement to X. In the following proof we will show that

$$Y = \{v \in Z \cap C^\infty(G) : \Delta v = 0 \text{ on } G\}. \tag{49}$$

This important result is due to Hermann Weyl (1940).

We make the following assumptions.

(H1) G is a bounded region in $\mathbb{R}^N$, $N \geq 1$, with $\partial G \in C^{0,1}$.

(H2) We choose linearly independent functions $u_{1n}, \ldots, u_{n'n} \in X$ and $v_{1n}, \ldots, v_{m'n} \in Y$, i.e.,

$$u_{kn} = 0 \quad \text{on } \partial G, \qquad k = 1, \ldots, n',$$

$$\Delta v_{kn} = 0 \quad \text{on } G, \qquad k = 1, \ldots, m',$$

and we consider the *Ritz equation*

$$\int_G \sum_i D_i(u_n - g) D_i u_{kn}\, dx = 0, \qquad k = 1, \ldots, n', \tag{50}$$

with $u_n = c_{1n}u_{1n} + \cdots + c_{n'n}u_{n'n}$ and the *Trefftz equation*

$$\int_G \sum_i D_i(v_n - g) D_i v_{kn}\, dx = 0, \qquad k = 1, \ldots, m', \tag{51}$$

with $v_n = d_{1n}v_{1n} + \cdots + d_{m'n}v_{m'n}$. Equation (50) (resp. (51)) is a linear system for the unknown real coefficients c_{kn} and d_{kn}, respectively. If $v_{kn} \in C^1(\bar{G})$ for all k, then integration by parts yields that the Trefftz equation (51) is equivalent to the equation

$$\int_{\partial G} (v_n - g)\frac{\partial v_{kn}}{\partial n}\, dO = 0, \qquad k = 1, \ldots, m'. \tag{51a}$$

Proposition 22.17 (Method of Orthogonal Projection of Weyl (1940)). *Assume* (H1), (H2). *Let $g \in Z$ be given, i.e., $g \in W_2^1(G)$ and $\int_{\partial G} g\, dO = 0$. Then*:

(a) Existence and uniqueness. *Let u be the orthogonal projection of g onto the Sobolev space $\mathring{W}_2^1(G)$. Then $v = g - u$ is the unique solution of the first boundary value problem* (48).

More precisely, we have $v \in W_2^1(G) \cap C^\infty(G)$, and v is a classical solution of the differential equation $\Delta v = 0$ on G. Furthermore, we get $v = g$ on ∂G in the sense of generalized boundary values.

(b) *A posteriori* error estimates. *For each $n \in \mathbb{N}$, the Ritz equation* (50) *and the Trefftz equation* (51) *have unique solutions u_n and v_n, respectively. For the solution v of the first boundary value problem* (48), *we obtain the error estimate*

$$\|g - v - u_n\|_{1,2,0}^2 \leq \|u_n - g\|_{1,2,0}^2 + \|v_n - g\|_{1,2,0}^2 - \|g\|^2.$$

By the inequality of Poincaré–Friedrichs, we get the additional error estimate:

$$c_0\|g - v - u_n\|_2^2 \leq \|g - v - u_n\|_{1,2,0}^2.$$

Proof. We use Theorem 22.B and the lemma of Weyl (Theorem 18.G). We first prove (49). Let $v \in Y$. Then, $(u|v) = 0$ for all $u \in X$, i.e.,

$$\int_G \sum_i D_i u D_i v\, dx = 0 \qquad \text{for all} \quad u \in \mathring{W}_2^1(G).$$

By Section 18.15, v is a weak solution of the Laplace equation $\Delta v = 0$ on G. By the lemma of Weyl, $v \in C^\infty(G)$, and v is a classical solution of the Laplace equation.

It follows from $g = u + v$ and $u \in \mathring{W}_2^1(G)$ that $u = 0$ on ∂G and hence $v = g$ on ∂G.

The remaining assertions follow from Theorem 22.B. □

22.5. Main Theorem on Linear Strongly Monotone Operators and the Galerkin Method

We consider the equation

$$a(u, v) = b(v) \qquad \text{for all} \quad v \in X. \tag{52}$$

We seek $u \in X$. The corresponding *dual* problem is given by

$$a(v, u^*) = b(v) \qquad \text{for all} \quad v \in X. \tag{52*}$$

Here, we seek $u^* \in X$. If $a(\cdot, \cdot)$ is symmetric, then (52) and (52*) coincide. However, in contrast to Section 22.1, we do not assume here that $a(\cdot, \cdot)$ is symmetric. In the *nonsymmetric* case, equation (52) does not correspond to a variational problem. The nonsymmetric case is related to boundary value problems for nonsymmetric elliptic differential equations. Along with the original problem (52) we consider the operator equation

$$Au = b, \qquad u \in X, \tag{53}$$

and the Galerkin equation

$$a(u_n, v) = b(v) \qquad \text{for all} \quad v \in X_n,$$

where we seek $u_n \in X_n$. From our assumptions below it follows that the Galerkin equation is equivalent to the following equation

$$a(u_n, w_k) = b(w_k), \qquad k = 1, \ldots, n, \tag{54}$$

with $u_n = c_{1n}w_1 + \cdots + c_{nn}w_n$. Equation (54) is a linear system for the unknown real coefficients c_{kn}.

We make the following assumptions.

(H1) Let X be a real separable H-space with the basis $\{w_1, w_2, \ldots\}$. Let $X_n = \operatorname{span}\{w_1, \ldots, w_n\}$.

(H2) The map $a\colon X \times X \to \mathbb{R}$ is bilinear, bounded, and strongly positive, i.e., there exist positive constants c and d such that

$$|a(u, v)| \le d\|u\| \|v\|,$$

$$a(u, u) \ge c\|u\|^2 \qquad \text{for all} \quad u, v \in X.$$

(H3) The functional $b\colon X \to \mathbb{R}$ is linear and continuous.

Theorem 22.C. *Assume* (H1) *through* (H3). *Then*:

(a) Existence and uniqueness. *The original problem* (52) *has a unique solution* $u \in X$, *and we have the a priori estimate*

$$\|u\| \leq c^{-1} \|b\|.$$

The dual problem (52*) *has a unique solution* $u^* \in X$, *and* $\|u^*\| \leq c^{-1} \|b\|$.

(b) Convergence of the Galerkin method. *Let* $\dim X = \infty$. *For each* $n \in \mathbb{N}$, *the Galerkin equation* (54) *has a unique solution* $u_n \in X_n$, *and*

$$u_n \to u \quad \text{in } X \qquad \text{as} \quad n \to \infty.$$

(c) *A priori* error estimates. *For each* $n \in \mathbb{N}$,

$$\|u - u_n\| \leq c^{-1} d \cdot \operatorname{dist}_X(u, X_n). \tag{55}$$

(d) *A posteriori* error estimates. *Suppose that, for fixed* $w \in X$, *there is a constant* $C > 0$ *such that*

$$|a(w, v) - b(v)| \leq C \|v\| \qquad \text{for all} \quad v \in X.$$

Then

$$c \|u - w\| \leq C. \tag{56}$$

Assertion (a) also holds true in the case where X is not separable.

Corollary 22.18 (Equivalent Operator Equation).

(i) *Assume* (H1) *through* (H3). *Then the original equation* (52) *is equivalent to the operator equation* (53). *The operator* $A: X \to X^*$ *given by*

$$\langle Au, v \rangle = a(u, v) \qquad \text{for all} \quad u, v \in X \tag{57}$$

has the following properties:

(P) *A is linear, continuous, and strongly monotone.*

(ii) *Conversely, let* $A: X \to X^*$ *be an operator on the real* H-space X *with* (P). *Then, for each* $b \in X^*$, *the equation* $Au = b$ *has a unique solution* $u \in X$, *and this equation is equivalent to* (52).

PROOF. Ad(a). The equivalence between (52) and the operator equation $Au = b$ follows from (1a).

The existence of a unique solution $u \in X$ of $Au = b$ follows from the main theorem on linear monotone operators (Theorem 18.F in Section 18.11).

The equivalence between (52) and $Au = b$ implies the existence of a unique solution of (52).

The *a priori* estimate $c\|u\| \leq \|b\|$ follows from

$$c\|u\|^2 \leq a(u, u) = b(u) \leq \|b\| \|u\|.$$

Since the dual problem (52*) has the same structure as (52), equation (52*) has also a unique solution.

Ad(b). We use the Identification Principle 21.18 and set $X = X^*$. Then the convergence of the Galerkin method follows from the main theorem on uniquely approximation-solvable linear operator equations (Theorem 21.G(c)).

Ad(c), (d). The error estimates are obtained as in the proof of Corollary 22.2. □

In the preceding proof of (a) we used the main theorem on linear strongly monotone operators (Theorem 18.F) whose proof was based on the theorem of Riesz and on the perpendicular principle, i.e., on the *idea of orthogonality*. In the following we want to give an alternative simple existence proof based on the Riesz theorem and on the *Banach fixed-point principle*. This argument will be used in Chapter 25 in order to prove the main theorem on Lipschitz continuous strongly monotone nonlinear operators and to prove the main theorem on quadratic variational *inequalities*.

EXISTENCE PROOF VIA THE BANACH FIXED-POINT THEOREM. Let $a: X \times X \to \mathbb{K}$ be bilinear, bounded, and strongly positive on the H-space X over $\mathbb{K} = \mathbb{R}, \mathbb{C}$, i.e.,

$$|a(u, v)| \leq d\,\|u\|\,\|v\|,$$

$$\operatorname{Re} a(u, u) \geq c\,\|u\|^2 \qquad \text{for all} \quad u, v \in X.$$

We want to prove that, for each given $b \in X^*$, the equation

$$a(u, v) = b(v) \qquad \text{for all} \quad v \in X \tag{58}$$

has a unique solution $u \in X$.

(I) Equivalent operator equation via the Riesz theorem. It follows from $|a(u, v)| \leq d\,\|u\|\,\|v\|$ for all $u, v \in X$ that

$$v \mapsto a(u, v)$$

is a linear continuous functional on X. By the Riesz theorem, there exists a unique element in X denoted by Au such that

$$a(u, v) = (Au|v) \qquad \text{for all} \quad u, v \in X.$$

Moreover, $\|Au\| \leq d\,\|u\|$. For all $u, v, w \in X$ and $\alpha, \beta \in \mathbb{K}$, we obtain

$$\begin{aligned}(A(\alpha u + \beta w)|v) &= \alpha a(u, v) + \beta a(w, v) \\ &= (\bar{\alpha} Au + \bar{\beta} Aw|v).\end{aligned}$$

This implies $A(\alpha u + \beta w) = \bar{\alpha} Au + \bar{\beta} Aw$, i.e., the operator $A: X \to X$ is antilinear.

Again by the Riesz theorem, there is an $f \in X$ such that

$$b(v) = (f|v) \qquad \text{for all} \quad v \in X.$$

Hence the original equation (58) is equivalent to the antilinear operator equation

$$Au = f, \qquad u \in X. \tag{58*}$$

(II) The equivalent fixed-point problem. Equation (58*) is equivalent to the fixed-point equation

$$u = u - t(Au - f), \qquad u \in X, \tag{59}$$

for fixed $t > 0$. This is the decisive *trick* of the proof.

(III) The Banach fixed-point theorem. For all $u \in X$,

$$\begin{aligned}\|(I - tA)u\|^2 &= \|u\|^2 + t^2\|Au\|^2 - 2t\,\mathrm{Re}(Au|u)\\ &\leq \|u\|^2(1 + t^2d^2 - 2ct).\end{aligned}$$

We now choose $t > 0$ in such a way that

$$k^2 = 1 + t^2d^2 - 2ct < 1.$$

For example, let $t = c/d^2$. Then $k^2 = 1 - c^2/d^2 < 1$. Note that $0 < c \leq d$.

Thus, the operator $u \mapsto u - t(Au - f)$ is a k-contraction on X. By the Banach fixed-point theorem (Theorem 1.A), equation (59) has a unique solution. □

22.5a. The Abstract Green Operator

Definition 22.19. Let Y be a real B-space. The operator $K: Y \to Y^*$ is called *angle-bounded* iff K is linear and monotone, and there is a constant $\gamma \geq 0$ such that

$$|\langle Ku, v\rangle - \langle Kv, u\rangle|^2 \leq \gamma\langle Ku, u\rangle\langle Kv, v\rangle \qquad \text{for all} \quad u, v \in X.$$

Let Y be a real H-space. The operator $K: Y \to Y$ is called angle-bounded iff K is linear and monotone, and there is a constant $\gamma \geq 0$ such that

$$|(Ku|v) - (Kv|u)|^2 \leq \gamma(Ku|u)(Kv|v) \qquad \text{for all} \quad u, v \in X.$$

In particular, each linear, monotone, and symmetric operator $K: Y \to Y$ is angle-bounded. As we will show in Chapter 28, angle-bounded operators play an important role in the theory of nonlinear Hammerstein equations. In particular, in connection with semilinear elliptic boundary value problems, the following result will be used in Chapter 28.

We set

$$b(v) = (f|v)_Y \qquad \text{for all} \quad v \in X$$

and consider again the equation

$$a(u, v) = b(v) \qquad \text{for fixed} \quad u \in X \quad \text{and all} \quad v \in X \tag{60}$$

together with its dual equation

$$a(v, u^*) = b(v) \qquad \text{for fixed} \quad u^* \in X \quad \text{and all} \quad v \in X. \tag{60*}$$

We make the following assumptions.

(H1) The map $a: X \times X \to \mathbb{R}$ is bilinear, bounded, and strongly positive on the real H-space X.
(H2) Let Y be a real H-space such that the embedding $X \subseteq Y$ is continuous.

As in Section 22.1b we obtain that $b \in X^*$ for given $f \in Y$. By Theorem 22.C, equation (60) (resp. (60*)) has a unique solution u (resp. u^*). We set

$$u = Kf$$

(resp. $u^* = Lf$), and we call $K: Y \to Y$ the *abstract Green operator* or also the solution operator to equation (60).

Corollary 22.20. *Assume* (H1), (H2). *Then*:

(a) *The solution operator $K: Y \to Y$ is linear, continuous, monotone, and angle-bounded.*
(b) *If the embedding $X \subseteq Y$ is compact, then $K: Y \to Y$ is compact.*
(c) *If X is dense in Y, then K is strictly monotone.*
(d) *The adjoint operator $K^*: Y \to Y$ coincides with the solution operator L of the dual equation* (60*).

PROOF. Ad(a). As in the proof of Corollary 22.4, we obtain that $K: Y \to Y$ is linear, continuous, and monotone. It follows from

$$|a(u, v)| \leq d\,\|u\|\,\|v\|,$$
$$a(u, u) \geq c\,\|u\|^2 \qquad \text{for all} \quad u, v \in X$$

that

$$\begin{aligned}|(f|Kg)_Y - (g|Kf)_Y|^2 &= |a(Kf, Kg) - a(Kg, Kf)|^2 \\ &\leq 4d^2\|Kf\|^2\|Kg\|^2 \leq 4d^2c^{-2}a(Kf, Kf)a(Kg, Kg) \\ &= 4d^2c^{-2}(f|Kf)_Y(g|Kg)_Y \qquad \text{for all} \quad f, g \in Y.\end{aligned}$$

Hence $K: Y \to Y$ is angle-bounded.

Ad(b). Compare the proof of Corollary 22.4.

Ad(c). Let $b(v) = (f|v)_Y$ for all $v \in X$. If X is dense in Y, then $f \neq 0$ implies $b \neq 0$ and hence $Kf \neq 0$, since (60) has a unique solution. Thus, $Kf = 0$ implies $f = 0$. Now suppose that

$$(g|Kg)_Y = 0.$$

This implies $(g|Kg)_Y = a(Kg, Kg) = 0$ and hence $Kg = 0$. Consequently, $g = 0$. This means that K is strictly monotone.

Ad(d). For all $f, g \in Y$,

$$(f|Kg)_Y = a(Kg, Lf) = (g|Lf)_Y.$$

Hence $K^* = L$. $\square$

22.5b. The Duality Trick and Refined Error Estimates

We want to improve substantially the *a priori* error estimates for the Galerkin method in Theorem 22.C. To this end, we set

$$b(v) = (f|v)_Y \qquad \text{for all} \quad v \in X,$$

and we consider the equation

$$a(u, v) = b(v) \qquad \text{for fixed} \quad u \in X \quad \text{and all} \quad v \in X \tag{61}$$

together with its *dual* equation

$$a(v, u^*) = b(v) \qquad \text{for fixed} \quad u^* \in X \quad \text{and all} \quad v \in X. \tag{61*}$$

In addition, we consider the Galerkin equation

$$a(u_n, v) = b(v) \qquad \text{for fixed} \quad u_n \in X_n \quad \text{and all} \quad v \in X_n, \tag{62}$$

where $n = 1, 2, \ldots$. We make the following assumptions.

(H1) X, Y, and Z are real H-spaces, where the embedding $X \subseteq Y$ is continuous. Let X_n be a finite-dimensional subspace of X.

(H2) The map $a: X \times X \to \mathbb{R}$ is bilinear, bounded, and strongly positive.

(H3) *Approximation.* There exists a constant α_n such that

$$\operatorname{dist}_X(z, X_n) \le \alpha_n \|z\|_Z \qquad \text{for all} \quad z \in X \cap Z.$$

(H4) *A priori estimates.* There are constants C and C^* such that

$$\|u\|_Z \le C\|f\|_Y, \qquad \|u^*\|_Z \le C^*\|f\|_Y$$

for all $f \in Y$ and all the corresponding solutions u (resp. u^*) of equation (61) (resp. (61*)).

It follows from Theorem 22.C that for fixed $f \in Y$, the equations (61), (61*), (62) have unique solutions u, u^*, u_n, respectively.

Corollary 22.21. *Assume* (H1) *through* (H4). *Then we obtain the following a priori error estimates for the Galerkin method*:

$$\|u - u_n\|_X \le \alpha_n (c^{-1} dC)\|f\|_Y \qquad \textit{for all} \quad f \in Y. \tag{63}$$

In the space Y, we get the following stronger a priori error estimates:

$$\|u - u_n\|_Y \le \alpha_n^2 (c^{-1} d^2 CC^*)\|f\|_Y \qquad \textit{for all} \quad f \in Y. \tag{64}$$

The point is that the factor α_n^2 appears in (64) in contrast to (63). This is responsible for the *quadratic* convergence of the Galerkin method in Y.

PROOF. Let $\|\cdot\|$ denote the norm on X. By Section 22.5a,

$$u = Kf \qquad \text{and} \qquad u^* = K^*f.$$

(I) Subtracting (62) from (61) we get the *first key* relation:

$$a(u - u_n, v) = 0 \qquad \text{for all} \quad v \in X_n.$$

Hence, for all $v \in X_n$,

$$\begin{aligned} c\|u - u_n\|^2 &\le a(u - u_n, u - u_n) = a(u - u_n, u - v) \\ &\le d\|u - u_n\|\,\|u - v\|. \end{aligned}$$

This implies

$$c\|u - u_n\| \le d \cdot \operatorname{dist}_X(u, X_n) \le \alpha_n d\|u\|_Z \le \alpha_n dC\|f\|_Y.$$

This is (63).

(II) *The duality trick.* For all $g \in Y$ and all $v \in X_n$, we obtain the *second key* relation:

$$\begin{aligned} (u - u_n|g)_Y &= a(u - u_n, K^*g) = a(u - u_n, K^*g - v) \\ &\le d\|u - u_n\|\,\|K^*g - v\|. \end{aligned}$$

Hence

$$\begin{aligned} (u - u_n|g)_Y &\le d\|u - u_n\| \cdot \operatorname{dist}_X(K^*g, X_n) \\ &\le (\alpha_n d^2 c^{-1} C\|f\|_Y)(\alpha_n C^*\|g\|_Y) \end{aligned}$$

for all $f \in Y$. This implies (64). □

The duality trick above is due to Aubin (1967) and Nitsche (1970).

22.6. Application to Boundary Value Problems

We consider the first boundary value problem

$$\begin{aligned} -\Delta u + \sum_{i=1}^{N} a_i D_i u + a_0 u = f \quad &\text{on } G, \\ u = 0 \quad &\text{on } \partial G \end{aligned} \tag{65}$$

together with the dual problem

$$\begin{aligned} -\Delta u^* - \sum_{i=1}^{N} D_i(a_i u^*) + a_0 u^* = f \quad &\text{on } G, \\ u^* = 0 \quad &\text{on } \partial G. \end{aligned} \tag{65*}$$

We make the following assumptions.

(H) G is a bounded region in $\mathbb{R}^N$, $N \geq 1$, and $a_j \in C(\bar{G})$ for all j. Let $X = \mathring{W}_2^1(G)$.

We set

$$a(u, v) = \int_G \left(\sum_{i=1}^N D_i u D_i v + (a_i D_i u)v + a_0 uv \right) dx,$$

$$b(v) = \int_G fv \, dx.$$

Definition 22.22. Assume (H). The *generalized problem* to (65) (resp. (65*)) reads as follows. For given $f \in L_2(G)$ we seek $u \in X$ (resp. $u^* \in X$) such that

$$a(u, v) = b(v) \qquad \text{for all} \quad v \in X \tag{66}$$

(resp.

$$a(v, u^*) = b(v) \qquad \text{for all} \quad v \in X). \tag{66*}$$

Formally, we obtain (66) from (65) by multiplying (65) with $v \in C_0^\infty(G)$ and by using subsequent integration by parts. Similarly, we obtain (66*) from (65*). Note that in contrast to the classical problem (65*), we do not need the differentiability of a_j in (66*).

Proposition 22.23. *Assume* (H). *Let* c_0 *be the constant in the Poincaré–Friedrichs inequality* (23) *and suppose that*

$$\beta + \gamma < c_0/(1 + c_0), \qquad \beta, \gamma \geq 0,$$

where $\min_{x \in \bar{G}} a_0(x) \geq -\gamma$ *and*

$$\max_{x \in \bar{G}} |a_j(x)| \leq \beta, \qquad j = 1, \ldots, N.$$

Then all the assertions of Theorem 22.C *in Section* 22.5 *are valid for the two generalized problems* (66) *and* (66*). *In particular,* (66) *and* (66*) *have unique solutions.*

PROOF.

(I) By (1b), $b \in X^*$.

(II) We investigate $a: X \times X \to \mathbb{R}$. By the Hölder inequality, for all $u, v \in X$,

$$|a(u, v)| \leq \sum_i \int_G \left(|D_i u D_i v| + \beta |(D_i u) v| \right) dx + \max|a_0| \int_G |uv| \, dx$$

$$\leq \sum_i \|D_i u\|_2 \|D_i v\|_2 + \beta \|D_i u\|_2 \|v\|_2$$

$$+ \max|a_0| \, \|u\|_2 \|v\|_2 \leq \text{const} \, \|u\|_{1,2} \|v\|_{1,2}.$$

Hence, $a(\cdot, \cdot)$ is bounded and bilinear.

To prove the *strong positivity* of $a(\cdot, \cdot)$, we use the Poincaré–Friedrichs inequality. By (24), we obtain for the principal part of $a(u, v)$:

$$\int_G \sum_i (D_i u)^2 \, dx \geq c \|u\|_{1,2}^2 \qquad \text{for all} \quad u \in X,$$

where $c = c_0/(1 + c_0)$. Clearly,

$$\int_G a_0 u^2 \, dx \geq -\gamma \int_G u^2 \, dx \geq -\gamma \|u\|_{1,2}^2 \qquad \text{for all} \quad u \in X.$$

By the Hölder inequality,

$$\sum_i \left| \int_G a_i (D_i u) u \, dx \right| \leq \beta \|u\|_{1,2}^2 \qquad \text{for all} \quad u \in X.$$

Putting these three inequalities together, we get

$$a(u, u) \geq (c - \gamma - \beta) \|u\|_{1,2}^2 \qquad \text{for all} \quad u \in X,$$

i.e., $a(\cdot, \cdot)$ is strongly positive.

Thus, all the assumptions of Theorem 22.C are fulfilled. □

The results of the next section will be used in Section 22.10 in order to consider the first boundary value problem (65) in the case where the coefficients $|a_j|$ are arbitrarily large. Then the unique solvability of (65) can be violated and we have to apply Fredholm alternatives.

22.7. Compact Perturbations of Strongly Monotone Operators, Fredholm Alternatives, and the Galerkin Method

We consider the operator equation

$$Au + Cu = b, \qquad u \in X, \tag{67}$$

together with the *dual* equation

$$A^* u^* + C^* u^* = 0, \qquad u^* \in X. \tag{67*}$$

The Galerkin equations to (67) have the following form:

$$\langle Au_n + Cu_n, v \rangle = \langle b, v \rangle \qquad \text{for fixed} \quad u_n \in X_n \quad \text{and all} \quad v \in X_n.$$

Explicitly, that means

$$\langle Au_n + Cu_n, w_j \rangle = \langle b, w_j \rangle, \qquad j = 1, \ldots, n. \tag{68}$$

We seek the unknown real coefficients c_{jn} of the Galerkin approximation $u_n = c_{1n} w_1 + \cdots + c_{nn} w_n$.

Theorem 22.D. *Suppose that the operators A, C: $X \to X^*$ are linear and continuous on the real* H-*space* X. *Furthermore, suppose that A is strongly monotone, and C is compact. Then*:

(a) Uniqueness implies existence. *If the original equation* (67) *with* $b = 0$ *has only the trivial solution* $u = 0$, *then for each* $b \in X^*$, *equation* (67) *has a unique solution* u, *and*

$$\|u\| \leq \text{const}\,\|b\|.$$

(b) Fredholm alternative. *For fixed* $b \in X^*$, *equation* (67) *has a solution iff*

$$\langle b, u^* \rangle = 0$$

for all solutions u^* *of* (67*).

(c) Index equals zero. *Equation* (67) *with* $b = 0$ *and equation* (67*) *have the same finite number of linearly independent solutions.*

(d) Convergence of the Galerkin method. *Suppose that* $\{w_1, w_2, \ldots\}$ *is a basis in the infinite-dimensional separable* H-*space* X, *and set* $X_n = \text{span}\{w_1, \ldots, w_n\}$. *Moreover, suppose that for fixed* $b \in X^*$, *equation* (67) *has a unique solution* u.

Then there is a number $n_0(b)$ *such that, for each* $n \geq n_0(b)$, *the Galerkin equation* (68) *has a unique solution, and*

$$u_n \to u \quad \textit{in } X \qquad \textit{as} \quad n \to \infty.$$

This theorem tells us that $A + C$ is a Fredholm operator of index zero. Fredholm alternatives for nonlinear operators will be considered in Chapter 29.

Proof. Using the Identification Principle 21.18, we set $X = X^*$. Now, the assertions follow immediately from Theorem 21.F (Riesz–Schauder theory) and Theorem 21.G(d). Note that A: $X \to X^*$ is bijective, by Theorem 22.C. $\square$

The following example shows that Theorem 22.D generalizes a well-known result for linear equations in $\mathbb{R}^N$.

Example 22.24. Let B: $\mathbb{R}^N \to \mathbb{R}^N$ be a linear operator, i.e., B is an $N \times N$-matrix, and B^* is the adjoint matrix. Then the equation

$$Bu = b, \qquad u \in \mathbb{R}^N,$$

has a solution iff $\langle b|u^* \rangle = 0$ for all solutions u^* of the dual equation $B^*u^* = 0$.

Proof. Use Theorem 22.D with $X = \mathbb{R}^N$ and $A = I$, $C = B - I$. Note that the continuous operator C is compact since $\dim X < \infty$. $\square$

In what follows we will apply Theorem 22.D to integral equations, bilinear forms, and boundary value problems.

22.8. Application to Integral Equations

We consider the Fredholm integral equation of the second kind

$$u(x) - \int_G k(x, y)u(y)\,dy = b(x), \qquad x \in G, \tag{69}$$

together with the adjoint equation

$$u^*(x) - \int_G k(y, x)u^*(y)\,dy = 0, \qquad x \in G. \tag{69*}$$

The Galerkin equations for $u_n = c_{1n}w_1 + \cdots + c_{nn}w_n$ are given by

$$\int_G u_n w_j\,dx - \int_{G\times G} k(x, y)u_n(y)w_j(x)\,dy\,dx = \int_G bw_j\,dx, \tag{70}$$

where $j = 1, \ldots, n$.

Proposition 22.25. *Suppose that:*

(i) *G is a nonempty measurable set in $\mathbb{R}^N$, $N \geq 1$ (e.g., G is open). Let $X = L_2(G)$.*
(ii) *We set $X_n = \operatorname{span}\{w_1, \ldots, w_n\}$, where $\{w_1, w_2, \ldots\}$ is a basis in X.*
(iii) *The function $k: G \times G \to \mathbb{R}$ is measurable (e.g., continuous) and square-integrable, i.e., $\int_{G\times G} k(x, y)^2\,dx\,dy < \infty$.*

Then:

(a) *For fixed $b \in X$, equation (69) has a solution $u \in X$ iff*

$$\int_G bu^*\,dx = 0$$

for all solutions $u^* \in X$ of the dual equation (69).*
(b) *Equation (69) with $b(x) \equiv 0$ and equation (69*) have the same finite number of linearly independent solutions in X.*
(c) *Suppose that equation (69) with $b(x) \equiv 0$ has only the trivial solution $u = 0$ in X. Then, for each $b \in X$, equation (69) has a unique solution u in X.*

Furthermore, for each fixed $b \in X$, there is a number $n_0(b)$ such that, for each $n \geq n_0(b)$, the Galerkin equation (70) has a unique solution $u_n \in X_n$, and

$$u_n \to u \quad \text{in } X \qquad \text{as} \quad n \to \infty.$$

Proof. We set

$$(Ku)(x) = \int_G k(x, y)u(y)\,dy, \qquad x \in G.$$

By Problem 21.11, the operator $K: X \to X$ is linear and compact. The adjoint operator $K^*: X \to X$ is given by

$$(K^*u)(x) = \int_G k(y, x)u(y)\, dy.$$

Then the assertions follow from Theorem 22.D by letting $X = X^*$, $A = I$, and $C = -K$. $\square$

22.9. Application to Bilinear Forms

With a view to elliptic boundary value problems we consider the equation

$$a(u, v) + c(u, v) = b(v) \qquad \text{for fixed} \quad u \in X \quad \text{and all} \quad v \in X, \tag{71}$$

together with the *dual* equation

$$a(v, u^*) + c(v, u^*) = 0 \qquad \text{for fixed} \quad u^* \in X \quad \text{and all} \quad v \in X. \tag{71*}$$

The Galerkin equations for $u_n = c_{1n}w_1 + \cdots + c_{nn}w_n$ are given by

$$a(u_n, w_j) + c(u_n, w_j) = b(w_j), \qquad j = 1, \ldots, n. \tag{72}$$

Proposition 22.26. *Suppose that*:

(i) *X is a real separable* H-*space. Set $X_n = \operatorname{span}\{w_1, \ldots, w_n\}$, where $\{w_1, w_2, \ldots\}$ is a basis in X.*

(ii) *The maps a, c: $X \times X \to \mathbb{R}$ are bilinear and bounded. Moreover, a is strongly positive, c is compact, and $b \in X^*$.*

Then:

(a) *Suppose that equation* (71) *with $b = 0$ has only the trivial solution $u = 0$. Then, for each $b \in X^*$, equation* (71) *has a unique solution u, and $\|u\| \le \text{const}\,\|b\|$.*

Let $\dim X = \infty$. Then, for each $b \in X^*$, there is a number $n_0(b)$ such that, for each $n \ge n_0(b)$, the Galerkin equation (72) *has a unique solution $u_n \in X_n$, and*

$$u_n \to u \quad \text{in } X \qquad \text{as} \quad n \to \infty.$$

(b) *Equation* (71) *has a solution iff*

$$b(u^*) = 0$$

for all solutions u^* of (71*).

(c) *Equation* (71) *with $b = 0$ and the dual equation* (71*) *have the same finite number of linearly independent solutions.*

PROOF. By (1a), there exist linear continuous operators A, C: $X \to X^*$ such that

$$\langle Au, v\rangle = a(u, v), \qquad \langle Cu, v\rangle = c(u, v) \qquad \text{for all} \quad u, v \in X.$$

Here, A is strongly monotone, and C is compact. Moreover, since $X^{**} = X$,

$$\langle u, A^*v\rangle = a(u,v), \qquad \langle u, C^*v\rangle = c(u,v) \qquad \text{for all} \quad u, v \in X.$$

Thus, (71) and (71*) is equivalent to (67) and (67*), respectively. Hence, Theorem 22.D yields the assertions. □

22.10. Application to Boundary Value Problems

We consider the first boundary value problem

$$\begin{aligned} -\Delta u + \sum_{i=1}^{N} a_i D_i u + a_0 u &= f \quad \text{on } G, \\ u &= 0 \quad \text{on } \partial G \end{aligned} \tag{73}$$

together with the dual problem

$$\begin{aligned} -\Delta u^* - \sum_{i=1}^{N} D_i(a_i u^*) + a_0 u^* &= 0 \quad \text{on } G, \\ u^* &= 0 \quad \text{on } \partial G. \end{aligned} \tag{73*}$$

We set

$$a(u,v) = \int_G \sum_{i=1}^{N} D_i u D_i v\,dx,$$

$$c(u,v) = \int_G \sum_{i=1}^{N} (a_i D_i u)v + a_0 uv\,dx, \qquad b(v) = \int_G fv\,dx,$$

i.e., $a(\cdot,\cdot)$ and $c(\cdot,\cdot)$ correspond to the principal part and to the lower order terms in (73), respectively.

Definition 22.27. Let G be a bounded region in $\mathbb{R}^N$, $N \geq 1$. Set $X = \mathring{W}_2^1(G)$. Suppose that $a_j \in C(\bar{G})$ for all j. Then the *generalized problem* to (73) (resp. (73*)) reads as follows. For given $f \in L_2(G)$, we seek $u \in X$ (resp. $u^* \in X$) such that

$$a(u,v) + c(u,v) = b(v) \qquad \text{for all} \quad v \in X \tag{74}$$

(resp.

$$a(v,u^*) + c(v,u^*) = 0 \qquad \text{for all} \quad v \in X). \tag{74*}$$

By the proof of Proposition 22.7, the map $a: X \times X \to \mathbb{R}$ is bilinear bounded, and strongly positive. Below we shall prove that $c: X \times X \to \mathbb{R}$ is linear, bounded, and compact. This corresponds to the following general principle for elliptic differential equations:

Lower order terms yield compactness.

This observation also remains true for quasi-linear elliptic differential equations. This will be the key to the theory of pseudomonotone operators in Chapter 27.

Proposition 22.28. *All the assertions of Proposition* 22.26 *are true for the generalized problems* (74) *and* (74*). *In particular, for* $f \in L_2(G)$, *the original problem* (73) *has a generalized solution* $u \in X$ *iff*

$$\int_G fu^* \, dx = 0$$

for all generalized solutions $u^* \in X$ *of the dual problem* (73*).

PROOF. By (1b), $b \in X$.*

As in the proof of Proposition 22.23, we obtain that $c: X \times X \to \mathbb{R}$ is linear and bounded. To prove the compactness of $c(\cdot, \cdot)$, assume

$$u_n \rightharpoonup u \quad \text{and} \quad v_n \rightharpoonup v \quad \text{in } X \quad \text{as} \quad n \to \infty.$$

Since the embedding $X \subseteq L_2(G)$ is compact, we get

$$u_n \to u \quad \text{and} \quad v_n \to v \quad \text{in } L_2(G) \quad \text{as} \quad n \to \infty.$$

This implies

$$a_i v_n \to a_i v \quad \text{in } L_2(G) \quad \text{as} \quad n \to \infty.$$

By Problem 21.4, it follows from $u_n \rightharpoonup u$ in X that

$$D_i u_n \rightharpoonup D_i u \quad \text{in } L_2(G) \quad \text{as} \quad n \to \infty,$$

and hence

$$(D_i u_n | a_i v_n)_2 \to (D_i u | a_i v)_2 \quad \text{as} \quad n \to \infty,$$

where $(\cdot|\cdot)_2$ denotes the scalar product in $L_2(G)$. Therefore,

$$c(u_n, v_n) \to c(u, v) \quad \text{as} \quad n \to \infty,$$

i.e., $c(\cdot, \cdot)$ is compact.

Consequently, all the assumptions of Proposition 22.26 are satisfied. □

22.11. Eigenvalue Problems and the Ritz Method

We want to study the eigenvalue problem

$$Bu = \lambda u. \tag{75}$$

Here, we seek $\lambda \in \mathbb{R}$ and $u \in X$ with $u \neq 0$. Along with (75) we consider the Ritz equation

$$(Bu | w_{jn}) = \lambda (u | w_{jn}), \quad j = 1, \ldots, n', \tag{76}$$

for $\lambda \in \mathbb{R}$ and $u = c_{1n}w_{1n} + \cdots + c_{n'n}w_{n'n}$ with $u \neq 0$. The Ritz equation (76) represents an eigenvalue problem for a symmetric matrix.

We make the following assumptions.

(H1) X is a real separable H-space, $X \neq \{0\}$.

(H2) The operator $B: X \to X$ is linear, symmetric, compact, and strictly monotone.

(H3) (X_n) is a Galerkin scheme in X. Let $X_n = \operatorname{span}\{w_{1n}, \ldots, w_{n'n}\}$.

22.11a. The Variational Approach

In order to solve the eigenvalue problem (75), we consider the variational problem

$$\max(Bu|u) = \lambda, \qquad (u|u) = 1, \qquad u \in X. \tag{77a}$$

The solution is denoted by λ_1 and u_1. For $m = 2, 3, \ldots,$ we now determine successively (λ_m, u_m) by solving the variational problems

$$\begin{gathered} \max(Bu|u) = \lambda, \qquad (u|u) = 1, \qquad u \in X, \\ (u_i|u) = 0, \qquad i = 1, \ldots, m-1. \end{gathered} \tag{77b}$$

In order to characterize directly the mth eigenvvalue λ_m, we will use the Courant maximum–minimum principle

$$\lambda_m = \max_{M \in \mathscr{L}_m} \min_{u \in M} (Bu|u). \tag{78}$$

Here, $\mathscr{L}_m$ is the class of all the sets $S \cap L$, where S is the surface of the unit ball in X, and L is an arbitrary m-dimensional linear subspace of X. This principle was used by Fischer (1905), Weyl (1911), and Courant (1920).

In 1911, Hermann Weyl used this principle in order to solve the famous problem on the asymptotic distribution of the eigenvalues of the Laplace equation.

Theorem 22.E. *Assume* (H1), (H2). *Then*:

(i) Complete orthonormal system. *The operator B has a complete orthonormal system of eigenvectors in X. The eigenvalues of B are positive, and they have finite multiplicity. The eigenvectors related to two different eigenvalues are orthogonal to each other.*

(ii) Construction of the eigensolutions. *For $m = 1, 2, \ldots,$ the variational problems* (77) *have successively the solutions (λ_m, u_m), which are eigensolutions of B. The set $\{u_1, u_2, \ldots\}$ forms a complete orthonormal system of eigenvectors to B. The set $\{\lambda_1, \lambda_2, \ldots\}$ contains all the eigenvalues of B counted according to their multiplicity. Furthermore,*

$$\lambda_1 \geq \lambda_2 \geq \cdots > 0.$$

If $\dim X = \infty$, then $\lambda_m \to 0$ as $m \to \infty$.

(iii) The Courant maximum–minimum principle. *For* $m = 1, 2, \ldots$, *the eigenvalue* λ_m *is obtained through* (78). *Moreover*, u_m *is solution of* (78).

The convergence of the Ritz method will be proved in Section 22.11c. The Ljusternik–Schnirelman theory generalizes the Courant maximum–minimum principle to nonlinear eigenvalue problems. This will be studied in detail in Part III.

PROOF. Ad(i), (ii). If $Bu = 0$, then $(Bu|u) = 0$ and hence $u = 0$, because B is strictly monotone. Consequently, assertions (i) and (ii) are special cases of the Hilbert–Schmidt theory considered in Section 19.6.

Ad(iii). We set

$$\mathscr{A} = \max_{M \in \mathscr{L}_m} \min_{u \in M} (Bu|u).$$

Let $M \in \mathscr{L}_m$. Since $\dim M = m$, there exists a $u \in M$ with

$$(u|u) = 1 \qquad \text{and} \qquad (u_i|u) = 0 \qquad \text{for } i = 1, \ldots, m-1,$$

where we set $u_0 = 0$. From (77) it follows that

$$(Bu|u) \leq \lambda_m,$$

and hence $\mathscr{A} \leq \lambda_m$.

In order to obtain $\mathscr{A} \geq \lambda_m$, set $M_0 = S \cap \operatorname{span}\{u_1, \ldots, u_m\}$. Then $M_0 \in \mathscr{L}_m$. If $u \in M_0$, then $u = b_1 u_1 + \cdots + b_m u_m$ and

$$\begin{aligned}(Bu|u) &= \lambda_1 b_1^2 + \cdots + \lambda_m b_m^2 \\ &\geq \lambda_m(b_1^2 + \cdots + b_m^2) = \lambda_m.\end{aligned}$$

Noting $(Bu_m|u_m) = \lambda_m$, we get $\min_{u \in M_0}(Bu|u) = \lambda_m$, and hence $\mathscr{A} \geq \lambda_m$. □

In order to solve the variational problem (77a), we used a simple special argument in Section 19.7, which is only applicable to linear problems. In the following, we give an alternative existence proof which can be generalized to nonlinear problems. In particular, we will use the following argument in Section 43.4 in order to prove the existence of a bifurcation point for nonlinear operator equations.

EXISTENCE PROOF FOR THE LARGEST EIGENVALUE VIA MAXIMUM TRICK. We study the maximum problem

$$\max(Bu|u) = \lambda, \qquad (u|u) = 1, \qquad u \in X. \tag{79}$$

Assume (H1), (H2). The trick is to consider first the modified maximum problem

$$\max_{u \in K} (Bu|u) = \lambda, \tag{80}$$

where $K = \{u \in X : \|u\| \leq 1\}$, and then to show that the solution u of (80) satisfies $u \in \partial K$, i.e., u is also a solution of (79).

(I) Existence for (80). Set $\alpha = \sup_{u \in K} (Bu|u)$. Then there is a sequence (u_n) in K such that

$$(Bu_n|u_n) \to \alpha \qquad \text{as} \quad n \to \infty.$$

Since (u_n) is a bounded sequence in the H-space X, there exists a subsequence, again denoted by (u_n), such that

$$u_n \rightharpoonup u \qquad \text{as} \quad n \to \infty.$$

Since B is compact, $(Bu_n|u_n) \to (Bu|u)$ as $n \to \infty$, by Corollary 21.33. Hence $(Bu|u) = \alpha$, i.e., u is a solution of (80).

(II) We show that $\|u\| = 1$. Otherwise, we have $\|u\| < 1$. Then, for each $v \in X$, the real function

$$\varphi(t) = (B(u + tv)|u + tv), \qquad t \in \mathbb{R},$$

has a local maximum at $t = 0$, i.e., $\varphi'(0) = 0$. Hence

$$(Bu|v) = 0 \qquad \text{for all} \quad v \in X.$$

This implies $\alpha = (Bu|u) = 0$. However, since B is strictly monotone, we obtain $(Bv|v) > 0$ for all $v \neq 0$. Hence $\alpha > 0$. This is a contradiction. □

22.11b. A Comparison Principle for Eigenvalues

The following important results are easy consequences of the Courant maximum–minimum principle.

Corollary 22.29. *Suppose that the operators B, C: $X \to X$ are linear, symmetric, compact, and strictly monotone on the real separable* H*-space $X \neq \{0\}$. Then:*

(a) *The operator B (resp. C) has the eigenvalues*

$$0 < \cdots \leq \lambda_2 \leq \lambda_1$$

(resp.

$$0 < \cdots \leq \mu_2 \leq \mu_1),$$

counted according to their multiplicity.

(b) *If $B \leq C$, i.e., $(Bu|u) \leq (Cu|u)$ for all $u \in X$, then $\lambda_m \leq \mu_m$ for all m.*

(c) *$|\lambda_m - \mu_m| \leq \|B - C\|$ for all m.*

Proof. Ad(a). Cf. Theorem 22.E.

Ad(b). This follows immediately from the Courant maximum–minimum principle (78).

Ad(c). By (78),

$$\lambda_m = \max_{u \in \mathscr{L}_m} \min_{u \in M} (Bu|u),$$

$$\mu_m = \max_{u \in \mathscr{L}_m} \min_{u \in M} (Cu|u).$$

Set $B = C + D$. Note that $(Du|u) \geq -\|Du\|\,\|u\| \geq -\|D\|\,\|u\|^2$, by the Schwarz inequality. Hence

$$\min_{u \in M} (Bu|u) \geq \min_{u \in M} (Cu|u) + \min_{u \in M} (Du|u)$$
$$\geq \min_{u \in M} (Cu|u) - \|D\|.$$

This implies

$$\lambda_m \geq \mu_m - \|B - C\|,$$

and hence $\mu_m \geq \lambda_m - \|C - B\|$, i.e., $|\lambda_m - \mu_m| \leq \|B - C\|$. □

22.11c. The Convergence of the Ritz Method

Let $P_n: X \to X_n$ be the orthogonal projection operator from X onto X_n. Then the Ritz equation (76) is equivalent to the equation

$$P_n Bu = \lambda u, \qquad \lambda \in \mathbb{R}, \quad u \in X_n, \quad n = 1, 2, \ldots. \tag{81}$$

Corollary 22.30. *Assume* (H1) *through* (H3) *of Theorem* 22.E. *Let* $\dim X = \infty$. *Then*:

(a) Ritz equation. *The Ritz eigenvalue problem* (81) *has the eigenvalues*

$$\lambda_{1n} \geq \lambda_{2n} \geq \cdots \geq \lambda_{n'n} > 0$$

counted according to their multiplicity. Moreover, equation (81) *has a complete orthonormal system of eigenvectors in* X_n. *For all* $m \leq n$,

$$\lambda_{mn} \leq \lambda_m, \tag{82}$$

where λ_m *is the* mth *eigenvalue of the original equation* $Bu = \lambda u$.

(b) Error estimates for the eigenvalues and convergence of the eigenvalues. *For all* $m \leq n$,

$$|\lambda_{mn} - \lambda_m| \leq \|B - P_n B\|, \tag{83}$$

where $\|B - P_n B\| \to 0$ *as* $n \to \infty$. *Hence the sequence* (λ_{mn}) *of the approximate eigenvalues converges to* λ_m *as* $n \to \infty$.

If $X_1 \subseteq X_2 \subseteq \cdots$, *then* (λ_{mn}) *is monotonically increasing as* $n \to \infty$.

(c) Error estimates for the eigenvectors. *Let* u_{mn} *be an eigenvector of the Ritz problem* (81) *corresponding to the eigenvalue* λ_{mn}. *Then there is an eigenvector* u *of the original problem* $Bu = \lambda_m u$ *such that*

$$\|u_{mn} - u\| \leq C_m \|B - P_n B\|, \tag{84}$$

where the constant C_m *is independent of* n.

In the following proof we will essentially make use of the Courant maximum–minimum principle.

Proof. Ad(a). It follows from $P_n^* = P_n$ that $(P_n Bu|v) = (Bu|P_n v)$ and hence

$$(P_n Bu|v) = (Bu|v) \qquad \text{for all} \quad u, v \in X_n.$$

Therefore, the operator $P_nB: X_n \to X_n$ is linear, symmetric, compact, and strictly monotone. We now apply Theorem 22.E to the operator P_nB. In particular, we obtain

$$\lambda_{mn} = \max_{M \in \mathscr{L}_m, M \subseteq X_n} \min_{u \in M} (P_nBu|u)$$
$$\leq \max_{M \in \mathscr{L}_m} \min_{u \in M} (Bu|u) = \lambda_m.$$

Moreover, if $X_1 \subseteq X_2 \subseteq \cdots$, then (λ_{mn}) is monotonically increasing as $n \to \infty$.

Ad(b). By Lemma 21.56, $\|B - P_nB\| \to 0$ as $n \to \infty$, since B is compact.

Relation (83) (resp. (84)) follows from Corollary 22.29 (resp. Problem 22.5a). $\square$

Remark 22.31 (Upper Bounds for the Eigenvalues). By (82), we obtain a *lower* bound for the eigenvalue λ_m of the original problem $Bu = \lambda u$. There exists a number of methods in the literature which provide *upper* bounds for λ_m. For example, we mention the methods of Bazley–Fox, Krylov–Bogoljubov, Knauer, Temple, Weinberger, and Weinstein–Aronszajn. Unfortunately, there is no universal method equally well suited in any situation. For example, the stability of mechanical systems in engineering depends substantially on the behavior of appropriate eigenvalues. Moreover, the energies of quantum systems (e.g., atoms, molecules, solid states, semiconductors) correspond to eigenvalues of the Schrödinger equation. Therefore, engineers and physicists are strongly interested in obtaining sharp two-sided error estimates for eigenvalues. In this connection we recommend Collatz (1963, M), Fichera (1965, L), (1978, M), Gould (1966, M), Weinstein and Stenger (1972, M), Weinberger (1974, L), Thirring (1983, M), Vol. 3 (quantum theory), Velte (1984, S) and Kuttler and Sigilito (1987, M) (cf. also Problem 22.5b).

In the following we describe a basic idea for obtaining upper bounds for λ_m. Suppose we construct a sequence (B_j) of operators $B_j: X \to X$ which are linear, symmetric, and compact such that

$$B \leq \cdots \leq B_1 \leq B_0.$$

Let $\lambda_m^{(j)}$ denote the mth eigenvalue of B_j. Then it follows from Corollary 22.29 and (82) that

$$\lambda_{mn} \leq \lambda_m \leq \lambda_m^{(j)} \qquad \text{for all } m,$$

and

$$|\lambda_m - \lambda_m^{(j)}| \leq \|B - B_j\| \qquad \text{for all } m \text{ and } j.$$

22.12. Application to Bilinear Forms

With a view to elliptic boundary value problems, we now reformulate the results of the preceding section in terms of bilinear forms. To this end, we consider the equation

$$\lambda a(u, v) = b(u, v) \qquad \text{for all} \quad v \in X. \tag{85}$$

We seek $\lambda \in \mathbb{R}$ and $u \in X$ with $u \neq 0$.

The corresponding Ritz equation reads as follows:

$$\lambda a(u, w_{jn}) = b(u, w_{jn}), \qquad j = 1, \ldots, n'. \tag{86}$$

Here, we seek $\lambda \in \mathbb{R}$ and $u = c_{1n}w_{1n} + \cdots + c_{n'n}w_{n'n}$ with $u \neq 0$. This is an eigenvalue problem for a symmetric matrix.

We make the following assumptions.

(H1) X is a real separable H-space with the scalar product $(\cdot|\cdot)_X$, where $X \neq \{0\}$.

(H2) The bilinear forms $a, b: X \times X \to \mathbb{R}$ are bounded and *symmetric*. Moreover, $a(\cdot, \cdot)$ is strongly positive, and $b(\cdot, \cdot)$ is *compact* and strictly positive.

(H3) (X_n) is a Galerkin scheme in X and $\dim X = \infty$. Let $X_n = \operatorname{span}\{w_{1n}, \ldots, w_{n'n}\}$.

By (1a), problem (85) is equivalent to the operator equation

$$\lambda Au = B_1 u, \qquad u \in X, \tag{87}$$

where

$$a(u, v) = \langle Au, v\rangle \qquad \text{and} \qquad b(u, v) = \langle B_1 u, v\rangle$$

for all $u, v \in X$. Here, the operators $A, B_1: X \to X^*$ are linear, continuous, and symmetric. Moreover, A is strongly monotone, and B_1 is compact and strictly monotone.

In order to reduce problems (85) and (87) to the situation considered in Section 22.11, we set

$$(u|v) = a(u, v) \qquad \text{for all} \quad u, v \in X.$$

Then, $(\cdot|\cdot)$ is an equivalent scalar product on X called the *energetic* scalar product. By (1a) and the Identification Principle 21.18, the original problem (85) and the operator equation (87) are equivalent to the equation

$$\lambda u = Bu, \qquad u \in X, \tag{88}$$

where

$$b(u, v) = (Bu|v) \qquad \text{for all} \quad u, v \in X.$$

The operator $B: X \to X$ is linear, symmetric, compact, and strictly monotone. Consequently, all the results of Section 22.11 can be applied to equation (88) and hence to the equivalent problems (85) and (87). However, in order to obtain the classical formulations for elliptic boundary value problems in the next section, we replace the variational problems and the Courant maximum–minimum principle by equivalent expressions. To this end, we first consider the minimum problem

$$\min a(u, u) = \lambda^{-1}, \qquad b(u, u) = 1, \qquad u \in X, \tag{89a}$$

with the solution u_1. For $m = 2, 3, \ldots$, we then consider successively the minimum problems

$$\begin{aligned} \min a(u,u) = \lambda^{-1}, \qquad & b(u,u) = 1, \qquad u \in X, \\ b(u_i, u) = 0, \qquad & i = 1, \ldots, m-1. \end{aligned} \tag{89b}$$

The Courant maximum–minimum principle is replaced by the following minimum–maximum principle:

$$\lambda_m^{-1} = \min_{M \in \mathscr{L}_m} \max_{u \in M} a(u,u). \tag{90}$$

Set $S = \{u \in X : b(u,u) = 1\}$. Then $\mathscr{L}_m$ denotes the class of all the sets $S \cap L$, where L is an arbitrary m-dimensional linear subspace of X.

Proposition 22.32. *Assume* (H1) *through* (H3). *Then*:

(i) Complete orthonormal system. *For* $m = 1, 2, \ldots$, *the variational problems* (89) *possess successively the solutions* (λ_m, u_m), *which are eigensolutions of the original problem* (85). *This way, we obtain all the eigenvalues*

$$\lambda_1 \geq \lambda_2 \geq \cdots > 0$$

of (85) *counted according to their multiplicity, and* $\lambda_m \to 0$ *as* $m \to \infty$. *Each eigenvalue* λ_m *has finite multiplicity.*

The eigenvectors $u_1, u_2, \ldots$ *form a complete orthogonal system in* X *with respect to the energetic scalar product* $(u|v) = a(u,v)$. *Furthermore,*

$$b(u_i, u_j) = \delta_{ij} \qquad \text{for all } i, j,$$

and for each $u \in X$, *the series*

$$u = \sum_{k=1}^{\infty} b(u, u_k) u_k$$

converges in X.

(ii) The Courant minimum–maximum principle. *For* $m = 1, 2, \ldots$, *the* mth *eigenvalue* λ_m *can be characterized by* (90).

(iii) Convergence of the Ritz method. *The Ritz equation* (86) *has the eigenvalues*

$$\lambda_{1n} \geq \lambda_{2n} \geq \cdots \geq \lambda_{n'n} > 0$$

counted according to their multiplicity. For each m,

$$\lim_{n \to \infty} \lambda_{mn} = \lambda_m,$$

and $\lambda_{mn} \leq \lambda_m$ *for all* $m \leq n$. *If* $X_1 \subseteq X_2 \subseteq \cdots$, *then*

$$\lambda_{mm} \leq \lambda_{m,m+1} \leq \lambda_{m,m+2} \leq \cdots \leq \lambda_m \qquad \text{for all } m.$$

For fixed m *and* $n = m, m+1, \ldots$, *let* (u_{mn}, λ_{mn}) *be an eigensolution of the Ritz equation* (86) *normalized by* $\|u_{mn}\| = 1$. *Then there exist eigenvectors* $v_1, v_2, \ldots$ *of the original problem* (85), *corresponding to the eigenvalue* λ_m, *such that*

$$\lim_{n\to\infty} \|u_{mn} - v_n\|_X = 0.$$

PROOF. This follows from Theorem 22.E in Section 22.11 and from Corollary 22.30. Note the following. By Theorem 22.E,

$$\lambda_i(u_i|u_j) = \delta_{ij} \qquad \text{for all } i, j.$$

Since $b(u_i, u_j) = \lambda_i a(u_i, u_j) = \lambda_i(u_i|u_j)$, we obtain the additional orthogonality relation

$$b(u_i, u_j) = \delta_{ij} \qquad \text{for all } i, j.$$

Moreover, by Theorem 22.E, the eigensolutions (u_m, λ_m) are obtained from the maximum problem

$$\max_{u\neq 0} \frac{b(u,u)}{(u|u)} = \lambda, \qquad (u_i|u) = 0, \qquad i = 1, \ldots, m-1.$$

Since $b(u_i, u) = \lambda_i(u_i|u)$ by (85), and since $\lambda_i \neq 0$, we obtain the equivalent minimum problem (89).

By Theorem 22.E, for each $u \in X$, there exists a convergent series

$$u = \sum_{j=1}^{\infty} \alpha_j u_j.$$

From $b(u_k, u_j) = \delta_{kj}$ we get $\alpha_k = b(u, u_k)$. □

In order to obtain strong error estimates for the Ritz method, we make the following assumptions.

(A1) X, Y, and Z are real H-spaces. The embedding $X \subseteq Y$ is continuous.

(A2) Assume (H1) through (H3) above. In particular, suppose that there are positive constants c and d such that

$$c\|u\|_X^2 \leq a(u,u),$$
$$|a(u,v)| \leq d\|u\|_X\|v\|_X \qquad \text{for all} \quad u, v \in X.$$

In addition, suppose that

$$b(u,v) = (u|v)_Y \qquad \text{for all} \quad u, v \in X.$$

(A3) *Approximation.* For each $n \in \mathbb{N}$, there is a constant α_n such that

$$\text{dist}_X(z, X_n) \leq \alpha_n\|z\|_Z \qquad \text{for all} \quad z \in X \cap Z.$$

(A4) *A priori estimate.* There is a constant C such that $u \in Z$ and

$$\|u\|_Z \leq C\|f\|_Y$$

for all $f \in Y$ and all solutions $u \in X$ of the equation

$$a(u,v) = (f|v)_Y \qquad \text{for all} \quad v \in Y.$$

Corollary 22.33 (Strong Error Estimates for the Ritz Method). *Assume* (A1) *through* (A4) *and use the notation of Proposition* 22.32(iii). *Then, for all* $m \le n$, *we obtain the error estimates*

$$|\lambda_{mn} - \lambda_m| \le \alpha_n^2(c^{-1}d^2C^2),$$

$$\|u_{mn} - v_n\|_Y \le \alpha_n^2 D_m,$$

where the constant D_m *depends on m, but not on n.*

Since α_n^2 appears in the error estimates, we speak of *quadratic* convergence of the Ritz method.

Applications of this result to the finite-element method will be considered in Section 22.13.

PROOF.

(I) Solution operators. Let $K: Y \to Y$ be the solution operator of the equation in (A4), i.e., we set $u = Kf$. Further let $K_n: Y \to Y$ be the solution operator of the corresponding Ritz equation

$$a(u, v) = (f|v)_Y \qquad \text{for fixed} \quad u \in X_n \quad \text{and all} \quad v \in X_n.$$

Then the proof of Corollary 22.20 shows that the operators $K, K_n: Y \to Y$ are linear, symmetric, compact, and monotone. By Corollary 22.21,

$$\|Kf - K_nf\|_Y \le \alpha_n^2 c^{-1}d^2C^2\|f\|_Y \qquad \text{for all} \quad f \in Y,$$

and hence $\|K - K_n\|_Y \le \alpha_n^2 c^{-1} d^2 C^2$.

(II) Equivalent operator equation. The original equation (85) (resp. the Ritz equation (86)) is equivalent to the operator equation

$$\lambda u = Ku, \qquad u \in X, \quad \lambda \in \mathbb{R}$$

(resp.

$$\lambda u = K_n u, \qquad u \in X, \quad \lambda \in \mathbb{R}).$$

By the comparison principle in Section 22.11b,

$$|\lambda_{mn} - \lambda_m| \le \|K - K_n\|_Y.$$

By Problem 22.5a,

$$\|u_{mn} - v_n\|_Y \le C_m\|K - K_n\|_Y. \qquad \square$$

22.13. Application to Boundary–Eigenvalue Problems

We consider the boundary–eigenvalue problem

$$\begin{aligned} -\lambda\Delta u &= u + f \quad \text{on } G,\\ u &= 0 \quad \text{on } \partial G, \end{aligned} \tag{91}$$

where λ is a real number. The case $f = 0$ corresponds to an eigenvalue problem. However, we also want to study the case where $f \neq 0$ in order to explain the interplay between eigenvalue problems and Fredholm alternatives. We set

$$a(u, v) = \int_G \sum_{i=1}^N D_i u D_i v \, dx,$$

$$b(u, v) = \int_G uv \, dx, \qquad b_1(v) = \int_G fv \, dx.$$

Definition 22.34. Let G be a bounded region in $\mathbb{R}^N$, $N \geq 1$. Set $X = \mathring{W}_2^1(G)$. The *generalized problem* to (91) reads as follows. For given $f \in L_2(G)$, we seek $u \in X$ such that

$$\lambda a(u, v) = b(u, v) + b_1(v) \qquad \text{for all} \quad v \in X. \tag{92}$$

In the case where $f = 0$, we seek $\lambda \in \mathbb{R}$ and $u \in X$ with $u \neq 0$.

Formally, we obtain (92) from (91) by multiplying (91) with $v \in C_0^\infty(G)$ and by using subsequent integration by parts.

Proposition 22.35. *The generalized problem* (92) *has the following properties:*

(i) Eigenvalue problem. *If $f = 0$, i.e., $b_1 = 0$, then all the assertions of Proposition* 22.32 *are valid for* (92).

In particular, there exist infinitely many eigenvalues λ. Each eigenvalue has a finite multiplicity. If we count the eigenvalues according to their multiplicity, then

$$\lambda_1 \geq \lambda_2 \geq \cdots > 0,$$

and $\lambda_m \to 0$ as $m \to \infty$.

There exists a complete orthonormal system of eigenvectors $\{u_m\}$ in the space $L_2(G)$, and for each $u \in X$ (resp. $u \in L_2(G)$), the series

$$u = \sum_{k=1}^\infty \alpha_k u_k,$$

with $\alpha_k = \int_G u u_k \, dx$, converges in X (resp. $L_2(G)$).

(ii) Unique solvability. *If λ is not an eigenvalue, then, for each $f \in L_2(G)$, equation* (92) *has a unique solution.*

(iii) Fredholm alternative. *Let λ be an eigenvalue. Then equation* (92) *has a solution* iff

$$\int_G fu \, dx = 0$$

for all eigenvectors u corresponding to λ.

PROOF. By (1b), $b_1 \in X^*$.

(I) The proof of Proposition 22.7 shows that $a: X \times X \to \mathbb{R}$ is bilinear, bounded, symmetric, and strongly positive. Furthermore, by the proof of Proposition 22.28, the bilinear map $b: X \times X \to \mathbb{R}$ is compact. Therefore, all the assumptions of Proposition 22.32 are fulfilled.

Note the following. By Proposition 22.32, $\{u_m\}$ is an orthonormal system with respect to $b(\cdot, \cdot)$, i.e., $\{u_m\}$ is an orthonormal system in $L_2(G)$. Let $v \in X$. Then it follows from

$$b(u_i, v) = \int_G u_i v \, dx = 0 \qquad \text{for all } i$$

that $a(u_i, v) = (u_i|v) = 0$ for all i. Since $\{u_m\}$ is complete in X, we get $v = 0$. Moreover, X is dense in $L_2(G)$. Thus, $\{u_m\}$ is also complete in $L_2(G)$, by Theorem 19.A(iv).

(II) The Fredholm alternatives follow from Proposition 22.26. □

Remark 22.36 (The Rapidity of Convergence of the Method of Finite Elements for Eigenvalue Problems). In order to explain the basic idea in a simple way, we consider the boundary–eigenvalue problem

$$\text{(P)} \qquad \begin{aligned} -\lambda\Delta u &= u \quad \text{on } G, \\ u &= 0 \quad \text{on } \partial G, \end{aligned}$$

where G is a bounded region in $\mathbb{R}^2$ with $\partial G \in C^\infty$. Then the corresponding generalized problem has eigensolutions (u, λ) in $\mathring{W}_2^1(G) \times \mathbb{R}$ with the eigenvalues

$$\lambda_1 \geq \lambda_2 \geq \cdots > 0,$$

counted according to their multiplicity. Here, $\lambda_m \to 0$ as $m \to \infty$.

We now proceed as in Section 22.2e. Consider a sequence of triangulations of G, (T_n), where h_n denotes the *maximal diameter* of the triangles corresponding to the triangulation T_n. Furthermore, consider the special finite elements w_{kn} as defined in Section 22.2e, i.e., the functions w_{kn} are piecewise linear. Then the Ritz equations are given by

$$\text{(R)} \qquad \lambda \int_G (D_1 u_n D_1 w_{kn} + D_2 u_n D_2 w_{kn}) \, dx = \int_G u_n w_{kn} \, dx$$

for $k = 1, \ldots, n'$. We seek $u_n = c_{1n} w_{1n} + \cdots + c_{n'n} w_{n'n}$, i.e., we seek the real coefficients $c_{1n}, \ldots, c_{n'n}$. Note that (R) is an eigenvalue problem for a symmetric matrix which has the eigenvalues

$$\lambda_{1n} \geq \lambda_{2n} \geq \cdots \geq \lambda_{n'n} > 0,$$

counted according to their multiplicity. Set

$$Z = W_2^2(G), \qquad X = \mathring{W}_2^1(G), \qquad Y = L_2(G).$$

Using the estimates given in Section 22.2e, we obtain the following error estimates from Corollary 22.33.

(i) *Eigenvalues.* For all $m \leq n$,

$$|\lambda_{mn} - \lambda_m| \leq \text{const} \cdot h_n^2,$$

where the constant is independent of m and n.

(ii) *Eigenvectors.* Let (u_{mn}, λ_m) be an eigensolution of the Ritz equation (R) normalized by $\|u_{mn}\|_X = 1$. Then there is a generalized eigensolution (v_n, λ_m) of the original problem (P) such that

$$\|u_{mn} - v_n\|_2 = \left(\int_G (u_{mn} - v_n)^2 \, dx\right)^{1/2} \leq \text{const} \cdot h_n^2$$

for all $m \leq n$. Here, the constant is independent of n.

This shows the *quadratic* convergence of the method of finite elements with respect to h_n.

22.14. Gårding Forms

Definition 22.37. Let X and Z be H-spaces over $\mathbb{K} = \mathbb{R}, \mathbb{C}$ with the continuous embedding $X \subseteq Z$. Then $g: X \times X \to \mathbb{K}$ is called a *Gårding form* iff g is bilinear and bounded, and there is a constant $c > 0$ and a real constant C such that

$$\operatorname{Re} g(u, u) \geq c\|u\|_X^2 - C\|u\|_Z^2 \qquad \text{for all} \quad u \in X. \tag{93}$$

This inequality is called a Gårding inequality. If $C = 0$, then we speak of a strict Gårding inequality. Then, g is called a *strict* Gårding form, i.e., g is strongly positive.

The Gårding form g is called *regular* iff the embedding $X \subseteq Z$ is compact.

Inequalities of type (93) for strongly elliptic operators of order $2m$ were first proved by Gårding (1953) in a fundamental paper (cf. Section 22.15). In this chapter and the following two chapters we shall show the following:

Gårding forms play a fundamental role in the modern theory of linear partial differential equations of elliptic, parabolic, and hyperbolic type.

The importance of Gårding forms for the calculus of variations will be explained in Section 29.12.

The following simple result is the *key* to the theory of Gårding forms and their applications to boundary value problems.

Lemma 22.38. *Let $g: X \times X \to \mathbb{K}$ be a regular Gårding form. Then there exists a decomposition of the form*

$$g = a + b,$$

where $a, b: X \times X \to \mathbb{K}$ are bilinear and bounded, $a(\cdot, \cdot)$ is strongly positive, and $b(\cdot, \cdot)$ is compact.

If g is symmetric, then so are a and b.

Consequently, regular Gårding forms are compact perturbations of strongly positive bilinear forms.

PROOF. We set

$$a(u, v) = g(u, v) + C(u|v)_Z$$

and $b(u, v) = -C(u|v)_Z$. By (93),

$$\operatorname{Re} a(u, u) \geq c\|u\|_X^2 \qquad \text{for all} \quad u \in X.$$

Since the embedding $X \subseteq Z$ is compact, it follows from

$$u_n \rightharpoonup u \qquad \text{and} \qquad v_n \rightharpoonup v \quad \text{in } X \qquad \text{as} \quad n \to \infty$$

that $u_n \to u$ and $v_n \to v$ in Z as $n \to \infty$ and hence

$$(u_n|v_n)_Z \to (u|v)_Z \qquad \text{as} \quad n \to \infty.$$

Thus, $b: X \times X \to \mathbb{K}$ is compact. □

Proposition 22.39 (Hestenes (1951)). *Let X be a real* H-*space. Then*:

(i) *Suppose that the bilinear forms*

$$a, b: X \times X \to \mathbb{R}$$

are bounded and symmetric, where, in addition, a is strongly positive and b is compact.

If $a + b$ is strictly positive, then $a + b$ is strongly positive.

(ii) *Suppose that $g: X \times X \to \mathbb{R}$ is a strictly positive, symmetric, regular Gårding form. Then, g is strongly positive.*

The proof of (i) will be given in Problem 22.4. Statement (ii) follows immediately from Lemma 22.38.

In Section 22.16, Proposition 22.39 will be applied to stability theory. In Section 29.12, this proposition will play a fundamental role in order to obtain sufficient criteria for the existence of minima via accessory quadratic eigenvalue problems (the functional analytic generalization of the classical Jacobi theory).

The following proposition is the key to our proof of the Gårding inequality for strongly elliptic differential equations in the next section.

Proposition 22.40 (The Abstract Ehrling Inequality). *Let X, Y, and Z be three* B-*spaces over $\mathbb{K} = \mathbb{R}, \mathbb{C}$ with*

$$X \subseteq Y \subseteq Z.$$

Suppose that the embedding $X \subseteq Y$ is compact, and that the embedding $Y \subseteq Z$ is continuous.

Then, for each $\varepsilon > 0$, there is a constant $c(\varepsilon) > 0$ such that

$$\|u\|_Y \leq \varepsilon\|u\|_X + c(\varepsilon)\|u\|_Z \qquad \textit{for all} \quad u \in X. \tag{94}$$

EXAMPLE 22.41. Let G be a bounded region in $\mathbb{R}^N$, $N \geq 1$. Then the inequality (94) holds in the case where

$$X = \mathring{W}_2^m(G), \qquad Y = \mathring{W}_2^{m-1}(G), \qquad Z = L_2(G), \qquad m = 1, 2, \ldots.$$

Inequalities of this type are due to Ehrling (1954).

PROOF OF PROPOSITION 22.40. Suppose that (94) is not true for a fixed $\varepsilon > 0$. Then there is a sequence (u_n) such that

$$\|u_n\|_Y > \varepsilon \|u_n\|_X + n\|u_n\|_Z \qquad \text{for all } n. \tag{95}$$

Hence $u_n \neq 0$. After division, if necessary, we may assume that $\|u_n\|_X = 1$ for all n. Since the embedding $X \subseteq Y$ is compact, there is a subsequence, again denoted by (u_n), such that $u_n \to u$ in Y as $n \to \infty$. This implies $u_n \to u$ in Z as $n \to \infty$. By (95), $\|u_n\|_Y > \varepsilon$ and hence $u \neq 0$. Again by (95), $u_n \to 0$ in Z as $n \to \infty$, i.e., $u = 0$. This is a contradiction. □

22.15. The Gårding Inequality for Elliptic Equations

We want to study boundary value problems for linear differential operators of order $2m$, i.e., we consider

$$Lu = \sum_{|\alpha|, |\beta| \leq m} (-1)^{|\alpha|} D^\alpha(a_{\alpha\beta} D^\beta u). \tag{96}$$

We make the following assumptions.

(H) Let G be a bounded region in $\mathbb{R}^N$, $N \geq 1$. Let $m = 1, 2, \ldots$. All the coefficient functions $a_{\alpha\beta}\colon G \to \mathbb{R}$ are measurable and bounded (e.g., $a_{\alpha\beta} \in C(\bar{G})$).

Recall that $x = (\xi_1, \ldots, \xi_N)$, $D_i = \partial/\partial\xi_i$, and

$$D^\alpha = D_1^{\alpha_1} D_2^{\alpha_2} \cdots D_N^{\alpha_N},$$

where $\alpha = (\alpha_1, \ldots, \alpha_N)$ and $|\alpha| = \alpha_1 + \cdots + \alpha_N$.

Definition 22.42. Assume (H). The differential operator L in (96) is called *regularly strongly elliptic* iff there is a number $c_0 > 0$ such that

$$\sum_{|\alpha|, |\beta| = m} a_{\alpha\beta}(x) D^\alpha D^\beta \geq c_0 \sum_{|\alpha| = m} |D^\alpha|^2 \tag{97}$$

for all $x \in G$ and $D^\alpha, D^\beta \in \mathbb{R}$.

The differential operator L is called *strongly elliptic* iff all the functions $a_{\alpha\beta}\colon \bar{G} \to \mathbb{R}$ are continuous for $|\alpha|, |\beta| = m$ and there is a $c_0 > 0$ such that

$$\sum_{|\alpha|, |\beta| = m} a_{\alpha\beta}(x) d^\alpha d^\beta \geq c_0 |d|^{2m} \tag{97*}$$

for all $x \in G$, $d \in \mathbb{R}^N$.

Moreover, L is called *symmetric* iff $a_{\alpha\beta}(x) = a_{\beta\alpha}(x)$ for all indices α and β, and all $x \in G$.

The *dual* differential operator L^* to L is defined to be

$$L^*u = \sum_{|\alpha|, |\beta| \le m} (-1)^{|\alpha|} D^\alpha (a_{\beta\alpha} D^\beta u).$$

Notice the following peculiarities. In (96), D^α denotes a differential operator, but in (97), D^α denotes a real number, where α is an index. Roughly speaking, condition (97) means that the principal part of L is positive definite. If L is symmetric, then $L^* = L$.

In (97*), we have $d = (d_1, \ldots, d_N)$ and

$$d^\alpha = d_1^{\alpha_1} d_2^{\alpha_2} \cdots d_N^{\alpha_N}.$$

Since

$$\sum_{|\alpha|=m} |d_1|^{2\alpha_1} \cdots |d_N|^{2\alpha_N} = |d|^{2m} \qquad \text{for all} \quad d \in \mathbb{R}^N,$$

condition (97) implies (97*). Therefore, we obtain the following:

Regularly strongly elliptic operators are also strongly elliptic in the case where $a_{\alpha\beta}$ is continuous and bounded on G for all α, β with $|\alpha|$, $|\beta| = m$.

In the following sections we want to prove:

(i) Fredholm alternatives for regularly strongly elliptic or strongly elliptic differential operators L, and
(ii) the existence of infinitely many eigensolutions in the case where L is symmetric.

Here, the key is the Gårding inequality below.

Example 22.43. The second-order differential operator

$$Lu = \sum_{i=1}^N -D_i^2 u + a_{0i} D_i u + a_{00}$$

is regularly strongly elliptic. The dual operator is given by

$$L^*u = \sum_{i=1}^N -D_i^2 u - D_i(a_{0i} u) + a_{00},$$

and L is symmetric in the case where $a_{0i}(x) \equiv 0$ for $i = 1, \ldots, N$.

Example 22.44. The fourth-order differential operator

$$Lu = \Delta^2 u$$

is regularly strongly elliptic and symmetric.

PROOF. Explicitly, we obtain that

$$Lu = \left(\sum_i D_i^2\right)^2 u = \sum_{|\alpha|=2} a_{\alpha\beta} D^\alpha D^\beta u,$$

where $a_{\alpha\beta} = 1$ if $\alpha = \beta$, and $a_{\alpha\beta} = 0$ if $\alpha \neq \beta$. Note that $D^\alpha = D_i D_j$ for $|\alpha| = 2$. □

Proposition 22.45 (The Gårding Inequality). *Assume* (H) *above, and assume that the differential operator L is regularly strongly elliptic or strongly elliptic. Then, there are real numbers $c > 0$ and C such that*

$$\int_G \sum_{|\alpha|, |\beta| \leq m} a_{\alpha\beta} D^\alpha u D^\beta u \, dx \geq c\|u\|_{m,2}^2 - C\|u\|_2^2 \tag{98}$$

for all $u \in \mathring{W}_2^m(G)$.

PROOF. We first assume that L is regularly strongly elliptic.

(I) The principal part. Let $u \in \mathring{W}_2^m(G)$. The strong ellipticity condition (97) implies

$$\int_G \sum_{|\alpha|, |\beta| = m} a_{\alpha\beta} D^\alpha u D^\beta u \, dx \geq c_0 \int_G \sum_{|\alpha| = m} (D^\alpha u)^2 \, dx$$

$$= c_0 \|u\|_{m,2,0}^2.$$

By $A_2(53)$, the norms $\|\cdot\|_{m,2,0}$ and $\|\cdot\|_{m,2}$ are equivalent on $\mathring{W}_2^m(G)$, i.e., there is a $c > 0$ such that

$$c_0 \|u\|_{m,2,0}^2 \geq 2c\|u\|_{m,2}^2.$$

(II) The lower order terms. Again let $u \in \mathring{W}_2^m(G)$. Note that $\sup_{x \in G} |a_{\alpha\beta}(x)| < \infty$ for all α, β. By the Hölder inequality, there is a $K > 0$ such that

$$\left| \int_G \sum_{|\alpha|, |\beta| \leq m-1} a_{\alpha\beta} D^\alpha u D^\beta u \, dx \right| \leq K\|u\|_{m-1,2}^2.$$

By the Ehrling inequality in Example 22.41,

$$\|u\|_{m-1,2} \leq \varepsilon\|u\|_{m,2} + c(\varepsilon)\|u\|_2.$$

Choosing $\varepsilon > 0$ sufficiently small and using $(a+b)^2 \leq 2a^2 + 2b^2$, we obtain

$$K\|u\|_{m-1,2}^2 \leq c\|u\|_{m,2}^2 + C\|u\|_2^2.$$

Putting these estimates together, we get (98).

The proof for strongly elliptic differential operators will be given in Problem 22.7. There, we will also prove the Gårding inequality for strongly elliptic systems. □

22.16. The Main Theorems on Gårding Forms

The following results will be applied to strongly elliptic differential equations of order $2m$ in the next section.

22.16a. Fredholm Alternative

We consider the equation

$$g(u, v) = b(v) \qquad \text{for fixed} \quad u \in X \quad \text{and all} \quad v \in X, \tag{99}$$

together with the *dual* problem

$$g(v, u^*) = 0 \qquad \text{for fixed} \quad u^* \in X \quad \text{and all} \quad v \in X. \tag{99*}$$

Theorem 22.F. *Let $g: X \times X \to \mathbb{R}$ be a regular Gårding form on the real* H-*space X, and let $b \in X^*$. Then:*

(a) *The original equation* (99) *has a solution u iff*

$$b(u^*) = 0$$

for all solutions u^* of the dual equation (99*).

(b) *If the homogeneous equation* (99) *with $b = 0$ has only the trivial solution $u = 0$, then, for each $b \in X^*$, the inhomogeneous equation* (99) *has a unique solution.*

Proof. By Lemma 22.38,

$$g(u, v) = a(u, v) + c(u, v) \qquad \text{for all} \quad u, v \in X,$$

where $a, c: X \to \mathbb{R}$ are bilinear and bounded. Moreover, a is strongly positive, and c is compact. Now, the assertions follow from Proposition 22.26. □

Note that if g satisfies a strict Gårding inequality, then the case (b) is at hand.

22.16b. Eigenvalue Problem

We now study the eigenvalue problem

$$g(u, v) = \mu(u|v)_Y \qquad \text{for all} \quad v \in X. \tag{100}$$

We seek $\mu \in \mathbb{R}$ and $u \in X$ with $u \neq 0$.

Theorem 22.G. *Let $g: X \times X \to \mathbb{R}$ be a symmetric regular Gårding form on the real separable infinite-dimensional* H-*space X, i.e.,*

$$g(u, u) \geq c\|u\|_X^2 - C\|u\|_Y^2 \qquad \textit{for all} \quad u \in X,$$

where Y is a real H-*space with the compact embedding $X \subseteq Y$. Here, $c > 0$ and $C \in \mathbb{R}$. Then:*

(a) *Problem* (100) *has infinitely many eigenvalues μ. All the eigenvalues of* (100) *have finite multiplicity. If we count the eigenvalues according to their multiplicity, then*

$$-C < \mu_1 \le \mu_2 \le \cdots,$$

and $\mu_m \to +\infty$ as $m \to \infty$.

(b) *The smallest eigenvalue μ_1 can be characterized by the minimum problem*

$$\mu_1 = \min g(u, u), \qquad \|u\|_Y = 1, \qquad u \in X. \tag{101}$$

(c) *The* m*th eigenvalue μ_m can be characterized by the minimum–maximum principle*

$$\mu_m = \min_{M \in \mathscr{L}_m} \max_{u \in M} g(u, u),$$

where $S = \{u \in X: \|u\|_Y = 1\}$, and $\mathscr{L}_m$ denotes the class of all the sets $S \cap L$, where L is an arbitrary m-dimensional linear subspace of X.

22.16c. Stability

Corollary 22.46. *Suppose that $\mu_1 > 0$ in Theorem* 22.G. *Then, for each $b \in X^*$, the variational problem*

$$\tfrac{1}{2}g(u, u) - b(u) = \min!, \qquad u \in X, \tag{102}$$

has a unique solution u, which corresponds to a strict minimum. Moreover, u is the unique solution of the Euler equation

$$g(u, v) = b(v) \qquad \text{for fixed} \quad u \in X \quad \text{and all} \quad v \in X, \tag{103}$$

and u depends on b in a stable manner, i.e.,

$$\|u_1 - u_2\|_Y \le \mu_1^{-1} \|b_1 - b_2\|_{Y^*}, \tag{104}$$

where the solution u_j corresponds to $b_j \in Y^*$. Note that $Y^* \subseteq X^*$.

Moreover, there is a constant $c > 0$ such that

$$g(u, u) \ge c\|u\|_X^2 \qquad \text{for all} \quad u \in X. \tag{105}$$

Remark 22.47 (Interpretation in Terms of Elasticity). In elasticity theory we have:

$$u = \text{displacement of the elastic body,}$$

$$\tfrac{1}{2}g(u, u) = \text{elastic potential energy of the body,}$$

$$b(u) = \text{work of the outer forces.}$$

The variational problem (102) corresponds to the principle of minimal potential energy. The solution u represents an equilibrium state of the body which corresponds to the outer force b. Relation (104) tells us that all the equilibrium states u are *stable*, i.e., small changes of the outer forces, $\|b_1 - b_2\|_{Y^*}$, cause only small changes of the displacements, $\|u_1 - u_2\|_Y$. In addition, we obtain from (104) the following fundamental principle:

The equilibrium states of the body are more stable, the larger the smallest eigenvalue μ_1 is.

Therefore, engineers are very interested in constructing such systems (e.g., buildings, bridges, etc.) for which μ_1 is large. Relation (105) says that the displacement u is small in the case where the elastic energy is small. In fact, the condition $\mu_1 > 0$ is equivalent to (105) (strong positivity of the elastic potential energy).

PROOF OF THEOREM 22.G. We consider the Gårding inequality

$$g(u, u) \geq c\|u\|_X^2 - C\|u\|_Y^2 \qquad \text{for all} \quad u \in X.$$

We set

$$a(u, v) = g(u, v) + C(u|v)_Y, \qquad b(u, v) = (u|v)_Y.$$

Then the original problem (100) is *equivalent* to the equation

$$a(u, v) = (\mu + C)b(u, v) \qquad \text{for all} \quad v \in X. \tag{106}$$

The *point* is that the Gårding inequality implies the strong positivity of $a(\cdot, \cdot)$, i.e.,

$$a(u, u) \geq c\|u\|^2 \qquad \text{for all} \quad u \in X.$$

The compact embedding $X \subseteq Y$ implies the compactness of $b: X \times X \to \mathbb{R}$ (cf. the proof of Lemma 22.38).

We now apply Proposition 22.32 to (106). □

PROOF OF COROLLARY 22.46. If $\mu_1 > 0$, then g is strictly positive by (101), and hence g is *strongly* positive by the theorem of Hestenes (Proposition 22.39).

Theorem 22.A yields the assertions. In this connection, note the following. If u is a solution of (103), then it follows from (101) that

$$\mu_1\|u\|_Y^2 \leq g(u, u) = b(u) \leq \|b\|_{Y^*}\|u\|_Y$$

and hence $\mu_1\|u\|_Y \leq \|b\|_{Y^*}$. This is (104). □

22.17. Application to Strongly Elliptic Differential Equations of Order $2m$

We consider the boundary value problem

$$\begin{gathered} Lu - \mu u = f \quad \text{on } G, \\ D^\alpha u = 0 \quad \text{on } \partial G \qquad \text{for all} \quad \alpha: |\alpha| \leq m - 1, \end{gathered} \tag{107}$$

together with the *dual* problem

$$\begin{gathered} L^*u - \mu u = 0 \quad \text{on } G, \\ D^\alpha u = 0 \quad \text{on } \partial G \qquad \text{for all} \quad \alpha\colon |\alpha| \le m-1. \end{gathered} \tag{107*}$$

Here, $\mu \in \mathbb{R}$ and

$$Lu = \sum_{|\alpha|, |\beta| \le m} (-1)^{|\alpha|} D^\alpha (a_{\alpha\beta} D^\beta u),$$

$$L^*u = \sum_{|\alpha|, |\beta| \le m} (-1)^{|\alpha|} D^\alpha (a_{\beta\alpha} D^\beta u).$$

In order to study these problems in detail, we make the following assumptions.

(H) Let G be a bounded region in $\mathbb{R}^N$, $N \ge 1$, and let $m = 1, 2, \ldots$. All the coefficient functions $a_{\alpha\beta}\colon G \to \mathbb{R}$ are measurable and bounded (e.g., $a_{\alpha\beta} \in C(\bar{G})$). The differential operator L is regularly strongly elliptic or strongly elliptic, and $f \in L_2(G)$. Let $X = \mathring{W}_2^m(G)$ and $Y = L_2(G)$.

We define

$$g(u, v) = \int_G \sum_{|\alpha|, |\beta| \le m} a_{\alpha\beta} D^\alpha v D^\beta u \, dx,$$

$$b(u, v) = \int_G uv \, dx, \qquad b_1(v) = \int_G fv \, dx.$$

By Proposition 22.45, we obtain the Gårding inequality

$$g(u, u) - \mu b(u, u) \ge c \|u\|_X^2 - (C + \mu) \|u\|_Y^2$$

for all $u \in X$ and $\mu \in \mathbb{R}$. Recall that the embedding $X \subseteq Y$ is compact.

Definition 22.48. The *generalized problem* to (107) (resp. (107*)) is given by

$$g(u, v) - \mu b(u, v) = b_1(v) \qquad \text{for fixed} \quad u \in X \quad \text{and all} \quad v \in X \tag{108}$$

(resp.

$$g(v, u) - \mu b(v, u) = 0 \qquad \text{for fixed} \quad u \in X \quad \text{and all} \quad v \in X). \tag{108*}$$

Formally, equation (108) is obtained from (107) by multiplying the first equation of (107) with $v \in C_0^\infty(G)$ and by using subsequent integration by parts. Analogously, we obtain (108*) from (107*). Note that $u \in \mathring{W}_2^m(G)$ implies that the boundary condition in (107) is satisfied in the sense of generalized boundary values.

In the following, solutions of (107) and (107*) are to be understood in the sense of generalized solutions. We say that μ is an *eigenvalue* of (107) iff equation (107), with $f = 0$, has a solution $u \in X$ with $u \ne 0$.

Proposition 22.49 (Fredholm Alternatives). *Assume* (H). *Then*:

(a) *Suppose that μ is not an eigenvalue of* (107) *(e.g., $\mu \le -C$). Then, for each $f \in L_2(G)$, equation* (107) *has a unique solution $u \in X$.*

If we set $u = Kf$, then the solution operator $K: L_2(G) \to L_2(G)$ of (107) *is linear and compact. The adjoint operator $K^*: L_2(G) \to L_2(G)$ is the solution operator to the dual problem* (107*), *i.e., K is symmetric if L is symmetric.*

If $\mu \leq -C$, then $K: L_2(G) \to L_2(G)$ is angle-bounded.

(b) *Suppose that μ is an eigenvalue of* (107). *Then μ is also an eigenvalue of* (107*). *Both eigenvalues have the same finite multiplicity.*

For given $f \in L_2(G)$, equation (107) *has a solution iff*

$$\int_G fu \, dx = 0$$

for all eigensolutions $u \in X$ of (107*).

(c) *The eigenvalues μ of* (107) *have no finite accumulation point.*

Proof. This follows from Theorem 22.F, Corollary 22.20, and Corollary 21.53. □

Proposition 22.50 (Eigenvalue Problem). *Assume* (H) *and assume that L is symmetric. Then:*

(a) *The eigenvalue problem* (107), *with $f = 0$, has infinitely many eigenvalues μ. All the eigenvalues have finite multiplicity. If we count the eigenvalues according to their multiplicity, then*

$$-C < \mu_1 \leq \mu_2 \leq \cdots$$

and $\mu_k \to +\infty$ as $k \to \infty$.

There exists a complete orthonormal system of eigenvectors $\{u_k\}$ in the space $L_2(G)$.

(b) *The smallest eigenvalue μ_1 can be characterized by*

$$\mu_1 = \min g(u, u), \qquad \int_G u^2 \, dx = 1, \qquad u \in X.$$

(c) *The* kth *eigenvalue μ_k can be characterized by*

$$\mu_k = \min_{M \in \mathscr{L}_k} \max_{u \in M} g(u, u),$$

where $S = \{u \in X: \int_G u^2 \, dx = 1\}$, and $\mathscr{L}_k$ is the class of all the sets $S \cap L$, where L is an arbitrary k-dimensional subspace of X.

Corollary 22.51 (Stability). *Suppose that we have $\mu_1 > 0$ in Proposition* 22.50. *Then for each $f \in L_2(G)$, Problem* (107), *with $\mu = 0$, has a unique solution $u \in X$, and*

$$\int_G u^2 \, dx \leq \mu_1^{-2} \int_G f^2 \, dx. \tag{109}$$

Moreover, u is the unique solution of the variational problem

$$\tfrac{1}{2}g(u,u) - \int_G fu\,dx = \min!, \qquad u \in X,$$

and u corresponds to a strict minimum.

PROOF. This is a consequence of Theorem 22.G. In order to prove the completeness of $\{u_k\}$, apply the same argument as in the proof of Proposition 22.35 to the equation

$$g(u,v) + Cb(u,v) = (\mu + C)b(u,v) \qquad \text{for fixed} \quad u \in X \quad \text{and all} \quad v \in X,$$

which is equivalent to the original problem (108).

Relation (109) follows from the following estimates:

$$\mu_1 \int_G u^2\,dx \le g(u,u) = \int_G fu\,dx \le \left(\int_G f^2\,dx\right)^{1/2} \left(\int_G u^2\,dx\right)^{1/2}. \qquad \square$$

22.18. Difference Approximations

The following results will be used to prove the regularity of generalized solutions in the next sections. Let $x \in \mathbb{R}^N$. Denote by $e_i = (0, \ldots, 0, 1, 0, \ldots)$ the unit vector in the ith direction. Define the translation operator

$$(T_i^h u)(x) = u(x + he_i)$$

and the difference operator

$$\nabla_i^h u = \frac{T_i^h u - u}{h}.$$

Hence

$$(\nabla_i^h u)(x) = \frac{u(x + he_i) - u(x)}{h}.$$

Proposition 22.52. *Let G be a bounded region in $\mathbb{R}^N$, $N \ge 1$, and let H be a proper subregion of G, i.e., $H \subset\subset G$. Furthermore, let $h_0 = \operatorname{dist}(H, \partial G)$ and let $1 \le p < \infty$, $m = 1, 2, \ldots$.*

Then, for all $h \in \mathbb{R}^N$ with $|h| < h_0$, the following are true:

(a) *The classical operators T_i^h and ∇_i^h can be uniquely extended to continuous linear operators from $L_p(G)$ to $L_p(H)$.*

(b) *For all $u \in W_p^m(G)$,*

$$\|\nabla_i^h u\|_{W_p^{m-1}(H)} \le c\,\|u\|_{W_p^m(G)}, \tag{110}$$

$$\lim_{h \to 0} \|\nabla_i^h u - D_i u\|_{W_p^{m-1}(H)} = 0. \tag{111}$$

(c) *Let* $1 < p < \infty$. *If* $u \in L_p(H)$ *and*

$$\|\nabla_i^h u\|_{L_p(H)} \leq a \qquad \text{for all} \quad h \in]0, h_1[,$$

where $h_1 > 0$ *and* $a \geq 0$ *are fixed real numbers, then the generalized derivative* $D_i u$ *exists on* H *and* $D_i u \in L_p(H)$ *with*

$$\|D_i u\|_{L_p(H)} \leq a.$$

(d) *Let* $1 < p < \infty$. *If* $u \in W_p^{m-1}(H)$ *and*

$$\|\nabla_i^h u\|_{W_p^{m-1}(H)} \leq a \qquad \text{for all} \quad h \in]0, h_1[,$$

where $h_1 > 0$ *and* $a \geq 0$ *are fixed real numbers, then* $u \in W_p^m(G)$ *and*

$$\|u\|_{W_p^m(H)} \leq a.$$

Statements (c) *and* (d) *remain true if we replace* ∇_i^h *by* ∇_i^{-h}.

Statement (c) above justifies the following principle:

A priori estimates for difference quotients yield the existence of generalized derivatives.

This fundamental principle will be used critically in our regularity proofs in the next sections.

PROOF. Ad(a). For all $u \in C_0^\infty(G)$,

$$\int_H |u(x + he_i)|^p \, dx \leq \int_G |u(x)|^p \, dx.$$

By the extension principle in Section 18.12, we obtain the continuous linear operator $T_i^h: L_p(G) \to L_p(H)$.

Ad(b). Suppose that $\partial G \in C^{0,1}$. This assumption can always be satisfied by passing to a smaller region G. Since $C^\infty(\bar{G})$ is dense in $W_p^m(G)$, it suffices to prove (b) for $u \in C^\infty(\bar{G})$. To simplify notation, let $N = 1$ and $m = 1$.

Let $u \in C^\infty(\bar{G})$. Then

$$\nabla^h u(x) = \int_0^1 u'(x + th) \, dt.$$

By the Hölder inequality with 1-trick,

$$|\nabla^h u(x)|^p \leq \int_0^1 |u'(x + th)|^p \, dt.$$

Hence

$$\begin{aligned} \int_H |\nabla^h u|^p \, dx &\leq \int_0^1 \left(\int_H |u'(x + th)|^p \, dx \right) dt \\ &\leq \int_G |u'(x)|^p \, dx. \end{aligned}$$

This yields (110). Moreover, we obtain that

$$\int_H |\nabla^h u - u'|^p \, dx \le \int_H \left| \int_0^1 (u'(x+th) - u'(x))\, dt \right|^p dx$$
$$\le \int_0^1 \left(\int_H |u'(x+th) - u'(x)|^p \, dx \right) dt \to 0 \qquad \text{as} \quad h \to 0,$$

according to $A_2(32)$. This yields (111).

Ad(c). Since the B-space $L_p(H)$ is reflexive, there exists a subsequence such that

$$\nabla_i^h u \rightharpoonup g \quad \text{in } L_p(H) \qquad \text{as} \quad h \to 0.$$

A simple transformation yields

$$\int_H \varphi \nabla_i^h u \, dx = -\int_H u \nabla_i^{-h} \varphi \, dx \qquad \text{for all} \quad \varphi \in C_0^\infty(H)$$

and small h. Letting $h \to 0$, we get

$$\int_H \varphi g \, dx = -\int_H u D_i \varphi \, dx \qquad \text{for all} \quad \varphi \in C_0^\infty(H),$$

and hence $D_i u = g$ on H. Finally, we obtain

$$\|D_i u\|_{L_p(H)} \le \underline{\lim}_{h \to 0} \|\nabla_i^h u\|_{L_p(H)}.$$

Ad(d) This follows from (c) by induction. □

22.19. Interior Regularity of Generalized Solutions

Our goal is to justify the following fundamental regularity principle:

The generalized solutions of linear elliptic equations are smoother, the smoother the data are.

By data we understand coefficients, right-hand terms, boundary values, and boundaries. We shall prove two important results on interior regularity (Theorem 22.H) and on regularity up to the boundary (Theorem 22.I).

We consider the differential equation

$$\sum_{|\alpha|, |\beta| \le m} (-1)^{|\alpha|} D^\alpha (a_{\alpha\beta} D^\beta u) = f \quad \text{on } G \tag{112}$$

together with the boundary condition

$$D^\alpha u = 0 \quad \text{on } \partial G \qquad \text{for all} \quad \alpha\colon |\alpha| \le m - 1, \tag{113}$$

where $m = 1, 2, \ldots$. Moreover, we set

$$a(u, v) = \int_G \sum_{|\alpha|, |\beta| \le m} a_{\alpha\beta} D^\alpha v D^\beta u \, dx,$$

$$b(v) = \int_G fv \, dx.$$

We make the following assumptions:

(H1) Let G be a bounded region in $\mathbb{R}^N$, $N \ge 1$.
(H2) All the coefficients $a_{\alpha\beta}$ belong to $C^\infty(\bar{G})$, and $a(\cdot, \cdot)$ is strongly positive, i.e., there is a constant $d > 0$ such that

$$a(u, u) \ge d \|u\|_{m,2}^2 \qquad \text{for all} \quad u \in \mathring{W}_2^m(G).$$

(H3) Let H be a proper subregion of G, i.e., $H \subset\subset G$.

By a *generalized* solution of (112) we understand a function $u \in W_2^m(G)$ such that

$$a(u, v) = b(v) \qquad \text{for all} \quad v \in C_0^\infty(G). \tag{114}$$

Theorem 22.H (Main Theorem of Regularity Theory). *Assume* (H1) *through* (H3). *Let*

$$f \in W_2^k(G), \qquad k = 0, 1, \ldots$$

be given. Then, each generalized solution $u \in W_2^m(G)$ of (112) *belongs to the Sobolev space $W_2^{2m+k}(H)$, and we have*

$$\|u\|_{W_2^{2m+k}(H)} \le c(\|u\|_{W_2^m(G)} + \|f\|_{W_2^k(G)}), \tag{115}$$

where the constant $c > 0$ depends on G, H, k, and m.

The same remains true if we replace the strong positivity of $a(\cdot, \cdot)$ in (H2) *by the strong ellipticity of* (112).

Note that this is a very natural result. In fact, since the differential operator in (112) is of order $2m$, we expect that the solution u of (112) has derivatives up to order $2m + k$ in case f has derivatives up to order k.

The proof of Theorem 22.H will be given in the next section.

Corollary 22.53 (Boundary Value Problem). *Assume* (H1) *through* (H3). *Let*

$$f \in W_2^k(G), \qquad k = 0, 1, \ldots$$

be given. Then, the generalized boundary value problem (114) *corresponding to* (112), (113) *has a unique solution $u \in \mathring{W}_2^m(G)$. In addition, u belongs to the Sobolev space $W_2^{2m+k}(H)$, i.e.,*

$$u \in \mathring{W}_2^m(G) \cap W_2^{2m+k}(H)$$

and

$$\|u\|_{W_2^{2m+k}(H)} + \|u\|_{W_2^m(G)} \le c\|f\|_{W_2^k(G)},$$

where the constant $c > 0$ depends on G, H, k, and m.

In particular, if $f \in C^\infty(\bar{G})$, then $u \in W_2^{2m+k}(H)$ for all k. Since H is an arbitrary proper subregion of G, this implies $u \in C^\infty(G)$, according to the Sobolev embedding theorems.

By Section 22.21, Corollary 22.53 remains true for $H = G$ in the case where the boundary of the region G is smooth, i.e., $\partial G \in C^\infty$.

PROOF. Let $X = \mathring{W}_2^m(G)$. By Theorem 22.C, problem (114) has a unique solution $u \in X$ and

$$\|u\|_X \le \text{const}\,\|b\|_{X^*} \le \text{const}\,\|f\|_{L_2(G)}.$$

Now use Theorem 22.H. □

EXAMPLE 22.54. The first boundary value problem for the Poisson equation,

$$-\Delta u = f \quad \text{on } G, \qquad u = 0 \quad \text{on } \partial G,$$

corresponds to (112), (113) with $m = 1$. Thus, for each given $f \in W_2^k(G)$, $k = 0, 1, \ldots$, we obtain a unique generalized solution $u \in \mathring{W}_2^1(G)$. In addition, we have

$$u \in W_2^{2+k}(H)$$

for each proper subregion H of G.

In Section 22.21 we will show that $u \in W_2^{2+k}(G)$ in the case where the boundary of G is smooth, i.e., $\partial G \in C^\infty$.

22.20. Proof of Theorem 22.H

We will essentially make use of *negative norms*. To this end, let

$$W_2^{-k}(G), \qquad k = 0, 1, \ldots,$$

denote the set of all linear functionals $b\colon C_0^\infty(G) \to \mathbb{R}$ with the property

$$|b(\varphi)| \le B\|\varphi\|_{k,2} \qquad \text{for all} \quad \varphi \in C_0^\infty(G) \quad \text{and fixed} \quad B \ge 0.$$

The smallest possible number B is denoted by $\|b\|_{-k,2}$. Hence

$$|b(\varphi)| \le \|b\|_{-k,2}\|\varphi\|_{k,2} \qquad \text{for all} \quad \varphi \in C_0^\infty(G).$$

STANDARD EXAMPLE 22.55. Let G be a nonempty bounded open set in $\mathbb{R}^N$, $N \ge 1$. Define

$$b(\varphi) = \int_G agD^\gamma\varphi\,dx \qquad \text{for all} \quad \varphi \in C_0^\infty(G),$$

where $a \in C^r(\bar{G})$, $g \in W_2^r(G)$, $k, r = 0, 1, \ldots$, and

$$|\gamma| \leq k + r.$$

Then $b \in W_2^{-k}(G)$ and $\|b\|_{-k,2} \leq \text{const}\,\|g\|_{r,2}$.

PROOF. Integration by parts yields

$$b(\varphi) = \int_G (-1)^{|\alpha|} D^\alpha(ag) D^\beta \varphi\, dx,$$

where $|\alpha| \leq r$ and $|\beta| \leq k$. By the product rule for $D^\alpha(ag)$ and the Hölder inequality,

$$|b(\varphi)| \leq \text{const}\,\|g\|_{r,2}\|\varphi\|_{k,2} \qquad \text{for all} \quad \varphi \in C_0^\infty(G). \qquad \square$$

22.20a. The Fundamental Regularity Lemma

Lemma 22.56. *Assume* (H1) *through* (H3) *from Section* 22.19. *Let* $b \in W_2^{-m+1}(G)$ *be given. Then each generalized solution* $u \in W_2^m(G)$ *of* (114) *belongs to the Sobolev space* $W_2^{m+1}(H)$, *and we have*

$$\|u\|_{W_2^{m+1}(H)} \leq c(\|u\|_{W_2^m(G)} + \|b\|_{-m+1,2}),$$

where the constant $c > 0$ *depends on* G, H, *and* m.

Before proving this, we want to show that Lemma 22.56 immediately implies Theorem 22.H.

PROOF OF THEOREM 22.H. The simple idea is to set

$$w = D_j u$$

and to consider the corresponding differential equation for the derivative w. Classically, differentiation of the original equation (112) yields

$$\sum_{|\alpha|,|\beta| \leq m} (-1)^{|\alpha|} D^\alpha(a_{\alpha\beta} D^\beta w) = g \quad \text{on } G, \tag{116}$$

where

$$g = D_j f - \sum_{|\alpha|,|\beta| \leq m} (-1)^{|\alpha|} D^\alpha(D_j a_{\alpha\beta} D^\beta u).$$

We want to translate this procedure to generalized problems. Note that problem (116*) below is precisely the generalized problem to (116).

We will improve the regularity of u step by step.

Case 1: Let $a(\cdot,\cdot)$ be strongly positive on $\mathring{W}_2^m(G)$.

(I) Let $f \in L_2(G)$, and let $u \in W_2^m(G)$ be a solution of the original problem (114), that is,

(E) $$a(u, v) = b(v) \qquad \text{for all} \quad v \in C_0^\infty(G),$$

where

$$b(v) = \int_G fv\,dx, \qquad a(u,v) = \int_G \sum_{|\alpha|,|\beta| \le m} a_{\alpha\beta} D^\alpha v D^\beta u\,dx.$$

By Example 22.55,

$$b \in W_2^{-m+1}(G) \qquad \text{and} \qquad \|b\|_{-m+1,2} \le \text{const}\,\|f\|_2.$$

By Lemma 22.56,

$$u \in W_2^{m+1}(H) \qquad \text{and} \qquad \|u\|_{m+1,2} \le c(\|u\|_{m,2} + \|f\|_2).$$

In the following we may set $G = H$, by making the region G a little smaller.

(II) Let $f \in L_2(G)$, $m \ge 2$, and let $u \in W_2^{m+1}(G)$ be a solution of (E). We set $v = D_j\varphi$ and $w = D_j u$. Then, using integration by parts, it follows from (E) that

$$a(w, \varphi) = b_1(\varphi) \qquad \text{for all} \quad \varphi \in C_0^\infty(G), \tag{116*}$$

where

$$b_1(\varphi) = -\int_G fD_j\varphi\,dx - \int_G \sum_{|\alpha|,|\beta| \le m} D_j a_{\alpha\beta} D^\alpha \varphi D^\beta u\,dx.$$

By Example 22.55,

$$b_1 \in W_2^{-m+1}(G) \qquad \text{and} \qquad \|b_1\|_{-m+1,2} \le \text{const}(\|f\|_2 + \|u\|_{m+1,2}).$$

Note that $D^\beta u \in W_2^1(G)$ for $|\beta| \le m$.

Applying Lemma 22.56 to (116*), we obtain $w \in W_2^{m+1}(H)$, and hence

$$u \in W_2^{m+2}(H).$$

(III) Let $f \in W_2^1(G)$, $m \ge 1$, and let $u \in W_2^{m+1}(G)$ be a solution of (E). Then we can use the same argument as in (II) in order to show that $u \in W_2^{m+2}(H)$, since again $b_1 \in W_2^{-m+1}(G)$, and

$$\|b_1\|_{-m+1,2} \le \text{const}(\|f\|_{1,2} + \|u\|_{m+1,2}).$$

(IV) From this we obtain Theorem 22.H by induction.

Case 2: Let the differential equation (112) be strongly elliptic.

Then problem (E) can be written as

(E*) $$a_1(u,v) = b_1(v) \qquad \text{for all} \quad v \in C_0^\infty(G),$$

where

$$a_1(u,v) = a(u,v) + C\int_G uv\,dx, \qquad b_1(v) = b(v) + C\int_G uv\,dx.$$

By the Gårding inequality in Section 22.15, $a_1(\cdot,\cdot)$ is strongly positive on $\mathring{W}_2^m(G)$ if $C > 0$ is sufficiently large. Now apply the arguments from Case 1 to (E*).

The proof of Theorem 22.H is complete. □

22.20b. Proof of Lemma 22.56

The basic idea of the proof is the following.

(i) We use the *strong positivity* of the bilinear form $a(\cdot, \cdot)$ in order to obtain estimates for the difference quotients $\nabla^h(D^\alpha u)$, $|\alpha| \leq m$.
(ii) By the limiting process $h \to 0$, we get the existence of the derivatives $D_j(D^\alpha u)$ via Section 22.18.

In order to make the simple idea of the proof as transparent as possible, we restrict ourselves to a special case. However, note that the proof of the general case proceeds completely analogously. We consider the differential equation

$$-(Au')' = f \quad \text{on } G, \tag{117}$$

where G is a bounded open interval in $\mathbb{R}^1$, i.e., we consider the special case $m = 1$ and $N = 1$. Let $A \in C^\infty(\bar{G})$.

In the following, all constants are denoted by c.

An elementary computation yields the following two key formulas:

$$\nabla^h(vw) = w\nabla^h v + T^h v \nabla^h w, \tag{118}$$

$$\int_G v\nabla^{-h} w\, dx = -\int_G w\nabla^h v\, dx. \tag{119}$$

In this connection we assume that the support of v or w is a *compact subset* of the region G and that $|h| > 0$ is sufficiently small. Instead of ∇^h, ∇^{-h}, T^h, we write briefly ∇, ∇^-, T, respectively. Moreover, we set $\int = \int_G$. Note that

$$a(u, v) = \int Au'v'\, dx.$$

Step 1: Discrete integration by parts of the bilinear form $a(\cdot, \cdot)$.

We choose a region Ω with $H \subset\subset \Omega \subset\subset G$ and $\partial\Omega \in C^{0,1}$. Furthermore, we choose a test function $\psi \in C_0^\infty(\Omega)$ with

$$\psi = 1 \quad \text{on } H.$$

For all $\varphi \in C_0^\infty(\Omega)$, we obtain that

$$\begin{aligned} a(\nabla^-(\psi u), \varphi) &= \int A\nabla^-(\psi u)'\varphi'\, dx \\ &= -\int (\psi u)'\nabla(A\varphi')\, dx = -a(u, \nabla(\varphi\psi)) + R. \end{aligned}$$

Step 2: Estimates for the remainder R.

The explicit form of R follows from (118). However, we only need the structure of R. It is *important* that the term

$$A\psi u'\nabla\varphi'$$

does *not* occur in R. More precisely, R contains only terms of the structure

$$\int \rho D^\alpha u \nabla \varphi \, dx, \qquad \int \rho D^\alpha u D^\beta \varphi \, dx,$$

where $|\alpha|, |\beta| \leq 1$. In addition, there appear similar terms, which contain the operator T. But the translation operator T does not change the following arguments. In this connection, note that $\varphi \in C_0^\infty(\Omega)$ and that $|h|$ is sufficiently small.

Let $\|\cdot\|$ and $\|\cdot\|_2$ denote the norm on $W_2^1(G)$ and $L_2(G)$, respectively. By Proposition 22.52,

$$\|\nabla \varphi\|_2 \leq c\|\varphi\|.$$

The Hölder inequality yields

$$|R| \leq c\|u\| \, \|\varphi\|. \tag{120}$$

Step 3: Estimates for the principal part $a(u, \nabla(\varphi\psi))$.

Since u is a generalized solution, we obtain that

$$a(u, \nabla(\varphi\psi)) = b(\nabla(\varphi\psi)).$$

Let $|b|$ denote the norm of $b \in W_2^{-m+1}(G)$ with $m = 1$. From

$$\begin{aligned} |b(\nabla(\varphi\psi))| &\leq |b| \, \|\nabla(\varphi\psi)\|_2 \\ &\leq c|b| \, \|\varphi\psi\| \leq c_1|b| \, \|\varphi\| \end{aligned}$$

it follows that

$$|a(u, \nabla(\varphi\psi))| \leq c_1|b| \, \|\varphi\|. \tag{121}$$

By (120) and (121), we get

$$a(\nabla^-(\psi u), \varphi) \leq c(\|u\| + |b|)\|\varphi\|$$

for all $\varphi \in C_0^\infty(\Omega)$, and hence for all $\varphi \in \mathring{W}_2^1(\Omega)$.

Step 4: Strong positivity of $a(\cdot, \cdot)$.

We choose $\varphi = \nabla^-(\psi u)$. From $a(\varphi, \varphi) \geq d\|\varphi\|^2$ it follows that

$$\|\nabla^-(\psi u)\| \leq c(\|u\| + |b|).$$

Since $\psi = 1$ on H, we obtain the *key inequality*

$$\|\nabla^{-h} u\|_{W_2^1(H)} \leq c(\|u\| + |b|), \tag{122}$$

for all sufficiently small $|h| > 0$. Recall that $\nabla^- = \nabla^{-h}$.

Step 5: The limiting process $h \to 0$.

From (122) and Proposition 22.52 it follows that $u \in W_2^2(H)$ and

$$\|u\|_{W_2^2(H)} \leq c(\|u\| + |b|). \tag{123}$$

The proof of Lemma 22.56 is complete. □

22.21. Regularity of Generalized Solutions up to the Boundary

Theorem 22.I. *Assume* (H1) *through* (H3) *from Section* 22.19. *In addition, assume that the boundary of the region G is smooth, i.e.,* $\partial G \in C^\infty$. *Let*

$$f \in W_2^k(G), \qquad k = 0, 1, \ldots,$$

be given. Then the generalized boundary value problem (114) *corresponding to the classical first boundary value problem* (112), (113) *of order* $2m$ *has a unique solution* $u \in \mathring{W}_2^m(G)$. *In addition,* u *belongs to the Sobolev space* $W_2^{2m+k}(G)$ *and*

$$\|u\|_{W_2^{2m+k}(G)} \leq c\|f\|_{W_2^k(G)}, \tag{124}$$

where the constant $c > 0$ *depends on* G, k, *and* m.

EXAMPLE 22.57. We consider the first boundary value problem

$$-\Delta u = f \quad \text{on } G, \qquad u = 0 \quad \text{on } \partial G, \tag{125}$$

where G is a bounded region in $\mathbb{R}^N$, $N \geq 1$, with $\partial G \in C^\infty$. Let $f \in L_2(G)$ be given. Then the corresponding generalized problem

$$\int_G \sum_{j=1}^N D_j u D_j \varphi \, dx = \int_G f\varphi \, dx \qquad \text{for all} \quad \varphi \in C_0^\infty(G) \tag{126}$$

has exactly one solution $u \in \mathring{W}_2^1(G)$, and u also belongs to the Sobolev space $W_2^2(G)$.

Using this result, it is possible to give a precise characterization of the *Friedrichs extension* of the negative Laplace operator. To this end, let $X = L_2(G)$ and let $A\colon D(A) \subseteq X \to X$ be the Friedrichs extension of the operator $-\Delta\colon C_0^\infty(G) \to X$. Then

$$D(A) = \mathring{W}_2^1(G) \cap W_2^2(G).$$

Moreover, we have

$$Au = -\Delta u \qquad \text{for all} \quad u \in D(A).$$

Here, the second derivatives are to be understood in the generalized sense. This is a very natural result.

PROOF. Let $u \in D(A)$. We set $f = Au$. Let X_E be the energetic space of A. By Section 19.11, $X_E = \mathring{W}_2^1(G)$ and $D(A) \subseteq X_E$. The self-adjointness of A yields

$$(Au|\varphi) = (u|A\varphi) \qquad \text{for all} \quad \varphi \in C_0^\infty(G).$$

Thus, u is a solution of (126). By Theorem 22.I, $u \in W_2^2(G)$.

Conversely, let $u \in \mathring{W}_2^1(G) \cap W_2^2(G)$. We set $f = -\Delta u$. Integration by parts yields (126), i.e.,

$$(u|A\varphi) = (f|\varphi) \qquad \text{for all} \quad \varphi \in C_0^\infty(G).$$

Since A is self-adjoint, we obtain $Au = f$. □

22.22. Proof of Theorem 22.I

The strategy of proof is the following.

(i) The regularity of the tangential derivatives at the boundary is proved similarly to the proof of Theorem 22.H.
(ii) The regularity of the normal derivatives at the boundary follows from the differential equation.

We first restrict ourselves to the Poisson equation in $\mathbb{R}^2$, i.e., we consider

$$-u_{\xi\xi} - u_{\eta\eta} = f \quad \text{on } G, \qquad u = 0 \quad \text{on } \partial G. \tag{127}$$

For given $f \in L_2(G)$, let $u \in \mathring{W}_2^1(G)$ be a generalized solution of (127).

Step 1: Interior regularity.

From Theorem 22.H it follows that $u \in W_2^2(H)$ for all proper subregions $H \subset\subset G$.

Step 2: Local regularity up to the boundary in the case of straight-line boundary parts.

Suppose that a part of the boundary ∂G corresponds to a part of the ξ-axis. Choose a half-ball as pictured in Figure 22.4(a). The same method of proof as in Section 22.20 yields the existence of the generalized tangential derivatives

$$(u_\xi)_\xi, \qquad (u_\eta)_\xi$$

and those derivatives belong to $L_2(H)$. Note that the application of the difference operators $\nabla^{\pm h}$ in the ξ-direction does not cause any difficulties.

In order to get the existence of the remaining derivative $u_{\eta\eta}$, we use the differential equation (127). More precisely, we have to use the generalized equation

$$\int_H (u_\xi \varphi_\xi + u_\eta \varphi_\eta)\, dx = \int_H f\varphi\, dx \qquad \text{for all} \quad \varphi \in C_0^\infty(H).$$

Integration by parts yields

$$\int_H u\varphi_{\eta\eta}\, dx = -\int_H (f + u_{\xi\xi})\varphi\, dx.$$

Hence we obtain that

$$u_{\eta\eta} = -f - u_{\xi\xi},$$

in the generalized sense. This implies $u_{\eta\eta} \in L_2(H)$. Therefore, $u \in W_2^2(H)$.

Step 3: Local regularity up to the boundary in the case of curved boundary parts.

We consider a boundary part as pictured in Figure 22.5. We choose a local (μ, ζ)-coordinate system, and we assume that the boundary part is given by

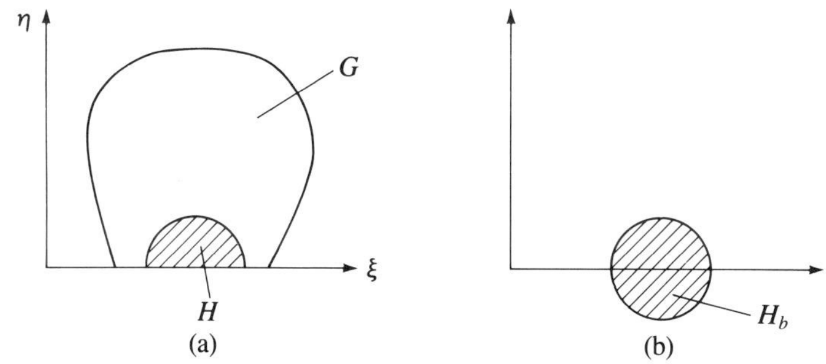

Figure 22.4

the equation $\zeta = g(\mu)$. The transformation

$$\xi^* = \mu, \qquad \eta^* = \zeta - g(\mu) \tag{128}$$

corresponds to a local C^∞-diffeomorphism. Note that the Jacobian is different from zero. By (128), we reduce our problem to Step 2. More precisely, by (128), the original differential equation (127) is transformed locally into a new differential equation having C^∞-coefficients with respect to the new variables ξ^*, η^*. An explicit calculation shows that the corresponding transformed bilinear form *remains strongly positive*. Therefore, we can use the same argument as in Step 2.

Step 4: Globalization.

We choose a sufficiently fine finite covering $\{U_j\}$ of $\bar{G}$, where all the sets U_j are open. Let $\{\varphi_j\}$ be a partition of unity subordinated to $\{U_j\}$. In particular, we have $\varphi_j \in C_0^\infty(U_j)$. We set

$$u_j = \varphi_j u.$$

By our previous local regularity results in Steps 1–3, we can choose U_j in such a way that $u \in W_2^2(G \cap U_j)$. This implies

$$u_j \in W_2^2(G).$$

It follows from $u = \sum_j u_j$ on G that $u \in W_2^2(G)$.

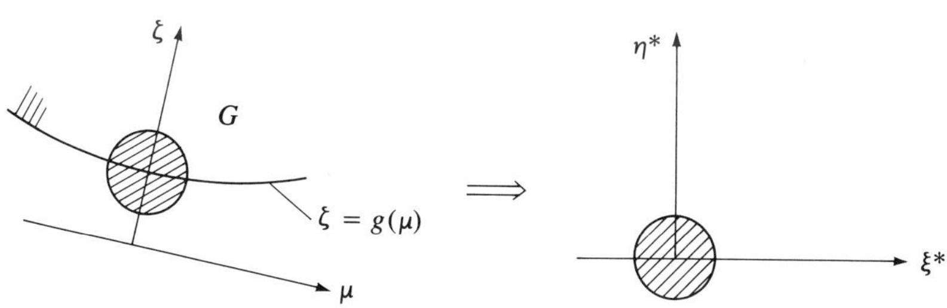

Figure 22.5

This proves Theorem 22.I in the special case of the Poisson equation.

In the general case of equations of order $2m$, one can use similar arguments. In Step 2 above one has to extend the function u from the half-ball H to the ball H_b in Figure 22.4, according to Problem 21.7. The details can be found in Friedman (1969, M), p. 61.

Problems

22.1. *General duality principle for quadratic minimum problems.* We consider the quadratic minimum problem

(P) $$F(u) \stackrel{\text{def}}{=} a(u,u) - 2b(u) = \min!, \qquad u \in u_0 + X,$$

together with the variational equation

(V) $$a(u,w) = b(w) \qquad \text{for all} \quad w \in X.$$

The dual problem reads as follows:

(P*) $$F^*(v) \stackrel{\text{def}}{=} -a(v,v) + 2a(u_0,v) - 2b(u_0) = \max!, \qquad v \in v_0 + Y.$$

In preparation we formulate the *key* relation

$$\|u - v\|_E^2 = F(u) - F^*(v) \tag{129}$$

for all $u \in u_0 + X$ and all $v \in v_0 + Y$. We make the following assumptions.

(H1) X is a closed linear subspace of the real H-space Z.

(H2) The map $a\colon Z \times Z \to \mathbb{R}$ is bilinear, bounded, symmetric, and positive.

(H3) $a\colon X \times X \to \mathbb{R}$ is strongly positive.

(H4) The functional $b\colon Z \to \mathbb{R}$ is linear and continuous.

(H5) We are given $u_0 \in Z$.

(H6) We set

$$Y = \{u \in Z\colon a(u,w) = 0 \text{ for all } w \in X\},$$

i.e., Y consists of all the solutions $u \in Z$ of the homogeneous variational equation (V). We choose a fixed solution $v_0 \in Z$ of (V). Furthermore, we set

$$(u|v)_E = a(u,v) \qquad \text{for all} \quad u, v \in Z.$$

Show:

(i) The original problem (P) has a unique solution $\bar{u}$. This solution $\bar{u}$ satisfies the variational equation (V).

(ii) Relation (129) holds.

(iii) The dual problem (P*) also has the solution $\bar{u}$, and the extremal values of (P) and (P*) are the same, i.e., $F(\bar{u}) = F^*(\bar{u})$.

(iv) *The Ritz–Trefftz method.* Let X_n and Y_n be a finite-dimensional subspace of X and Y, respectively. Let u (resp. v) be a solution of the Ritz problem

$$F(u) = \min!, \qquad u \in u_0 + X_n \tag{130}$$

(resp. Trefftz problem

$$F^*(v) = \max!, \qquad v \in v_0 + Y_n). \tag{130*}$$

Then we obtain the fundamental error estimates

$$F^*(v) \leq F^*(\bar{u}) = F(\bar{u}) \leq F(u)$$

and

$$\|\bar{u} - v\|_E^2 \leq F(u) - F^*(v).$$

(v) *The hypercircle method.* For all $u \in u_0 + X$ and all $v \in v_0 + Y$,

$$\|2^{-1}(u + v) - \bar{u}\|_E = 2^{-1}\|u - v\|_E. \tag{131}$$

This relation tells us that the solution $\bar{u}$ of the original problem (P) lies on a hypercircle with center $2^{-1}(u + v)$ and radius $2^{-1}\|u - v\|_E$. Let (u, v) be a solution of the minimum problem

$$\|u - v\|_E^2 = \min!, \qquad u \in u_0 + X_n, \quad v \in v_0 + Y_n.$$

Then equation (131) yields an error estimate for $\bar{u}$. This is the so-called hypercircle method of Prager and Synge (cf. Synge (1957, **M**)), which is frequently used in elasticity.

Solution: Ad(i). Cf. Theorem 18.A.

Ad(ii). By construction of v_0 and Y, we get

$$a(v, u - u_0) = b(u - u_0) \qquad \text{for all} \quad u \in u_0 + X, \quad v \in v_0 + Y,$$

and hence

$$\begin{aligned}\|u - v\|_E^2 &= a(u, u) - 2a(u, v) + a(v, v)\\ &= F(u) - F^*(v) + 2a(u_0 - u, v) - 2b(u_0 - u)\\ &= F(u) - F^*(v).\end{aligned}$$

Ad(iii). By (129),

$$F^*(v) \leq F(\bar{u}) \qquad \text{for all} \quad v \in v_0 + Y.$$

It follows from the variational equation (V) that

$$a(\bar{u}, w) = b(w) \qquad \text{for all} \quad w \in X.$$

By the construction of v_0,

$$a(v_0, w) = b(w) \qquad \text{for all} \quad w \in X.$$

Hence $a(\bar{u} - v_0, w) = 0$ for all $w \in X$, i.e., $\bar{u} \in v_0 + Y$. Letting $u = v = \bar{u}$ in (129), we obtain $F(\bar{u}) = F^*(\bar{u})$.

Ad(iv). This follows from $X_n \subseteq X$ and $Y_n \subseteq Y$, by (129).

Ad(v). Let $u \in u_0 + X$ and $v \in v_0 + Y$. From

$$a(y, x) = 0 \qquad \text{for all} \quad y \in Y, \quad x \in X,$$

we get $a(\bar{u} - v, \bar{u} - u) = 0$. This implies

$$\begin{aligned}\|u + v - 2\bar{u}\|_E^2 &= \|(u - \bar{u}) + (v - \bar{u})\|_E^2\\ &= \|(u - \bar{u}) - (v - \bar{u})\|_E^2 = \|u - v\|_E^2.\end{aligned}$$

22.1a. *Application to the first boundary value problem.* Apply the preceding results to:

$$-\Delta u = f \quad \text{on } G, \qquad u = u_0 \quad \text{on } \partial G. \tag{132}$$

Here, G is a bounded region in $\mathbb{R}^N$, $N \geq 1$. We are given $f \in L_2(G)$ and $u_0 \in W_2^1(G)$.

Solution: We set

$$a(u, v) = \int_G \sum_{i=1}^N D_i u D_i v \, dx, \qquad b(v) = \int_G fv \, dx.$$

Moreover, let

$$Z = W_2^1(G), \qquad X = \mathring{W}_2^1(G).$$

In this case, v_0 is a generalized solution of the equation

$$-\Delta v_0 = f \quad \text{on } G,$$

and the functions v in Y are generalized solutions of the homogeneous equation

$$-\Delta v = 0 \quad \text{on } G.$$

By the lemma of Weyl (Theorem 18.G),

$$Y = \{v \in C^\infty(G) \colon \Delta v = 0 \text{ on } G\}.$$

Thus, the functions in X and Y satisfy the homogeneous boundary condition $u = 0$ on ∂G and the homogeneous equation $\Delta v = 0$ on G, respectively.

Now, we can apply the results (i) through (v) above to the first boundary value problem (132).

22.1b. *Application to the method of orthogonal projection.* Show that problems (P) and (P*) above generalize the method of orthogonal projection in Section 22.3.

Solution: Choose X and Z as above. We set

$$a(u, v) = (u|v)_Z, \qquad b = 0,$$

and we are given $u_0 \in Z$. Let $v_0 = 0$. Then $Y = X^\perp$, i.e., $Z = X \oplus Y$. Thus, we obtain

$$F(u) = \|u\|^2, \qquad F^*(v) = -\|v\|^2 + 2(u_0|v).$$

Hence the original problem (P) (resp. the dual problem (P*)) above are identical with

$$\|u_0 - u\|^2 = \min!, \qquad u \in X$$

(resp.

$$-\|u_0 - v\|^2 + \|u_0\|^2 = \max!, \qquad v \in X^\perp).$$

This corresponds to the method of orthogonal projection in Section 22.3.

22.2. *Weak sequential lower semicontinuity of positive quadratic functionals.* Let $a \colon X \times X \to \mathbb{R}$ be bilinear, bounded, symmetric, and positive on the real H-space X. Show that

$$u_n \rightharpoonup u \quad \text{in } X \qquad \text{as} \quad n \to \infty$$

implies

$$a(u, u) \leq \varliminf_{n \to \infty} a(u_n, u_n). \tag{133}$$

Solution: For fixed $u \in X$, we set

$$b(v) = a(u, v) \qquad \text{for all} \quad v \in X.$$

Then $b \in X^*$, since $|b(v)| \leq d\|u\| \|v\|$. Thus, $u_n \rightharpoonup u$ as $n \to \infty$ implies $b(u_n) \to b(u)$, i.e.,

$$\lim_{n\to\infty} a(u_n, u) = a(u, u).$$

It follows from $a(u - u_n, u - u_n) \geq 0$ that

$$a(u, u) \leq a(u_n, u_n) + 2a(u, u) - 2a(u_n, u).$$

Letting $n \to \infty$, we get (133).

22.3. *Alternative proof of the main theorem on quadratic minimum problems.* We consider the problem

$$F(u) \stackrel{\text{def}}{=} \tfrac{1}{2}a(u, u) - b(u) = \min!, \qquad u \in X. \tag{134}$$

Suppose that the map $a: X \times X \to \mathbb{R}$ is bilinear, bounded, symmetric, and strongly positive on the real H-space X, and let $b \in X^*$. By means of weak compactness (Theorem 21.D), show that (134) has a solution.

Solution: Since $a(\cdot, \cdot)$ is strongly positive, there is a $c > 0$ such that $a(u, u) \geq c\|u\|^2$ for all $u \in X$. We set

$$\alpha \stackrel{\text{def}}{=} \inf_{u \in X} F(u).$$

Thus, there exists a sequence (u_n) with

$$F(u_n) \to \alpha \qquad \text{as} \quad n \to \infty.$$

It follows from

$$\tfrac{1}{2}c\|u_n\|^2 - \|b\| \|u_n\| \leq F(u_n) \leq \text{constant} \qquad \text{for all } n$$

that (u_n) is bounded. By Theorem 21.D, there is a subsequence, again denoted by (u_n), such that

$$u_n \rightharpoonup u \qquad \text{as} \quad n \to \infty.$$

Hence $b(u_n) \to b(u)$ as $n \to \infty$. By Problem 22.2,

$$F(u) \leq \underline{\lim}_{n\to\infty} F(u_n) = \alpha.$$

This implies $F(u) = \alpha$, i.e., u is a solution of (134).

22.4. *The theorem of Hestenes.* Let $a, b: X \times X \to \mathbb{R}$ be bilinear, bounded, and symmetric on the real H-space X. In addition, let a be strongly positive, and let b be compact.

Show that if $a + b$ is strictly positive, then $a + b$ is strongly positive.

Solution: If $a + b$ is *not* strongly positive, then there exists a sequence (u_n) with $\|u_n\| = 1$ for all n and

$$a(u_n, u_n) + b(u_n, u_n) \to 0 \qquad \text{as} \quad n \to \infty. \tag{135}$$

Since (u_n) is bounded, there exists a subsequence, again denoted by (u_n), such that

$$u_n \rightharpoonup u \quad \text{in } X \qquad \text{as} \quad n \to \infty,$$

by Theorem 21.D. The compactness of b implies

$$b(u_n, u_n) \to b(u, u) \qquad \text{as} \quad n \to \infty.$$

Since $a(\cdot, \cdot)$ is strongly positive, there is a $c > 0$ such that $a(u, u) \geq c\|u\|^2$ for all $u \in X$. By (135),

$$c + b(u, u) \leq 0.$$

By Problem 22.2,

$$a(u, u) + b(u, u) \leq 0.$$

Since $a + b$ is strictly positive, $u = 0$. Hence $c \leq 0$. This contradicts $c > 0$.

22.5. *Eigenvalue problems.* Let M and N be linear subspaces of a B-space X. We set

$$\alpha = \sup\{\text{dist}(u, N) \colon u \in M, \ \|u\| = 1\},$$

$$\beta = \sup\{\text{dist}(u, M) \colon u \in N, \ \|u\| = 1\}$$

and $\delta(M, N) = \max\{\alpha, \beta\}$ (Fig. 22.6).

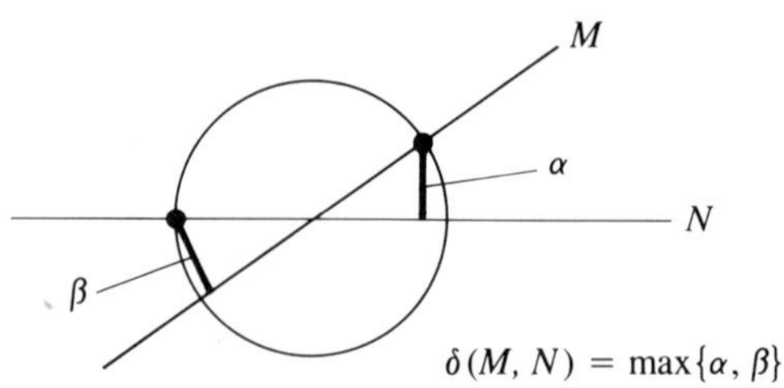

Figure 22.6

22.5a.* *Estimates for eigenvalues and eigenvectors.* Suppose that

$$\lim_{n \to \infty} \|A - A_n\| = 0,$$

where the operators $A, A_n \colon X \to X$ are linear, strictly monotone, symmetric, and compact on the real H-space X. Let

$$\lambda_1 \geq \lambda_2 \geq \cdots > 0$$

(resp.

$$\lambda_{1n} \geq \lambda_{2n} \geq \cdots > 0)$$

be the eigenvectors of A (resp. A_n) counted according to their multiplicity. The eigenvector space to λ_k (resp. λ_{kn}) is denoted by X_k (resp. X_{kn}).

(i) By Corollary 22.29,

$$|\lambda_k - \lambda_{kn}| \leq \|A - A_n\| \qquad \text{for all } k, n.$$

(ii) Show that for each k, there is a constant C_k such that

$$\delta(X_k, X_{kn}) \leq C_k\|A - A_n\| \qquad \text{for all } n.$$

Hint: Cf. Osborn (1975). This paper also contains similar results for compact operators in B-spaces.

22.5b.* *Two-sided error estimates for eigenvalues* (*the method of Knauer*). We consider the eigenvalue problem

$$Lu = \lambda u, \qquad u \in D(L). \tag{136}$$

We want to obtain the two-sided error estimate

$$\mu_k \le \lambda_k \le \rho_k \le q \tag{137}$$

for the kth eigenvalue λ_k of (136). The key to (137) is the estimate (138) below. We make the following assumptions.

(H1) The operator $L: D(L) \subseteq X \to X$ is linear, symmetric, and strictly monotone on the H-space X.

(H2) There exists a sequence of eigenvalues of L with

$$0 < \lambda_1 \le \lambda_2 \le \cdots, \qquad \lim_{k\to\infty} \lambda_k = \infty$$

and an associated sequence (u_n) of eigenvectors which form a complete orthonormal system in X.

(H3) Let $X_n = \operatorname{span}\{v_1, \ldots, v_n\}$ be an n-dimensional linear subspace of $D(L)$. The eigenvalues of the Ritz method

$$(Lu|v_j) = \rho(u|v_j), \qquad u \in X_n, \quad j = 1, \ldots, n,$$

are denoted by $0 < \rho_1 \le \rho_2 \le \cdots \le \rho_n$.

(H4) Suppose that we know a number $q > 0$ such that

$$\frac{(Lu|u)}{(u|u)} \ge q \qquad \text{for all nonzero} \quad u \in X_n^\perp \cap D(L), \tag{138}$$

where $X_n^\perp$ denotes the orthogonal complement to X_n in X.

Let μ_k denote the lowest eigenvalue of the matrix

$$\left(\begin{array}{ccc|c} \rho_k & & 0 & r_k \\ & \ddots & & \vdots \\ 0 & & \rho_n & r_n \\ \hline r_k & \cdots & r_n & q \end{array}\right)$$

where $r_j = (\|Lu_j\|^2 - \rho_j^2)^{1/2}$.

Show that (137) holds.

Hint: Cf. Knauer (1971) and Velte (1984, S). Further important methods for obtaining two-sided error estimates for eigenvalues can be found in Remark 22.31.

22.5c. *An example.* Apply (137) to the following boundary–eigenvalue problem:

$$((1 + x)u'')'' = \lambda u, \qquad 0 < x < \pi, \tag{139a}$$

$$u(0) = u(\pi) = u''(0) = u''(\pi) = 0. \tag{139b}$$

Solution: Set $X = L_2(0, \pi)$ and $Lu = ((1 + x)u'')''$ with

$$D(L) = \{u \in C^4[0, \pi]: u \text{ satisfies (139b)}\}.$$

Along with (139), we consider

$$\begin{aligned} u^{(4)} &= \lambda u, \qquad 0 < x < \pi, \\ u(0) &= u(\pi) = u''(0) = u''(\pi) = 0. \end{aligned} \tag{140}$$

This auxiliary problem has the eigensolutions

$$v_n(x) = \sin nx, \qquad \lambda_n^{(1)} = n^4, \qquad n = 1, 2, \ldots.$$

Set $L_1 u = u^{(4)}$ with $D(L_1) = D(L)$, and let $X_n = \operatorname{span}\{v_1, \ldots, v_n\}$. Since $\lambda_{n+1}^{(1)}$ is the lowest eigenvalue of L_1 on $X_n^\perp$, we obtain

$$\frac{(L_1 u|u)}{(u|u)} \geq \lambda_{n+1}^{(1)} \qquad \text{for all nonzero } u \in X_n^\perp \cap D(L).$$

For all $u \in D(L)$, integration by parts yields

$$\begin{aligned} (Lu|u) &= \int_0^\pi (1 + x)u''^2\, dx \geq \int_0^\pi u''^2\, dx \\ &= \int_0^\pi u^{(4)} u\, dx = (L_1 u|u). \end{aligned}$$

This implies the key estimate

$$(Lu|u) \geq (L_1 u|u) \geq \lambda_{n+1}^{(1)}\, (u|u)$$

for all $u \in X_n^\perp \cap D(L)$, i.e., $q = (n + 1)^4$.

Let λ_k be the kth eigenvalue of (139). Using the nine basis functions $v_1, \ldots, v_9$, we obtain from (137) the following error estimates:

$$\begin{aligned} 2.42587 &\leq \lambda_1 \leq 2.42602 \\ 38.07887 &\leq \lambda_2 \leq 38.08746 \\ 191.4660 &\leq \lambda_3 \leq 191.5835 \\ 603.2088 &\leq \lambda_4 \leq 603.9993 \\ 1468.176 &\leq \lambda_5 \leq 1473.626. \end{aligned}$$

22.6.* *The numerical stability of the Ritz method.* Examples show that the Ritz method may behave badly if the system of basis functions is *not* strongly minimal. We want to explain this *dangerous* effect. Our main result will be that it is important to use strongly minimal systems.

We consider the minimum problem

$$\tfrac{1}{2}a(u, u) - b(u) = \min!, \qquad u \in X,$$

and make the following assumptions.

(H1) Let X be a real separable H-space. The map $a\colon X \times X \to \mathbb{R}$ is bilinear, bounded, symmetric, and strongly positive. Let $b \in X^*$.

(H2) Let $X_n = \operatorname{span}\{v_1, \ldots, v_n\}$, and suppose that $X = \overline{\bigcup_n X_n}$.

We consider the Ritz equation

$$a(v_k, u_n) = b(v_k), \qquad k = 1, \ldots, n, \tag{141}$$

where

$$u_n = \sum_{m=1}^{n} c_{mn} v_m.$$

Then the Ritz equation (141) means that

$$\sum_{m=1}^{n} a(v_k, v_m) c_{mn} = b(v_k), \qquad k = 1, \ldots, n. \tag{142}$$

We write this linear system in the following matrix form:

$$A_n c_n = b_n. \tag{143}$$

Together with this equation, we consider the perturbed equation

$$\bar{A}_n \bar{c}_n = \bar{b}_n. \tag{143*}$$

We introduce the energetic norm, $\|u\|_E = a(u, u)^{1/2}$, and we set

$$\|B\|_\infty = \max_{1 \le k, m \le n} |b_{km}|$$

for the real $n \times n$-matrix B.

The Ritz method is called *numerically stable* iff there are positive constants α, β, γ such that

$$\|u_n - \bar{u}_n\|_E \le \alpha \|A_n - \bar{A}_n\|_\infty + \beta \|b_n - \bar{b}_n\|_\infty \qquad \text{for all } n$$

and all real $n \times n$-matrices $\bar{A}_n$ with $\|A_n - \bar{A}_n\|_\infty \le \gamma$. This relation tells us that, uniformly for all n, sufficiently small perturbations of the Ritz equation lead to small perturbations of the Ritz solutions u_n.

The system of basis elements $\{v_1, v_2, \ldots\}$ is called *strongly minimal* in the energetic space iff

$$\inf_n \lambda_1^{(n)} > 0,$$

where $\lambda_1^{(n)}$ denotes the lowest eigenvalue of the matrix $A_n = (a(v_k, v_m))_{k,m=1,\ldots,n}$.

For example, if (v_k) is a complete orthonormal system with respect to the energetic scalar product $(u|v)_E = a(u, v)$ on X, then A_n equals the identity, and hence (v_k) is strongly minimal in the energetic space, since $\lambda_1^{(n)} = 1$ for all n.

Show that the Ritz method is numerically stable iff (v_k) is strongly minimal in the energetic space.

Hint: Cf. Michlin (1969, M), Chapter 2, Section 2. Chapter 5 of this book also contains similar results for eigenvalue problems.

22.7. *The Gårding inequality for strongly elliptic systems via Fourier transform.* In order to make the proof for the general case below as transparent as possible, we first consider a very simple situation.

22.7a. *A typical special case.* We consider the following differential operator in divergence form:

$$Lu = -(au')' + bu' + du. \tag{144}$$

Let G be a nonempty bounded open set in $\mathbb{R}^1$. We set

$$g(u, v) = \int_G (au'v' + bu'v + duv)\, dx.$$

If the coefficients a, b, d are sufficiently smooth, then integration by parts yields

$$g(u, v) = \int_G (Lu)v\,dx \qquad \text{for all} \quad u, v \in C_0^\infty(G).$$

Obviously, the more general differential operator

$$L_1 u = -a_1 u'' + b_1 u' + d_1 u$$

can be written in the divergence form (144) with

$$a = a_1, \qquad b = a_1' + b_1, \qquad d = d_1,$$

in the case where the derivative a_1' exists.

We assume that the functions $a, b, d\colon G \to \mathbb{R}$ are measurable and bounded. Furthermore, suppose that there is a number $c_0 > 0$ such that

$$a(x) \geq c_0 \qquad \text{for all} \quad x \in G. \tag{145}$$

Recall that

$$\|u\|_{1,2,0}^2 = \int_G u'^2\,dx, \qquad \|u\|_{1,2}^2 = \int_G u^2 + u'^2\,dx,$$

$$\|u\|_2^2 = \int_G u^2\,dx.$$

Show that there are real constants $c > 0$ and C such that

$$g(u, u) \geq c\|u\|_{1,2}^2 - C\|u\|_2^2 \qquad \text{for all} \quad u \in \mathring{W}_2^1(G). \tag{146}$$

Solution: Since the norms $\|\cdot\|_{1,2,0}$ and $\|\cdot\|_{1,2}$ are equivalent on $\mathring{W}_2^1(G)$, and since $C_0^\infty(G)$ is dense in $\mathring{W}_2^1(G)$, it is sufficient to prove that

$$g(u, u) \geq c\|u\|_{1,2,0}^2 - C\|u\|_2^2 \qquad \text{for all} \quad u \in C_0^\infty(G). \tag{147}$$

(I) The ε-trick for the lower order terms. By the Hölder inequality, we obtain

$$\left|\int_G u'u\,dx\right| \leq \left(\int_G u'^2\,dx\right)^{1/2}\left(\int_G u^2\,dx\right)^{1/2}.$$

Using the inequality $|2\alpha\beta| \leq \varepsilon^2\alpha^2 + \varepsilon^{-2}\beta^2$ for all $\alpha, \beta \in \mathbb{R}$, $\varepsilon > 0$, we get

$$2\left|\int_G bu'u\,dx\right| \leq \sup_G |b|\left(\varepsilon^2\int_G u'^2\,dx + \varepsilon^{-2}\int_G u^2\,dx\right).$$

Since we can choose $\varepsilon > 0$ arbitrarily small, it is sufficient for the proof of (147) to show that

$$\int_G au'^2\,dx \geq c\int_G u'^2\,dx - C\int_G u^2\,dx, \tag{148}$$

for all $u \in C_0^\infty(G)$ and appropriate constants $c > 0$ and C. This follows immediately from (145). However, we will give another proof of (148) which can also be applied to strongly elliptic systems below. To this end, we assume that the function a is *continuous* on $\overline{G}$.

(II) The Fourier transform. Let $F(u)$ denote the Fourier transform of u (cf. A_2(74)). For all $u, v \in C_0^\infty(G)$ we have

$$\int_G uv\,dx = \int_{\mathbb{R}} F(u)\overline{F(v)}\,dx \tag{149}$$

and

$$F(u') = ixF(u).$$

(III) Proof of (148) for constant coefficients. Let x_j be a fixed point in G. For all $u \in C_0^\infty(G)$, we obtain that

$$\begin{aligned}\int_G a(x_j)u'^2\,dx &= \int_{\mathbb{R}} a(x_j)F(u')\overline{F(u')}\,dx\\ &\geq \int_{\mathbb{R}} c_0|F(u)|^2x^2\,dx = \int_G c_0u'^2\,dx.\end{aligned}$$

(IV) Localization of the principal part and freezing of the coefficient of the principal part. We want to describe an important technique which *reduces* the case of variable coefficients to the case (III) of constant coefficients. The idea is to use the values $a(x_j)$ at a finite number of suitable points x_j. This procedure is called "*freezing of the coefficients.*"

(IV-1) Basic estimate. Let U be a nonempty open subset of G and suppose that

$$|a(x) - a(y)| < \eta \qquad \text{for all} \quad x, y \in U.$$

We fix $\varphi \in C_0^\infty(U)$ and $y \in U$. Then, for all $u \in C_0^\infty(G)$, we obtain that

$$\begin{aligned}\left|\int_G (a(x) - a(y))(u\varphi)'^2\,dx\right| &\leq \eta \int_G (u\varphi)'^2\,dx\\ &\leq \eta\cdot\text{const}\int_G (u'^2 + u^2)\,dx,\end{aligned}$$

noting that $(u\varphi)' = u'\varphi + u\varphi'$ and $2|\alpha\beta| \leq \alpha^2 + \beta^2$.

(IV-2) Partition of unity. Let $\eta > 0$ be given. Since a is continuous on $\bar{G}$, there exists a finite number of nonempty open sets $U_1, \ldots, U_k$ such that $\bar{G} \subseteq \bigcup_j U_j$ and

$$|a(x) - a(y)| < \eta \qquad \text{for all} \quad x, y \in U_j \cap G \quad \text{and all } j.$$

By Section 18.16, there exists a partition of unity subordinated to $\{U_j\}$, i.e., there is a family $\{\varphi_j\}$ of functions $\varphi_j \in C_0^\infty(U_j), j = 1, \ldots, k$, such that

$$\sum_j \varphi_j(x)^2 = 1 \qquad \text{for all} \quad x \in G, \tag{150}$$

where we set $\varphi_j(x) = 0$ for $x \notin U_j$. Let $u \in C_0^\infty(G)$ be given. We localize the function u by setting

$$u_j = \varphi_j u.$$

Then $u_j \in C_0^\infty(U_j)$, and from (150) we obtain the *key relation*:

$$\sum_j au_j'^2 = \sum_j a\varphi_j^2u'^2 + A = au'^2 + A, \tag{151}$$

where A contains lower order terms, i.e., A contains only terms of the form uu' and u^2.

(IV-3) In order to prove (148) it is now sufficient to show that

$$J \stackrel{\text{def}}{=} \int_G \sum_j a(x_j)u_j'^2\,dx \geq c\int_G u'^2\,dx - C\int_G u^2\,dx \tag{152}$$

in the case where we choose $\eta > 0$ sufficiently small and we fix the points x_j such that $x_j \in U_j$ for all j. In fact, by (151), we can estimate the expression A according to (I) for small $\varepsilon > 0$, and we can use the argument (IV-1) for small $\eta > 0$.

(VI-4) Proof of (152). By (III) and (151),

$$J \geq c_0 \int_G \sum_j u_j'^2\,dx = c_0\int_G u'^2\,dx + \int_G B\,dx,$$

where B denotes lower order terms. Using (I) with respect to B, for small $\varepsilon > 0$, we obtain (152).

22.7b. *Strongly elliptic systems.* We consider the system

$$\begin{aligned} L_j u = f_j \quad &\text{on } G, \qquad j = 1, \ldots, M,\\ D^\gamma u_j = 0 \quad \text{on } \partial G \qquad &\text{for all} \quad \gamma\colon |\gamma| \leq m-1, \end{aligned} \tag{153}$$

where

$$L_j u = \sum_{i=1}^{M} \sum_{|\alpha|,|\beta|\leq m} (-1)^{|\alpha|} D^\alpha(a_{\alpha\beta ij}D^\beta)u_i,$$

and we set

$$g(u,v) = \int_G a_{\alpha\beta ij}D^\alpha v_j D^\beta u_i\,dx,$$

where we sum over $i, j = 1, \ldots, M$ and α, β with $|\alpha|, |\beta| \leq m$. For $u, v \in C_0^\infty(G)^M$, integration by parts yields

$$g(u,v) = \int_G \sum_{j=1}^{M} (L_j u)v_j\,dx.$$

The system (153) is called *strongly elliptic* iff all the functions $a_{\alpha\beta ij}$ with $i, j = 1, \ldots, M$ and $|\alpha|, |\beta| = m$ are *continuous* on $\bar{G}$ and there is a $c_0 > 0$ such that

$$\sum_{i,j=1}^{M} a_{\alpha\beta ij}(x)d^\alpha d^\beta w_i w_j \geq c_0|d|^{2m}|w|^2, \tag{154}$$

for all $x \in G$, $d \in \mathbb{R}^N$, $w \in \mathbb{R}^M$, where $d^\alpha = d_1^{\alpha_1}d_2^{\alpha_2}\cdots d_N^{\alpha_N}$. We assume:

(H1) G is a nonempty bounded open set in $\mathbb{R}^N$, where $N, M, m \geq 1$.

(H2) All the functions $a_{\alpha\beta ij}\colon G \to \mathbb{R}$ are measurable and bounded for $i, j = 1, \ldots, M$ and α, β with $|\alpha|, |\beta| \leq m$.

(H3) The system (153) is strongly elliptic.

Show that $g(\cdot,\cdot)$ satisfies the *Gårding inequality*, i.e., there are real numbers $c > 0$ and C such that

$$g(u,u) \geq c\|u\|_{m,2}^2 - C\|u\|_2^2 \qquad \text{for all} \quad u \in \mathring{W}_2^m(G)^M. \tag{155}$$

Here, $u \in \mathring{W}_2^m(G)^M$ means $u_i \in \mathring{W}_2^m(G)$ for all i, and

$$\|u\|_{m,2}^2 = \sum_i \|u_i\|_{m,2}^2, \qquad \|u\|_2^2 = \sum_i \|u_i\|_2^2.$$

Recall that

$$\|u_i\|_{m,2}^2 = \int_G \sum_{|\alpha| \leq m} |D^\alpha u_i|^2\, dx,$$

$$\|u_i\|_{m,2,0}^2 = \int_G \sum_{|\alpha| = m} |D^\alpha u_i|^2\, dx.$$

Solution: Since the two norms $\|\cdot\|_{m,2}$ and $\|\cdot\|_{m,2,0}$ are equivalent on $\mathring{W}_2^m(G)$, and since $C_0^\infty(G)$ is dense in $\mathring{W}_2^m(G)$, it is sufficient to show that

$$g(u,u) \geq c\|u\|_{m,2,0}^2 - C\|u\|_2^2 \qquad \text{for all} \quad u \in C_0^\infty(G)^M. \tag{156}$$

As in the proof of Proposition 22.45, the *lower order terms* of g with $|\alpha| + |\beta| < 2m$ can be estimated via Ehrling's inequality and the ε-trick.

Thus it is sufficient to prove (156) for the *principal part*, i.e., we may assume that $a_{\alpha\beta ij} = 0$ for $|\alpha| + |\beta| < 2m$. The proof now proceeds completely analogously to the proof of Problem 22.7a. In this connection, observe the following. It follows from (154) that

$$\operatorname{Re} \sum_{i,j=1}^{M} \sum_{|\alpha|,|\beta|=m} a_{\alpha\beta ij}(x) d^\alpha d^\beta w_i \overline{w}_j \geq c_0 |d|^{2m} |w|^2,$$

for all *complex* tupels $w \in \mathbb{R}^M$, since

$$\operatorname{Re}(w_i \overline{w}_j) = \operatorname{Re} w_i \operatorname{Re} w_j + \operatorname{Im} w_i \operatorname{Im} w_j,$$

and $a_{\alpha\beta ij}(x)$, d are real. Hence, for constant $a_{\alpha\beta ij}$ and $u \in C_0^\infty(G)^M$, we obtain the following *key inequality* via the Fourier transform:

$$\begin{aligned}\int_G a_{\alpha\beta ij} D^\alpha u_i D^\beta u_j\, dx &= \int_{\mathbb{R}^N} \operatorname{Re} a_{\alpha\beta ij} x^\alpha x^\beta F(u_i)\overline{F(u_j)}\, dx \\ &\geq c_0 \int_{\mathbb{R}^N} |x|^{2m} F(u_i)\overline{F(u_i)}\, dx = c_0 \int_G |D^\alpha u_i|^2\, dx,\end{aligned}$$

where we sum over $i, j = 1, \ldots, M$ and α, β with $|\alpha|$, $|\beta| = m$. Note that the left-hand integral is real, since $a \ldots, u, v$ are real, and note that $\operatorname{Re} \int \cdots = \int \operatorname{Re} \ldots.$

22.8. *Existence theory for strongly elliptic systems.* By definition, the *generalized problem* to the boundary value problem (153) reads as follows.

Let $X = \mathring{W}_2^m(G)^M$. We are given $f_i \in L_2(G)$ for all i, and we are looking for a function $u \in X$ such that

$$g(u,v) = b(v) \qquad \text{for all} \quad v \in X. \tag{157}$$

Assume (H1) to (H3) from Problem 22.7b. Then, all the abstract results from Section 22.16 can be applied to equation (157). In particular, equation (157) has a solution iff

$$b(u^*) = 0$$

for all solutions $u^* \in X$ of the dual equation

$$g(v, u^*) = 0 \qquad \text{for all} \quad v \in X. \tag{157*}$$

The number of linearly independent solutions u^* of (157*) is finite.

If (157) has at most one solution, then it possesses a unique solution.

22.9. *The method of homogenization and delicate convergence processes for functionals and operators.* We consider the operator equation

$$(\mathrm{E}_\varepsilon) \qquad L_\varepsilon u_\varepsilon = f,$$

which depends on the small parameter $\varepsilon > 0$. We are looking for an asymptotic expansion of the solution of the form

$$u_\varepsilon = v_0 + \varepsilon v_1 + \varepsilon^2 v_2 + \cdots,$$

and we are looking for an operator equation

$$(\mathrm{E}_0) \qquad L_0 v_0 = f,$$

which yields the approximation v_0 of u_ε. In engineering, for example, equation (E_ε) describes processes in material of the small period $\varepsilon > 0$ (e.g., heat conduction in granulated material). The limiting process $\varepsilon \to 0$ corresponds to a *homogenization* of the material (Fig. 22.7). The idea of engineers is to use equation (E_0) as an approximation for (E_ε). In this connection, note that the parameters in (E_ε) and (E_0) correspond to the material constants of the periodic (granulated) and the homogenized material, respectively (e.g., the heat conductivity).

To explain the typical difficulty let us consider the following simple problem:

$$(\mathrm{P}_\varepsilon) \qquad \begin{aligned} -(a_\varepsilon(x)u_\varepsilon'(x))' &= f(x), \qquad 0 < x < 1, \\ u_\varepsilon(0) = u_\varepsilon(1) &= 0, \end{aligned}$$

where we set

$$a_\varepsilon(x) = a(x/\varepsilon),$$

and the function a_ε has the period $\varepsilon > 0$. More precisely, assume that the continuous function $a\colon \mathbb{R} \to \mathbb{R}$ has period one and $\inf_{x \in \mathbb{R}} a(x) > 0$. The mean value of a is denoted by

$$M(a) = \int_0^1 a(x)\,dx.$$

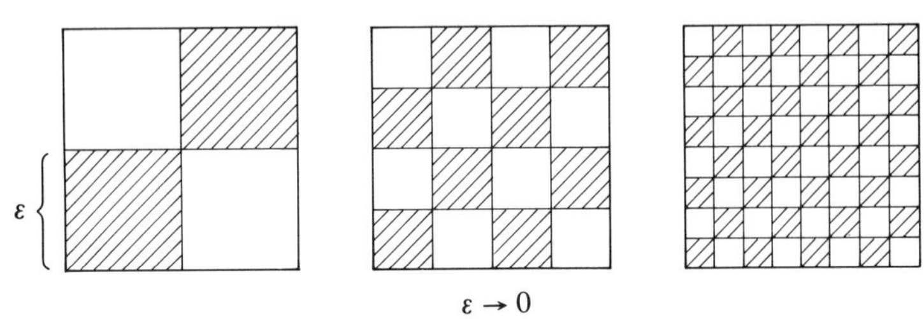

Figure 22.7

Let $f \in C[0,1]$. One can show that

$$a_\varepsilon(x) \overset{*}{\rightharpoonup} M(a) \quad \text{in } L_\infty(0,1) \qquad \text{as} \quad \varepsilon \to 0.$$

Furthermore, the generalized solution $u \in \mathring{W}_2^1(0,1)$ of (P_ε) satisfies the estimate $\|u_\varepsilon\|_{1,2} \leq$ const for all small $\varepsilon > 0$. Hence, there is a subsequence such that

$$u_\varepsilon \rightharpoonup u \quad \text{in } \mathring{W}_2^1(0,1) \qquad \text{as} \quad \varepsilon \to 0.$$

We now come to the *point*. Intuitively, we would expect that the limit function u satisfies the equation

$$(\mathrm{P}_0) \qquad \begin{aligned} -(bu')' &= f, \qquad 0 < x < 1, \\ u(0) &= u(1) = 0 \end{aligned}$$

with $b = M(a)$. However, this is *wrong*. In fact, we have to set

$$b = 1/M(a^{-1}).$$

This has been discovered by engineers. Similar difficulties arise for variational problems.

As an introduction to the method of homogenization, we recommend Bensoussan, Lions, and Papanicolaou (1978, M). From the physical point of view, the homogenization of material leads to new physical effects. From the mathematical point of view, the method of homogenization corresponds to *delicate convergence processes* for operators and functionals in connection with differential equations and variational problems, respectively (e.g., the G-convergence). A general theory for both linear and nonlinear problems can be found in Attouch (1984, M).

22.10.** *The end of a long struggle—the complete description of the Laplacian and of general elliptic equations in terms of functional analysis.* We consider the first boundary value problem:

$$-\Delta u = f \quad \text{on } G, \tag{158a}$$

$$u = g \quad \text{on } \partial G. \tag{158b}$$

Let G be a bounded region in $\mathbb{R}^N$, $N \geq 1$, and let $\partial G \in C^\infty$. Let $m = 0, 1, \ldots,$ and $1 < p < \infty$. We set

$$Au = (f, g),$$

where f and g are given by (158). The following fundamental result *finishes* the long struggle of mathematicians against the Dirichlet problem.

The operator

$$A\colon W_p^{m+2}(G) \to W_p^m(G) \times W_p^{m+2-1/p}(\partial G)$$

is a linear homeomorphism.

The definition of the boundary spaces $W_p^k(\partial G)$ of fractional order k will be given in the Appendix $\mathrm{A}_2(51)$. This very natural result allows the following interpretation.

(i) If $u \in W_p^{m+2}(G)$, then u satisfies (158) with

$$f \in W_p^m(G), \qquad g \in W_p^{m+2-1/p}(\partial G). \tag{159}$$

In this connection, note that u has at least generalized second-order derivatives. Thus, equations (158a) and (158b) are to be understood in the sense of generalized derivatives and generalized boundary values, respectively.

(ii) Conversely, for each given f and g with (159), the Dirichlet problem (158) has a unique solution $u \in W_p^{m+2}(G)$, and there is a constant $c > 0$ depending on G, m, and p such that

$$\|u\|_{m+2,p} \leq c(\|f\|_{m,p} + \|g\|_{m+2-1/p,p}).$$

Hint: For $p = 2$, see Lions and Magenes (1968, M), Vol. 1.

For $p > 1$, see Triebel (1978, M), (1983, M). The proof relies on interpolation theory.

More general results in terms of Sobolev spaces can be found in Problem 6.8.

In Triebel (1983, M), Section 4.3, one finds the following fundamental result via interpolation theory:

Broad classes of elliptic equations with fairly general boundary conditions represent isomorphic mappings with respect to appropriate $B^s_{p,q}$-spaces and $F^s_{p,q}$-spaces.

These two important classes of function spaces coincide, for special values s, p, q, with Sobolev spaces or Hölder spaces (cf. A_2(116b)).

A detailed study of these spaces can be found in Triebel (1983, M). There it is also shown that the $B^s_{p,q}$-spaces and $F^s_{p,q}$-spaces comprehend the following important special cases: Hölder–Zygmund spaces, Sobolev–Slobodeckii spaces, Besov spaces, Lebesgue spaces, Bessel potential spaces, and Hardy spaces. In some sense, one can say that the $B^s_{p,q}$-spaces and $F^s_{p,q}$-spaces are related to elliptic equations in a natural manner.

References to the Literature

Classical papers: Poincaré (1890), (1916, M) (collected works), Hilbert (1900a), (1912, M), (1932, M) (collected works), Weyl (1911), (1940), Courant (1920), Friedrichs (1934), (1944), (1953), Courant and Hilbert (1937, M), Sobolev (1937), (1938), (1950, M), Višik (1949), (1951), (1952), Gårding (1953), Browder (1954) (strongly elliptic systems), Lax and Milgram (1954), Nirenberg (1955), (1959), Agmon, Douglis, and Nirenberg (1959) (fundamental *a priori* estimates), Hörmander (1955), (1964, M), (1983, M) (general theory of linear partial differential operators), Lions and Magenes (1960), (1968, M) (interpolation theory and linear elliptic equations).

Modern standard work on linear partial differential equations: Dautray and Lions (1984, M), Vols. 1–3 (theory, applications, and numerical methods).

Handbook of numerical analysis: Ciarlet and Lions (1988, M), Vols. 1ff (finite element method, difference method, etc.).

Collection of problems in the equations of mathematical physics with solutions: Vladimirov (1986a, M).

The partial differential equations of theoretical physics and their explicit solutions: Morse and Feshbach (1953, M), Frank and v. Mises (1961, M), Babič (1967, M), Myint and Debnath (1987, M).

Introduction to the theory of linear partial differential equations: Courant and

Hilbert (1953, M) (classical standard work), Tychonov and Samarskii (1959, M), Garabedian (1964, M), Babič (1967, M) (handbook without proofs), John (1982, M).

Introduction to the modern functional analytic approach to linear elliptic differential equations: Friedman (1969, M) (especially recommended), Bers, John, and Schechter (1964, M) (basic ideas), Agmon (1965, L), Fichera (1965, L), Nečas (1967, M), Wloka (1982, M), Ladyženskaja (1985, M).

Elementary expositions: Velte (1976, M), Rektorys (1977, M), Showalter (1977, M).

Further important monographs: Morrey (1966), Maurin (1967), Simader (1972), Triebel (1972), Ladyženskaja and Uralceva (1973), Treves (1975), Schechter (1977), Gilbarg and Trudinger (1983), Grisvard (1985) (nonsmooth domains).

Survey: Miranda (1970, M).

Pseudodifferential operators: Hörmander (1983, M), Vols. 1–4 (standard work), Taylor (1981, M), Treves (1982, M), Vols. 1, 2, Rempel and Schulze (1982, M) (modern index theory for elliptic boundary value problems), Chazarain and Pirion (1982, M) (introduction).

Interpolation theory and linear elliptic differential equations: Lions and Magenes (1968, M), Vol. 1 (L_2-theory), Triebel (1978, M), (1983, M) (general L_p-theory).

Ritz method: Cf. the References to the Literature for Chapter 18.

Difference method: Cf. the References to the Literature for Chapter 20.

Finite elements: Ciarlet (1977, M), Babuška and Szabo (1988, M), Ciarlet and Lions (1988, M), Vol. 1. Cf. also the References to the Literature for the Appendix.

Boundary element method and its applications in engineering: Feng Kang (1983, S) (abstract approach), Bannerjee (1981, M), Crouch (1983, M), Brebbia (1984, M), Wendland (1984, S), Tanaka (1986, M).

Multigrid methods: Hackbusch (1985, M).

Software system ELLPACK for solving elliptic differential equations: Rice and Boisvert (1984).

Software system QUADPACK for integration: Piessens (1983).

Software system PDE/PROTRAN for the finite element method: Sewell (1985).

Approximation methods for eigenvalues: Gould (1966, M) (introduction), Collatz (1963, M) (many applications in engineering), Collatz (1964, M) (Newton method), Michlin (1969, M) (stability of the Ritz method), Krasnoselskii (1973, M) and Vainikko (1977) (Galerkin method), Bramble and Osborn (1973) and Osborn (1975) (nonsymmetric compact operators), Kolomý (1978) (iterative methods).

Two-sided error estimates for eigenvalues: Velte (1984) (survey), Collatz (1963, M), (1964, M) (method of Temple), Fichera (1965, L) (methods of Rayleigh–Ritz, Bazley–Fox, and Weinstein–Aronszajn), Fichera (1978, M) (method of orthogonal invariants), Knauer (1971), Weinstein and Stenger (1972, M), Weinberger (1974, M), Kuttler and Sigilito (1987, M) (estimating eigenvalues with *a posteriori/a priori* inequalities).

Eigenvalues and spectra of molecules: Reed and Simon (1971, M), Vol. 4, Thirring (1983, M), Vol. 3.

Eigenvalue criteria for minima in the calculus of variations: Klötzler (1971, M).

Eigenvalue distributions of compact operators: König (1985, M), Pietsch (1987, M).

Pseudodifferential operators and eigenvalue problems: Shubin (1986, M).

Random eigenvalue problems: v. Scheidt and Purkert (1983, M).

Homogenization: Bensoussan, Lions, Papanicolau (1978, M), Oleinik (1979, S), Attouch (1984, L), Ericksen (1986, P).

G-convergence and weak variational convergence and their applications to differential equations: De Giorgi (1983, S), Attouch (1984, L) (linear and nonlinear problems).

Supercomputing: Murman (1985, S), Lichnewsky and Saguez (1987, S), Martin (1988, S).

CHAPTER 23

Hilbert Space Methods and Linear Parabolic Differential Equations

> While studying parabolic equations it is useful to introduce and make use of several function spaces.
>
> Olga Aleksandrovna Ladyženskaja (1985)

23.1. Particularities in the Treatment of Parabolic Equations

In order to obtain existence theorems for linear parabolic differential equations one has two different possibilities:

(i) the theory of semigroups, and
(ii) the Galerkin method.

In Chapter 19 we considered (i). In this chapter we want to study (ii). Let $Q_T = G \times]0, T[$. In order to explain the basic problems and methods we consider the classical initial–boundary value problem for the heat equation:

$$\begin{aligned} u_t - \Delta u = f \quad & \text{on } Q_T, \\ u = 0 \quad & \text{on } \partial G \times [0, T], \\ u(x, 0) = u_0(x) \quad & \text{on } G. \end{aligned} \tag{1}$$

Here, the functions f and u_0 are given and u is to be found.

The *generalized problem* corresponding to this is obtained formally by multiplying the differential equation by $v \in C_0^\infty(G)$ and subsequent integration by parts; therefore, for all $v \in C_0^\infty(G)$,

$$\frac{d}{dt}\int_G u(x,t)v(x)\,dx + \int_G \sum_{i=1}^N D_i u(x,t) D_i v(x)\,dx$$

$$= \int_G f(x,t)v(x)\,dx \qquad \text{for all} \quad t \in\,]0,T[, \tag{2}$$

$$u(x,0) = u_0(x) \qquad \text{for all} \quad x \in G,$$

where $x = (\xi_1, \ldots, \xi_N)$ and $D_i = \partial/\partial\xi_i$. We set

$$V = \mathring{W}_2^1(G), \qquad H = L_2(G)$$

and refer to four particularities which arise in the functional analytic treatment of parabolic equations.

(I) The *first* important particularity of (2) consists in that we deal with the space variable x and the time variable t in *different ways.*

(a) For fixed time t, we think of the function

$$x \mapsto u(x,t)$$

of the space variable x as an element of the Sobolev space V. We briefly denote the element of V that arises this way by $u(t)$; therefore,

$$u(t) \in V.$$

(b) If we now vary the time t in the interval $[0,T]$, then we obtain a function

$$t \mapsto u(t).$$

There thus arises from the real function $(x,t) \mapsto u(x,t)$ the function $t \mapsto u(t)$ with *values in the* B-*space* V.

We now write equation (2) in the form

$$\frac{d}{dt}(u(t)|v)_H + a(u(t),v) = (f(t)|v)_H \quad \text{on }]0,T[\qquad \text{for all} \quad v \in V, \tag{3}$$

$$u(0) = u_0 \in H.$$

We are looking for the function $t \mapsto u(t)$ with $u(t) \in V$ for all $t \in [0,T]$. Here, for all $w, v \in V$, we set

$$a(w,v) = \int_G \sum_{i=1}^N D_i w(x) D_i v(x)\,dx,$$

$$(f(t)|v)_H = \int_G f(x,t)v(x)\,dx.$$

(II) Equation (3) shows the *second* particularity that the generalized problem requires, in a natural way, the use of *two* spaces H and V. Here, H is obtained in connection with the time derivative, and V results from the elliptic term $-\Delta u$ in (1) and the boundary condition $u = 0$ on $\partial G \times [0, T]$. In the present case we have $V \subseteq H$ and V is dense in H. This leads us to the concept of the *evolution triple*

$$\text{“}V \subseteq H \subseteq V^*\text{”}$$

in Section 23.4.

(III) The *third* particularity consists in that we think of the time derivative d/dt in (3) as the *generalized derivative* on $]0, T[$. For this reason, equation (3) needs to be satisfied only for almost all $t \in\,]0, T[$.

(IV) The *fourth* particularity results from the choice of the space for the solution $t \mapsto u(t)$. For (3) we shall choose the Sobolev space

$$W_2^1(0, T; V, H).$$

We give the precise definition of this space in Section 23.6. Here, we note only the corresponding norm

$$\|u\| = \left(\int_0^T \|u(t)\|_V^2 \, dt \right)^{1/2} + \left(\int_0^T \|u'(t)\|_{V^*}^2 \, dt \right)^{1/2},$$

where u' is the generalized derivative of u. The definition of u' is so made that:

$$\langle u'(t), v \rangle_V = \frac{d}{dt} (u(t)|v)_H \qquad \text{for all} \quad v \in V.$$

In order to obtain an operator equation, we introduce the operator $A: V \to V^*$ and the functional $b(t) \in V^*$ by means of

$$\langle Aw, v \rangle_V = a(w, v),$$

$$\langle b(t), v \rangle_V = (f(t)|v)_H \qquad \text{for all} \quad v \in V.$$

Then from the basic equation (3) there arises

$$\langle u'(t) + Au(t) - b(t), v \rangle_V = 0 \quad \text{on }]0, T[\qquad \text{for all} \quad v \in V$$

and hence we get the desired *operator equation*:

$$\begin{aligned} u'(t) + Au(t) &= b(t) \qquad \text{for almost all} \quad t \in\,]0, T[, \\ u(0) &= u_0 \in H. \end{aligned} \tag{4}$$

Here, we are looking for the function

$$u \in W_2^1(0, T; V, H).$$

This is an abstract first-order differential equation that we call a first-order evolution equation. The point is that

$$u(t) \in V \qquad \text{and} \qquad u'(t) \in V^*,$$

i.e., because $V \subset V^*$, the time derivative $u'(t)$ lies in a more comprehensive space than $u(t)$. The existence theorems concerning abstract differential equations from Chapter 3 can therefore *not* be applied. To solve equation (4), we will use the Galerkin method in Section 23.7.

In Section 23.4 we shall elucidate in what sense one is to understand $V \subseteq H \subseteq V^*$. Analogously to the regularity theorems in Sobolev spaces of real functions there results in Section 23.6 the *continuity of the embedding*

$$W_2^1(0, T; V, H) \subseteq C([0, T], H).$$

That is, u in equation (4) is continuous as the function $u\colon [0, T] \to H$. It is only then that the initial condition $u(0) = u_0 \in H$ is meaningful.

We emphasize again that this approach works in the *two* spaces V and H. For the investigation of (1) there is still another possibility which works only in *one* space, namely, in $H = L_2(G)$. To this end, we write the original problem (1) in the form

$$\begin{aligned} u'(t) + A_F u(t) &= f(t) \qquad \text{for all} \quad t \in \,]0, T[, \\ u(0) &= u_0. \end{aligned} \tag{4*}$$

The main difference is the following:

(α) We set

$$Bu = -\Delta u, \qquad D(B) = C_0^\infty(G).$$

Then the operator $A_F\colon D(A_F) \subseteq H \to H$ in (4*) is the Friedrichs extension of B. This operator is *unbounded.*

(β) In contrast to A_F, the operator $A\colon V \to V^*$ in (4) is *continuous.* More precisely, A is the so-called energetic extension of B, i.e., the space $V = \mathring{W}_2^1(G)$ is the energetic space of B and A is an extension of A_F. We have that

$$D(B) \subseteq D(A_F) \subseteq V \subseteq H, \qquad B \subseteq A_F \subseteq A,$$

where $D(B) = C_0^\infty(G)$, $V = \mathring{W}_2^1(G)$, and $H = L_2(G)$.

Equation (4*) was used in Chapter 19 in the context of semigroups. The difficulties which arise in (4) and (4*), in different ways, have their origin in the fact that in equation (2) the first integral originating from u_t contains *no* space derivatives, and the second integral originating from the elliptic part $-\Delta$ does contain space derivatives.

In equation (4), we need

$$b(t) \in V^* \qquad \text{for almost all} \quad t \in \,]0, T[,$$

whereas, in equation (4*), we need the *stronger* condition

$$f(t) \in H \qquad \text{for all} \quad t \in [0, T].$$

Note that $H \subseteq V^*$. Since equations (4) and (4*) correspond to the Galerkin method and the method of semigroups, respectively, we obtain the following:

The Galerkin method allows us more general right-hand sides of evolution equations than the method of semigroups.

In Chapter 3 we have considered, parallel to the classical calculus, the Riemann integral and the derivative of functions $t \mapsto u(t)$ of the real variable t with values in a B-space. For the investigation of the evolution equation (4), we need the following more general concepts for functions of this sort, namely:

(a) measurability,
(b) the Lebesgue integral and Lebesgue spaces,
(c) generalized derivatives and Sobolev spaces.

In Appendix $A_2(1)$, measures, measurability, and the Lebesgue integral are introduced in a unified way for real functions and functions with values in B-spaces. The reader who is not familiar with these concepts should look at $A_2(1)$ through $A_2(13)$ before studying this chapter.

We provide the remaining tools (c) in the following sections. In principle, all the proofs corresponding to (c) proceed parallel to the classical proofs for real functions. We attach importance to presenting results in such a form, and to clarifying them by means of examples so that later treatment of linear and nonlinear evolution equations and evolution inequalities will be as transparent as possible for the reader. In particular, we emphasize the clear working out of a number of space identifications in the context of evolution triples "$V \subseteq H \subseteq V^*$." In the proofs, e.g., in Example 23.4, one must always pay careful attention to the fact that the Lebesgue integral is only defined for measurable functions. For that reason, one must also always verify the *measurability* of the functions under consideration.

The main result of this chapter is Theorem 23.A in Section 23.7. For linear evolution equations, this theorem contains:

existence and uniqueness of the solution,
convergence of the Galerkin method, and
continuous dependence of the solution on the data.

Generalization to nonlinear problems will be considered

in Chapter 30 (the Galerkin method), and
in Chapter 31 (semigroups).

23.2. The Lebesgue Space $L_p(0, T; X)$ of Vector-Valued Functions

Definition 23.1. Let X be a B-space and $0 < T < \infty$.

(a) The space $C^m([0, T], X)$ with $m = 0, 1, \ldots$ consists of all continuous functions $u: [0, T] \to X$ that have continuous derivatives up to order m on $[0, T]$ with the norm

$$\|u\| \stackrel{\text{def}}{=} \sum_{i=0}^{m} \max_{0 \le t \le T} |u^{(i)}(t)|. \tag{5}$$

Here, only the right-hand and the left-hand derivatives need exist at the boundary points $t = 0$ and $t = T$, respectively. In (5), $u^{(0)}$ means u.

We write $C([0, T], X)$ instead of $C^0([0, T], X)$.

(b) The space $L_p(0, T; X)$ with $1 \le p < \infty$ consists of all measurable functions $u\colon]0, T[\to X$ for which

$$\|u\|_p \overset{\text{def}}{=} \left(\int_0^T \|u(t)\|_X^p \, dt \right)^{1/p} < \infty \tag{6}$$

holds.

The space $L_\infty(0, T, X)$ will be introduced in Problem 23.12.

Proposition 23.2 (Properties of Lebesgue Spaces). *Let* $m = 0, 1, \ldots$ *and* $1 \le p < \infty$. *Let* X *and* Y *be* B-*spaces over* $\mathbb{K}$. *Then*:

(a) $C^m([0, T], X)$ *with the norm* (5) *is a* B-*space over* $\mathbb{K}$.

(b) $L_p(0, T; X)$ *with the norm* (6) *is a* B-*space over* $\mathbb{K}$ *in the case where one identifies functions that are equal almost everywhere on* $]0, T[$.

Moreover, the set of all step functions $u\colon [0, T] \to X$ *is dense in* $L_p(0, T; X)$ (*cf.* $A_2(4)$).

(c) $C([0, T], X)$ *is dense in* $L_p(0, T; X)$, *and the embedding*

$$C([0, T], X) \subseteq L_p(0, T; X)$$

is continuous.

(d) *The set of all polynomials* $w\colon [0, T] \to X$, *i.e.*,

$$w(t) = a_0 + a_1 t + \cdots + a_n t^n$$

with $a_i \in X$ *for all* i *and* $n = 0, 1, \ldots$ *is dense in* $C([0, T], X)$ *and* $L_p(0, T; X)$.

(e) *If* X *is an* H-*space with scalar product* $(\cdot|\cdot)_X$, *then* $L_2(0, T; X)$ *is also an* H-*space with the scalar product*

$$(u|v) = \int_0^T (u(t)|v(t))_X \, dt.$$

(f) $L_p(0, T; X)$ *is separable in the case where* X *is separable and* $1 \le p < \infty$.

(g) $L_p(0, T; X)$ *is uniformly convex in the case where* X *is uniformly convex and* $1 < p < \infty$.

(h) *If the embedding* $X \subseteq Y$ *is continuous, then the embedding*

$$L_r(0, T; X) \subseteq L_q(0, T; Y), \qquad 1 \le q \le r \le \infty,$$

is also continuous.

The proof will be given in Problem 23.2.

EXAMPLE 23.3. Let X be a B-space. If the function

$$u\colon]0, T[\to X$$

is continuous almost everywhere and bounded, i.e.,

$$\sup_{0<t<T} \|u(t)\| < \infty,$$

then $u \in L_p(0, T; X)$ for all $1 \le p \le \infty$.

PROOF.

(I) Let $p < \infty$. By $A_2(8)$, the function $t \mapsto u(t)$ is measurable. The measurability of $t \mapsto \|u(t)\|$ results from $A_2(9)$. Due to the assumption that this function is bounded, condition (6) is valid.

(II) In the case where $p = \infty$ use the definition of $L_\infty(0, T; X)$ in Problem 23.12. □

The following example is important for the functional analytic interpretation of the right side f of parabolic differential equations. Our goal is the next two formulas:

$$\langle b(t), v \rangle_Y = \int_G f(x, t) v(x)\, dx \qquad \text{for all} \quad v \in Y, \tag{7a}$$

$$\langle b_1(t), v \rangle_V = \int_G f(x, t) v(x)\, dx \qquad \text{for all} \quad v \in V. \tag{7b}$$

For this purpose we make the following assumptions:

(H) Let G be a bounded region in $\mathbb{R}^N$, $N \ge 1$. For $0 < T < \infty$, let $Q_T = G \times]0, T[$, i.e.,

$$Q_T = \{(x, t): x \in G, t \in]0, T[\}$$

(Fig. 23.1). Moreover, we set

$$Y = L_p(G), \qquad V = \mathring{W}_p^m(G)$$

with $1 < p < \infty$, $p^{-1} + q^{-1} = 1$, $m = 0, 1, \ldots$.

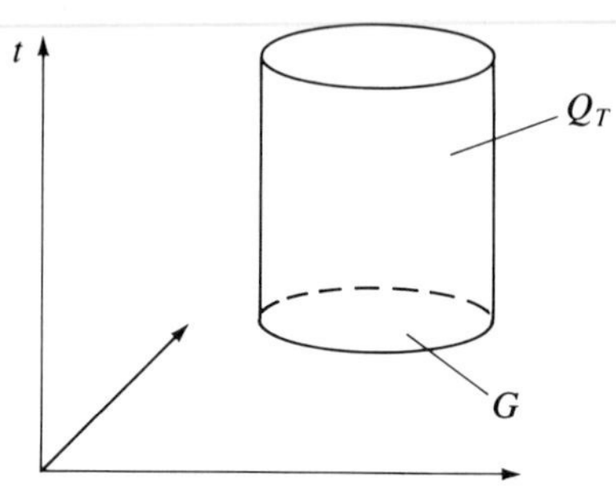

Figure 23.1

EXAMPLE 23.4. Under the assumption (H), suppose the real function $(x, t) \mapsto f(x, t)$ belongs to $L_q(Q_T)$. Then the following hold:

(a) For all $t \in [0, T]$, there exists a functional $b(t) \in Y^*$ such that (7a) is valid for almost all $t \in [0, T]$ and $t \mapsto b(t)$ belongs to $L_q(0, T; Y^*)$.

To be precise, for almost all $t \in]0, T[$, the function

$$x \mapsto f(x, t)$$

belongs to the space $L_q(G)$, and this function can be identified with a functional $b(t) \in Y^*$ such that we have (7a) and

$$\|b(t)\|_{Y^*}^q = \int_G |f(x, t)|^q \, dx, \tag{8a}$$

$$\|b\|_{L_q(0, T; Y^*)}^q = \int_0^T \left(\int_G |f(x, t)|^q \, dx \right)$$

$$= \int_{Q_T} |f(x, t)|^q \, dx \, dt. \tag{8b}$$

(b) For all $t \in [0, T]$, there exists a functional $b_1(t) \in V^*$ such that (7b) is valid for almost all $t \in [0, T]$ and $t \mapsto b_1(t)$ belongs to $L_q(0, T; V^*)$.

Here, the linear continuous functional $b_1(t): V \to \mathbb{R}$ is the restriction of the linear continuous functional $b(t): Y \to \mathbb{R}$ to the space V.

Moreover, we get

$$\|b\|_{L_q(0, T; V^*)}^q \leq \int_{Q_T} |f(x, t)|^q \, dx \, dt.$$

Remark 23.5. Because of this result one frequently says briefly that every function

$$f \in L_q(Q_T)$$

belongs to the spaces $L_q(0, T; Y^*)$ and $L_q(0, T; V^*)$.

By $A_2(34)$, we may identify $Y^* = L_p(G)^*$ with $L_q(G)$. Pictorially, $b(t_0)$ corresponds to the function $f = f(x, t)$ on the cross section of the cylinder Q_T with the plane $t = t_0$ (Fig. 23.1).

PROOF. Ad(a). By assumption, f, as a function of (x, t), is an element of $L_q(Q_T)$. Together with the Fubini theorem $A_2(23)$ this means that

$$\int_{Q_T} |f(x, t)|^q \, dx \, dt = \int_0^T \left(\int_G |f(x, t)|^q \, dx \right) dt. \tag{9}$$

Here, the inner integral on the right-hand side exists for all $t \in [0, T] - N$ where N is a set with meas $N = 0$. For $t \in N$, we set

$$f(x, t) = 0 \qquad \text{for all} \quad x \in G.$$

By $A_2(34)$, the functions from $L_q(G)$ can be identified with the elements of Y^*.

Thereby, the element $b(t) \in Y^*$ corresponding to $x \mapsto f(x, t)$ is given precisely by relation (7a).

Moreover, (8a) holds. Then the relation (8b) follows immediately from (8a) and (9).

It remains to show the measurability of the function

$$b\colon [0, T] \to Y^*.$$

By the crucial *Pettis theorem* $A_2(10)$ and the reflexivity of the B-space Y, for this it suffices to show the measurability of the *real* function

$$t \mapsto \langle b(t), v\rangle_Y \quad \text{on } [0, T],$$

for each $v \in Y$. Recall that $Y = L_p(G)$. Let $v \in Y$. By the Hölder inequality, it follows from $v \in L_p(Q_T)$ and $f \in L_q(Q_T)$ that $vf \in L_1(Q_T)$. The Fubini theorem yields the relation

$$\int_{Q_T} v(x)f(x,t)\,dx\,dt = \int_0^T \left(\int_G v(x)f(x,t)\,dx\right)dt$$

and the measurability of the right-hand inner integral as a function of t. By (7a), that is exactly the measurability of $t \mapsto \langle b(t), v\rangle_Y$.

Ad(b). For almost all $t \in [0, T]$, the Hölder inequality,

$$\begin{aligned}\left|\int_G f(x,t)v(x)\,dx\right| &\le \left(\int_G |f(x,t)|^q\,dx\right)^{1/q}\left(\int_G |v(x)|^p\,dx\right)^{1/p}\\ &\le \left(\int_G |f(x,t)|^q\,dx\right)^{1/q}\|v\|_V \qquad \text{for all} \quad v \in V,\end{aligned}$$

implies the existence of a functional $b_1(t) \in V^*$ which satisfies (7b). Moreover, we get

$$\|b_1(t)\|_{V^*}^q \le \int_G |f(x,t)|^q\,dx$$

and hence $b_1 \in L_q(0, T; V^*)$, since

$$\int_0^T \|b_1(t)\|_{V^*}^q\,dt \le \int_0^T \left(\int_G |f(x,t)|^q\,dx\right)dt < \infty.$$

The existence of the left-hand integral follows from the majorant criterion $A_2(17)$. In this connection, note that the measurability of $t \mapsto b_1(t)$ follows as the measurability of $t \mapsto b(t)$ above. □

23.3. The Dual Space to $L_p(0, T; V)$

We first justify the *Hölder inequality* that is basic for many applications:

$$\int_0^T |\langle v(t), u(t)\rangle_V|\,dt \le \left(\int_0^T \|v(t)\|_{V^*}^q\,dt\right)^{1/q}\left(\int_0^T \|u(t)\|_V^p\,dt\right)^{1/p}. \tag{10}$$

Proposition 23.6. *Let V be a* B*-space. Then the Hölder inequality* (10) *holds for all*

$$u \in L_p(0, T; V), \qquad v \in L_q(0, T; V^*)$$

with $1 < p < \infty$, $p^{-1} + q^{-1} = 1$. In particular, all the integrals in (10) *do exist.*

PROOF. Let $J =]0, T[$. Since the functions u and v are measurable on J, there exist step functions $u_n: J \to V$ and $v_n: J \to V^*$ such that

$$u_n(t) \to u(t) \qquad \text{and} \qquad v_n(t) \to v(t) \qquad \text{as} \quad n \to \infty,$$

for almost all $t \in J$. This implies

$$\langle v_n(t), u_n(t) \rangle_V \to \langle v(t), u(t) \rangle_V \qquad \text{as} \quad n \to \infty,$$

for almost all $t \in J$. Thus the real function $t \mapsto \langle v(t), u(t) \rangle_V$ is measurable on J. Because of

$$|\langle v(t), u(t) \rangle_V| \leq \|v(t)\|_{V^*} \|u(t)\|_V$$

the assertion follows now from the classical Hölder inequality in Section 18.6. □

We now justify the important *duality relation*:

$$L_p(0, T; V)^* = L_q(0, T; V^*), \qquad \frac{1}{p} + \frac{1}{q} = 1.$$

We set $X = L_p(0, T; V)$. Then we have to show that $X^* = L_q(0, T; V^*)$. For this, we consider the two key formulas:

$$\langle \bar{v}, u \rangle_X = \int_0^T \langle v(t), u(t) \rangle_V \, dt \qquad \text{for all} \quad u \in X, \tag{11}$$

$$\|\bar{v}\|_{X^*} = \|v\|_{L_q(0, T; V^*)}. \tag{12}$$

Proposition 23.7. *Let V be a reflexive and separable* B*-space and let $1 < p < \infty$, $p^{-1} + q^{-1} = 1$. Then the following hold:*

(a) *To each function $v \in L_q(0, T; V^*)$ there corresponds a unique functional $\bar{v} \in X^*$ with* (11).
(b) *Conversely, to each $\bar{v} \in X^*$ there corresponds exactly one $v \in L_q(0, T; V^*)$ with* (11). *Moreover,* (12) *holds.*
(c) *The* B*-space $L_p(0, T; V)$ is reflexive and separable.*

PROOF. Assertion (a) follows from the Hölder inequality (10).

Assertion (b) is a standard result of the theory of function spaces. Proofs may be found in Edwards (1965, M), 8.20.5, and in Gajewski, Gröger, and Zacharias (1974, M), Chapter 4.

The reflexivity of $Y = L_p(0, T; V)$ in (c) is a direct consequence of (b). The separability of Y follows from Proposition 23.2(f). □

Convention 23.8 (First Identification). Let $X = L_p(0, T; V)$. By Proposition 23.7, the two real B-spaces

$$X^* = L_p(0, T; V)^* \qquad \text{and} \qquad L_q(0, T; V^*)$$

are *normisomorphic*, i.e., there exists a linear bijective isometric mapping j: $L_q(0, T; V^*) \to X^*$ given by $j(v) = \overline{v}$. Normisomorphic B-spaces can be identified with each other. Therefore, we may identify X^* with $L_q(0, T; V^*)$. Henceforth, we write

$$X^* = L_q(0, T; V^*).$$

The same way, we identify $\overline{v} = j(v)$ with v. Therefore, in future, we write equations (11) and (12) in the following form:

$$\langle v, u \rangle_X = \int_0^T \langle v(t), u(t) \rangle_V \, dt, \tag{13}$$

$$\|v\|_{X^*} = \left(\int_0^T \|v(t)\|_{V^*}^q \, dt \right)^{1/q} \tag{14}$$

for all $u \in X$ and $v \in X^*$.

The following results will be used later on in an essential way in order to prove the convergence of the Galerkin method for linear and nonlinear evolution equations of first and second order.

Proposition 23.9 (Limit Relations for Integrals). *Let V be a reflexive and separable* B-*space. Furthermore, let $1 < p, q < \infty$, $p^{-1} + q^{-1} = 1$, and $0 \le t \le T < \infty$. Then we have the following statements*:

(a) *If $u \in L_p(0, T; V)$, then*

$$\left\langle v^*, \int_0^t u(s) \, ds \right\rangle_V = \int_0^t \langle v^*, u(s) \rangle_V \, ds \qquad \textit{for all} \quad v^* \in V^*.$$

(b) *If $u \in L_p(0, T; V^*)$, then*

$$\left\langle \int_0^t u(s) \, ds, v \right\rangle_V = \int_0^t \langle u(s), v \rangle_V \, ds \qquad \textit{for all} \quad v \in V.$$

(c) *From $u_n \to u$ in $L_p(0, T; V)$ as $n \to \infty$ it follows that*

$$\int_0^t u_n(s) \, ds \to \int_0^t u(s) \, ds \quad \textit{in } V \qquad \textit{as} \quad n \to \infty.$$

(d) *From*

$$u_n \to u \quad \textit{in } L_p(0, T; V) \qquad \textit{as} \quad n \to \infty,$$

$$v_n \rightharpoonup v \quad \textit{in } L_q(0, T; V^*) \qquad \textit{as} \quad n \to \infty,$$

if follows that

$$\int_0^t \langle v_n(s), u_n(s)\rangle_V \, ds \to \int_0^t \langle v(s), u(s)\rangle_V \, ds \quad \text{as } n \to \infty.$$

(e) *From*

$$u_n \rightharpoonup u \quad \text{in } L_p(0, T; V) \qquad \text{as} \quad n \to \infty,$$

$$v_n \to v \quad \text{in } L_q(0, T; V^*) \qquad \text{as} \quad n \to \infty,$$

it follows that

$$\int_0^t \langle v_n(s), u_n(s)\rangle_V \, ds \to \int_0^t \langle v(s), u(s)\rangle_V \, ds.$$

Assertions (a), (b), (c) *are also valid for arbitrary* B-*spaces* V.

Further results concerning the weak* convergence in $L_\infty(0, T; V^*)$ can be found in Problem 23.12.

PROOF.

(I) First we set $t = T$. Moreover, let $X = L_p(0, T; V)$. Then we have that $X^* = L_q(0, T; V^*)$.

Ad(a). Let $u \in X$. This implies $u \in L_1(0, T; V)$ by Proposition 23.2(h). According to the definition of the integral in $A_2(14)$, there exists a sequence (u_n) of step functions such that, as $n \to \infty$, we have that

$$u_n(s) \to u(s) \quad \text{in } V \quad \text{almost everywhere on }]0, T[,$$

$$\int_0^T u_n(s)\, ds \to \int_0^T u(s)\, ds \quad \text{in } V.$$

Moreover, we get

$$\int_0^T \|u_n(s) - u_m(s)\| \, ds < \varepsilon$$

for all n, $m \geq n_0(\varepsilon)$. Let $v^* \in V^*$. By definition, the integral of a step function is a sum. Hence we obtain that

$$\langle v^*, \int_0^T u_n(s)\, ds\rangle = \int_0^T \langle v^*, u_n(s)\rangle \, ds.$$

Now set $f_n(s) = \langle v^*, u_n(s)\rangle$ and $f(s) = \langle v^*, u(s)\rangle$. Then (f_n) is a sequence of real step functions with

$$f_n(s) \to f(s) \qquad \text{as} \quad n \to \infty \qquad \text{for almost all} \quad s \in]0, T[.$$

Furthermore, we get

$$\int_0^T |f_n(s) - f_m(s)| \, ds \leq \int_0^T \|v^*\| \, \|u_n(s) - u_m(s)\| \, ds < \varepsilon$$

for all $n, m \geq n_1(\varepsilon)$. Consequently, the integral $\int_0^T f(s)\,ds$ exists. Now a passage to the limit as $n \to \infty$ yields assertion (a).

Analogously, one obtains (b).

Ad(c). We set

$$A(u) = \int_0^T u(s)\,ds \qquad \text{for all} \quad u \in X.$$

The Hölder inequality together with the 1-trick yields

$$\|A(u)\| \leq \int_0^T \|u(s)\|\,ds \leq c\left(\int_0^T \|u(s)\|^p\,ds\right)^{1/p}.$$

Hence the operator $A: X \to V$ is linear and continuous. This implies (c) for $t = T$.

Ad(d). From $u_n \to u$ in X and $v_n \rightharpoonup v$ in X^* as $n \to \infty$ it follows that

$$\langle v_n, u_n \rangle \to \langle v, u \rangle_X.$$

Note that such a relation holds for each reflexive B-space X. Hence assertion (d) follows from

$$\langle v, u \rangle_X = \int_0^T \langle v(s), u(s) \rangle\,ds.$$

Analogously, one obtains assertion (e).

(II) Now let us consider the more general case $0 \leq t < T$. The decisive trick is to use the characteristic function

$$\chi(s) = \begin{cases} 1 & \text{for} \quad 0 \leq s \leq t, \\ 0 & \text{otherwise,} \end{cases}$$

and to replace u and v with χu and χv, respectively. This way, we reduce this general case (II) to the more special case (I). Note that multiplication by χ does not effect strong and weak convergence in X or X^*. To see this consider, for example, a sequence (u_n) with

$$u_n \rightharpoonup u \quad \text{in } X \qquad \text{as} \quad n \to \infty.$$

We want to show that

$$\chi u_n \rightharpoonup \chi u \quad \text{in } X \qquad \text{as} \quad n \to \infty.$$

In fact, it follows from $u_n \rightharpoonup u$ in X as $n \to \infty$ that

$$\langle v, u_n \rangle_X = \int_0^T \langle v(s), u_n(s) \rangle\,ds \to \langle v, u \rangle_X \qquad \text{as} \quad n \to \infty,$$

for all $v \in X^*$, where $X^* = L_q(0, T; V^*)$. Since

$$v \in L_q(0, T; V^*) \qquad \text{implies} \qquad \chi v \in L_q(0, T; V^*),$$

we obtain that, as $n \to \infty$,

$$\int_0^T \langle v(s), \chi(s)u_n(s)\rangle \, ds = \langle \chi v, u_n\rangle_X \to \langle \chi v, u\rangle_X = \int_0^T \langle v(s), \chi(s)u(s)\rangle \, ds.$$

Hence $\chi u_n \rightharpoonup \chi u$ in X as $n \to \infty$. □

In order to obtain a simple proof for the uniqueness of generalized derivatives in Section 23.5, we now generalize the crucial variational lemma from Section 18.2.

Proposition 23.10 (Variational Lemma). *Let X be a* B-*space. Then, it follows from $u \in L_1(0, T; X)$ and*

$$\int_0^T \varphi(t)u(t)\, dt = 0 \qquad \text{for all} \quad \varphi \in C_0^\infty(0, T)$$

that $u = 0$ in $L_1(0, T; X)$, i.e.,

$$u(t) = 0 \qquad \text{for almost all} \quad t \in \,]0, T[.$$

PROOF.

(I) First we consider the special case $X = V^*$, where V is a separable B-space. For each $v \in V$ we have that

$$\int_0^T \varphi(s)\langle u(s), v\rangle \, ds = 0 \qquad \text{for all} \quad \varphi \in C_0^\infty(0, T),$$

by Proposition 23.9(b). The classical variational lemma (Proposition 18.2) implies that

$$\langle u(s), v\rangle = 0 \qquad \text{for all} \quad s \in \,]0, T[\, - Z(v),$$

where $Z(v)$ is a set of measure zero. Let $\{v_i\}$ be a countable set which is dense in V. Then the set $Z = \bigcup_i Z(v_i)$ also has the measure zero and we get

$$\langle u(s), v\rangle = 0 \qquad \text{for all} \quad s \in \,]0, T[\, - Z,$$

and *all* $v \in V$. Hence we obtain that

$$u(s) = 0 \qquad \text{for all} \quad s \in \,]0, T[\, - Z.$$

This is the assertion.

(II) The special case (I) is sufficient for the proof of Proposition 23.20 below. The proof of Proposition 23.10 in the general case will be given in Problem 23.6. □

23.4 Evolution Triples

Definition 23.11. We understand an *evolution triple*

$$\text{“}V \subseteq H \subseteq V^*\text{”} \tag{15}$$

to be the following:

(i) V is a real, separable, and reflexive B-space.
(ii) H is a real, separable H-space.
(iii) The embedding $V \subseteq H$ is continuous, i.e.,

$$\|v\|_H \leq \text{const}\,\|v\|_V \qquad \text{for all} \quad v \in V, \tag{16}$$

and V is dense in H.

With evolution triples we describe the fact that *two* spaces V and H appear in evolution equations. We shall later apply the following example to concrete parabolic and hyperbolic equations.

EXAMPLE 23.12. Let G be a bounded region of $\mathbb{R}^N$ with $N \geq 1$. We set

$$V = \mathring{W}_p^m(G), \qquad H = L_2(G),$$

with $2 \leq p < \infty$ and $m \geq 1$. Then "$V \subseteq H \subseteq V^*$" is an evolution triple.

We will now justify that in a suitable sense the inclusion

$$H \subseteq V^*$$

is valid. The point of departure is the formula

$$\langle \bar{h}, v \rangle_V = (h|v)_H \qquad \text{for all} \quad v \in V. \tag{17a}$$

Proposition 23.13. *Let "$V \subseteq H \subseteq V^*$" be an evolution triple. Then the following hold*:

(a) *To each $h \in H$ there corresponds, via* (17a), *a linear continuous functional $\bar{h}$: $V \to \mathbb{R}$, i.e., $\bar{h} \in V^*$.*
(b) *The mapping $h \mapsto \bar{h}$ from H into V^* is linear, injective, and continuous.*

PROOF. Let $h \in H$. By (16),

$$|(h|v)_H| \leq \|h\|_H \|v\|_H \leq \text{const}\,\|h\|_H \|v\|_V$$

holds for all $v \in V$. Therefore, there exists an $\bar{h} \in V^*$ with (17a) and

$$\|\bar{h}\|_{V^*} \leq \text{const}\,\|h\|_H. \tag{18a}$$

The mapping $h \mapsto \bar{h}$ is obviously linear; it is continuous from H to V^*, by (18a). However, it is also injective for it follows from $\bar{h} = 0$ that

$$(h|v)_H = 0 \qquad \text{for all} \quad v \in V.$$

The space V is dense in H and therefore $h = 0$. □

Convention 23.14 (Second Identification). By Proposition 23.13, we may identify $\bar{h}$ with h. In this sense,

$$H \subseteq V^*.$$

Henceforth, we shall write h instead of $\bar{h}$. Then the following are valid:

$$\langle h, v \rangle_V = (h|v)_H \qquad \text{for all} \quad h \in H, \quad v \in V, \tag{17}$$

$$\|h\|_{V^*} \leq \text{const}\|h\|_H \qquad \text{for all} \quad h \in H. \tag{18}$$

In the following, the relation

$$V \subseteq H \subseteq V^*,$$

for evolution triples is to be understood in the sense of Convention 23.14. Observe that the embeddings $V \subseteq H$ and $H \subseteq V^*$ are continuous. In particular, we get

$$\langle v, w \rangle_V = \langle w, v \rangle_V \qquad \text{for all} \quad v, w \in V.$$

This follows from $V \subseteq H$ and (17), since $(w|v)_H = (v|w)_H$.

Formulas (17) and (18) will be used frequently.

Using Problem 18.6, Convention 23.14 can briefly be summarized as follows:

(a) *It follows from $V \subseteq H$ and from the density of V in H that $H^* \subseteq V^*$.*
(b) *We identify H with H^*, i.e., we have $H^* = H$, and hence $H \subseteq V^*$.*
(c) *Since V is reflexive, the space H^* is dense in V^*, and hence H is dense in V^*.*

23.5. Generalized Derivatives

The following definition is *basic* for understanding evolution equations. We proceed from the classical formula

$$\int_0^T \varphi^{(n)}(t)u(t)\,dt = (-1)^n \int_0^T \varphi(t)u^{(n)}(t)\,dt \qquad \text{for all} \quad \varphi \in C_0^\infty(0, T) \tag{19}$$

and choose as a starting point for the definition of the nth generalized derivative

$$w = u^{(n)}$$

the formula

$$\int_0^T \varphi^{(n)}(t)u(t)\,dt = (-1)^n \int_0^T \varphi(t)w(t)\,dt \qquad \text{for all} \quad \varphi \in C_0^\infty(0, T). \tag{20}$$

Definition 23.15. Let $u \in L_1(0, T; Y)$ and $w \in L_1(0, T; Z)$. Then, the function w is called the nth *generalized derivative* of the function u on $]0, T[$ iff (20) is valid. We write $w = u^{(n)}$.

We call the reader's attention to the fact that the values of the functions

$$u\colon]0, T[\to Y \qquad \text{and} \qquad w\colon]0, T[\to Z$$

may lie in different spaces Y and Z. Then (20) means that the integrals appearing in the right and left members belong to $Y \cap Z$. Since

$$\|\varphi^{(n)}(t)u(t)\| \leq \text{const}\|u(t)\|,$$

it follows from $u \in L_1(0, T; Y)$ and from the majorant criterion $A_2(17)$ that the integral appearing in the left member of (20) exists for all $\varphi \in C_0^\infty(0, T)$. Analogously, one obtains the existence of the integral appearing in the right member of (20).

EXAMPLE 23.16. In order to consider the simplest situation with $Y = Z$, let

$$u \in C^n([0, T], Y), \qquad n \geq 1.$$

Then the continuous nth derivative $u^{(n)}\colon [0, T] \to Y$ is also, at the same time, the generalized nth derivative of u on $]0, T[$.

PROOF. Let $v \in C^1([0, T]; Y)$ and $\varphi \in C_0^\infty(0, T)$. From $(\varphi v)' = \varphi' v + \varphi v'$ we obtain the classical integration by parts formula

$$\int_0^T \varphi' v \, dt = -\int_0^T \varphi v' \, dt.$$

Repeated application of this formula yields (19). □

Remark 23.17 (Generalized Derivatives and Distributions). We explain the connection of Definition 23.15 with the general distribution theory which we summarize in $A_2(62)$. Let $u \in L_1(0, T; Y)$. Then one can assign to the function u a *distribution* U via

$$U(\varphi) = \int_0^T \varphi(t)u(t)\, dt \qquad \text{for all} \quad \varphi \in C_0^\infty(0, T).$$

For each n, this distribution has an nth distributive derivative $U^{(n)}$ which is defined by

$$U^{(n)}(\varphi) = (-1)^n U(\varphi^{(n)}) \qquad \text{for all} \quad \varphi \in C_0^\infty(0, T). \tag{21a}$$

If (20) holds, then $U^{(n)}$ can be obviously represented in the form

$$U^{(n)}(\varphi) = \int_0^T \varphi(t)u^{(n)}(t)\, dt \qquad \text{for all} \quad \varphi \in C_0^\infty(0, T). \tag{21b}$$

The advantage of the distribution conception consists in that each function

$$u \in L_1(0, T; Y)$$

possesses derivatives of *every order* in the distributional sense. Definition 23.15

singles out the cases in which by (21b) the nth distributional derivative of u can be represented by a *function* $w \in L_1(0, T; Z)$. In this case, we set $u^{(n)} = w$ and we write briefly

$$u \in L_1(0, T; Y), \qquad u^{(n)} \in L_1(0, T; Z).$$

Proposition 23.18 (Uniqueness of Generalized Derivatives). *Let X and Z be* B-*spaces. Moreover, suppose that $u \in L_1(0, T; Y)$ and $v, w \in L_1(0, T; Z)$. If*

$$u^{(n)} = v \qquad \text{and} \qquad u^{(n)} = w$$

in the sense of generalized derivatives, then we obtain

$$v(t) = w(t) \qquad \text{almost everywhere on} \quad]0, T[,$$

i.e., we have $v = w$ in $L_1(0, T; Z)$.

PROOF. The assertion follows from

$$\int_0^T \varphi(v - w)\,dt = 0 \qquad \text{for all} \quad \varphi \in C_0^\infty(0, T),$$

by Proposition 23.10. □

The following proposition shows that generalized derivatives are compatible with weak limits. This fact is important for proving the convergence of the Galerkin method for evolution equations.

Proposition 23.19 (Generalized Derivatives and Weak Convergence). *Let Y and Z be* B-*spaces over $\mathbb{K} = \mathbb{R}, \mathbb{C}$ with the continuous embedding $Y \subseteq Z$. Then it follows from*

$$u_k^{(n)} = v_k \quad \text{on }]0, T[\qquad \text{for all } k \text{ and fixed} \quad n \geq 1$$

and

$$u_k \rightharpoonup u \quad \text{in } L_p(0, T; Y) \qquad \text{as} \quad k \to \infty,$$

$$v_k \rightharpoonup v \quad \text{in } L_q(0, T; Z) \qquad \text{as} \quad k \to \infty, \qquad 1 \leq p, q < \infty,$$

that

$$u^{(n)} = v \quad \text{on }]0, T[.$$

PROOF. Let A and B be B-spaces over $\mathbb{K}$. If the embedding $A \subseteq B$ is continuous, then, as $k \to \infty$,

$$a_k \rightharpoonup a \quad \text{in } A \qquad \text{implies} \qquad a_k \rightharpoonup a \quad \text{in } B.$$

This implies that, as $k \to \infty$,

$$u_k \rightharpoonup u \qquad \text{and} \qquad v_k \rightharpoonup v \quad \text{in } L_1(0, T; Z),$$

by Proposition 23.2(h). Let $\varphi \in C_0^\infty(0, T)$. From

$$\int_0^T \varphi^{(n)} u_k \, dt = (-1)^n \int_0^T \varphi v_k \, dt$$

and $k \to \infty$ we obtain that

$$\int_0^T \varphi^{(n)} u \, dt = (-1)^n \int_0^T \varphi v \, dt,$$

by Proposition 23.9. □

We now consider, with a view to evolution equations, the situation in evolution triples.

Proposition 23.20 (Existence of $u^{(n)}$). *Let "$V \subseteq H \subseteq V^*$" be an evolution triple and let $1 \leq p, q \leq \infty, 0 < T < \infty$. Then the following are valid.*

(a) Uniqueness. *For $u \in L_p(0, T; V)$, the generalized derivative $u^{(n)}$ is unique as an element of $L_q(0, T; V^*)$, that is, $t \mapsto u^{(n)}(t)$ can be modified only on a subset of $]0, T[$ of measure zero.*

(b) Existence. *Let $u \in L_p(0, T; V)$. Then there exists the generalized derivative*

$$u^{(n)} \in L_q(0, T; V^*)$$

iff there is a function $w \in L_q(0, T; V^*)$ such that

$$\int_0^T (u(t)|v)_H \varphi^{(n)}(t) \, dt = (-1)^n \int_0^T \langle w(t), v \rangle_V \varphi(t) \, dt \tag{22}$$

for all $v \in V$ and all $\varphi \in C_0^\infty(0, T)$.

Then $u^{(n)} = w$ and

$$\frac{d^n}{dt^n}(u(t)|v)_H = \langle u^{(n)}(t), v \rangle_V \tag{23}$$

holds for all $v \in V$ and almost all $t \in]0, T[$. Here, d^n/dt^n means the nth generalized derivative of real functions on $]0, T[$.

By the definition of generalized derivatives of real functions, (22) and (23) with $w = u^{(n)}$ are equivalent.

We shall later frequently use formula (22) in connection with parabolic and hyperbolic differential equations. Their advantage consists in that one has to investigate only the generalized derivatives of real functions. Observe that

$$L_p(0, T; V) \subseteq L_1(0, T; V) \qquad \text{and} \qquad L_q(0, T; V^*) \subseteq L_1(0, T; V^*)$$

for $1 \leq p, q \leq \infty$ as well as $V \subseteq V^*$.

PROOF. Ad(a). Use Proposition 23.18.

Ad(b). Formula (20) is equivalent to

$$\left\langle \int_0^T \varphi^{(n)}(t)u(t)\,dt, v\right\rangle_V = (-1)^n \left\langle \int_0^T \varphi(t)w(t)\,dt, v\right\rangle_V,$$

for all $v \in V$. Therefore, by Proposition 23.9, we get

$$\int_0^T \langle \varphi^{(n)}(t)u(t), v\rangle_V\,dt = (-1)^n \int_0^T \langle \varphi(t)w(t), v\rangle_V\,dt,$$

for all $v \in V$. Now, (22) follows from (17). □

EXAMPLE 23.21. Assume:

(i) "$V \subseteq H \subseteq V^*$" is an evolution triple and $0 < T < \infty$;
(ii) the function $u: [0, T] \to V$ is continuous and the derivative

$$u'(t) = \lim_{h\to 0} \frac{u(t+h) - u(t)}{h}$$

exists for all $t \in [0, T]$ as a limiting value in the space H; and
(iii) the function $u': [0, T] \to H$ is continuous.

Then, u' is the generalized derivative of u on $]0, T[$. Furthermore,

$$u \in L_p(0, T; V) \quad \text{and} \quad u' \in L_q(0, T; V^*) \quad \text{for all } p, q\colon 1 \le p,\ q \le \infty.$$

The peculiarity here consists in that the values of u lie in the space V, nonetheless the limit value $u'(t)$ exists only in the space H.

PROOF. The mapping $u: [0, T] \to H$ is continuous by (ii) and (16); therefore $u \in C^1([0, T]; H)$. The assertion then follows from Example 23.16. □

Corollary 23.22. *Let H be a real* H-*space, and let $u: [0, T] \to H$ be Lipschitz continuous, that is*

$$\|u(t) - u(s)\| \le L|t - s| \qquad \text{for all} \quad t, s \in [0, T] \quad \text{and fixed} \quad L \ge 0.$$

Then the following holds:

(a) *For almost all $t \in [0, T]$, the function u has a derivative,*

$$u'(t) = \lim_{h\to 0} \frac{u(t+h) - u(t)}{h}$$

and

$$u(t) = u(0) + \int_0^t u'(s)\,ds \qquad \text{for all} \quad t \in [0, T].$$

(b) *For almost all* $t \in [0, T]$,

$$\|u'(t)\| \leq L,$$

and u' *is the generalized derivative of* u *on* $]0, T[$.

PROOF. Ad(a). Compare Problems 23.4 and 23.5c.

Ad(b). It follows immediately from the Lipschitz condition that $\|u'(t)\| \leq L$ Integration of

$$(\varphi u)' = \varphi' u + \varphi u'$$

yields

$$\int_0^T \varphi' u \, dt = -\int_0^T \varphi u' \, dt \qquad \text{for all} \quad \varphi \in C_0^\infty(0, T). \qquad \square$$

23.6. The Sobolev Space $W_p^1(0, T; V, H)$

The following proposition, in particular the integration by parts formula (25) below, will play a central role in the treatment of evolution equations. Let $X = L_p(0, T; V)$. The basic definition is given by

$$W_p^1(0, T; V, H) = \{u \in X : u' \in X^*\}, \qquad 1 < p < \infty.$$

Recall that $X^* = L_q(0, T; V^*)$, where $p^{-1} + q^{-1} = 1$.

Proposition 23.23. *Let "*$V \subseteq H \subseteq V^*$*" be an evolution triple, and let* $1 < p < \infty$, $p^{-1} + q^{-1} = 1$, $0 < T < \infty$. *Then the following hold*:

(i) The space $W_p^1(0, T; V, H)$. *The set of all* $u \in L_p(0, T; V)$ *that have generalized derivatives*

$$u' \in L_q(0, T; V^*)$$

forms a real B*-space with the norm*

$$\|u\|_{W_p^1} = \|u\|_{L_p(0,T;V)} + \|u'\|_{L_q(0,T;V^*)}.$$

We denote this B*-space by* $W_p^1(0, T; V, H)$.

(ii) Regularization. *The embedding*

$$W_p^1(0, T; V, H) \subseteq C([0, T], H)$$

is continuous. More precisely, we have the following. If $u \in W_p^1(0, T; V, H)$, *then there exists a uniquely determined continuous function* $u_1 : [0, T] \to H$ *which coincides almost everywhere on* $[0, T]$ *with the initial function* u. *Henceforth, we write* u *instead of* u_1. *Further, in this sense,*

$$\max_{0 \leq t \leq T} \|u(t)\|_H \leq \text{const} \, \|u\|_{W_p^1}. \tag{24}$$

(iii) Dense subset. *The set of all polynomials* $w\colon [0, T] \to V$, *that is*

$$w(t) = \sum_i t^i a_i \qquad \text{with} \quad a_i \in V \quad \text{for all } i,$$

is dense in the spaces $W_p^1(0, T; V, H)$, $L_p(0, T; V)$, *and* $L_p(0, T; H)$.

(iv) Integration by parts. *For all* $u, v \in W_p^1(0, T; V, H)$ *and arbitrary* t, s, $0 \le s \le t \le T$, *the following generalized integration by parts formula holds*:

$$(u(t)|v(t))_H - (u(s)|v(s))_H = \int_s^t \langle u'(\tau), v(\tau)\rangle_V + \langle v'(\tau), u(\tau)\rangle_V \, d\tau. \quad (25)$$

Here, the values $u(t), v(t), u(s), v(s)$ *are the values of the continuous functions* $u, v\colon [0, T] \to H$ *in the sense of* (ii).

The proof will be given in Problem 23.10. Observe the following important fact. In applications the space V contains smoother functions than the space H. For example, we will have $V = \mathring{W}_2^1(G)$ and $H = L_2(G)$. In Proposition 23.23(ii) above, the functions from $W_p^1(0, T; V, H)$ are not continuous with respect to the space V, but only with respect to the worse space H.

23.7. Main Theorem on First-Order Linear Evolution Equations and the Galerkin Method

We study the initial value problem

$$\frac{d}{dt}(u(t)|v)_H + a(u(t), v) = \langle b(t), v\rangle_V, \quad (26a)$$

$$u(0) = u_0 \in H, \quad (26b)$$

$$u \in W_2^1(0, T; V, H), \quad (26c)$$

where we assume that (26a) is valid for all $v \in V$ and almost all $t \in]0, T[$.

To be precise, we assume that there exists a subset Z of $]0, T[$ of measure zero such that equation (26a) holds for all $v \in V$ and all $t \in]0, T[- Z$. Note that Z is independent of v. Furthermore, d/dt in (26a) means the generalized derivative of real functions on $]0, T[$, i.e., (26a) means that

$$-\int_0^T (u(t)|v)_H \varphi'(t)\, dt + \int_0^T a(u(t), v)\varphi(t)\, dt$$

$$= \int_0^T \langle b(t), v\rangle_V \varphi(t)\, dt \qquad \text{for all} \quad \varphi \in C_0^\infty(0, T).$$

Let $u \in W_2^1(0, T; V, H)$. The initial condition (26b) is meaningful because, by

Proposition 23.23, after a modification on a subset of $[0, T]$ of measure zero, u is a uniquely determined continuous mapping of the form $u: [0, T] \to H$. Condition (26b) is to be understood in this sense.

We make the following assumptions:

(H1) "$V \subseteq H \subseteq V^*$" is an evolution triple with $\dim V = \infty$, $0 < T < \infty$. The spaces V and H are real H-spaces.

(H2) The mapping $a: V \times V \to \mathbb{R}$ is bilinear, bounded, and strongly positive. Moreover, we are given $u_0 \in H$ and $b \in L_2(0, T; V^*)$.

(H3) $\{w_1, w_2, \dots\}$ is a basis in V, and (u_{n0}) is a sequence from H with

$$u_{n0} \to u_0 \quad \text{in } H \qquad \text{as} \quad n \to \infty,$$

where

$$u_{n0} \in \operatorname{span}\{w_1, \dots, w_n\} \qquad \text{for all } n.$$

In order to formulate the Galerkin method, we set

$$u_n(t) = \sum_{k=1}^{n} c_{kn}(t) w_k, \qquad u_{n0} = \sum_{k=1}^{n} \alpha_{kn} w_k.$$

By definition, for almost all $t \in]0, T[$, the *Galerkin equations* read as follows:

$$\sum_{k=1}^{n} c'_{kn}(t)(w_k | w_j)_H + c_{kn}(t) a(w_k, w_j) = \langle b(t), w_j \rangle_V, \tag{27a}$$

$$c_{jn}(0) = \alpha_{jn}, \qquad j = 1, \dots, n. \tag{27b}$$

Here, (27a) arises formally from (26a) if one replaces u by u_n and v by w_j. Equation (27b) results from $u_n(0) = u_{n0}$.

Since $w_1, \dots, w_n$ are linearly independent, we get

$$\det((w_k | w_j)_H) \neq 0, \qquad k, j = 1, \dots, n,$$

for the Gram determinant. Therefore, equation (27a) can be solved for c'_{kn} and thus represents a system of linear differential equations. Recall that

$$\|u\|_{W_2^1} = \|u\|_{L_2(0,T;V)} + \|u'\|_{L_2(0,T;V^*)}.$$

Theorem 23.A. *If the assumptions* (H1), (H2), (H3) *hold, then*:

(a) Existence and uniqueness. *The original equation* (26) *has exactly one solution* u.

(b) Continuous dependence on the data. *The map*

$$(u_0, b) \mapsto u$$

is linear and continuous from $H \times L_2(0, T; V^*)$ *to* $W_2^1(0, T; V, H)$, *i.e., there is a constant* $D > 0$ *such that*

$$\|u\|_{W_2^1} \leq D(\|u_0\|_H + \|b\|_{L_2(0,T;V^*)}),$$

for all $u_0 \in H$ *and* $b \in L_2(0, T; V^*)$.

(c) Convergence of the Galerkin method. *For all* $n = 1, 2, \dots,$ *the Galerkin equation* (27) *has exactly one solution*

$$u_n \in W_2^1(0, T; V, H).$$

The sequence (u_n) *converges as* $n \to \infty$ *to the solution* u *of* (26) *in the following sense*:

$$u_n \to u \quad in\ L_2(0, T; V), \tag{28a}$$

$$\max_{0 \le t \le T} \|u_n(t) - u(t)\|_H \to 0. \tag{28b}$$

Corollary 23.24. *The original equation* (26) *is equivalent to the following operator equation*:

$$\begin{aligned} u'(t) + Au(t) &= b(t) \qquad \text{for almost all} \quad t \in]0, T[, \\ u(0) &= u_0 \in H, \qquad u \in W_2^1(0, T; V, H). \end{aligned} \tag{29}$$

Here, the operator $A\colon V \to V^*$ *is linear, continuous, and strongly monotone with*

$$\langle Au, v \rangle_V = a(u, v) \qquad \text{for all} \quad u, v \in V.$$

The proofs of Theorem 23.A and Corollary 23.24 will be given in Section 23.9.

Remark 23.25 (Gårding's Inequality and a Transformation Trick). In assumption (H2) of Theorem 23.A, we required the strong positivity of the bilinear form a. We want to show that it is sufficient to postulate the validity of the more general Gårding inequality

$$a(u, u) \ge c\|u\|_V^2 - d\|u\|_H^2 \qquad \text{for all} \quad u \in V, \tag{30}$$

where $c > 0$ and $d \ge 0$ are constants. To this end, we use the transformation

$$u(t) = e^{dt} w(t).$$

From the original equation (26) we obtain the transformed equation

$$\begin{aligned} &\frac{d}{dt}(w(t)|v)_H + a_1(w(t), v) = \langle b_1(t), v \rangle \qquad \text{for all} \quad v \in V, \\ &w(0) = u_0, \qquad w \in W_2^1(0, T; V, H) \end{aligned} \tag{31}$$

where $a_1(w, v) = a(w, v) + d(w|v)_H$ and $b_1(t) = e^{-dt} b(t)$. By (30),

$$a_1(u, u) \ge c\|u\|_V^2 \qquad \text{for all} \quad u \in V.$$

That is, the transformed bilinear form a_1 is *strongly positive*. Thus we can apply Theorem 23.A to the transformed problem (31).

We now replace the original problem (26) with the more general problem

$$\frac{d}{dt}(u(t)|v)_H + a(u(t), v; t) = \langle b(t), v \rangle_V, \tag{32a}$$

$$u(0) = u_0, \qquad u \in W_2^1(0, T; V, H). \tag{32b}$$

In contrast to (26), the bilinear form a *depends* on time t. Here, (32a) must hold for all $v \in V$ and almost all $t \in]0, T[$. In this case, we replace the assumption (H2) with the following assumption.

(H2*) For all $t \in]0, T[$, the mapping

$$a(\cdot, \cdot; t)\colon V \times V \to \mathbb{R}$$

is bilinear, bounded, and satisfies an abstract Gårding inequality, where the constants are *independent* of time t. That is, there exist constants $C, c > 0$ and $d \geq 0$ such that

$$|a(u, v; t)| \leq C \|u\|_V \|v\|_V,$$

$$a(u, u) \geq c \|u\|_V^2 - d \|u\|_H^2,$$

for all $u, v \in V$ and $t \in]0, T[$. Moreover, the function

$$t \mapsto a(u, v; t)$$

is measurable on $]0, T[$ for all $u, v \in V$.

We are given $u_0 \in H$ and $b \in L_2(0, T; V^*)$.

Corollary 23.26. *Suppose that* (H1), (H2*), (H3) *hold true. Then all the assertions of Theorem* 23.A *are also valid for equation* (32).

Moreover, (32) *is equivalent to the following operator equation*:

$$\begin{aligned} u'(t) + A(t)u(t) &= b(t) \qquad \text{for almost all} \quad t \in]0, T[, \\ u(0) &= u_0, \qquad u \in W_2^1(0, T; V, H), \end{aligned} \tag{33}$$

where the operator $A(t)\colon V \to V^*$ *is defined by*

$$\langle A(t)u, v \rangle_V = a(u, v; t) \qquad \text{for all} \quad u, v \in V.$$

The proof will be given in Section 23.9.

23.8. Application to Parabolic Differential Equations

Let $Q_T = G \times]0, T[$. We consider the following initial–boundary value problem:

$$\begin{aligned} u_t(x, t) - \Delta u(x, t) &= f(x, t) \quad \text{on } Q_T, \\ u(x, t) &= 0 \text{ on } \partial G \times [0, T], \\ u(x, 0) &= u_0(x) \quad \text{on } G. \end{aligned} \tag{34}$$

Definition 23.27. Let G be a bounded region in $\mathbb{R}^N$ with $N \geq 1$. We set

$$V = \mathring{W}_2^1(G), \qquad H = L_2(G).$$

The *generalized problem* corresponding to (34) reads as follows: Let $f \in L_2(Q_T)$ and $u_0 \in H$ be given. We seek $u \in W_2^1(0, T; V, H)$ with $u(0) = u_0$ so that

$$\frac{d}{dt}(u(t)|v)_H + a(u(t), v) = \langle b(t), v \rangle_V$$

holds for all $v \in V$ and almost all $t \in]0, T[$. In this connection, for all $u, v \in V$, we set

$$a(u, v) = \int_G \sum_{i=1}^N D_i u D_i v \, dx,$$

$$\langle b(t), v \rangle_V = \int_G f(x, t) v(x) \, dx.$$

The last equation is justified by Example 23.4. Here, $x = (\xi_1, \ldots, \xi_N)$ and $D_i = \partial/\partial \xi_i$. Recall that

$$(u|v)_H = \int_G u(x) v(x) \, dx.$$

The generalized problem and the corresponding Galerkin method (27) are motivated by Section 19.2. As the basis elements $\{w_1, w_2, \ldots\}$ in $V = \mathring{W}_2^1(G)$ one can choose polynomials or finite elements (see $A_2(56)$–$A_2(60)$).

Proposition 23.28. *All the assertions of Theorem* 23.A *and Corollary* 23.24 *hold for the generalized problem corresponding to* (34).

PROOF. By Example 23.4, $b \in L_2(0, T; V^*)$ and

$$\|b\|_{L_2(0,T;V^*)} \leq \|f\|_{L_2(Q_T)}.$$

It results from the proof of Proposition 22.7 that $a: V \times V \to \mathbb{R}$ is bilinear, bounded, and strongly positive. Thus, all the assumptions of Theorem 23.A are fullfilled. $\square$

We now replace the classical heat equation (34) with the following more general initial–boundary value problem:

$$\begin{aligned} u_t - \sum_{i,j=1}^N D_i(a_{ij} D_j u) + \sum_{i=1}^N \alpha_i D_i u + \beta u = f \quad &\text{on } Q_T, \\ u(x, t) = 0 \qquad &\text{on } \partial G \times [0, T], \\ u(x, 0) = u_0(x) \quad &\text{on } G. \end{aligned} \tag{35}$$

Here, the coefficients a_{ij}, α_i, and β may depend on time t. Such equations are related to general heat conduction processes and diffusion processes.

Our assumptions are the following.

(A1) Let G be a bounded region in $\mathbb{R}^N$ with $N \geq 1$, and let $0 < T < \infty$. We set $Q_T = G \times]0, T[$.

(A2) All the functions $a_{ij}, \alpha_i, \beta\colon Q_T \to \mathbb{R}$ are measurable and bounded, i.e., they belong to $L_\infty(Q_T)$. In particular, this condition is satisfied if all these functions are continuous on $\bar{Q}_T$.

(A3) Uniform ellipticity. There exists a constant $\gamma > 0$ such that

$$\sum_{i,j=1}^{N} a_{ij}(x,t)\zeta_i\zeta_j \geq \gamma \sum_{i=1}^{N} \zeta_i^2$$

for all $(x,t) \in Q_T$ and all $(\zeta_1, \ldots, \zeta_N) \in \mathbb{R}^N$.

Definition 23.29. The *generalized problem* to (35) is defined analogously to Definition 23.27 with

$$a(u,v;t) = \int_G \left(\sum_{i,j=1}^{N} a_{ij} D_i u D_j v + \sum_{i=1}^{N} \alpha_i v D_i u + \beta u v \right) dx.$$

Formally, the generalized problem is obtained in the following way. Multiply the first equation of (35) by $v \in C_0^\infty(G)$ and use integration by parts.

Proposition 23.30. *Assume that* (A1)–(A3) *are satisfied. Then all the assertions of Corollary* 23.26 *hold true for the generalized problem to* (35).

In particular, for given $u \in L_2(G)$, $f \in L_2(Q_T)$, *the generalized problem to* (35) *has a unique solution u in the space* $W_2^1(0, T; V, H)$ *and*

$$\|u\|_{W_2^1} \leq \text{const}(\|u_0\|_{L_2(G)} + \|f\|_{L_2(Q_T)}),$$

where $V = \mathring{W}_2^1(G)$ *and* $H = L_2(G)$.

PROOF. We use Corollary 23.26.

(I) The Hölder inequality yields

$$|a(u,v;t)| \leq C\|u\|_V \|v\|_V \qquad \text{for all} \quad u, v \in V, \quad t \in \,]0, T[.$$

Observe that

$$\left|\int \alpha_i v D_i u\, dx\right| \leq \sup|\alpha_i| \left(\int (D_i u)^2\, dx\right)^{1/2} \left(\int v^2\, dx\right)^{1/2}$$
$$\leq \sup|\alpha_i|\, \|u\|_V \|v\|_V \tag{36}$$

and

$$\left|\int \beta u v\, dx\right| \leq \sup|\beta|\, \|u\|_V \|v\|_V.$$

(II) The Poincaré–Friedrichs inequality yields

$$\int \sum_{i,j} a_{ij} D_i u D_i v\, dx \geq \gamma \int \sum_i (D_i u)^2\, dx$$
$$\geq c\|u\|_V^2 \qquad \text{for all} \quad u \in V \quad \text{and fixed} \quad c > 0.$$

Moreover, we have

$$\|u\|_H \leq \|u\|_V \qquad \text{for all} \quad u \in V.$$

From (I) we obtain the decisive *Gårding inequality*:

$$a(u, u, t) \geq c\|u\|_V^2 - d\|u\|_H^2 \qquad \text{for all} \quad u \in V, \quad t \in]0, T[,$$

where the constants $c > 0$ and $d \geq 0$ are *independent* of time t.

(III) The function

$$t \mapsto a(u, v; t)$$

is measurable on $]0, T[$ for all u, $v \in V$. This follows from a_{ij}, α_i, $\beta \in L_\infty(Q_T)$, by Example 23.4.

(IV) Let $f \in L_2(Q_T)$. From

$$\langle b(t), v\rangle_V = \int_G f(x, t)v(x)\,dx \qquad \text{for all} \quad v \in V,$$

it follows that $\|b\|_{L_2(0,T;V^*)} \leq \|f\|_{L_2(Q_T)}$, again by Example 23.4.

The assertion is now a consequence of Corollary 23.26. □

Let

$$L(t)u(x, t) = \sum_{|\alpha|, |\beta| \leq m} (-1)^{|\alpha|} D^\alpha(a_{\alpha\beta}(x, t)D^\beta u(x, t)),$$

where the partial derivatives D^α, D^β refer to the space variable x. We now study the following initial–boundary value problem of order $2m$:

$$\begin{aligned} u_t(x, t) + L(t)u(x, t) &= f(x, t) \quad \text{on } Q_T, \\ D^\alpha u(x, t) &= 0 \qquad\quad \text{on } \partial G \times [0, T] \qquad \text{for all} \quad \alpha\colon |\alpha| \leq m - 1, \\ u(x, 0) &= u_0(x) \quad\;\; \text{on } G. \end{aligned} \tag{37}$$

We assume:

(B1) Let G be a bounded region in $\mathbb{R}^N$, $N \geq 1$, and let $0 < T < \infty$. We set $Q_T = G \times]0, T[$. Moreover, let $m = 1, 2, \ldots$.

(B2) All the coefficients $a_{\alpha\beta}\colon Q_T \to \mathbb{R}$ are measurable and bounded.

(B3) *Strong ellipticity.* For each $t \in]0, T[$, the differential operator $L(t)$ of order $2m$ is regularly strong elliptic or strong elliptic in the sense of Definition 22.42, where the ellipticity constant c_0 is *independent* of time t.

Definition 23.31. The *generalized problem* to (37) reads as follows. Let

$$V = \mathring{W}_2^m(G), \qquad H = L_2(G),$$

and let $f \in L_2(Q_T)$ and $u_0 \in H$ be given. We seek $u \in W_2^1(0, T; V, H)$ with $u(0) = u_0$ such that

$$\frac{d}{dt}(u(t)|v)_H + a(u(t), v; t) = \langle b(t), v\rangle_V$$

holds for all $v \in V$ and almost all $t \in]0, T[$. Here, for all $u, v \in V$, we set

$$a(u, v; t) = \int_G \sum_{|\alpha|, |\beta| \le m} a_{\alpha\beta}(x, t) D^\alpha v D^\beta u \, dx,$$

$$\langle b(t), v \rangle_V = \int_G f(x, t) v(x) \, dx.$$

Proposition 23.32. *Assume* (B1) *through* (B3). *Then, all the assertions of Corollary* 23.26 *hold true for the generalized problem to* (37). *In particular, the generalized problem to* (37) *has a unique solution u and*

$$\|u\|_{W_2^1} \le \text{const}(\|u_0\|_{L_2(G)} + \|f\|_{L_2(Q_T)}).$$

PROOF. Observe the Gårding inequality from Section 22.15 and use the same argument as in the proof of Proposition 23.30. □

Remark 23.33 (Generalizations). Using the Gårding inequality for strongly elliptic systems from Problem 22.7, one can immediately generalize Proposition 23.32 to parabolic systems.

Applications of the theory of analytic semigroups to parabolic equations will be considered in the Appendix, $A_2(110)$. However, note that this approach needs stronger assumptions with respect to $a_{\alpha\beta}$ and f than in Proposition 23.32.

Combining Theorem 23.A with the Banach fixed-point theorem, one can obtain existence theorems for semilinear parabolic equations. This method will be used in Section 72.5 in order to obtain existence theorems for the nonstationary Navier–Stokes equations for viscous flows.

23.9. Proof of the Main Theorem

The following proof of Theorem 23.A contains a number of methods of deduction that are typical for existence and uniqueness proofs for time-dependent differential equations by means of the Galerkin method. We recommend a careful study of this proof inasmuch as it contains important elements which facilitate the understanding of more complicated existence proofs for nonlinear evolution equations which will be considered later on.

The key to the proof is:

(i) the integration by parts formula (25); and
(ii) the Gronwall lemma (Proposition 3.10), which yields the necessary *a priori* estimates for the solutions of the Galerkin equations.

For brevity we set

$$\langle u, v \rangle = \langle u, v \rangle_V, \qquad (u|v) = (u|v)_H,$$

$$\|u\| = \|u\|_V, \qquad |u| = \|u\|_H.$$

Moreover, we set $\int = \int_0^T$ and

$$W = W_2^1(0, T; V, H).$$

Observe that the embeddings $V \subseteq H \subseteq V^*$ are continuous, that is,

$$|u| \leq \text{const}\,\|u\| \qquad \text{for all} \quad u \in V,$$

$$\|u\|_{V^*} \leq \text{const}\,|u| \qquad \text{for all} \quad u \in H.$$

According to (17) we have

$$(u|v) = \langle u, v \rangle \qquad \text{for all} \quad u \in H, \quad v \in V.$$

By assumption (H2),

$$a(u, u) \geq c\,\|u\|_V^2 \qquad \text{for all} \quad u \in V \quad \text{and fixed} \quad c > 0. \tag{38}$$

We set

$$\langle Au, v \rangle = a(u, v) \qquad \text{for all} \quad u, v \in V.$$

Then the operator $A: V \to V^*$ is linear, continuous, and strongly monotone.

Step 1: An equivalent operator equation.

Our goal is to prove Corollary 23.24, i.e., we want to show that the original equation (26) is equivalent to the operator equation (29). To this end, we use the formula

$$\frac{d}{dt}(u(t)|v) = \langle u'(t), v \rangle \qquad \text{for all} \quad v \in V, \tag{39}$$

and almost all $t \in \,]0, T[$. This follows from Proposition 23.20. Here, d/dt denotes the generalized derivative on $]0, T[$.

(I-1) Suppose that $u \in W$ and that

$$\frac{d}{dt}(u(t)|v) + a(u(t), v) = \langle b(t), v \rangle \qquad \text{for all} \quad v \in V, \tag{40}$$

and almost all $t \in \,]0, T[$. This implies

$$\langle u'(t) + Au(t) - b(t), v \rangle = 0 \qquad \text{for all} \quad v \in V,$$

and almost all $t \in \,]0, T[$. Hence

$$u'(t) + Au(t) - b(t) = 0 \qquad \text{for almost all} \quad t \in \,]0, T[. \tag{41}$$

(I-2) Conversely, (41) and $u \in W$ imply (40).

Step 2: Uniqueness.

For $j = 1, 2$, suppose that

$$u_j'(t) + Au_j(u) = b(t) \qquad \text{for almost all} \quad t \in \,]0, T[,$$

$$u_j(0) = u_0, \qquad u_j \in W.$$

The difference $u = u_1 - u_2$ then satisfies the equation

$$u'(t) + Au(t) = 0 \qquad \text{for almost all} \quad t \in \,]0, T[,$$

$$u(0) = 0, \qquad u \in W.$$

We have to show that

$$u(t) = 0 \qquad \text{for almost all} \quad t \in]0, T[. \tag{42}$$

By the integration by parts formula (25) and by (38), we get

$$\begin{aligned} \tfrac{1}{2}|u(T)|^2 &= \int \langle u'(t), u(t) \rangle \, dt \\ &= -\int \langle Au(t), u(t) \rangle \, dt = -\int a(u(t), u(t)) \, dt \\ &\le -c \int \|u(t)\|^2 \, dt. \end{aligned}$$

Note that $u(0) = 0$. This implies

$$\tfrac{1}{2}|u(T)|^2 + c \int \|u(t)\|^2 \, dt \le 0,$$

and hence (42).

Step 3: Existence proof via the Galerkin method.
For the sake of simplicity, we assume that the function

$$t \mapsto \langle b(t), v \rangle$$

is continuous on $[0, T]$ for all $v \in V$. The more general case that $b \in L_2(0, T; V^*)$ will be considered at the end of this section in connection with the proof of Corollary 23.26.

(I) Solution of the Galerkin equations. The Galerkin equations (27) are equivalent to

$$\begin{gathered} (u_n'(t)|w_j) + a(u_n(t), w_j) = \langle b(t), w_j \rangle, \qquad j = 1, \ldots, n, \\ u_n(0) = u_{n0}, \end{gathered} \tag{43}$$

where

$$u_n(t) = \sum_{j=1}^{n} c_{jn}(t) w_j,$$

and $w_j \in V$ for all j. Equation (43) represents a linear system of ordinary differential equations. According to Corollary 3.8, for each $n = 1, 2, \ldots$, equation (43) has a unique classical solution on $[0, T]$.

(II) *A priori* estimates for the Galerkin solutions. We show that

$$\int \|u_n(t)\|^2 \, dt \le K\left(|u_n(0)|^2 + \int \|b\|_{V^*}^2 \, dt\right) \tag{44}$$

holds for all n, where the constant K is independent of n.

Multiplying (43) by c_j and summing over j, we obtain that

$$(u_n'(t)|u_n(t)) + a(u_n(t), u_n(t)) = \langle b(t), u_n(t) \rangle.$$

Because of the well-known product formula

$$\frac{d}{dt}(u_n(t)|u_n(t)) = 2(u_n'(t)|u_n(t)),$$

we get

$$\frac{d}{dt}|u_n|^2 + 2a(u_n, u_n) = 2\langle b, u_n\rangle.$$

Integration yields

$$|u_n(T)|^2 - |u_n(0)|^2 + 2\int a(u_n, u_n)\,dt = 2\int \langle b, u_n\rangle\,dt.$$

We now use the strong positivity condition

$$a(u_n, u_n) \geq c\|u_n\|^2$$

and the classical inequality

$$2|xy| \leq c^{-1}x^2 + cy^2 \qquad \text{for all} \quad x, y \in \mathbb{R}.$$

This implies

$$\begin{aligned}|u_n(T)|^2 + 2c\int \|u_n\|^2\,dt &\leq |u_n(0)|^2 + 2\int \|b\|_{V^*}\|u_n\|\,dt\\ &\leq |u_n(0)|^2 + c^{-1}\int \|b\|_{V^*}^2\,dt + c\int \|u_n\|^2\,dt,\end{aligned}$$

and hence (44).

(III) Weak convergence of the Galerkin method in $L_2(0, T; V)$. By (44), the sequence (u_n) of the Galerkin solutions is bounded in the H-space $L_2(0, T; V)$, since $u_n(0) \to u_0$ in H as $n \to \infty$. Thus there exists a weakly convergent subsequence $(u_{n'})$, i.e.,

$$u_{n'} \rightharpoonup u \quad \text{in } L_2(0, T; V) \qquad \text{as} \quad n \to \infty. \tag{45}$$

Below we shall show that from (45) we obtain:

$$u'(t) + Au(t) = b(t) \qquad \text{for almost all} \quad t \in]0, T[, \tag{46a}$$

$$u \in W, \tag{46b}$$

$$u(0) = u_0. \tag{46c}$$

By Step 2, equation (46) determines uniquely the solution u. Consequently, all the weakly convergent subsequences of (u_n) have the same limit u. Thus we obtain that

$$u_n \rightharpoonup u \quad \text{in } L_2(0, T; V) \qquad \text{as} \quad n \to \infty, \tag{47}$$

according to Proposition 21.23(i).

In order to prove (46), we will use the following *idea*. By means of generalized derivatives, we write the Galerkin equation as an integral identity. It is important that this identity does not contain any derivatives with respect to u_n. Applying the weak limit (45) to this identity, we get (46). The *key* equation is given by

$$-(u_0|v)\varphi(0) - \int \langle u(t), v\rangle \varphi'(t)\,dt + \int \langle Au(t), v\rangle \varphi(t)\,dt = \int \langle b(t), v\rangle \varphi(t)\,dt, \quad (48)$$

for all $v \in V$ and all real functions

$$\varphi \in C^1[0, T] \qquad \text{with} \quad \varphi(T) = 0. \quad (49)$$

(III-1) Proof of (48). Choose φ as in (49). Multiplying the Galerkin equation (43) by φ and using integration by parts, we obtain the *integral identity*:

$$-(u_n(0)|w_j)\varphi(0) - \int \langle u_n(t), w_j\rangle \varphi'(t)\,dt + \int a(u_n, w_j)\varphi\,dt = \int \langle b, w_j\rangle \varphi\,dt, \quad (50)$$

for all $n \geq j$. Observe that

$$(u_n(t)|w_j) = \langle u_n(t), w_j\rangle.$$

Applying the weak limit (45) to equation (50), we get

$$-(u_0|w_j)\varphi(0) - \int \langle u(t), w_j\rangle \varphi'(t)\,dt + \int a(u, w_j)\varphi\,dt = \int \langle b, w_j\rangle\,\varphi\,dt, \quad (51)$$

for all j.

In order to justify this limit we need the fact that the second and third term in (50) represent linear continuous functionals on the space $X = L_2(0, T; V)$, with respect to u_n. Indeed, we have that

$$\begin{aligned}\left|\int \langle u_n(t), w_j\rangle \varphi'(t)\,dt\right| &\leq C_1 \int \|u_n(t)\|\,\|w_j\|\,|\varphi'(t)|\,dt \\ &\leq C_2 \|w_j\| \left(\int \|u_n(t)\|^2\,dt\right)^{1/2} \\ &= C_2 \|w_j\|\,\|u_n\|_X, \end{aligned} \quad (52)$$

where C_1 and C_2 are constants, by the Hölder inequality and $\|v\|_{V^*} \leq$

$C_1\|v\|$ for all $v \in V$. Analogously,

$$\left|\int a(u_n(t), w_j)\varphi(t)\,dt\right| \le C_1 \int \|u_n(t)\|\,\|w_j\|\,|\varphi(t)|\,dt$$
$$\le C_2\|w_j\|\,\|u_n\|_X. \tag{53}$$

In order to finish our argument let $v \in V$. By assumption (H3), there exists a sequence (v_n) with

$$v_n \to v \quad \text{in } V \qquad \text{as} \quad n \to \infty,$$

where each v_n is a finite linear combination of certain basis elements w_j. Letting $n \to \infty$ we see that equation (51) is also valid if we replace w_j with v. This is the assertion (48).

To justify this limit we need the fact that the terms in (51) represent linear continuous functionals on the space V, with respect to w_j. This follows from (52), (53) and

$$\left|\int \langle b, w_j\rangle\,dt\right| \le C\left(\int \|b(t)\|_{V^*}^2\,dt\right)^{1/2} \|w_j\|.$$

(III-2) Proof of (46a). From (48) we obtain that

$$\left\langle -\int \varphi' u\,dt + \int \varphi(Au - b)\,dt, v\right\rangle = 0,$$

for all $\varphi \in C_0^\infty(0, T)$ and all $v \in V$. Hence

$$-\int \varphi' u\,dt + \int \varphi(Au - b)\,dt = 0 \qquad \text{for all} \quad \varphi \in C_0^\infty(0, T).$$

This shows that $u(\cdot)$ has a *generalized derivative* on $]0, T[$ with

$$u' + Au = b \qquad \text{on }]0, T[.$$

(III-3) Proof of (46b). From $u \in L_2(0, T; V)$ and $\|Av\|_{V^*} \le C\|v\|$ for all $v \in V$, we obtain that

$$\int \|Au(t)\|_{V^*}^2\,dt \le C^2 \int \|u(t)\|^2\,dt,$$

i.e., $Au \in L_2(0, T; V^*)$. From $b \in L_2(0, T; V^*)$ we get $u' \in L_2(0, T; V^*)$. Hence $u \in W$.

(III-4) Proof of (46c). Since $u \in W$, we can apply the integration by parts formula (25). This yields

$$(u(T)|\varphi(T)v) - (u(0)|\varphi(0)v) = \int \langle u'(t), \varphi(t)v\rangle + \langle u(t), \varphi'(t)v\rangle\,dt,$$

for all $\varphi \in C^1[0, T]$ and all $v \in V$. In particular, if $\varphi(0) = 1$ and $\varphi(T) =$

0, then equation (48) along with $u' + Au = b$ yields

$$(u(0) - u_0|v) = 0 \qquad \text{for all} \quad v \in V.$$

Since V is dense in H, we get $u(0) = u_0$.

The existence proof of Theorem 23.A is complete.

Step 4: Continuous dependence on the data.

From

$$u_n \rightharpoonup u \quad \text{in } L_2(0, T; V) \qquad \text{as} \quad n \to \infty$$

it follows that

$$\|u\|_{L_2(0,T;V)} \leq \varliminf_{n\to\infty} \|u_n\|_{L_2(0,T;V)},$$

by Proposition 21.23(c). Thus, letting $n \to \infty$, we obtain from the *a priori* estimate (44) and from $u_n(0) \to u_0$ as $n \to \infty$ that

$$\|u\|_{L_2(0,T;V)} \leq K^{1/2}\left(|u_0|^2 + \int \|b(t)\|_{V^*}^2\, dt\right)^{1/2}.$$

By (III-3),

$$\|Au\|_{L_2(0,T;V^*)} \leq C\|u\|_{L_2(0,T;V)}. \tag{54}$$

Finally, from $u' + Au = b$ we get

$$\|u\|_W \leq D\left(|u_0|^2 + \int \|b(t)\|_{V^*}^2\, dt\right)^{1/2}.$$

Step 5: Convergence of the Galerkin method in $C([0, T], H)$.

We show that

$$\max_{0 \leq t \leq T} |u_n(t) - u(t)| \to 0 \qquad \text{as} \quad n \to \infty. \tag{55}$$

To this end, we will use the approximation argument (56a) and (56b) below.

We have $u \in W$ and also $u_n \in W$ for all n. The embedding $W \subseteq C([0, T], H)$ implies

$$u, u_n \in C([0, T], H) \qquad \text{for all } n.$$

(I) Approximation. Set $H_n = \operatorname{span}\{w_1, \ldots, w_n\}$. Hence $H_n \subseteq V \subseteq H$. By Proposition 23.23, for each $u \in W$ and $\varepsilon > 0$ there exists a polynomial

$$p(t) = \sum_i t^i a_i$$

with coefficients $a_i \in V$ and

$$\|u - p\|_W = \left(\int \|u(t) - p(t)\|_V^2\, dt\right)^{1/2} + \left(\int \|u'(t) - p'(t)\|_{V^*}^2\, dt\right)^{1/2} < \varepsilon.$$

This inequality remains true if we change the coefficients a_i of p a little with respect to the norm of the space V. Since the set $\bigcup_n H_n$ is dense in V, we obtain the stronger result that the set of all polynomials with

coefficients in $\bigcup_n H_n$ is *dense* in the space W. Thus there exists a sequence (p_n) of polynomials p_n: $[0, T] \to H_n$ with

$$p_n \to u \quad \text{in } W \qquad \text{as} \quad n \to \infty.$$

The continuity of the embedding $W \subseteq C([0, T], H)$ implies

$$\max_{0 \le t \le T} |u(t) - p_n(t)| \le \text{const} \|u - p_n\|_W \to 0 \qquad \text{as} \quad n \to \infty. \tag{56a}$$

Below we will show that $u_n \rightharpoonup u$ in $L_2(0, T; V)$ as $n \to \infty$ implies that

$$\max_{0 \le t \le T} |u_n(t) - p_n(t)| \to 0 \qquad \text{as} \quad n \to \infty. \tag{56b}$$

From (56a) and (56b) we obtain the desired relation (55).

(II) Proof of (56b) via integration by parts.

(II-1) From $u_n(0) \to u(0)$ in H as $n \to \infty$ and (56a), it follows that

$$|u_n(0) - p_n(0)| \to 0 \qquad \text{as} \quad n \to \infty.$$

(II-2) From the Galerkin equations we get

$$\begin{aligned}\langle u_n', u_n - p_n \rangle &= \langle b - Au_n, u_n - p_n \rangle \\ &= \langle u' + A(u - u_n), u_n - p_n \rangle.\end{aligned}$$

The positivity of $a(\cdot, \cdot)$ yields

$$\langle A(u - u_n), u - u_n \rangle \ge 0.$$

(II-3) We set $X = L_2(0, T; V)$, and hence $X^* = L_2(0, T; V^*)$. Using (II-1) and (II-2), we obtain from the integration by parts formula (25) and the Hölder inequality that

$$\begin{aligned}&2^{-1}|u_n(t) - p_n(t)|^2 - 2^{-1}|u_n(0) - p_n(0)|^2 \\ &\quad = \int_0^t \langle u_n' - p_n', u_n - p_n \rangle \, ds \\ &\quad = \int_0^t \langle u' + A(u - u_n) - p_n', (u_n - u) + (u - p_n) \rangle \, ds \\ &\quad \le \int_0^t \langle u' - p_n', u_n - p_n \rangle + \langle A(u - u_n), u - p_n \rangle \, ds \\ &\quad \le \|u' - p_n'\|_{X^*} \|u_n - p_n\|_X + \|Au - Au_n\|_{X^*} \|u - p_n\|_X \\ &\quad \le \text{const} \|u - p_n\|_W \to 0 \qquad \text{as} \quad n \to \infty.\end{aligned}$$

Note that the sequences $(u_n), (p_n)$ are bounded in X, and (Au_n) is bounded in X^*. This follows from the weak convergence of (u_n) and (p_n) in X as well as from (54).

Step 6: Strong convergence of the Galerkin method in the space $X = L_2(0, T; V)$.

(I) From $u_n \rightharpoonup u$ in X as $n \to \infty$ it follows that

$$Au_n \rightharpoonup u \quad \text{in } X^* \qquad \text{as} \quad n \to \infty,$$

since $A: X \to X^*$ is linear and continuous by (54). This yields

$$\int \langle Au_n, u\rangle\, dt \to \int \langle Au, u\rangle\, dt \qquad \text{as} \quad n \to \infty,$$

and

$$\int \langle b, u_n\rangle\, dt \to \int \langle b, u\rangle\, dt \qquad \text{as} \quad n \to \infty,$$

by Proposition 23.9. Letting $n \to \infty$, we get

$$\int \langle b - Au_n, u\rangle + \langle b - Au, u_n\rangle\, dt - (u_n(T)|u(T)) + (u_n(0)|u(0))$$

$$\to 2\int \langle b - Au, u\rangle\, dt - |u(T)|^2 + |u(0)|^2 = 0. \quad (57)$$

Observe $u' = b - Au$ and the integration by parts formula (25).

(II) We show that $u_n \to u$ in X as $n \to \infty$. By (25), we obtain the following two relations:

$$\begin{aligned} 2^{-1}|u(T) - u_n(T)|^2 - 2^{-1}|u(0) - u_n(0)|^2 &= \int \langle u'(t) - u_n'(t), u(t) - u_n(t)\rangle\, dt \\ &= \int \langle b - Au - u_n', u - u_n\rangle\, dt \end{aligned}$$

and

$$\begin{aligned} (u_n(T)|u(T)) - (u_n(0)|u(0)) &= \int \langle u_n'(t), u(t)\rangle + \langle u'(t), u_n(t)\rangle\, dt \\ &= \int \langle u_n', u\rangle + \langle b - Au, u_n\rangle\, dt. \end{aligned}$$

The Galerkin equation (43) yields

$$\langle u_n'(t), u_n(t)\rangle + \langle Au_n(t), u_n(t)\rangle = \langle b(t), u_n(t)\rangle.$$

Now to the *key* of our proof. The strong positivity of $a(\cdot, \cdot)$ yields

$$\begin{aligned} c\|u - u_n\|_X^2 \equiv c \int \|u(t) - u_n(t)\|^2\, dt \\ \leq \int \langle A(u - u_n), u - u_n\rangle\, dt + 2^{-1}|u(T) - u_n(T)|^2. \end{aligned}$$

Hence

$$
\begin{aligned}
c\|u-u_n\|_X^2 & \\
&\leq 2^{-1}|u(0)-u_n(0)|^2+\int\langle b-Au_n-u_n', u-u_n\rangle\,dt \\
&= 2^{-1}|u(0)-u_n(0)|^2+\int\langle b-Au_n-u_n', u\rangle\,dt \\
&= 2^{-1}|u(0)-u_n(0)|^2+\int\langle b-Au_n, u\rangle\,dt \\
&\quad+\int\langle b-Au, u_n\rangle\,dt-(u_n(T)|u(T))+(u_n(0)|u(0))\to 0 \qquad \text{as} \quad n\to\infty,
\end{aligned}
$$

according to (57). This implies $u_n \to u$ in X as $n \to \infty$.

The proof of Theorem 23.A is complete.

PROOF OF COROLLARY 23.26. This proof proceeds analogously to the proof of Theorem 23.A. Observe the following two pecularities.

(i) The bilinear functional $(u, v) \mapsto a(u, v; t)$ depends on time t, and hence the operator $A(t)$ also depends on t. But this does not change our estimates, since all the constants are independent of t according to assumption (H2*).

(ii) The Galerkin equations represent a linear system of ordinary differential equations, where the coefficients are measurable and *bounded*. In order to solve this system, we transform it into a system of Volterra integral equations, which has a unique continuous solution (see Section 3.3). By the main theorem of calculus A_2(25c), we may differentiate this integral equation with respect to t almost everywhere on $]0, T[$. This way we obtain a solution of the Galerkin equations with bounded derivatives, which exist almost everywhere on $]0, T[$. □

PROBLEMS

Without stating explicitly the opposite, we assume that

$$1 \leq p, q < \infty \qquad \text{and} \qquad 0 < T < \infty.$$

Problems 23.1 to 23.10 form a logical chain.

23.1. *Generalized approximation theorem of Weierstrass.* Let X be a B-space. Show that the set of all polynomials $p: [0, T] \to X$ with coefficients in X is dense in $C([0, T], X)$.

Solution: We translate the classical proof of Bernstein (1880–1968) for $X = \mathbb{R}$ to our more general case. Let $\sum = \sum_{k=0}^n$.

(I) Two identities. For $b_k(t) = \binom{n}{k} t^k(1-t)^{n-k}$, we have

$$\sum b_k(t) = 1, \tag{58a}$$

$$\sum b_k(t)(nt-k)^2 = nt(t-1), \tag{58b}$$

for all $t \in \mathbb{R}$ and $n = 1, 2, \ldots$. To prove this, we begin with

$$(t+s)^n = \sum \binom{n}{k} t^k s^{n-k}.$$

Differentiation with respect to t yields

$$nt(t+s)^{n-1} = \sum \binom{n}{k} k t^k s^{n-k},$$

$$n(n-1)t^2(t+s)^{n-2} = \sum \binom{n}{k} k(k-1) t^k s^{n-k}.$$

Setting $s = 1 - t$ we obtain (58), by summation.

(II) The Bernstein polynomials B_n. After a similarity transformation we may assume that $T = 1$. Let $u \in C([0,1], X)$ with the norm $\|u\|_C = \max_{0 \le t \le 1} \|u(t)\|$. We set

$$B_n(t) = \sum u\left(\frac{k}{n}\right) b_k(t).$$

By (58),

$$\|u(t) - B_n(t)\| \le \sum \left\| u(t) - u\left(\frac{k}{n}\right) \right\| b_k(t).$$

For $\varepsilon > 0$ we choose $\delta > 0$ such that

$$\left\| u(t) - u\left(\frac{k}{n}\right) \right\| \le \varepsilon \quad \text{if} \quad \left| t - \frac{k}{n} \right| < \delta.$$

Obviously,

$$\left\| u(t) - u\left(\frac{k}{n}\right) \right\| \le 2\|u\|_C \left(t - \frac{k}{n}\right)^2 \Big/ \delta^2 \quad \text{if} \quad \left| t - \frac{k}{n} \right| \ge \delta.$$

From (58) we obtain the desired result:

$$\|u(t) - B_n(t)\| \le \sum \left(\varepsilon + 2\|u\|_C \left(t - \frac{k}{n}\right)^2 \Big/ \delta^2 \right) b_k(t)$$

$$\le \varepsilon + 2\|u\|_C t(1-t)/\delta^2 n \le 2\varepsilon,$$

for all $n \ge n_0(\varepsilon)$ with suitable $n_0(\varepsilon)$.

23.2. *Proof of Proposition* 23.2. The proof proceeds along the lines of the classical proof for real functions, which may be found, for example, in Kufner, John, and Fučik (1977, M). Also compare Gajewski, Gröger, and Zacharias (1974, M), Chapter 4. Let us consider some typical arguments.

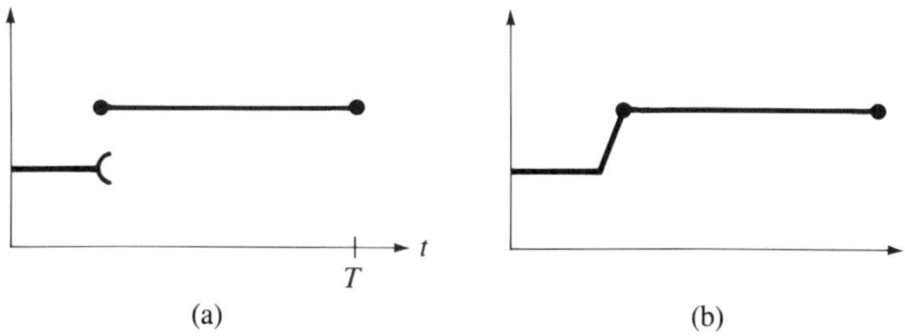

Figure 23.2

(I) *Density*.

(I-1) Prove first that the set of all step functions is dense in $L_p(0, T; X)$. In fact, it follows from the definition of the Lebesgue integral $A_2(14)$ that for each $u \in L_p(0, T; X)$, there exists a sequence (u_n) of step functions such that

$$u_n(t) \to u(t) \qquad \text{almost everywhere on }]0, T[\qquad \text{as} \quad n \to \infty.$$

Setting

$$v_n(t) = \begin{cases} u_n(t) & \text{if } \|u_n(t)\| \leq 2\|u(t)\|, \\ 0 & \text{otherwise,} \end{cases}$$

we get

$$\int_0^T \|u(t) - v_n(t)\|^p \, dt \to 0 \qquad \text{as} \quad n \to \infty,$$

by the principle of majorized convergence $A_2(19)$.

(I-2) A simple approximation argument of measure theory shows that all those special step functions are also dense in $L_p(0, T; X)$, which are constant on intervals (Fig. 23.2(a)). Each such special step function can be arbitrarily closely approximated by continuous functions (Fig. 23.2(b)). Hence $C([0, T], X)$ is dense in $L_p(0, T; X)$.

(I-3) By Problem 23.1, the set of all polynomials p: $[0, T] \to X$, with coefficients in X, is also dense in the space $L_p(0, T; X)$.

(II) *Separability*. Let $Z = L_p(0, T; X)$ and suppose that X is separable. Choose an at most countable set M, which is dense in X. Let P be the set of all the polynomials p: $[0, T] \to X$ with coefficients in M. A simple approximation argument shows that P is dense in Z by (I-3). Hence Z is separable.

(III) *Embeddings*. The continuity of the embedding

$$C([0, T], X) \subseteq L_p(0, T; X)$$

follows from

$$\left(\int_0^T \|u(t)\|^p \, dt \right)^{1/p} \leq T^{1/p} \max_{0 \leq t \leq T} \|u(t)\|.$$

The continuity of the embeddings in Proposition 23.2(h) is an immediate consequence of

$$\left(\int_0^T \|u(t)\|_Y^q \,dt\right)^{1/q} \le c\left(\int_0^T \|u(t)\|_X^q \,dt\right)^{1/q}$$
$$\le c_1\left(\int_0^T \|u(t)\|_X^r \,dt\right)^{1/r}.$$

Here, use the Hölder inequality and note that $1 \le q \le r < \infty$.

23.3. *Dense subsets of the space* $Y = C^m([0, T], X)$, $m \ge 1$. Let M be a dense set in the B-space X. Show that the set of all the polynomials $p: [0, T] \to X$ with coefficients in M is dense in Y.

Solution:

(I) Let $M = X$ and $m = 1$. We set $C = C([0, T], X)$. By Problem 23.1, for each $u \in Y$, there exists a sequence of polynomials $q_n: [0, T] \to X$ with

$$q_n \to u' \quad \text{in } C \qquad \text{as} \quad n \to \infty.$$

Integration yields

$$u(t) = u(0) + \int_0^t u'(s)\,ds.$$

If we set

$$p_n(t) = u(0) + \int_0^t q_n(s)\,ds,$$

then $p_n' = q_n$ and

$$\|u(t) - p_n(t)\| \le T\|u' - q_n\|_C.$$

Hence we obtain $p_n \to u$ in $C^1([0, T], X)$ as $n \to \infty$.

(II) In the case where $M = X$ and $m \ge 2$, integrate the mth derivative. If $M \subset X$, then use a simple approximation argument with respect to the coefficients of the polynomials.

23.4. *Absolutely continuous functions.* The function $w: [0, T] \to X$ with values in the B-space X is called *absolutely continuous* iff, for each $\varepsilon > 0$, there exists a $\delta > 0$ so that

$$\sum_n \|u(t_n) - u(s_n)\| < \varepsilon$$

holds for all finite systems of pairwise disjoint intervals $]t_n, s_n[$ of total length $< \delta$.

Show that each Lipschitz continuous function $w: [0, T] \to X$ is absolutely continuous.

Solution: Observe that $\|u(t_n) - u(s_n)\| \le L|t_n - s_n|$.

23.5. *Generalized main theorem of calculus.* Let X be a B-space. For a given function $u \in L_1(0, T; X)$, we set

$$v(t) = \int_0^t u(s)\,ds. \tag{59}$$

23.5a. Show that $v \in C([0, T], X)$ and that

$$v' = u \quad \text{on }]0, T[,$$

in the generalized sense. Furthermore, show that v is absolutely continuous.

Solution: We will use a typical approximation argument. The set of all the polynomials $q\colon [0, T] \to X$ is dense in $L_1(0, T; X)$. Hence there exists a sequence (q_n) of such polynomials with

$$q_n \to u \quad \text{in } L_1(0, T; X) \qquad \text{as} \quad n \to \infty.$$

We set

$$p_n(t) = \int_0^t q_n(s)\,ds.$$

From

$$\|v(t) - p_n(t)\| \le \int_0^T \|u(s) - q_n(s)\|\,ds$$

it follows that

$$p_n \to v \quad \text{in } C([0, T], X) \qquad \text{as} \quad n \to \infty.$$

The continuity of the embedding $C([0, T], X) \subseteq L_1(0, T; X)$ implies that

$$p_n \to v \quad \text{in } L_1(0, T; X) \qquad \text{as} \quad n \to \infty.$$

By Proposition 3.7, we have $p_n'(t) = q_n(t)$ on $[0, T]$, in the classical sense. By Example 23.16,

$$\int_0^T \varphi q_n\,dt = -\int_0^T \varphi' p_n\,dt \qquad \text{for all} \quad \varphi \in C_0^\infty(0, T).$$

Letting $n \to \infty$ we obtain that

$$\int_0^T \varphi u\,dt = -\int_0^T \varphi' v\,dt,$$

according to Proposition 23.9. This yields $v' = u$ on $]0, T[$.

The absolute continuity of v follows from the absolute continuity of the integral (see $A_2(20)$).

23.5b.* Consider the function v in (59), where $u \in L_1(0, T; X)$. Show that

$$v'(t) = u(t) \qquad \text{for almost all} \quad t \in]0, T[, \tag{60}$$

where the derivative exists in the classical sense.

Hint: This is a special case of $A_2(25)$. Compare Dunford and Schwartz (1958, M), Vol. 1, III.12.8. In particular, equation (60) holds true if u is continuous at the point t.

23.5c.* *The generalized theorem of Lebesgue.* Let $v\colon [0, T] \to X$ be an absolutely continuous function with values in the reflexive B-space X. Then the classical derivative $v'(t)$ exists at almost all points $t \in [0, T]$. Moreover, we have $v' \in L_1(0, T; X)$ and

$$v(t) = v(0) + \int_0^t v'(s)\,ds \qquad \text{for all} \quad t \in [0, T].$$

Hint: Cf. Brézis (1973, L), p. 145.

23.6. *Generalized variational lemma.* Let X be a B-space. Show that if $u \in L_1(0, T; X)$ and

$$\int_0^T \varphi u\, ds = 0 \qquad \text{for all} \quad \varphi \in C_0^\infty(0, T),$$

then $u = 0$ in $L_1(0, T; X)$, that is,

$$u(t) = 0 \qquad \text{for almost all} \quad t \in]0, T[.$$

Solution: Let $0 < t \leq T$ and let $\varepsilon > 0$ be sufficiently small. We define a function $\varphi \in C_0^\infty(0, T)$ such that $0 \leq \varphi \leq 1$ and

$$\varphi(s) = \begin{cases} 1 & \text{if} \quad s \in [\varepsilon, t - \varepsilon], \\ 0 & \text{if} \quad s \notin [0, t]. \end{cases}$$

Hence

$$\begin{aligned} \left\| \int_0^t u\, ds \right\| &= \left\| \int_0^t u\, ds - \int_0^T \varphi u\, ds \right\| \\ &= \left\| \int_0^\varepsilon u(1 - \varphi)\, ds + \int_{t-\varepsilon}^t u(1 - \varphi)\, ds \right\| \\ &\leq \int_0^\varepsilon \|u(s)\|\, ds + \int_{t-\varepsilon}^t \|u(s)\|\, ds. \end{aligned}$$

Letting $\varepsilon \to 0$ we obtain that

$$\int_0^t u\, ds = 0 \qquad \text{for all} \quad t \in [0, T],$$

by $A_2(20)$. The assertion follows now from Problem 23.5b, by differentiation.

23.7. *Test functions.*

23.7a. Show that the real function $\chi \in C_0^\infty(0, T)$ allows a representation of the form

$$\chi = \varphi', \qquad \varphi \in C_0^\infty(0, T)$$

iff $\int_0^T \chi\, dt = 0$.

Solution: Set $\varphi(s) = \int_0^s \chi\, dt$.

23.7b. Choose a fixed function $\rho \in C_0^\infty(0, T)$ with

$$\int_0^T \rho\, dt = 1.$$

Show that each function $\psi \in C_0^\infty(0, T)$ allows a representation of the form

$$\psi = \rho \int_0^T \psi\, dt + \varphi', \qquad \varphi \in C_0^\infty(0, T). \tag{61}$$

Solution: This follows from Problem 23.7a.

23.8. *Characterization of constant functions by generalized derivatives.* Let Y be a B-space. If $u \in L_p(0, T; Y)$, $1 \leq p < \infty$, and if

$$u' = 0 \quad \text{on }]0, T[, \tag{62}$$

in the generalized sense, then u = constant, i.e., there exists a $c \in Y$ with

$$u(t) = c \qquad \text{for almost all} \quad t \in]0, T[.$$

Conversely, each such constant function belongs trivially to $L_p(0, T; Y)$, and we have $u' = 0$ on $]0, T[$.

Solution: We set $c = \int_0^T \rho u \, dt$. From (62) it follows that

$$\int_0^T \varphi' u \, dt = 0 \qquad \text{for all} \quad \varphi \in C_0^\infty(0, T).$$

Using (61), we get

$$\int_0^T \psi u \, dt = \int_0^T \psi \, dt \int_0^T \rho u \, dt,$$

that is,

$$\int_0^T \psi(u - c) \, dt = 0 \qquad \text{for all} \quad \psi \in C_0^\infty(0, T).$$

The assertion follows now from Problem 23.6.

23.9. *p-mean continuity.* Let X be a B-space. For each $u \in L_p(0, T; X)$ with $1 \le p < \infty$, we have

$$\lim_{h \to 0} \int_0^T \|u(t + h) - u(t)\|^p \, dt = 0, \tag{63}$$

where we set $u(s) = 0$ for $s \notin [0, T]$.

Solution: Define the translation operator A_h by

$$(A_h u)(t) = u(t + h),$$

and denote the norm on $L_p(0, T; X)$ by $\|\cdot\|$. Obviously,

$$\begin{aligned} \|A_h u\| &= \left(\int_0^T \|u(t + h)\|^p \, dt \right)^{1/p} \\ &\le \left(\int_0^T \|u(t)\|^p \, dt \right)^{1/p} = \|u\| \qquad \text{for all} \quad h \in \mathbb{R}. \end{aligned} \tag{64}$$

(I) Step functions. Let M be a measurable subset of $[0, T]$ with the characteristic function

$$\chi(t) = \begin{cases} 1 & \text{if} \quad t \in M, \\ 0 & \text{if} \quad t \notin M. \end{cases}$$

Setting $u = \chi x$ for fixed $x \in X$, we get

$$\begin{aligned} \left(\int_0^T \|u(t + h) - u(t)\|^p \, dt \right)^{1/p} &\le 2|h| \, \|x\|_X \\ &= 2|h| T^{-1/p} \|u\|. \end{aligned}$$

The linearity of A_h implies that, for all *step functions u*, we get

$$\|A_h u - u\| \le 2|h| T^{-1/p} \|u\| \qquad \text{for all} \quad h \in \mathbb{R}. \tag{65}$$

(II) Approximation. Since the set of all the step functions is dense in $L_p(0, T; X)$, relation (65) also holds true for all functions $u \in L_p(0, T; X)$. This is (63).

23.10. *The Sobolev space* $W = W_p^1(0, T; V, H)$, $1 < p < \infty$. Choose q with $q^{-1} + p^{-1} = 1$.

23.10a. Show that W is a B-space.

Solution: If (u_n) is a Cauchy sequence in W, then

$$u_n \to u \quad \text{in } L_p(0, T; V) \qquad \text{as} \quad n \to \infty,$$

$$u_n' \to v \quad \text{in } L_q(0, T; V^*) \qquad \text{as} \quad n \to \infty,$$

with suitable points u and v. Observe that $L_p(0, T; V)$ and $L_q(0, T; V^*)$ are B-spaces. The continuity of the embedding $V \subseteq V^*$ implies $v = u'$, according to Proposition 23.19. Hence

$$u_n \to u \quad \text{in } W \qquad \text{as} \quad n \to \infty.$$

23.10b. Show that $C^1([0, T], V)$ is dense in W.

Solution: The proof proceeds analogously to the corresponding proof for classical Sobolev spaces in Section 21.4. Observe that, because of the p-mean continuity in Problem 23.9, the smoothing operator for $u \in W$ has the same properties as the classical smoothing operator from Section 18.14.

From Problem 23.3 it follows additionally that the set of all the polynomials p: $[0, T] \to V$ with coefficients in V is dense in the space W.

23.10c. Show that the embedding $W \subseteq C([0, T], V^*)$ is continuous.

Solution: Denote the norm on V^* and $L_p(0, T; V^*)$ by $\|\cdot\|$ and $\|\cdot\|_p$, respectively. Let $u \in W$. The idea of the proof is to consider the function

$$v(t) = \int_0^t u'(s)\, ds.$$

Then, $v \in C([0, T], V^*)$. To see this, observe that

$$\|v(t) - v(s)\| \leq \int_s^t \|u'(z)\|\, dz$$

and use the absolute continuity of the integral, $A_2(20)$.

By Problems 23.5a and 23.8, we obtain $v' = u'$, and hence

$$v(t) = u(t) + c \qquad \text{for almost all } t \in\,]0, T[, \tag{66}$$

where $c \in V^*$. The Hölder inequality yields

$$\begin{aligned} \max_{0 \leq t \leq T} \|v(t)\| &\leq T^{1/p} \|u'\|_q, \\ \|c\| &= T^{-1/p} \|c\|_p. \end{aligned} \tag{67}$$

From (66) and (67) it follows that

$$\begin{aligned} \max_{0 \leq t \leq T} \|u(t)\| &\leq d(\|u'\|_q + \|v - u\|_p) \\ &\leq d(\|u'\|_q + \|v\|_p + \|u\|_p) \leq d_1 \|u\|_W. \end{aligned}$$

23.10d. Show that the embedding $W \subseteq C([0, T], H)$ is continuous.

Solution:

(I) Integration by parts. Let $u, v \in C^1([0,T], H)$. Integrating

$$(u(t)|v(t))' = (u'(t)|v(t)) + (u(t)|v'(t)),$$

we get the integration by parts formula

$$(u(t)|v(t)) - (u(s)|v(s)) = \int_s^t (u'(z)|v(z)) + (u(z)|v'(z))\,dz, \tag{68}$$

for all $0 \le s \le t \le T$. By (17),

$$(u|v) = \langle u, v\rangle \qquad \text{for all} \quad u, v \in V.$$

For $u, v \in C^1([0,T], V)$, this implies

$$(u(t)|v(t)) - (u(s)|v(s)) = \int_s^t \langle u'(z), v(z)\rangle + \langle v'(z), u(z)\rangle\,dz, \tag{69}$$

for all $0 \le s < t \le T$.

(II) In order to cancel the term $(u(s)|v(s))$ in (69), we choose a test function $\varphi \in C^1(\mathbb{R})$ with $\varphi(s) = 0$ and $\varphi(t) = 1$. Moreover, let $|\varphi| + |\varphi'| \le 1$ on $\mathbb{R}$. Set $v = \varphi u$. Equation (69) implies that

$$(u(t)|u(t)) \le \text{const}\,\|u\|_W^2 \qquad \text{for all} \quad u \in C^1([0,T], V). \tag{70}$$

To see this, observe that $v' = \varphi' u + \varphi u'$ and use the Hölder inequality. Equation (70) yields

$$\|u\|_{C([0,T],H)} \le \text{const}\,\|u\|_W \qquad \text{for all} \quad u \in C^1([0,T], V). \tag{71}$$

The set $C^1([0,T], V)$ is dense in W. By (71) and the extension principle from Section 18.12, the embedding operator

$$j\colon C^1([0,T], V) \subseteq W \to C([0,T], H)$$

has a unique continuous extension $j\colon W \to C([0,T], H)$. In this sense, the embedding

$$W \subseteq C([0,T], H)$$

is continuous.

23.10e. Show that the integration by parts formula (69) holds for all $u, v \in W$.

Solution: Use the density of $C^1([0,T], V)$ in W and a limiting process in (69).

23.11. *Weak* convergence.* Let X be a B-space. Recall that a sequence (x_n^*) from X^* is called weakly* convergent to the point x^* in X^* iff

$$\langle x_n^*, x\rangle \to \langle x^*, x\rangle \qquad \text{as} \quad n \to \infty \qquad \text{for all} \quad x \in X.$$

We write $x_n^* \overset{*}{\rightharpoonup} x^*$.

Show that if the embedding $X \subseteq Y$ is continuous, then as $n \to \infty$:

$$x_n \rightharpoonup x \quad \text{in } X \qquad \text{implies} \qquad x_n \rightharpoonup x \quad \text{in } Y,$$

$$y_n^* \overset{*}{\rightharpoonup} y^* \quad \text{in } Y^* \qquad \text{implies} \qquad y_n^* \overset{*}{\rightharpoonup} y^* \quad \text{in } X^*.$$

Here, X and Y are B-spaces over $\mathbb{K}$.

Solution: Observe that $Y^* \subseteq X^*$.

23.12. *The space $Y = L_\infty(0, T; X)$ and weak* convergence.* Let X be a B-space. By definition, the space Y consists of precisely all the measurable functions $u\colon]0, T[\to X$, which are essentially bounded, that is, there exists a number B such that

$$\|u(t)\| \leq B \qquad \text{for almost all} \quad t \in]0, T[.$$

Precisely all the numbers B with this property are called essential bounds of u. Moreover, we set

$$\|u\|_\infty = \inf\{B\},$$

where the infimum is taken over all the essential bounds of u.

23.12a. Show that Y equipped with the norm $\|\cdot\|_\infty$ is a B-space if we identify any two functions which are identical almost everywhere on $]0, T[$.

Solution: The union of a countable number of sets of measure zero is again a set of measure zero. Let (u_n) be a Cauchy sequence in Y, i.e.,

$$\|u_n - u_m\|_\infty < \varepsilon \qquad \text{for all} \quad n, m \geq n_0(\varepsilon).$$

Then there exists a set Z of measure zero such that

$$|u_n(t) - u_m(t)| < 1/k \qquad \text{for all} \quad t \in]0, T[- Z,$$

and all $n, m \geq n_0(1/k)$. Further there exists a number B with

$$|u_n(t)| \leq B \qquad \text{for all} \quad t \in]0, T[- Z.$$

Here, Z is independent of n, m, and k. Thus the sequence (u_n) converges uniformly on $]0, T[- Z$ to a function which belongs to the space Y. Observe that the limit of a sequence of measurable functions is again a measurable function.

23.12b. *An embedding theorem.* Show that the embedding

$$L_\infty(0, T; X) \subseteq L_p(0, T; X)$$

is continuous for all $1 \leq p \leq \infty$.

Solution: For all $u \in L_p(0, T; X)$, $p < \infty$, we get

$$\|u\|_p = \left(\int_0^T \|u(t)\|^p \, dt\right)^{1/p} \leq T^{1/p}\|u\|_\infty.$$

23.12c. *Functionals.* Let V be a B-space and let $v \in L_\infty(0, T; V^*)$. We set

$$\bar{v}(u) = \int_0^T \langle v(t), u(t)\rangle_V \, dt \qquad \text{for all} \quad u \in L_1(0, T; V).$$

Show that $\bar{v} \in L_1(0, T; V)^*$.

Solution: From $|\langle v, u\rangle| \leq \|v\| \, \|u\|$ it follows that

$$|\bar{v}(u)| \leq \|v\|_\infty \int_0^T \|u(t)\| \, dt = \|v\|_\infty \|u\|_1.$$

This is the assertion.

The following result shows that this way one obtains precisely all the linear continuous functionals on $L_1(0, T; V)$.

23.12d.* *Duality*. Let V be a reflexive separable B-space. Then the space $L_1(0, T; V)$ is separable and

$$L_1(0, T; V)^* = L_\infty(0, T; V^*).$$

More precisely, there exists a bijective linear mapping $\bar{v} \mapsto v$ from $L_1(0, T; V)^*$ onto $L_\infty(0, T; V^*)$ with

$$\langle \bar{v}, u \rangle = \int_0^T \langle v(t), u(t) \rangle_V \, dt \qquad \text{for all} \quad u \in L_1(0, T; V)$$

and $\|\bar{v}\| = \|v\|_\infty$. Thus, the B-spaces $L_1(0, T; V)^*$ and $L_\infty(0, T; V^*)$ are *normisomorphic*.

One identifies $\bar{v}$ with v and writes briefly

$$\langle v, u \rangle = \int_0^T \langle v(t), u(t) \rangle_V \, dt.$$

Hint: Cf. Edwards (1965, M), 8.18.3.

23.12e. *Weak* sequential compactness*. Let the B-space V be given as in Problem 23.12d. By Theorem 21.E, each bounded sequence (v_n) in $L_\infty(0, T; V^*)$ has a subsequence with

$$v_{n'} \overset{*}{\rightharpoonup} v \quad \text{in } L_\infty(0, T; V^*) \qquad \text{as} \quad n \to \infty,$$

that is, for all $u \in L_1(0, T; V)$,

$$\int_0^T \langle v_{n'}(t), u(t) \rangle \, dt \to \int_0^T \langle v(t), u(t) \rangle \, dt \qquad \text{as} \quad n \to \infty. \tag{72}$$

23.12f. *Limits with respect to integrals*. Let V be a B-space. Show that

$$v_n \overset{*}{\rightharpoonup} v \quad \text{in } L_\infty(0, T; V^*) \qquad \text{as} \quad n \to \infty$$

implies

$$\int_0^T v_n(t) \, dt \overset{*}{\rightharpoonup} \int_0^T v(t) \, dt \qquad \text{as} \quad n \to \infty.$$

Solution: For all $u \in V$, as $n \to \infty$, we get

$$\left\langle \int_0^T v_n(t) \, dt, u \right\rangle = \int_0^T \langle v_n(t), u \rangle \, dt$$

$$\to \int_0^T \langle v(t), u \rangle \, dt = \left\langle \int_0^T v(t) \, dt, u \right\rangle.$$

This follows from Problem 23.12c.

23.12g. *Generalized derivatives*. Let V be a separable reflexive B-space. Show that from

$$v_n' = u_n \quad \text{on }]0, T[\qquad \text{for all } n,$$

in the generalized sense, and from

$$\begin{aligned} u_n &\overset{*}{\rightharpoonup} u \quad \text{in } L_\infty(0, T; V^*) \qquad \text{as} \quad n \to \infty, \\ v_n &\overset{*}{\rightharpoonup} v \quad \text{in } L_\infty(0, T; V^*) \qquad \text{as} \quad n \to \infty, \end{aligned} \tag{73}$$

it follows that

$$v' = u \quad \text{on }]0, T[.$$

Solution: Let $\varphi \in C_0^\infty(0, T)$. By Problem 23.12d it follows that (73) remains valid if we replace u_n, u, v_n, v with φu_n, φu, $\varphi' v_n$, $\varphi' v$, respectively. Letting $n \to \infty$ we obtain from

$$\int_0^T \varphi u_n \, dt = -\int_0^T \varphi' v_n \, dt \qquad \text{for all} \quad \varphi \in C_0^\infty(0, T)$$

the equation

$$\int_0^T \varphi u \, dt = -\int_0^T \varphi' v \, dt \qquad \text{for all} \quad \varphi \in C_0^\infty(0, T),$$

according to Problem 23.12f.

23.13. *Two general Sobolev embedding theorems for evolution equations.* Let X, Y, Z be real B-spaces. We consider functions $u: [0, T] \to X$. Let

$$W = \{u: u \in L_p(0, T; X), u' \in L_q(0, T; Z)\},$$

where u' denotes the generalized derivative of $u = u(t)$ on $]0, T[$.

23.13a. *Continuity.* Show that

$$W \subseteq C([0, T], Z) \tag{74}$$

in the case where the following conditions are satisfied:

(i) The embedding $X \subseteq Z$ is continuous (e.g., $X = Z$).

(ii) $0 < T < \infty$, $1 \leq q, p \leq \infty$.

Solution: Use the same argument as in Problem 23.10c. Observe that $u \in W$ implies $u, u' \in L_1(0, T; Z)$.

As usual, relation (74) is to be understood in the following sense. If $u \in W$, then there exists a continuous function $v: [0, T] \to Z$ such that $u(t) = v(t)$ for almost all $t \in [0, T]$.

23.13b.* *Compactness.* Show that the embedding

$$W \subseteq L_p(0, T; Y)$$

is compact in the case where the following conditions are satisfied:

(i) The embeddings

$$X \subseteq Y \subseteq Z$$

are continuous, and the embedding $X \subseteq Y$ is compact.

(ii) The spaces X and Z are reflexive.

(iii) $0 < T < \infty$, $1 < p, q < \infty$.

Here, we equip the space W with the norm

$$\|u\|_W = \|u\|_{L_p(0,T;X)} + \|u'\|_{L_q(0,T;Z)}.$$

This way W becomes a B-space.

Hint: Cf. Lions (1969, M), Chapter 1, 5.2.

23.14. *The intersection and the sum of* B-*spaces.* Let X, Y, and Z be B-spaces over $\mathbb{K} = \mathbb{R}, \mathbb{C}$ such that the embeddings

$$X \subseteq Z \qquad \text{and} \qquad Y \subseteq Z$$

are continuous, and $X \cap Y$ is dense in both X and Y. Show that:

(i) The set $X \cap Y$ becomes a B-space over $\mathbb{K}$ with the norm $\max\{\|u\|_X, \|u\|_Y\}$.

(ii) The set $X + Y$ becomes a B-space over $\mathbb{K}$ with the norm

$$\|w\| = \inf_{w=u+v} \{\|u\|_X + \|v\|_Y\},$$

where the infimum is taken over all the possible decompositions $w = u + v, u \in X, v \in Y$.

(iii) $(X \cap Y)^* = X^* + Y^*$.

(iv) $(X + Y)^* = X^* \cap Y^*$.

Hint: Use the Hahn–Banach theorem. Cf. Gajewski, Gröger, and Zacharias (1974, M), p. 14.

The statements above remain true if Z is a locally convex space.

References to the Literature

Calculus: Dunford and Schwartz (1958, M), Vol. 1, Gajewski, Gröger and Zacharias (1974, M), Brézis (1973, L).

Modern standard work on linear partial differential equations: Dautray and Lions (1984, M), Vols. 1–3 (theory, applications, and numerical methods).

Handbook of numerical mathematics: Ciarlet and Lions (1988, M), Vols. 1ff (finite element method, difference method, etc.).

Linear parabolic differential equations of mathematical physics: Babič (1967, M, B) (general overview without proofs), Courant and Hilbert (1953, M), Tychonov and Samarskii (1959, M), Garabedian (1964, M), Widder (1975, M), Carrier and Pearson (1976, M), John (1982, M).

Introduction to the functional analytic treatment of linear parabolic equations: Friedman (1969, M) (theory of semigroups), Lions (1969, M) (Galerkin method), Triebel (1972, M), Showalter (1977, M), Wloka (1982, M), Ladyženskaja (1985, M), Leis (1986, M).

Further important monographs: Friedman (1964, M) and Ladyženskaja (1967, M) (classical standard works), Lions and Magenes (1968, M) (interpolation theory and parabolic equations), Vejvoda (1981, M), Fattorini (1983, M), Hörmander (1983, M).

Difference methods: Smith (1965, M) (elementary introduction), Richtmyer and Morton (1967, M), Meis and Marcowitz (1978, M), Marchuk and Shaidurov (1983, M), Ciarlet and Lions (1988, M), Vol. 2.

Galerkin method and finite elements: Thomée (1984, L) (recommended as an introduction), Glowinski, Lions and Trémolières (1976, M), Ciarlet (1977, M), Fletcher (1984, M), Ciarlet and Lions (1988, M), Vol. 1.

Method of discretization in time: Rektorys (1982, M).

Multistep multiderivative methods: Gekeler (1984, L).

Multigrid methods: Hackbusch (1985, M).

CHAPTER 24

Hilbert Space Methods and Linear Hyperbolic Differential Equations

If you want to keep a matter secret, publish it.

Folclore

At a talk which I gave at a celebration of the twenty-fifth anniversary of the construction of von Neumann's computer in Princeton a few years ago, I suddenly started estimating silently in my mind how many theorems are published yearly in mathematical journals. I made a quick mental calculation and came up with a number like 100,000 per year. I mentioned this and my audience gasped.

The next day, two of the younger mathematicians in the audience came to tell me that, impressed by this enormous figure, they undertook a more systematic and detailed search in the Institute library.... Their estimate came to nearly 200,000 theorems a year.

Stanislaw Ulam (1976)

We may distinguish between a useful theorem, i.e., one for which an application has been found, and a very useful theorem, i.e., one for which many applications have been found....

The mathematical literature contains millions of theorems and very likely most of them are useless.

Philip Davis and Reuben Hersch (1981)

In the sense of "publish or perish" many correct but mathematically uninteresting papers are written, and have led to the flood of mathematical publications which makes it difficult for the mathematician to find the significant contributions for his work in the literature. By extrapolation, as I recently heard, it follows that in the year 2500 the length of bookrows in the libraries will increase with *overlight* velocity.[1]

Friedrich Hirzebruch (1974)

[1] Recently, it has been proved that long before the year 2500 the weight of the existing publications will be equal to the weight of the earth. Today we also have environmental pollution caused by publications. The *Collected Works* of David Hilbert (1932) consist of three volumes of moderate size (without his books). Nowadays there exist many mathematicians and other scientists whose

In 1956, after about one month of work, I was ordered to write up my results and as a consequence of which I swore to myself that I would choose a subject for research where it would take at least five years before I had anything worth writing about.

Kenneth G. Wilson (1982)
(Winner of the Nobel Prize in Physics in 1982 for his fundamental work on critical phenomena in connection with phase transitions, born in 1936)

If your work doesn't please the experts, that is a dangerous sign. If, in addition, it is praised by fools, then it should be forgotten.

Christian Fürchtegott Gellert (1715–1769)
(Professor of Poetry and Humanities in Leipzig)

Even if we have, for many reasons, abandoned the eighteenth-century ideal of universal knowledge, it is still appropriate to sometimes remember its advantages, when we consider the typical academic life of today. In every developed country there are hundreds of active mathematicians, out of which each one masters only a very *small corner* of his science, which then, understandably, appears to him to outshine all others in importance. He publishes the fruits of his work in abrupt, single papers in several journals, often written in various languages and scattered all over the world. The presentation of the material is meant only for a few specialists and contains no hint of a connection with more general questions, and is, therefore, for colleagues of other interests, very difficult to understand; but for an even greater circle of mathematicians it is completely inedible.

Felix Klein (1926)

24.1. Main Theorem on Second-Order Linear Evolution Equations and the Galerkin Method

We study the following initial value problem:

$$\frac{d^2}{dt^2}(u(t)|v)_H + a(u(t), v) = \langle b(t), v\rangle_V, \tag{1a}$$

$$u(0) = u_0 \in V, \qquad u'(0) = u_1 \in H, \tag{1b}$$

$$u \in L_2(0, T; V), \qquad u' \in L_2(0, T; H), \qquad u'' \in L_2(0, T; V^*), \tag{1c}$$

where we assume that (1a) is valid for all $v \in V$ and almost all $t \in]0, T[$.

To be precise, we assume that there exists a subset Z of $]0, T[$ of measure zero such that (1a) is valid for all $v \in V$ and all $t \in]0, T[- Z$. Note that Z is independent of v. Furthermore, d^2/dt^2 in (1a) means the generalized derivative

publications in journals would fill many more volumes. All over the world bureaucrats promote this tendency. The easiest way to estimate the value of a scientist is to count the number of his publications. Time is ripe to pull the emergency brake. But this must be done by the scientists themselves.

of real functions on $]0, T[$, that is,

$$\int_0^T (u(t)|v)_H \varphi''(t)\,dt + \int_0^T a(u(t), v)\varphi(t)\,dt = \int_0^T \langle b(t), v\rangle_V \varphi(t)\,dt \qquad \text{for all} \quad \varphi \in C_0^\infty(0, T).$$

We make the following assumptions:

(H1) "$V \subseteq H \subseteq V^*$" is an evolution triple with $\dim V = \infty$ and $0 < T < \infty$. The spaces V and H are real H-spaces.

Let $b \in L_2(0, T; H)$, $u_0 \in V$ and $u_1 \in H$ be given.

(H2) The functional $a: V \times V \to \mathbb{R}$ is bilinear, *symmetric*, bounded and strongly positive.

(H3) $\{w_1, w_2, \ldots\}$ is a basis in V. We are given sequences (u_{n0}) and (u_{n1}) with

$$u_{n0} \to u_0 \quad \text{in } V \qquad \text{as} \quad n \to \infty,$$

$$u_{n1} \to u_1 \quad \text{in } H \qquad \text{as} \quad n \to \infty.$$

The u_{n0} and u_{n1} belong to $\operatorname{span}\{w_1, \ldots, w_n\}$ for all $n \in \mathbb{N}$.

In order to formulate the Galerkin method, we set

$$u_n(t) = \sum_{k=1}^n c_{kn}(t) w_k,$$

$$u_{n0} = \sum_{k=1}^n \alpha_{kn} w_k, \qquad u_{n1} = \sum_{k=1}^n \beta_{kn} w_k.$$

By definition, for almost all $t \in]0, T[$, the *Galerkin equations* read as follows:

$$\sum_{k=1}^n c''_{kn}(t)(w_k|w_j)_H + c_{kn}(t) a(w_k, w_j) = \langle b(t), w_j\rangle_V, \tag{2a}$$

$$c_{jn}(0) = \alpha_{jn}, \qquad c'_{jn}(0) = \beta_{jn}, \qquad j = 1, \ldots, n. \tag{2b}$$

Here, (2a) results formally from (1a), when one replaces u by u_n and v by w_j. Equation (2b) results from the initial condition

$$u_n(0) = u_{n0}, \qquad u'_n(0) = u_{n1}.$$

Since $w_1, w_2, \ldots, w_n$ are linearly independent, $\det((w_k|w_j)_H) \neq 0$. Therefore, equation (2a) can be solved for the second derivatives c''_{kn}, and there arises a linear system of second-order differential equations for $c_{1n}, \ldots, c_{nn}$.

Remark 24.1 (Justification). We now show that the initial conditions (1b) are formulated in a meaningful way. The triple "$V \subseteq H \subseteq V^*$" is an evolution triple. By Problem 24.1, H is dense in the H-space V^*. Consequently, H and V^* generate the evolution triple

$$\text{“}H \subseteq V^* \subseteq H^*.\text{”}$$

It follows from the corresponding continuous embeddings that

$$L_2(0, T; H) \subseteq L_2(0, T; V^*) \subseteq L_2(0, T; H^*).$$

By Problem 24.2, $u'' = (u')'$. Thus, condition (1c) yields $u' \in L_2(0, T; H)$ and $(u')' \in L_2(0, T; H^*)$; therefore

$$u' \in W_2^1(0, T; H, V^*)$$

by the definition of this Sobolev space in Section 23.6. The generalized derivative u' is determined only up to changes on a set of measure zero on $[0, T]$. However, by Proposition 23.23(ii) there exists a uniquely determined representative

$$u' \in C([0, T]; V^*).$$

We use this representative in the initial condition (1b). Because of $u_1 \in H$ and $H \subseteq V^*$ the requirement $u'(0) = u_1$ makes sense.

Analogously to the above, from (1c) there results

$$u \in W_2^1(0, T; V, H);$$

therefore,

$$u \in C([0, T]; H).$$

Because of $u_0 \in V$ and $V \subseteq H$ the condition $u(0) = u_0$ is meaningful.

Theorem 24.A. *Under the assumptions* (H1), (H2), (H3), *the following hold*:

(a) Existence and uniqueness. *The original problem* (1) *has a unique solution u.*
(b) Continuous dependence on the data. *The map*

$$(u_0, u_1, b) \mapsto (u, u')$$

is a linear continuous map of the form

$$V \times H \times L_2(0, T; H) \to L_2(0, T; V) \times L_2(0, T; H),$$

that is, there exists a constant $D > 0$ such that

$$\|u\|_{L_2(0,T;V)} + \|u'\|_{L_2(0,T;H)} \leq D(\|u_0\|_V + \|u_1\|_H + \|b\|_{L_2(0,T;H)}).$$

(c) Convergence of the Galerkin method. *For each $n = 1, 2, \ldots$, the Galerkin equation* (2) *has a unique solution u_n that satisfies condition* (1c), *and*

$$u_n \rightharpoonup u \quad \text{in } L_2(0, T; V) \qquad \text{as} \quad n \to \infty,$$

$$u_n' \rightharpoonup u' \quad \text{in } L_2(0, T; H) \qquad \text{as} \quad n \to \infty.$$

Corollary 24.2. *Assume* (H1)–(H3). *Then the original problem* (1) *is equivalent to the following operator equation*:

$$u''(u) + Au(t) = b(t) \qquad \text{for almost all} \quad t \in]0, T[,$$

$$u(0) = u_0, \qquad u'(0) = u_1, \tag{3}$$

$$u \in L_2(0, T; V), \qquad u' \in L_2(0, T; H), \qquad u'' \in L_2(0, T; V^*).$$

Here, the operator $A: V \to V^*$ *results from*

$$\langle Au, v\rangle_V = a(u, v) \qquad \textit{for all} \quad u, v \in V.$$

The operator A *is linear, continuous, symmetric, and strongly monotone.*

PROOF OF COROLLARY 24.2. The existence of A follows from Proposition 21.31. By Proposition 23.20(b), equation (1a) is equivalent to

$$\langle u''(t), v\rangle_V + \langle Au(t), v\rangle_V = \langle b(t), v\rangle_V,$$

for all $v \in V$ and almost all $t \in]0, T[$. This is equivalent to the operator equation (3). □

We deal with the proof of Theorem 24.A and of Corollary 24.3 below in Section 24.3.

We now consider the more general case that the bilinear form a in (1) depends also on time t and satisfies a Gårding inequality, where the corresponding constants are *independent* of t. To this end, we replace the assumption (H2) with (H2*).

(H2*) The map

$$t \mapsto a(u, v; t)$$

is C^1 on $[0, T]$ for all $u, v \in V$. For each $t \in [0, T]$, the map $a(\cdot, \cdot; t)$: $V \times V \to \mathbb{R}$ is bilinear, symmetric, and bounded. There are constants $C, c > 0$, and $d \geq 0$ such that

$$|a(u, v; t)| \leq C\|u\|_V\|v\|_V,$$

$$a(u, v; t) \geq c\|u\|_V^2 - d\|u\|_H^2 \qquad \text{for all} \quad u, v \in V, \quad t \in [0, T].$$

Corollary 24.3. *Assume* (H1), (H2*), (H3). *Then all the assertions of Theorem* 24.A *remain valid.*

24.2. Application to Hyperbolic Differential Equations

Let $Q_T = G \times]0, T[$. We consider the initial–boundary value problem:

$$\begin{aligned} u_{tt}(x, t) - \Delta u(x, t) &= f(x, t) \quad \text{on } Q_T, \\ u(x, t) &= 0 \qquad \text{on } \partial G \times]0, T[, \\ u(x, 0) &= u_0(x) \quad \text{on } G, \\ u_t(x, 0) &= u_1(x) \quad \text{on } G. \end{aligned} \tag{4}$$

Definition 24.4. Let G be a bounded region in $\mathbb{R}^N$, $N \geq 1$. We set

$$V = \mathring{W}_2^1(G), \qquad H = L_2(G).$$

The *generalized problem* corresponding to (4) reads as follows: Let $f \in L_2(Q_T)$, $u_0 \in V$ and $u_1 \in H$ be given. We seek a function u so that equation (1) holds with

$$a(u, v) = \int_G \sum_{i=1}^N D_i u D_i v \, dx,$$

$$\langle b(t), v \rangle_V = \int_G f(x, t) v(x) \, dx,$$

where $x = (\xi_1, \ldots, \xi_N)$ and $D_i = \partial/\partial \xi_i$. Recall that

$$(u|v)_H = \int_G u(x) v(x) \, dx.$$

The generalized problem and the corresponding Galerkin method (2) are motivated by Section 19.3. As basis elements $\{w_1, w_2, \ldots\}$ in $V = \mathring{W}_2^1(G)$, one can choose the finite elements or polynomials given in $A_2(56)$ through $A_2(60)$.

Proposition 24.5. *For the generalized problem corresponding to* (4), *all the assertions of Theorem* 24.A *and Corollary* 24.2 *are satisfied, where* $\|b\|_{L_2(0,T;H)} = \|f\|_{L_2(Q_T)}$.

In particular, there exists a unique solution u of the generalized problem to (4) *such that*

$$u \in L_2(0, T; V), \qquad u' \in L_2(0, T; H), \qquad u'' \in L_2(0, T; V^*).$$

Moreover, we obtain that

$$\|u\|_{L_2(0,T;V)} + \|u'\|_{L_2(0,T;H)} \leq \text{const}(\|u_0\|_{\mathring{W}_2^1(G)} + \|u_1\|_{L_2(G)} + \|f\|_{L_2(Q_T)}).$$

Proof. As in the proof of Proposition 23.28, it follows that all the assumptions (H1)–(H3) of Theorem 24.A are satisfied. □

Remark 24.6 (Energy). We equip the space $V = \mathring{W}_2^1(G)$ with the equivalent energetic scalar product $(u|v)_E = a(u, v)$, and we set

$$E(t) = \frac{1}{2} \int_G \sum_{i=1}^N (D_i u(x, t))^2 + u_t(x, t)^2) \, dx.$$

Then,

$$\|u\|_{L_2(0,T;V)}^2 + \|u'\|_{L_2(0,T;H)}^2 = 2 \int_0^T E(t) \, dt.$$

In the special case $N = 2$, we may regard $u(x, t)$ as the displacement of a membrane at place x and time t (cf. Fig. 18.8). Then, equation (4) describes the motion of the membrane under the influence of the "outer force f." Moreover, $E(t)$ represents the total *energy* of the membrane at time t. From Proposition 24.5 we obtain that

$$\int_0^T E(t) \, dt \leq \text{const}(\|u_0\|_{\mathring{W}_2^1(G)}^2 + \|u_1\|_{L_2(G)}^2 + \|f\|_{L_2(Q_T)}^2).$$

This shows clearly that our approach can be considered as an *energetic method*.

Instead of (4) we now consider the more general initial–boundary value problem of order $2m$:

$$\begin{aligned} u_{tt}(x,t) + L(t)u(x,t) &= f(x,t) \quad \text{on } Q_T, \\ D^\alpha u(x,t) &= 0 \quad \text{on } \partial G \times [0,T] \quad \text{for all} \quad \alpha\colon |\alpha| \le m-1, \\ u(x,0) &= u_0(x) \quad \text{on } G, \\ u_t(x,0) &= u_1(x) \quad \text{on } G, \end{aligned} \tag{4*}$$

where

$$L(t)u(x,t) = \sum_{|\alpha|,|\beta| \le m} (-1)^{|\alpha|} D^\alpha(a_{\alpha\beta}(x,t)D^\beta u(x,t)).$$

(A) Let $m = 1, 2, \ldots$, and let G be a bounded region in $\mathbb{R}^N$, $N \ge 1$. All the coefficients

$$a_{\alpha\beta}, \frac{\partial}{\partial t} a_{\alpha\beta}\colon Q_T \to \mathbb{R}$$

are measurable and bounded. For each $t \in [0,T]$, the differential operator $L(t)$ is regularly strongly elliptic or *strongly elliptic* of order $2m$ in the sense of Definition 22.42, and $L(t)$ is *symmetric*, i.e., $a_{\alpha\beta} = a_{\beta\alpha}$ on Q_T for all α, β. Finally, for each $x \in G$, all the functions $t \mapsto a_{\alpha\beta}(x,t)$ are C^1 on $[0,T]$.

Definition 24.7. Assume (A), and set

$$V = \mathring{W}_2^m(G), \qquad H = L_2(G).$$

The *generalized problem* to (4*) reads as follows. Let $f \in L_2(Q_T)$, $u_0 \in V$ and $u_1 \in H$ be given. We seek a function u such that equation (1) holds, where

$$a(u,v;t) = \int_G \sum_{|\alpha|,|\beta| \le m} a_{\alpha\beta}(x,t) D^\alpha v D^\beta u \, dx,$$

$$\langle b(t), v\rangle_V = \int_G f(x,t)v(x)\,dx,$$

for all $u, v \in V$, $t \in [0,T]$.

Proposition 24.8. *Assume* (A). *Then, for the generalized problem to* (4*), *all the assertions of Theorem* 24.A *hold true, where* $\|b\|_{L_2(0,T;H)} = \|f\|_{L_2(Q_T)}$.

PROOF. Use the Gårding inequality from Section 22.15 and use Example 23.4 with respect to b. Then, the assertion follows from Corollary 24.3. □

Remark 24.9 (Generalizations). Using the Gårding inequality on strongly elliptic systems from Problem 22.7, one can immediately apply Theorem 24.A to second-order hyperbolic systems. In this connection, applications to elastodynamics will be considered in Section 61.10 of Part IV.

The important class of symmetric hyperbolic first-order systems will be studied in Chapter 83 in connection with Maxwell's equations in electrodynamics.

24.3. Proof of the Main Theorem

The following proof of Theorem 24.A is based on:

(i) the integration by parts formulas; and
(ii) *a priori* estimates for the solutions of the Galerkin equations by means of the Gronwall lemma (Proposition 3.10).

The integration by parts formulas are obtained in a simple way for smooth functions. However, note that we are dealing here with generalized derivatives. Therefore, the identities in (i) require a special justification which we take care of by approximating the vector-valued functions and their generalized derivatives by polynomials, and follow this by passing to the limit. In this connection, one must always pay careful attention concerning which spaces the approximations are carried out in. Apparently very short proofs for evolution problems in the literature frequently have their origin in the fact that only the formal arguments are given, and their justification is left to the reader.

Instead of $\langle\cdot,\cdot\rangle_V$ we write briefly $\langle\cdot,\cdot\rangle$.

24.3a. Proof of Uniqueness

Suppose that u_1 and u_2 are solutions of the original problem (1). Then the difference

$$u = u_1 - u_2$$

satisfies equation (1) with $b(t) \equiv 0$ and $u(0) = u'(0) = 0$. By (3),

$$u'' + Au = 0.$$

Remark 24.1 shows that $u \in W_2^1(0, T; V, H)$. Integration by parts (23.25) yields

$$\|u(S)\|_H^2 = 2\int_0^S \langle u'(t), u(t)\rangle\, dt, \tag{5}$$

for all $S \in]0, T[$. For fixed S, we define

$$\chi_S(t) = \begin{cases} 1 & \text{for} \quad 0 \le t \le S, \\ 0 & \text{for} \quad S < t \le T, \end{cases}$$

and

$$v(t) = -\int_0^t \chi_S(s)u(s)\,ds + c, \tag{6}$$

where the constant c is chosen so that $v(T) = 0$. Then $v(t) = 0$ for $t \geq S$. It follows from $u \in L_2(0, T; V)$ that $\chi_S u \in L_2(0, T; V)$. Problem 23.5b yields

$$v' = \chi_S u.$$

Now, our uniqueness proof will be based on the following two formulas:

$$\langle u'(T), v(T)\rangle - \langle u'(0), v(0)\rangle = \int_0^T \langle u'', v\rangle + \langle u', v'\rangle\,dt \tag{7}$$

and

$$\int_0^S \langle Av'(t), v(t)\rangle\,dt \leq 0. \tag{8}$$

The proofs of (7) and (8) will be given below.

It follows from (7) and $u'(0) = v(T) = 0$, $u'' + Au = 0$, $v' = \chi_S u$ that

$$\int_0^S -\langle Av'(t), v(t)\rangle + \langle u'(t), u(t)\rangle\,dt = 0.$$

By (5) and (8), $u(S) = 0$. Because S is arbitrary,

$$u(t) \equiv 0.$$

This shows the uniqueness of the solution.

Proof of (7). As in Remark 24.1, we consider the evolution triple

$$\text{“}H \subseteq V^* \subseteq H^*.\text{”}$$

Because of the corresponding continuous embeddings, the embeddings

$$L_2(0, T; H) \subseteq L_2(0, T; V^*) \subseteq L_2(0, T; H^*)$$

are also continuous.

By (1c), $u' \in L_2(0, T; H)$ and $u'' \in L_2(0, T; V^*)$. According to Problem 24.3, there exist polynomials p_n: $[0, T] \to H$ with

$$p_n \to u' \quad \text{in } L_2(0, T; H) \qquad \text{as} \quad n \to \infty,$$
$$p_n' \to u'' \quad \text{in } L_2(0, T; V^*) \qquad \text{as} \quad n \to \infty.$$

Then the above-mentioned continuous embeddings also yield

$$p_n \to u' \text{ in } L_2(0, T; H) \quad \text{and} \quad p_n' \to u'' \text{ in } L_2(0, T; H^*) \quad \text{as } n \to \infty \tag{9a}$$

as well as $p_n \to u'$ in $L_2(0, T; V^*)$ as $n \to \infty$. Now (9a) means that

$$p_n \to u' \quad \text{in } W_2^1(0, T; H, V^*) \qquad \text{as} \quad n \to \infty.$$

By Proposition 23.23, the embedding $W_2^1(0, T; H, V^*) \subseteq C([0, T], V^*)$ is continuous. This yields further that

$$p_n \to u' \quad \text{in } C([0, T], V^*) \qquad \text{as} \quad n \to \infty.$$

In particular, note that

$$p_n \to u' \quad \text{and} \quad p_n' \to u'' \quad \text{in } L_2(0, T; V^*) \quad \text{as} \quad n \to \infty, \tag{9}$$

$$p_n(0) \to u'(0) \quad \text{and} \quad p_n(T) \to u'(T) \quad \text{in } V^* \quad \text{as} \quad n \to \infty. \tag{10}$$

Since $u \in L_2(0, T; V)$, there exist polynomials q_n: $[0, T] \to V$ such that

$$q_n \to u \quad \text{in } L_2(0, T; V) \quad \text{as} \quad n \to \infty.$$

Parallel to (6) we construct the function

$$v_n(t) = -\int_0^t \chi_S(s) q_n(s)\, ds + c.$$

We then obtain $v_n \to v$ in $C([0, T], V)$ as $n \to \infty$, by the Hölder inequality. Moreover, $v_n' = -\chi_S q_n$. Thus, we get

$$v_n \to v \quad \text{and} \quad v_n' \to v' \quad \text{in } L_2(0, T; V) \quad \text{as} \quad n \to \infty, \tag{11}$$

$$v_n(0) \to v(0) \quad \text{and} \quad v_n(T) \to v(T) \quad \text{in } V \quad \text{as} \quad n \to \infty. \tag{12}$$

From this we obtain that, as $n \to \infty$,

$$\begin{gathered} \int_0^T \langle p_n', v_n \rangle + \langle p_n, v_n' \rangle\, dt \to \int_0^T \langle u'', v \rangle + \langle u', v' \rangle\, dt, \\ \langle p_n(T), v_n(T) \rangle - \langle p_n(0), v_n(0) \rangle \to \langle u'(T), v(T) \rangle - \langle u'(0), v(0) \rangle. \end{gathered} \tag{13}$$

The first integral relation of (13) follows from (9), (11) and Proposition 23.9. According to the classical formula for integration by parts, the left-hand sides in (13) are equal. This implies (7).

Proof of (8): Classical formulas yield

$$\begin{aligned} \int_0^S 2\langle Av_n'(t), v_n(t) \rangle\, dt &= \int_0^S \frac{d}{dt} \langle Av_n(t), v_n(t) \rangle\, dt \\ &= -\langle Av_n(0), v_n(0) \rangle \leq 0. \end{aligned} \tag{14}$$

To prove this, observe that v_n is a polynomial and note the *symmetry* and positiveness of A: $V \to V^*$ as well as $v_n(S) = 0$. In fact,

$$\begin{aligned} \frac{d}{dt} \langle Av_n(t), v_n(t) \rangle &= \langle Av_n'(t), v_n(t) \rangle + \langle Av_n(t), v_n'(t) \rangle \\ &= 2\langle Av_n'(t), v_n(t) \rangle. \end{aligned}$$

From (11) and the continuity of A: $V \to V^*$ it follows that

$$\begin{aligned} v_n &\to v && \text{in } L_2(0, T; V) && \text{as} \quad n \to \infty, \\ Av_n' &\to Av' && \text{in } L_2(0, T; V^*) && \text{as} \quad n \to \infty. \end{aligned}$$

We thus obtain (8) by means of passing to the limit in (14) as $n \to \infty$.

24.3b. Proof of Existence

We agree to use the abbreviations

$$(u|v) = (u|v)_H, \qquad |u| = \|u\|_H,$$
$$\langle u, v\rangle = \langle u, v\rangle_V, \qquad \|u\| = \|u\|_V,$$

and recall via Section 23.4 the continuity of the embeddings $V \subseteq H \subseteq V^*$ with

$$\begin{aligned} (h|v) &= \langle h, v\rangle \qquad \text{for all} \quad h \in H, \quad v \in V, \\ |v| &\le C\|v\| \qquad \text{for all} \quad v \in V, \end{aligned} \tag{15a}$$

where C is a constant. It thus follows that

$$|\langle w, v\rangle| = |(w|v)| \le C^2\|w\|\,\|v\| \qquad \text{for all} \quad w, v \in V. \tag{15b}$$

Step 1: *A priori* estimates for the solutions of the Galerkin equations by means of the Gronwall lemma.

Because of $b \in L_2(0, T; H)$ and

$$|\langle b(t), v\rangle|^2 = |(b(t)|v)|^2 \le C|b(t)|^2\|v\|^2,$$

the real function $t \mapsto \langle b(t), v\rangle$ belongs to the space $L_2(0, T)$ for all $v \in V$. For the sake of simplicity, we first assume that

$$t \mapsto \langle b(t), v\rangle \text{ is continuous on } [0, T] \qquad \text{for all} \quad v \in V.$$

Then the Galerkin equation (2) possesses a unique classical solution $u_n \in C^2[0, T]$. It immediately follows from (2a) upon multiplication by c'_{jn} and subsequent summation over j that

$$\frac{d}{dt}\left(|u'_n(t)|^2 + a(u_n(t), u_n(t))\right) = 2\langle b(t), u'_n(t)\rangle. \tag{16}$$

In this connection, note the *symmetry* of a, i.e., $a(w_k, w_j) = a(w_j, w_k)$ for all j, k. Furthermore, by (15a), we get

$$\begin{aligned} 2|\langle b(t), u'_n(t)\rangle| &\le 2|b(t)|\,|u'_n(t)| \\ &\le (|b(t)|^2 + |u'_n(t)|^2). \end{aligned}$$

Moreover, by assumption (H2),

$$c\|w\|^2 \le a(w, w), \qquad |a(w, v)| \le d\|w\|\,\|v\|, \tag{17}$$

for all $v, w \in V$ with positive constants c, d. Therefore, integration of (16) over $[0, t]$ yields

$$\begin{aligned} |u'_n(t)|^2 + c\|u_n(t)\|^2 \le{}& \left(|u'_n(0)|^2 + d\|u_n(0)\|^2 + \int_0^t |b(s)|^2\,ds\right) \\ &+ \int_0^t |u'_n(s)|^2 + c\|u_n(s)\|^2\,ds. \end{aligned} \tag{18}$$

By the Gronwall lemma (Proposition 3.10), there follows from (18) that

$$|u_n'(t)|^2 + c\|u_n(t)\|^2 \leq \mathrm{const}(|u_n'(0)|^2 + \|u_n(0)\|^2 + \int_0^t |b(s)|^2\,ds),$$

and hence, for all n, we obtain the decisive *a priori estimate*:

$$\|u_n'\|^2_{L_2(0,T;H)} + c\|u_n\|^2_{L_2(0,T;V)} \leq \mathrm{const}(|u_n'(0)|^2 + \|u_n(0)\|^2 + \|b\|^2_{L_2(0,T;H)}), \tag{19}$$

by using integration over $[0, T]$.

Step 2: Weak convergence of a subsequence of the Galerkin sequence (u_n). It follows from (19) that (u_n) (resp. (u_n')) is bounded in the H-space $L_2(0, T; V)$ (resp. $L_2(0, T; H)$). Thus there exist subsequences, again denoted by (u_n) and (u_n'), such that

$$\begin{aligned} u_n &\rightharpoonup u \quad \text{in } L_2(0,T;V) \quad \text{as} \quad n \to \infty, \\ u_n' &\rightharpoonup v \quad \text{in } L_2(0,T;H) \quad \text{as} \quad n \to \infty. \end{aligned} \tag{20}$$

By Proposition 23.19,

$$u' = v.$$

Noting that

$$\|u\| \leq \varliminf_{n\to\infty} \|u_n\| \qquad \left(\text{resp. } \|u'\| \leq \varliminf_{n\to\infty} \|u_n'\|\right)$$

in $L_2(0, T; V)$ (resp. $L_2(0, T; H)$), we obtain from (19) that, as $n \to \infty$,

$$\|u'\|^2_{L_2(0,T;H)} + c\|u\|^2_{L_2(0,T;V)} \leq \mathrm{const}(|u_1|^2 + \|u_0\|^2 + \|b\|^2_{L_2(0,T;H)}).$$

Step 3: We show that u is a solution of the original equation (1).

Proof of (1a). It follows from the Galerkin equation (2a), after multiplication by φ and subsequent integration by parts twice, that

$$\int_0^T \varphi''(u_n|w_j) + a(u_n, w_j)\,dt = \int_0^T \langle b, w_j\rangle\,dt \qquad \text{for all} \quad \varphi \in C_0^\infty(0,T), \tag{21}$$

where φ, u_n, and b depend on t. As $n \to \infty$, this yields for all j that

$$\int_0^T \varphi''(u|w_j) + a(u, w_j)\,dt = \int_0^T \langle b, w_j\rangle\,dt \qquad \text{for all} \quad \varphi \in C_0^\infty(0,T). \tag{22}$$

This limit will be justified below.

Since $\operatorname{span}\{w_1, w_2, \ldots\}$ is dense in V, equation (22) implies that

$$\int_0^T \varphi''(u|v) + a(u,v)\,dt = \int_0^T \langle b, v\rangle\,dt \qquad \text{for all} \quad \varphi \in C_0^\infty(0,T), \quad v \in V. \tag{23}$$

This yields (1a), that is,

$$\frac{d^2}{dt^2}(u(t)|v) + a(u(t), v) = \langle b(t), v\rangle \qquad \text{for all} \quad v \in V,$$

according to the definition of generalized derivatives. By Corollary 24.2,

$$u'' + Au = b.$$

In order to justify the passage to the limit as $n \to \infty$ in (21), we note that $u_n \rightharpoonup u$ in $L_2(0, T; V)$ as $n \to \infty$. Thus it is sufficient to prove that all expressions in (21) are linear continuous functionals on $X = L_2(0, T; V)$, with respect to u_n. This follows from (15) and the Hölder inequality. In fact, we get, for example,

$$\left| \int_0^T \varphi''(t)(u_n(t)|w_j)\,dt \right| \le \text{const} \int_0^T \|u_n(t)\|\,\|w_j\|\,dt \le \text{const}\,\|u_n\|_X \|w_j\|.$$

For the terms with $a(\cdot, \cdot)$, use (17). Equation (23) is obtained from (22) in the same way.

Proof of (1c). We have to show that $u'' \in L_2(0, T; V^*)$. It follows from $u \in L_2(0, T; V)$ and the boundedness of $A: V \to V^*$ that

$$Au \in L_2(0, T; V^*).$$

Because $b \in L_2(0, T; H)$, and because of the continuous embedding $H \subseteq V^*$, we also have $b \in L_2(0, T; V^*)$. Finally, from

$$u'' + Au = b$$

we get $u'' \in L_2(0, T; V^*)$.

Proof of (1b). In order to show that

$$u(0) = u_0, \qquad u'(0) = u_1,$$

we choose a function

$$\psi \in C^1[0, T] \qquad \text{with} \qquad \psi(T) = 0, \qquad \psi(0) = 1.$$

Analogously to (7) with $v(t) = \psi(t)w_j$, there results from integration by parts that

$$\begin{aligned} \langle u_n'(0), \psi(0)w_j \rangle &= -\int_0^T \langle u_n'', \psi w_j \rangle + \langle u_n', \psi' w_j \rangle\,dt, \\ \langle u'(0), \psi(0)w_j \rangle &= -\int_0^T \langle u'', \psi w_j \rangle + \langle u', \psi' w_j \rangle\,dt. \end{aligned} \tag{24}$$

The Galerkin equation (2) and $u'' + Au = b$ yield

$$\begin{aligned} \langle u_n'', \psi w_j \rangle &= \langle -Au_n + b, \psi w_j \rangle, \\ \langle u'', \psi w_j \rangle &= \langle -Au + b, \psi w_j \rangle. \end{aligned}$$

It follows from the convergence statement (20) that the right members of (24) tend to one another as $n \to \infty$. This passage to the limit can be justified as for (21). Note that, by (15), the right members of (24) are linear continuous functionals on $L_2(0, T; V)$ (resp. $L_2(0, T; H)$) with respect to u_n (resp. u_n'). Thus, we obtain from (24) that

$$\langle u_n'(0), w_j \rangle \to \langle u'(0), w_j \rangle \qquad \text{as} \quad n \to \infty \qquad \text{for all } j.$$

By assumption (H3), $u_n'(0) \to u_1$ holds in H as $n \to \infty$, therefore also in V^*. This yields

$$\langle u_1, w_j \rangle = \langle u'(0), w_j \rangle \qquad \text{for all } j.$$

Since $\operatorname{span}\{w_1, w_2, \ldots\}$ is dense in V, we get

$$u_1 = u'(0).$$

Analogously, we obtain that

$$u_0 = u(0),$$

by considering the following integration by parts formula that follows from (23.25):

$$(u_n(0)|\psi(0)w_j) = -\int_0^T \langle u_n', \psi w_j \rangle + \langle \psi' w_j, u_n \rangle \, dt,$$

$$(u(0)|\psi(0)w_j) = -\int_0^T \langle u', \psi w_j \rangle + \langle \psi' w_j, u \rangle \, dt.$$

Step 4: Weak convergence of the total sequence (u_n) of the Galerkin approximations.

We saw in Step 2 that all subsequences of (u_n) that converge weakly in $L_2(0, T; V)$ possess as limit element the *uniquely* determined solution u of equation (1). It then follows, according to Proposition 21.23(i), that the total sequence (u_n) converges weakly to u in $L_2(0, T; V)$, and (u_n') converges weakly to u' in $L_2(0, T; H)$.

Step 5: The general case $b \in L_2(0, T; H)$.

If $t \mapsto \langle b(t), v \rangle$ is not assumed to be continuous on $[0, T]$ as in Step 1, then one has to apply the Carathéodory theorem $A_2(61)$ to solve the Galerkin equations (cf. Problem 30.3). For this purpose, one writes equation (2) as a first-order system of ordinary differential equations. Then, analogous to the above, one can infer taking into account the lessened regularity of u_n. Now the components $c_{nj}(\cdot)$ of u_n do not lie in $C^2[0, T]$ but rather in $W_2^2(0, T)$. Therefore, $u_n \in L_2(0, T; V)$ and $u_n' \in L_2(0, T; H)$.

The proof of Theorem 24.A is complete. □

PROOF OF COROLLARY 24.3. The proof proceeds analogously to the proof of Theorem 24.A (cf. Problem 24.4). □

PROBLEMS

24.1. *Evolution triple.* Show that in an evolution triple "$V \subseteq H \subseteq V^*$," the space H is dense in V^*.

Solution: This follows immediately from statement (c) in Section 23.4. A direct proof goes as follows.

If H is not dense in V^*, then there exists a point $v^* \in V^*$ with

$$v^* \notin \bar{H},$$

therefore, $v^* \neq 0$. By the Hahn–Banach theorem, one can construct a functional $F \in V^{**}$ with

$$\langle F, h\rangle_{V^*} = 0 \qquad \text{for all} \quad h \in \bar{H} \qquad \text{and} \qquad \langle F, v^*\rangle_{V^*} = 1.$$

The B-space V is reflexive, therefore $V = V^{**}$. Thus we get $F \in V$ and

$$\langle h, F\rangle_V = (h|F)_H = 0 \qquad \text{for all} \quad h \in H.$$

This implies $F = 0$ in contradiction to $\langle v^*, F\rangle_V = 1$.

24.2. *Iterated derivatives* $(u')' = u''$. Show that if u' and $(u')'$ exist as generalized derivatives on $]0, T[$, then u'' also exists and $u'' = (u')'$.

Solution: It follows from

$$\int_0^T \varphi' u' \, dt = -\int_0^T \varphi (u')' \, dt \qquad \text{for all} \quad \varphi \in C_0^\infty(0, T)$$

and

$$\int_0^T \psi' u \, dt = -\int_0^T \psi u' \, dt \qquad \text{for all} \quad \psi \in C_0^\infty(0, T)$$

that

$$\int_0^T \varphi'' u \, dt = \int_0^T \varphi (u')' \, dt \qquad \text{for all} \quad \varphi \in C_0^\infty(0, T),$$

by setting $\psi = \varphi'$. This means $u'' = (u')'$.

Conversely, the existence of $(u')'$ along with $(u')' = u''$ follows from the existence of u' and u''.

24.3. *A density theorem*. Let "$V \subseteq H \subseteq V^*$" be an evolution triple, and let

$$u \in L_p(0, T; V), \qquad u' \in L_p(0, T; H)$$

with $1 \le p < \infty$ and $0 < T < \infty$.

Show that there exists a sequence (p_n) of functions from $C^1([0, T], V)$ such that

$$p_n \to u \quad \text{in } L_p(0, T; V) \qquad \text{as} \quad n \to \infty,$$
$$p_n' \to u' \quad \text{in } L_p(0, T; H) \qquad \text{as} \quad n \to \infty.$$

Since the set of polynomials is dense in $C^1([0, T], V)$, one may assume that all $p_n \colon [0, T] \to V$ are polynomials with coefficients in V.

Hint: Use the smoothing operator in Section 18.14.

24.4. *Proof of Corollary* 24.3.

Hint: Use the same arguments as in the proof of Theorem 24.A. Cf. Lions (1968, M), Chapter 4, Section 1.

References to the Literature

Linear hyperbolic differential equations in mathematical physics: Babič (1967, M, B) (general survey without proofs), Courant and Hilbert (1953, M), Tychonov and Samarskii (1959, M), Garabedian (1964, M), Friedlander (1976, M), John (1982, M), Sakamoto (1982, M), Günther (1988, M) (Huygens' principle).

Introduction to the functional analytic treatment of hyperbolic differential equations: Lions (1968, M) (Galerkin method), Triebel (1972, M), Wloka (1982, M), Fattorini (1983, M), Ladyženskaja (1985, M), Leis (1986, M).

Further important monographs: Sobolev (1950), Ladyženskaja (1953), Lions and Magenes (1968), Hörmander (1983), Smoller (1983).

Standard work on linear partial differential equations: Dautray and Lions (1984, M), Vols. 1–3 (theory, applications, and numerical methods).

Handbook of numerical analysis: Ciarlet and Lions (1988, M), Vols. 1ff (finite element method, difference method, etc.).